Signal Design for Good Correlation

For Wireless Communication, Cryptography, and Radar

This book provides a comprehensive, up-to-date description of the methodologies and the application areas, throughout the range of digital communication, in which individual signals and sets of signals with favorable correlation properties play a central role. The necessary mathematical background is presented to explain how these signals are generated and to show how they satisfy the appropriate correlation constraints. All the known methods to obtain balanced binary sequences with two-valued autocorrelation, many of them only recently discovered, are presented in depth.

Important applications include Code Division Multiple Access (CDMA) signals, such as those already in widespread use for cell-phone communication and planned for universal adoption in the various approaches to third-generation (3G) cell-phone use; systems for coded radar and sonar signals; communication signals to minimize mutual interference in multiuser environments; and pseudorandom sequence generation for secure authentication and for stream cipher cryptology.

SOLOMON W. GOLOMB is professor of mathematics and of electrical engineering at the University of Southern California.

GUANG GONG is professor of electrical and computer engineering at the University of Waterloo, Ontario.

Signal Design for Good Correlation
For Wireless Communication, Cryptography, and Radar

SOLOMON W. GOLOMB
University of Southern California

GUANG GONG
University of Waterloo, Ontario

CAMBRIDGE
UNIVERSITY PRESS

32 Avenue of the Americas, New York NY 10013-2473, USA

Cambridge University Press is part of the University of Cambridge.

It furthers the University's mission by disseminating knowledge in the pursuit of education, learning and research at the highest international levels of excellence.

www.cambridge.org
Information on this title: www.cambridge.org/9780521821049

First published 2005

A catalogue record for this publication is available from the British Library

Library of Congress Cataloguing in Publication data

Golomb, Solomon W. (Solomon Wolf)
Signal design for good correlation for wireless communication, cryptography, and radar / Solomon W. Golomb, Guang Gong.
p.   cm.
Includes bibliographical references and index.
ISBN 0-521-82104-5 (hardcover)
1. Signal theory (Telecommunication)   2. Signal processing – Digital techniques.
I. Gong, Guang, 1956–   II. Title.
TK5102.92.G65   2005
621.382'23 – dc22   2005002719

ISBN 978-0-521-82104-9 Hardback

Dedicated to Andrew and Erna Viterbi

Contents

Preface

This book is the product of a fruitful collaboration between one of the earliest developers of the theory and applications of binary sequences with favorable correlation properties and one of the currently most active younger contributors to research in this area. Each of us has taught university courses based on this material and benefited from the feedback obtained from the students in those courses. Our goal has been to produce a book that achieves a balance between the theoretical aspects of binary sequences with nearly ideal autocorrelation functions and the applications of these sequences to signal design for communications, radar, cryptography, and so on. This book is intended for use as a reference work for engineers and computer scientists in the applications areas just mentioned, as well as to serve as a textbook for a course in this important area of digital communications. Enough material has been included to enable an instructor to make some choices about what to cover in a one-semester course. However, we have referred the reader to the literature on those occasions when the inclusion of further detail would have resulted in a book of inordinate length.

We plan to maintain a Web site at http://calliope.uwaterloo.ca/~ ggong/book/ book.htm for additions, corrections, and the continual updating of the material in this book.

Solomon W. Golomb, Los Angeles, CA, USA
Guang Gong, Waterloo, ON, Canada
August 31, 2004

Acknowledgments

Many people contributed significantly to the development of the material presented in this book. To the best of our ability we have acknowledged these contributions where they occur, as well as in the Bibliography; but inevitably some references have surely gone unattributed, for which we apologize in advance.

Colleagues as well as both current and former doctoral students of the authors have reviewed portions of the text, but we assume full responsibility for any deficiencies that remain.

Among those deserving special thanks for their assistance are Wensong Chu, Zongduo Dai, Tor Helleseth, Katrin Hoeper, Shaoquan Jiang, Khoongming Khoo, P. Vijay Kumar, Charles Lam, Heekwan Lee, Oscar Moreno, Reza Omrani, Susana Sin, Hong-Yeop Song, Douglas Stinson, Herbert Taylor, Lloyd R. Welch, Amr Youssef, and Nam Yul Yu. Our gratitude for help in preparing the manuscript goes to Mayumi Thatcher.

We further acknowledge reliance on articles we have previously published, either together or separately, in such journals as the *IEEE Transactions on Information Theory* and in conference proceedings, including

(a) *Surveys in Combinatorics, 1991*, A. D. Keedwell (Ed.), Cambridge University Press, 1991.
(b) *Difference Sets, Sequences and Their Correlation Properties (Bad Windsheim, 1998)*, A. Pott et al. (Eds.), NATO Adv. Sci. Inst. Ser. C, Math. Phys. Sci., Vol. 542, Kluwer Acad. Publ., Dordrecht, 1999.
(c) *Sequences and Their Applications — Proceedings of SETA'98, Discrete Mathematics and Theoretical Computer Science*, T. Helleseth et al. (Eds.), London, Springer-Verlag, 1999.

(d) *Sequences and Their Applications – Proceedings of SETA'01, Discrete Mathematics and Theoretical Computer Science*, V. Kumar, T. Helleseth, and K. Yang (Eds.), Berlin, Springer-Verlag, 2001.

Finally, we are grateful to Lauren Cowles of Cambridge University Press for her encouragement and forebearance with this project.

– S. W. Golomb and G. Gong

Historical Introduction

The prehistory of our subject can be backdated to 1202, with the appearance of Leonardo Pisano's *Liber Abaci* (Fibonacci 1202), containing the famous problem about breeding rabbits that leads to the linear recursion $f_{n+1} = f_n + f_{n-1}$ for $n \geq 2$, $f_1 = f_2 = 1$, which yields the Fibonacci sequence. Additional background can be attributed to Euler, Gauss, Kummer, and especially Edouard Lucas (Lucas 1876). For the history proper, the earliest milestones are papers by O. Ore (Ore 1934), R. E. A. C. Paley (Paley 1933), and J. Singer (Singer 1938). Ore started the systematic study of linear recursions over finite fields (including $GF(2)$), Paley inaugurated the search for constructions yielding Hadamard matrices, and Singer discovered the Singer difference sets that are mathematically equivalent to binary maximum length linear shift register sequences (also known as pseudorandom sequences, pseudonoise (PN) sequences, or m-sequences).

It appears that by the early 1950s devices that performed the modulo 2 sum of two positions on a binary delay line were being considered as key generators for stream ciphers in cryptographical applications. The question of what the periodicity of the resulting output sequence would be seemed initially mysterious. This question was explored outside the cryptographic community by researchers at a number of locations in the 1953–1956 time period, resulting in company reports by E. N. Gilbert at Bell Laboratories, by N. Zierler at Lincoln Laboratories, by L. R. Welch at the Jet Propulsion Laboratory, by S. W. Golomb at the Glenn L. Martin Company (now part of Lockheed-Martin), and probably by others as well. These earliest reports independently arrived at the correspondence between binary linear recurrence relations and polynomials over $GF(2)$, with the m-sequences corresponding to primitive irreducible polynomials. Golomb may have been the first to point out the correspondence between binary sequences with 2-level autocorrelation and cyclic (v, k, λ) difference sets (Golomb 1955) and even earlier (Golomb 1954) to recognize that quadratic residue sequences share the 2-level autocorrelation property of the PN-sequences and to formulate the objective of finding all binary sequences with this autocorrelation function (i.e., identifying all the constructions that yield $(4t - 1, 2t - 1, t - 1)$ cyclic difference sets, also called cyclic Hadamard difference sets).

Beyond the Singer difference sets (equivalent to m-sequences) and the quadratic residue sequences (also called Legendre sequences), additional cyclic Hadamard examples were discovered occasionally: the sextic residue sequences of Marshall Hall, Jr. (see Hall 1956), the twin prime sequences of R. G. Stanton and D. A. Sprott (Stanton and Sprott 1958), and the GMW sequences, with generalizations, of B. Gordon, W. H. Mills, and L. R. Welch (Gordon, Mills, and Welch 1962). This was the state of knowledge when L. D. Baumert's book (Baumert 1971) appeared, except that by exhaustive search at $(v, k, \lambda) = (127, 63, 31)$, Baumert had found *six* inequivalent examples, of which only *three* came from known constructions. More unexplained examples turned up when U. Cheng performed the complete search at $(v, k, \lambda) = (255, 127, 63)$ (Cheng 1983) and still more when R. B. Dreier and K. W. Smith exhaustively searched the case $(v, k, \lambda) = (511, 255, 127)$ (Dreier and Smith 1991).

As mentioned in Baumert (1971), all known examples of cyclic Hadamard difference sets with parameters (v, k, λ), where $k = 2\lambda - 1$ and $v = 2k - 1$, have v belonging to one of three types: (i) primes of the form $4t - 1$, (ii) products pq where $q = p + 2$ and both p and q are primes, and (iii) numbers of the form $2^n - 1$. This conjecture (that v must be of one of these three types) looks much stronger now than when Golomb suggested it to Baumert around 1960. All known examples of type (ii) come from the Stanton–Sprott construction. All known examples of type (i) that are not also of type (iii), that is, primes of the form $4t - 1$ that are not Mersenne primes, come either from the quadratic residue construction or from Hall's sextic residue constructions when $p = 4a^2 + 27$. The great multiplicity of examples are of type (iii) and are related, in some way or other, to trace mappings from $GF(2^n)$ to $GF(2)$.

By 1955, Golomb had found all examples of type (iii) through $v = 2^5 - 1$. It was from studying their exhaustive list of examples at $v = 2^6 - 1$ that Gordon, Mill, and Welch (1962) discovered the GMW construction. Starting at $2^5 - 1$ in the 1950s, each decade has seen one more value of n, in $2^n - 1$, subjected to a complete search. It will be a challenge to programming skill and ingenuity to perform a complete search of $2^{11} - 1$ before the year 2020.

Golomb's book *Shift Register Sequences* first appeared in 1967, including the old Martin Company report (Golomb 1955) as its Chapter 3 and further developing the theory of both linear and nonlinear shift register sequences, based on Jet Propulsion Laboratory (JPL) reports he had written from 1956 to 1961, as the subsequent chapters. An enlarged second edition of this book appeared in 1982 (see Golomb 1967) and is still in print. The collaboration of the present authors began in 1996, when Dr. Gong began a two-year postdoctoral fellowship at the University of Southern California, visiting Dr. Golomb.

After decades of very slow progress, there was a sudden profusion of newly discovered constructions for cyclic Hadamard difference sets, starting in 1997. When the complete search was carried out at $(v, k, \lambda) = (1023, 511, 255)$ in (Gaal and Golomb 2001), ten inequivalent examples were found, but all belonged to families that by then had been discovered. These families also included all the previously unexplained examples at $v = 127$, $v = 255$, and $v = 511$. The recent paper by J. Dillon and H. Dobbertin (Dillon and Dobbertin 2004) summarizes and completes the validation of all the constructions now known for cyclic Hadamard difference sets and lends credence to the belief that the identification of all such constructions (the task proposed in Golomb 1954) is finally complete. It is therefore timely for the present book, which describes all these constructions in reasonable detail, to make its appearance. We also discuss the more general question of constructing $4t \times 4t$ Hadamard matrices, which are conjectured to exist for all positive integers t (the first unknown case is $t = 167$), and the numerous ways in which these matrices are applied, to form advantageous sets of signals for communication and in Hadamard transforms. Our final chapter concerns the application of sequences with favorable autocorrelation properties to problems of radar, sonar, and synchronization. The only previous book describing applications of this type is Hans Dieter Lüke's Korrelationssignale (Lüke 1992), in German, which appeared before the discovery of all the new constructions.

Interest in sequences with favorable correlation properties, and in the communication signals based on these sequences, has increased dramatically in recent years. In addition to the radar and sonar applications, there are important cryptographic and security system applications (see, e.g., Beker and Piper 1982) and there is intense interest in the applications to Code Division Multiple Access (CDMA) signals for mobile and wireless communications (see Viterbi 1995). In fact, essentially all the standards for third-generation (3G) cellular telephony are based on CDMA, which in turn uses signals with the correlation properties described in the present book. It is interesting to note that many books, including those just cited, by Beker and Piper on secure communications and by Viterbi on CDMA, faithfully reproduce (with appropriate attribution) the derivation of the three randomness properties of pseudorandom sequences from Golomb's original (1955) Martin Company report.

In recent years, special international conferences on sequences (such as the series "Sequences and Their Applications," or SETA) have become frequent. Starting in 1998, the *Transactions on Information Theory* of the IEEE has had an associate editor for sequences.

For all of these reasons, we believe the appearance of our book to be highly useful, relevant, and timely.

1

General Properties of Correlation

1.1 What is correlation?

Correlation is a measure of the similarity, or relatedness, between two phenomena. When properly normalized, the correlation measure is a real number between -1 and $+1$, where a correlation value of $+1$ indicates that the two phenomena are identical, a correlation value of -1 means that they are diametrically opposite, and a correlation value of 0 means that they are uncorrelated, that is, that they agree exactly as much as they disagree.

In statistics, the correlation between two sets of data is called their covariance. In linear algebra, the correlation between two vectors is their (normalized) dot product. Specifically, let $\alpha = (a_1, a_2, \ldots, a_n)$ and $\beta = (b_1, b_2, \ldots, b_n)$ be two n-dimensional vectors of real numbers, which could represent two sets of experimental data. The magnitudes of these vectors are $|\alpha| = \left(\sum_{i=1}^{n} a_i^2\right)^{\frac{1}{2}}$ and $|\beta| = \left(\sum_{i=1}^{n} b_i^2\right)^{\frac{1}{2}}$. The normalized vectors are $\alpha' = \frac{\alpha}{|\alpha|}$ and $\beta' = \frac{\beta}{|\beta|}$. The correlation between the a_i's and the b_i's is the covariance of the two data sets

$$C(\alpha, \beta) = \frac{(\alpha \cdot \beta)}{|\alpha|\,|\beta|} = \frac{\sum_{i=1}^{n} a_i b_i}{\left(\sum_{i=1}^{n} a_i^2\right)^{\frac{1}{2}}\left(\sum_{i=1}^{n} b_i^2\right)^{\frac{1}{2}}}, \tag{1.1}$$

which is also the normalized dot product of the two vectors, that is, the dot product of the two vectors:

$$(\alpha' \cdot \beta') = \left(\frac{\alpha}{|\alpha|} \cdot \frac{\beta}{|\beta|}\right) = \frac{(\alpha \cdot \beta)}{|\alpha|\,|\beta|} = C(\alpha, \beta).$$

Geometrically, $(\alpha \cdot \beta) = |\alpha||\beta| \cos\theta$, so that

$$C(\alpha, \beta) = \frac{(\alpha \cdot \beta)}{|\alpha||\beta|} = \cos\theta,$$

where θ is the angle between the vectors α and β. When the vectors are orthogonal (i.e., perpendicular), we have

$$C(\alpha, \beta) = \cos 90° = 0,$$

so that uncorrelated data sets correspond to orthogonal vectors. Note that $-1 \leq \cos \theta \leq +1$ for all angles θ.

The only real vector that cannot be normalized is the zero vector, $\mathbf{0} = (0, 0, \ldots, 0)$. However, since the (unnormalized) dot product of this vector with any other has the (scalar) value zero, the zero vector is usually regarded as uncorrelated with all vectors.

1.2 Continuous correlation

Suppose that $f(x)$ and $g(x)$ are real-valued functions that are square-integrable on the interval $[0, T]$. That is, the definite integrals $\int_0^T f^2(x)dx$ and $\int_0^T g^2(x)dx$ are well defined. If the expressions $\phi = \left(\int_0^T f^2(x)dx\right)^{\frac{1}{2}}$ and $\gamma = \left(\int_0^T g^2(x)dx\right)^{\frac{1}{2}}$ are nonzero, then the correlation between $f(x)$ and $g(x)$ on $[0, T]$ is defined to be

$$C(f, g) = \frac{\int_0^T f(x)g(x)dx}{\phi\gamma}. \tag{1.2}$$

This is the natural extension of covariance, or of (normalized) dot product, to the continuous case. If we regard the vector $\alpha = (a_1, a_2, \ldots, a_n)$ as a step function $\alpha(x)$ on the interval $(0, n]$, where $\alpha(x) \equiv a_i$ on the subinterval $(i - 1, i]$, and similarly regard $\beta = (b_1, b_2, \ldots, b_n)$ as the step function $\beta(x)$ on $(0, n]$ where $\beta(x) \equiv b_i$, for $i - 1 < x \leq i$, then clearly, $\int_0^n \alpha(x)\beta(x)dx = \sum_{i=1}^n a_i b_i$, so that $C(\alpha, \beta)$ has the same value whether computed by (1.1) or (1.2).

1.3 Binary correlation

Suppose that $\alpha = (a_1, a_2, \ldots, a_n)$ and $\beta = (b_1, b_2, \ldots, b_n)$ are both binary vectors and specifically that the a_i's and b_i's are restricted to the two values $+1$ and -1. Then both $|\alpha| = \left(\sum_{i=1}^n a_i^2\right)^{\frac{1}{2}} = \sqrt{n} = \left(\sum_{i=1}^n b_i^2\right)^{\frac{1}{2}} = |\beta|$, from which

$$C(\alpha, \beta) = \frac{1}{n}\sum_{i=1}^n a_i b_i = \frac{1}{n}(A - D) = \frac{A - D}{A + D},$$

where A is the number of times, for i from 1 to n, that a_i and b_i agree, and D is the number of times that a_i and b_i disagree. Clearly, $A + D = n$, since a_i and b_i either agree or disagree at each value of i.

Because $(+1)(+1) = (-1)(-1) = +1$, whereas $(+1)(-1) = (-1)(+1) = -1$, each agreement between a_i and b_i contributes $+1$, and each disagreement contributes -1, to the sum $\sum_{i=1}^{n} a_i b_i$. If a_i and b_i agree completely, then $A = n$, $D = 0$, and $C(\alpha, \beta) = \frac{A-0}{A+0} = \frac{n}{n} = +1$. If a_i and b_i disagree everywhere, then $A = 0$, $D = n$, and $C(\alpha, \beta) = \frac{0-D}{0+D} = \frac{-n}{n} = -1$. If agreements and disagreements occur equally often, then $A = D = \frac{n}{2}$, $A - D = 0$, and $C(\alpha, \beta) = \frac{A-D}{A+D} = \frac{0}{n} = 0$.

When binary vectors are used where the two values taken on are r and s, with $r > s$, the correlation defined as a normalized dot product is in general less useful than the modified correlation obtained from the comparison of agreements and disagreements: $\bar{C}(\alpha, \beta) = \frac{A-D}{A+D}$.

1.4 Complex correlation

When working with complex-valued vectors or functions, the appropriate notion of dot product is the Hermitian dot product, for which, if $\alpha = (a_1, a_2, \ldots, a_n)$ and $\beta = (b_1, b_2, \ldots, b_n)$, we have $|\alpha| = \sqrt{(\alpha \cdot \alpha)} = \left(\sum_{i=1}^{n} a_i a_i^*\right)^{\frac{1}{2}} = \left(\sum_{i=1}^{n} |a_i|^2\right)^{\frac{1}{2}}$, $|\beta| = \sqrt{(\beta \cdot \beta)} = \left(\sum_{i=1}^{n} b_i b_i^*\right)^{\frac{1}{2}} = \left(\sum_{i=1}^{n} |b_i|^2\right)^{\frac{1}{2}}$, $(\alpha \cdot \beta) = \sum_{i=1}^{n} a_i b_i^*$, and the correlation $C(\alpha, \beta)$ is given by

$$C(\alpha, \beta) = \frac{(\alpha \cdot \beta)}{|\alpha||\beta|} = \frac{\sum_{i=1}^{n} a_i b_i^*}{\left(\sum_{i=1}^{n} |a_i|^2\right)^{\frac{1}{2}} \left(\sum_{i=1}^{n} |b_i|^2\right)^{\frac{1}{2}}},$$

where z^* denotes the complex conjugate of z. That is, if $z = x + y\sqrt{-1}$ then $z^* = x - y\sqrt{-1}$, so that $zz^* = x^2 + y^2 = |z|^2$.

If the complex numbers happen to be real numbers, these definitions reduce to the previous ones for real-valued vectors/data/functions, since $z = z^*$ if and only if z is real. However, in order for a complex number to have a correlation of $+1$ with itself (which is clearly what we desire), it is necessary to use the Hermitian dot product. Note that in the simple case $z = \sqrt{-1}$, if we used $(z \cdot z) = z^2$ we would get $(z \cdot z) = -1$, whereas with $(z \cdot z) = zz^* = |z|^2$ we get $(z \cdot z) = +1$.

1.5 Mutual orthogonality

It is well known that there are many possible infinite sets of pairwise orthogonal functions on a given interval $[0, T]$. For example, the Fourier functions $\{1, \cos nx, \sin nx \mid 1 \leq n < \infty\}$ are mutually orthogonal on the interval $[0, 2\pi]$. The Walsh functions provide an infinite set of mutually orthogonal

binary-valued functions on [0, 1]. Since orthogonal means a correlation value
of 0, it is possible to have an arbitrarily large number of mutually uncorrelated
vectors, if we do not limit the dimensionality (i.e., the number of components)
of the vectors.

The Walsh functions are square-wave functions on [0, 1], which jump between the values -1 and $+1$, and there are 2^n such functions when the unit
interval is divided into 2^n subintervals. That is, there are 2^n orthogonal vectors
of $+1$'s and -1's with 2^n components. (More generally, with d-component
real vectors, there are at most d mutually orthogonal vectors, because mutually orthogonal vectors are linearly independent, and the maximum number of linearly independent vectors it contains is the dimension of a vector
space.) The Walsh functions will be properly defined, and further discussed, in
Chapter 2.

In a similar way, the number of mutually orthogonal Fourier sine and cosine
functions on $[0, 2\pi]$ is finite if one imposes a bandwidth constraint, that is, an
upper limit on the frequency of the sine and cosine functions that can be used.

1.6 The simplex bound on mutual negative correlation

It is possible for any finite number n of vectors (data, sets, functions) to be
mutually negatively correlated, if there is no dimensionality limitation on the
vectors. Two vectors can point in opposite directions, in which case the angle
between them is $180°$, and $\cos 180° = -1$. If α is any unit vector, that is, a
vector for which $|\alpha| = 1$, and $\beta = -\alpha$, then $C(\alpha, \beta) = (\alpha \cdot -\alpha) = -|\alpha|^2 =$
-1, the lowest possible value of correlation. The vectors α and β lie on a
single line through the origin, so that one dimension suffices to provide this
example.

Three vectors in the plane can be picked with a mutual separation of $120°$
between each pair of them. Since $\cos 120° = -\frac{1}{2}$, we can find three unit vec-
tors $\alpha_1, \alpha_2, \alpha_3$ for which $(\alpha_1 \cdot \alpha_2) = (\alpha_1 \cdot \alpha_3) = (\alpha_2 \cdot \alpha_3) = -\frac{1}{2}$. For example,
the vectors $\alpha_1 = (1, 0), \alpha_2 = \frac{1}{2}(-1, \sqrt{3}), \alpha_3 = \frac{1}{2}(-1, -\sqrt{3})$ have this property. These three vectors define the vertices of an equilateral triangle inscribed
in the unit circle of the xy-plane.

Similarly, four vectors in three-dimensional space can be located at the
vertices of a regular tetrahedron inscribed in the unit sphere. The angle between
each pair of these vectors is $109°28'16.493\dots''$, the cosine of which is $-\frac{1}{3}$.

In general, n unit vectors can be placed at the n vertices of the $(n-1)$-
dimensional regular simplex inscribed in the $(n-1)$-dimensional unit hy-
persphere, achieving a mutual negative correlation of $\frac{-1}{(n-1)}$ for all $n \geq 2$.

Fortunately, the fact that no better set of vectors exists can be proved by simple arithmetic, with no need of geometric knowledge or intuition.

Theorem 1.1 (SIMPLEX BOUND). *Let $A = \{\alpha_1, \alpha_2, \ldots, \alpha_n\}$ be a set of n unit vectors (i.e., $(\alpha_i, \alpha_i) = 1$ for all i, $1 \le i \le n$) and let $c_{ij} = (\alpha_i \cdot \alpha_j)$ for all i, j. Then*

$$\max_{i \ne j} c_{ij} \ge \frac{-1}{n-1}.$$

(Note that $\max_{i \ne j} c_{ij}$ is the worst (largest) correlation value between any two vectors in the set A.)

Proof. For the members of any finite set S, the maximum value of some parameter must be at least as big as the average value. Hence

$$\max_{i \ne j} c_{ij} \ge \text{average}_{i \ne j} c_{ij} = \frac{1}{n(n-1)} \sum_{i \ne j} c_{ij}$$

$$= \frac{1}{n(n-1)} \left\{ \sum_{i=1}^{n} \sum_{j=1}^{n} c_{ij} - \sum_{i=1}^{n} c_{ii} \right\}$$

$$= \frac{1}{n(n-1)} \left\{ \sum_{i=1}^{n} \sum_{j=1}^{n} (\alpha_i \cdot \alpha_j) - \sum_{i=1}^{n} |\alpha_i|^2 \right\}$$

$$= \frac{1}{n(n-1)} \left\{ \left(\sum_{i=1}^{n} \alpha_i \cdot \sum_{j=1}^{n} \alpha_j \right) - \sum_{i=1}^{n} 1 \right\}$$

$$= \frac{1}{n(n-1)} \left\{ \left| \sum_{i=1}^{n} \alpha_i \right|^2 - n \right\} \ge \frac{1}{n(n-1)} (0 - n) = \frac{-1}{n-1}.$$

$\square$

Selecting n vectors that correspond to the vertices of the $(n-1)$-dimensional simplex is the basic way to achieve this bound. If the m components of the vectors are restricted to the binary values $\frac{+1}{\sqrt{m}}$ and $\frac{-1}{\sqrt{m}}$, the same proof idea yields a slightly stronger result. (This is equally true if the vectors are in fact step functions on $[0, T]$ that take on only the values $\frac{+1}{\sqrt{T}}$ and $\frac{-1}{\sqrt{T}}$.)

Theorem 1.2 *Let $A = \{\alpha_1, \alpha_2, \ldots, \alpha_n\}$ be a set of n binary unit vectors (i.e., the m components of α_i are restricted to the values $\pm\frac{1}{\sqrt{m}}$ for each i, $1 \le i \le n$) and let $c_{ij} = (\alpha_i \cdot \alpha_j)$ for all i, j. Then*

$$\max_{i \ne j} c_{ij} \ge \begin{cases} \frac{-1}{n-1}, & n \text{ is even,} \\ \frac{-1}{n}, & n \text{ is odd.} \end{cases}$$

Proof. We proceed as in the proof of Theorem 1.1, up to

$$\max_{i \neq j} c_{ij} \geq \frac{1}{n(n-1)} \left\{ \left| \sum_{i=1}^{n} \alpha_i \right|^2 - n \right\}.$$

We now observe that each of the m components of $\sum_{i=1}^{n} \alpha_i$ is a sum of n terms, each of which is either $+\frac{1}{\sqrt{m}}$ or $-\frac{1}{\sqrt{m}}$. If n is *odd*, the sum of n such terms cannot be 0 and must have an absolute value of at least $\frac{1}{\sqrt{m}}$. (For the sum to be 0, half the n terms would have to be $+\frac{1}{\sqrt{m}}$ and half $-\frac{1}{\sqrt{m}}$, which cannot happen for odd n.) Hence, we have $\left| \sum_{i=1}^{n} \alpha_i \right|^2 \geq \left| \left(\frac{1}{\sqrt{m}}, \frac{1}{\sqrt{m}}, \ldots, \frac{1}{\sqrt{m}} \right) \right|^2 = m \cdot \left(\frac{1}{\sqrt{m}} \right)^2 = 1$, from which

$$\max_{i \neq j} c_{ij} \geq \frac{1}{n(n-1)}(1-n) = -\frac{1}{n}.$$

Hence in this case,

$$\max_{i \neq j} c_{ij} \geq \begin{cases} \frac{-1}{n-1}, & n \text{ is even,} \\ \frac{-1}{n}, & n \text{ is odd.} \end{cases}$$

$\square$

1.7 Autocorrelation

The autocorrelation function $C(\tau)$ of a function $f(t)$ is the correlation between $f(t)$ and $f(t + \tau)$, regarded as a function of τ. It is customary to distinguish three cases:

1. Finite or aperiodic autocorrelation is computed on the assumption that $f(t)$ is identically 0 outside some interval $[0, T]$. This leads to

$$C_f^F(\tau) = \frac{\int_0^{T-\tau} f(t) f(t+\tau) dt}{\int_0^T |f(t)|^2 \, dt}.$$

2. Infinite autocorrelation is computed on the assumption that $f(t)$ is defined for all t, $-\infty < t < \infty$, with $\int_{-\infty}^{\infty} |f(t)|^2 \, dt < \infty$, and we can calculate

$$C_f^I(\tau) = \frac{\int_{-\infty}^{\infty} f(t) f(t+\tau) dt}{\int_{-\infty}^{\infty} |f(t)|^2 \, dt}.$$

3. Periodic autocorrelation is computed on the assumption that $f(t)$ is periodic with some period P, so that $f(t + P) = f(t)$ for all t. In this case, we compute

$$C_f^P(\tau) = \frac{\int_0^P f(t) f(t+\tau) dt}{\int_0^P |f(t)|^2 \, dt}.$$

The same concept of autocorrelation applies to sequences.

1. If $S = \{s_i\}$ is a sequence defined for $1 \leq i \leq n$, the finite autocorrelation of S is given by

$$C_S^F(\tau) = \frac{\sum_{i=1}^{n-\tau} s_i s_{i+\tau}}{\sum_{i=1}^{n} |s_i|^2}.$$

2. If $S = \{s_i\}$ is defined for all i, $-\infty < i < \infty$, and $\sum_{i=-\infty}^{\infty} |s_i|^2 < \infty$, then the infinite autocorrelation of S is given by

$$C_S^I(\tau) = \frac{\sum_{i=-\infty}^{\infty} s_i s_{i+\tau}}{\sum_{i=-\infty}^{\infty} |s_i|^2}.$$

3. If $S = \{s_i\}$ is periodic with period P, so that $s_i = s_{i+P}$ for all i, then the periodic autocorrelation of S is given by

$$C_S^P(\tau) = \frac{\sum_{i=1}^{P} s_i s_{i+\tau}}{\sum_{i=1}^{P} |s_i|^2}.$$

Note that in all the cases considered, $C(0) = 1$ and $C(-\tau) = C(\tau)$. However, if we are working with complex values, and therefore use the Hermitian dot product, we find that $C(-\tau) = C^*(\tau)$.

1.8 Crosscorrelation

Suppose that $f(t)$ and $g(t)$ are two functions of the continuous variable t. The crosscorrelation function $R_{f,g}(\tau)$ between f and g is defined, analogous to the autocorrelation function, on one of the following three assumptions:

1. Finite crosscorrelation is computed on the assumption that $f(t)$ and $g(t)$ are identically 0 outside of some interval $[0, T]$, leading to

$$R_{f,g}^F(\tau) = \frac{\int_0^{T-\tau} f(t)g(t+\tau)dt}{\left(\int_0^T |f(t)|^2 \, dt\right)^{\frac{1}{2}} \left(\int_0^T |g(t)|^2 \, dt\right)^{\frac{1}{2}}}.$$

2. Infinite crosscorrelation is computed on the assumption that $f(t)$ and $g(t)$ are defined and square-integrable on $(-\infty, \infty)$, leading to

$$R_{f,g}^I(\tau) = \frac{\int_{-\infty}^{\infty} f(t)g(t+\tau)dt}{\left(\int_{-\infty}^{\infty} |f(t)|^2 \, dt\right)^{\frac{1}{2}} \left(\int_{-\infty}^{\infty} |g(t)|^2 \, dt\right)^{\frac{1}{2}}}.$$

3. Periodic crosscorrelation assumes that both $f(t)$ and $g(t)$ are periodic with a common periodicity P, so that $f(t + P) \equiv f(t)$ and $g(t + P) \equiv g(t)$ for all t. In this case, we have

$$R_{f,g}^P(\tau) = \frac{\int_0^P f(t)g(t + \tau)dt}{\left(\int_0^P |f(t)|^2\, dt\right)^{\frac{1}{2}} \left(\int_0^P |g(t)|^2\, dt\right)^{\frac{1}{2}}}.$$

Similarly,

1. If we have two sequences $S = \{s_i\}$ and $T = \{t_i\}$, both defined for $1 \le i \le n$, the finite crosscorrelation between S and T is defined by

$$R_{S,T}^F(\tau) = \frac{\sum_{i=1}^{n-\tau} s_i t_{i+\tau}}{\left(\sum_{i=1}^n |s_i|^2\right)^{\frac{1}{2}} \left(\sum_{i=1}^n |t_i|^2\right)^{\frac{1}{2}}}.$$

2. If $S = \{s_i\}$ and $T = \{t_i\}$ are defined for all i, $-\infty < i < \infty$, and if both $\sum_{i=-\infty}^{\infty} |s_i|^2 < \infty$ and $\sum_{i=-\infty}^{\infty} |t_i|^2 < \infty$, then the infinite crosscorrelation between S and T is given by

$$R_{S,T}^I(\tau) = \frac{\sum_{i=-\infty}^{\infty} s_i t_{i+\tau}}{\left(\sum_{i=-\infty}^{\infty} |s_i|^2\right)^{\frac{1}{2}} \left(\sum_{i=-\infty}^{\infty} |t_i|^2\right)^{\frac{1}{2}}}.$$

3. Finally, if both $S = \{s_i\}$ and $T = \{t_i\}$ are periodic with period P, so that $s_i = s_{i+P}$ and $t_i = t_{i+P}$ for all i, then the periodic crosscorrelation between S and T is defined to be

$$R_{S,T}^P(\tau) = \frac{\sum_{i=1}^{P} s_i t_{i+\tau}}{\left(\sum_{i=1}^P |s_i|^2\right)^{\frac{1}{2}} \left(\sum_{i=1}^P |t_i|^2\right)^{\frac{1}{2}}}.$$

We note that, in all cases, the crosscorrelation reduces to the autocorrelation if the two functions, or the two sequences, being correlated are in fact the same. That is,

$$R_{f,f}^X(\tau) = C_f^X(\tau), \quad \text{where } X \text{ is any of } F, I, \text{ or } P,$$

and

$$R_{S,S}^X(\tau) = C_S^X(\tau), \quad \text{where } X \text{ is any of } F, I, \text{ or } P.$$

Moreover, we mention that the Hermitian dot product has to be used in the complex case. When we correlate $f(t)$ against $g(t + \tau)$, the effect is basically the same as correlating $g(t)$ against $f(t - \tau)$, except in the complex case, where we must remember the effect of the Hermitian dot product. Similarly,

s_i correlated with $t_{i+\tau}$ is essentially the same as t_i correlated with $s_{i-\tau}$. In the most general case (i.e., the complex case), we therefore have

$$R^X_{f,g}(-\tau) = \left(R^X_{g,f}(\tau)\right)^*, \quad \text{where } X \text{ is any of } F, I, \text{ or } P,$$

and

$$R^X_{S,T}(-\tau) = \left(R^X_{T,S}(\tau)\right)^*, \quad \text{where } X \text{ is any of } F, I, \text{ or } P.$$

We should also note that the convolution V between $f(t)$ and $g(t)$ (or between $\{s_i\}$ and $\{t_i\}$) is in fact the unnormalized crosscorrelation between $f(t)$ and $g(-t)$ (or between $\{s_i\}$ and $\{t_{-i}\}$). Thus

$$V^F_{f,g}(\tau) = \int_0^T f(t)g(\tau - t)dt$$

$$V^I_{f,g}(\tau) = \int_{-\infty}^{\infty} f(t)g(\tau - t)dt$$

$$V^P_{f,g}(\tau) = \int_0^P f(t)g(\tau - t)dt$$

and

$$V^F_{S,T}(\tau) = \sum_{i=1}^{n-\tau} s_i t_{\tau-i}$$

$$V^I_{S,T}(\tau) = \sum_{i=-\infty}^{\infty} s_i t_{\tau-i}$$

$$V^P_{S,T}(\tau) = \sum_{i=1}^{P} s_i t_{\tau-i}.$$

2

Applications of Correlation to the Communication of Information

2.1 The maximum likelihood detector

In order to communicate information from a sender to a receiver, there must be more than one possible message that the sender is able to transmit and the receiver is able to detect. If only one message were possible, its receipt would be a foregone conclusion, and it would convey no information.

It is important to distinguish here between a signal and a message. If the sender is capable of transmitting only one signal, but has the choice of whether or not to send it at a given time, then *signal* and *no signal* are two distinct messages, and the decision of which one to send does indeed convey information.

According to Claude Shannon's *Theory of Information* (Shannon, 1948), if there are N possible messages, $m_1, m_2, \ldots, m_N$, which might be sent, and the a priori probabilities of these N messages are $p_1, p_2, \ldots, p_N$, respectively, then the amount of information conveyed by knowing which one of these messages was actually sent is given by the expression

$$H(p_1, p_2, \ldots, p_N) = -\sum_{i=1}^{N} p_i \log_2 p_i, \tag{2.1}$$

where the information measure H is called the entropy of the probability distribution $\{p_1, p_2, \ldots, p_N\}$ and measures both the a priori uncertainty concerning what will occur and the a posteriori information gained as a result of removing this uncertainty. In Eq. (2.1), the use of logarithms to the base 2 has the effect of measuring information in bits, where one bit of information is the amount of uncertainty removed when one learns the outcome of an experiment that, a priori, had two equally likely possible outcomes.

Most real communication channels are noisy. That is, the signals that are received do not look identical to the signals that are sent. As a result, the receiver must decide which signal was actually sent, given the actual signal that was

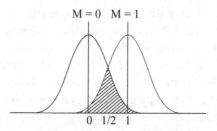

Figure 2.1. Ideal model for the gaussian binary symmetric channel.

received. As a result of noise in the channel, there is some probability, hopefully small, that an incorrect decision will have been made. The remaining uncertainty as to what was sent, given what was received, is called the equivocation in the channel. If the set of possible transmitted signals is represented by X, and the set of possible received signals is denoted by Y, then the a priori uncertainty, as calculated in Eq. (2.1), is written as $H(X)$, and the a posteriori uncertainty, or equivocation, is denoted $H_Y(X)$, the uncertainty regarding X given Y. The amount of information actually communicated in such a case is

$$I(X|Y) = H(X) - H_Y(X).$$

(For a fuller treatment of these basic concepts of information theory, see Shannon (1948), Gallager (1968), or Golomb, Peile, and Scholtz (1994).)

The basic problem that serves as a model of detection theory concerns the situation in which there are two possible transmitted signals, represented by the real numbers 0 and 1, and these are similarly corrupted by gaussian noise. That is, the receiver does not receive 0 or 1, but instead receives a sample from a gaussian distribution having mean M and standard deviation σ, where M is either 0 or 1. The larger the value of σ, the noisier the channel and the greater the probability that the receiver will make an incorrect decision as to what was sent. A picture of this idealized detection theory situation (in which both distributions have the same standard deviation σ) is shown in Figure 2.1. By the symmetry of the diagram, it is evident that the optimum detection strategy is to decide that if the received sample has a value less than one-half, then 0 was sent, and if the value is greater than one-half, then 1 was sent. Note, however, that the shaded portion of the figure corresponds to small probability regions where this strategy will, unavoidably, lead to an incorrect decision. (For further discussion of this model, see Selin (1965).)

A considerably more general result is that when the receiver is trying to decide which one of a set of N signals was actually sent, over a channel corrupted by gaussian noise, the optimum decision process is to perform correlation detection, that is, to calculate the correlation between the actual received signal and ideal models of each of the possible transmitted signals, and to decide that the highest value of the correlation corresponds to the signal that was actually sent. For a proof of this theorem, see Fano (1961). The optimum detector for a given channel is known as the matched filter for that channel, and the result we have just mentioned is frequently described as follows: the matched filter for the gaussian channel is a *correlation detector*.

2.2 Coherent versus incoherent detection

Most forms of electronic communication involve a carrier signal, which is a high-frequency sine wave, which is modulated (i.e., modified) by some lower frequency process that somehow embodies the information to be conveyed. In the classic forms of modulation, the sine wave is written as a function of time:

$$f(t) = A \sin(\omega t + \phi),$$

and then one of the three parameters – amplitude A, frequency ω, or phase ϕ – is made to vary with time to convey information. These are the familiar AM (amplitude modulation), FM (frequency modulation), and PM (phase modulation) systems for radio communication; see Black (1953). More recent digital communication systems are usually based on some form of PCM (pulse code modulation). In PCM systems, it is typical that some parameter (usually amplitude or phase) is switched back and forth between only two values, where the switching can occur only at multiples of a certain time period (whose reciprocal is called the chip rate) that is usually quite long compared to the period of the carrier sine wave. In some PCM systems, more than two values of a parameter are used.

Typical examples of the type of modulation that may occur are

(i) The amplitude A may be switched between the values 1 and 0. (This is equivalent to having the signal turned on and off.)

(ii) The amplitude A may be switched between the values $+1$ and -1.

(iii) The phase ϕ may be switched between the values 0 and π (i.e., $0°$ and $180°$). Note that this system of phase reversal is indistinguishable from (ii) above, where the amplitude undergoes sign reversal, since $\sin(x + \pi) = -\sin(x)$.

(iv) The phase ϕ may be switched between two values ϕ_1 and ϕ_2 that are not necessarily $180°$ apart. (Any such system is refered to as biphase modulation.)

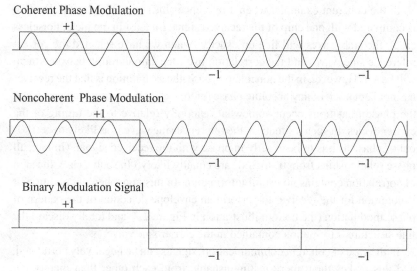

Figure 2.2. Coherent vs. noncoherent phase modulation.

(v) The phase ϕ may be switched among a finite set of values, $\phi_1, \phi_2, \ldots, \phi_n$, which are usually equally spaced modulo $360°$. For example, when $n = 4$ (called quadriphase modulation), it is customary to use $\phi_1 = 0°$, $\phi_2 = 90°$, $\phi_3 = 180°$, $\phi_4 = 270°$.

(vi) The frequency ω may be switched among a set with finite values.

To obtain the maximum information rate from the kinds of phase modulation enumerated above, it is necessary to maintain a fixed rational relationship between the chip rate (the frequency with which phase shifts are allowed to occur) and the frequency of the sine wave carrier. In such systems, which are called coherent communication systems, a phase change is allowed to occur every M cycles of the carrier sine wave, for some fixed integer M, and at no other times. In coherent systems, the clocking of the phase changes must be rigidly related to the timing of the underlying carrier sine wave.

The distinction between coherent and noncoherent communication clearly matters in the context of correlation detection. When phase coherence has been maintained, the distinction between a correlation value (between the incoming signal and a locally generated model of it at the receiver) of $+\delta$ (where $\delta > 0$) and a correlation value of $-\delta$ is meaningful and can be used to convey information. When phase coherence is not maintained, it is impossible to distinguish between correlation values of $+\delta$ and $-\delta$. This is illustrated in Figure 2.2.

In the coherent example, when a reference sine wave of one chip duration is compared with one chip of the received signal (at least in the ideal, noiseless case), the sine waves will line up exactly (when we have selected the correct reference sine wave) and the correlation over one chip duration between them will be +1. However, in the noncoherent case, the assumption is that the receiver has not kept track of any absolute phase reference for the carrier sine wave, and the phase transitions occur somewhat randomly relative to the timing of the carrier cycles. In such a situation, the reference sine wave is as likely to be 180° out of phase as it is to be exactly in phase with the received signal. (In fact, all phase relationships from 0° to 360° are equally likely.) In such a case, the sign of correlation contains no useful information. In this noncoherent condition, it is common for the receiver to perform an envelope detection of the pattern of phase modulation (the bottom illustration in Figure 2.2) and totally discard the fine structure information contained in the carrier sine wave.

In the case of coherent communication, signals can be negatively correlated, and this makes them more distinguishable from each other than merely being uncorrelated. The greatest distinguishability occurs between diametrically opposite signals, $s(t)$ and $-s(t)$, which have a normalized crosscorrelation of -1. On the other hand, in the case of noncoherent communications, maximum distinguishablity occurs when two signals are orthogonal, that is, when their crosscorrelation has the value 0.

2.3 Orthogonal, biorthogonal, and simplex codes

By Theorem 1.1 in Chapter 1, a set of $n > 1$ normalized signals (all having unit energy and unit duration) are *maximally uncorrelated* iff

$$c_{ij} = -1/(n - 1) \text{ for all } 1 \le i \ne j \le n. \tag{2.2}$$

Furthermore, by Theorem 1.2 in Chapter 1, if the n signals are binary, a necessary condition for the bound in Eq. (2.2) to be achieved is that n must be even.

Signal sets achieving the bound (2.2) are called simplex codes, since the bound is attained by the n signals corresponding to the vectors from the center to the n vertices of an $(n - 1)$-dimensional simplex. (See Section 1.6 in Chapter 1.) From the discussion in Section 2.2, we see that simplex codes are useful if and only if coherent detection is employed at the receiver. Otherwise, the best that can be achieved (when c_{ij} cannot be distinguished from $-c_{ij}$) is the case that $c_{ij} = 0$ for all $1 \le i \ne j \le n$. A set of n signals that achieves this is called an orthogonal code, since the signals in the set are pairwise orthogonal.

Suppose we have an orthogonal signal set $\{\alpha_1, \alpha_2, \ldots, \alpha_n\}$ and we adjoin to it the signals $\{-\alpha_1, -\alpha_2, \ldots, -\alpha_n\}$. The enlarged signal set is called a bi-orthogonal code. If we let $\{\beta_1, \beta_2, \ldots, \beta_{2n}\} = \{\alpha_1, \alpha_2, \ldots, \alpha_n, -\alpha_1, -\alpha_2, \ldots, -\alpha_n\}$, we have

$$(\beta_i \cdot \beta_j) = \begin{cases} 1 & \text{if } i = j, \\ -1 & \text{if } |i - j| = n, \\ 0 & \text{otherwise.} \end{cases}$$

(Obviously, in order to distinguish α_i from $-\alpha_i$, this is meaningful only in the case of coherent detection.) For the set of $2n$ signals β_i we see that

$$\text{average}_{i \neq j} c_{ij} = \frac{1}{(2n)^2 - (2n)} \cdot (-2n) = \frac{-1}{2n - 1}, \tag{2.3}$$

since for each of the $2n$ choices of β_i there is exactly one choice of β_j with $(\beta_i \cdot \beta_j) = -1$, and all the other choices of $j \neq i$ give $(\beta_i \cdot \beta_j) = 0$. Since the size of the signal set $\{\beta_i : i = 1, \ldots, 2n\}$ is $2n$, Eq. (2.3) achieves the simplex bound of Theorem 1.1 (in Chapter 1) for the average value of crosscorrelation, though not for the maximum value.

2.4 Hadamard matrices and code construction

A Hadamard matrix H of order n is an $n \times n$ matrix whose entries are restricted to the values $+1$ and -1, with the property

$$HH^T = nI,$$

where H^T is the transpose of H, and I is the $n \times n$ identity matrix. (If the rows of H are $\alpha_1, \alpha_2, \ldots, \alpha_n$, this equation requires $(\alpha_i \cdot \alpha_j) = 0$ for all $i \neq j$.)

Theorem 2.1 *The order n of a Hadamard matrix H is a member of the set $\{1, 2, 4t\}$ where t runs through the positive integers.*

Proof. $[+1]$ and $\begin{bmatrix} +1 & +1 \\ +1 & -1 \end{bmatrix}$ are Hadamard matrices of order 1 and 2. For $n > 2$, let the rows of H be $\alpha_1, \alpha_2, \ldots, \alpha_n$. Multiplying any of the columns of H by -1 has no effect on the pairwise orthogonality of the rows, nor does any permutation of the columns of H. For convenience, we multiply those columns of H by -1 where α_1 has a -1, so that in the normalized matrix H', $\alpha_1' = (+1, +1, \ldots, +1)$. We also permute the columns so that $\alpha_2' = (+1, +1, \ldots, +1, -1, -1, \ldots, -1)$. Since $(\alpha_1' \cdot \alpha_2') = 0$, α_2' must consist of equally many $+1$'s and -1's, hence $n/2$ of each, and n must be even. We further permute the columns, without affecting the appearance of α_1' or α_2', to get $\alpha_3' = (+1, \ldots, +1, -1, \ldots, -1, +1, \ldots, +1, -1, \ldots, -1)$ where we have r times $+1$ followed by $(\frac{n}{2} - r)$ times -1 and then s times $+1$ followed by $(\frac{n}{2} - s)$ times -1. Because $(\alpha'_1 \cdot \alpha'_3) = 0$, we have $(\alpha'_1 \cdot \alpha'_3) = (r + s) - (\frac{n}{2} - r) - (\frac{n}{2} - s) = 0$ from which $r + s = \frac{n}{2}$. Because $(\alpha'_2 \cdot \alpha'_3) = 0$, we have $(\alpha'_2 \cdot \alpha'_3) = r - (\frac{n}{2} - r) - s + (\frac{n}{2} - s) = 0$, from which $r - s = 0$. Hence $r = s = \frac{n}{4}$, and n must be a multiple of 4. □

It has long been conjectured that Hadamard matrices exist for all $n = 4t$, but this is still far from proved. However, the smallest value of $4t$ for which no Hadamard matrix was known (for many years) was $4t = 428$. But by Kharaghani and Tayfeh-Rezaie (2004) that case was solved, and now the smallest unknown case is $4t = 668$.

For a description of many of the systematic methods for constructing Hadamard matrices, see Hall (1986).

Theorem 2.2 *If there is a Hadamard matrix H of order $n > 1$, then there is a binary orthogonal code consisting of n vectors each of length n, there is a binary bi-orthogonal code consisting of $2n$ vectors each of length n, and a binary simplex code consisting of n vectors each of length $n - 1$.*

Proof. The row vectors of H need only to be normalized by multiplication of each row by $\frac{1}{\sqrt{n}}$ to obtain the n vectors of an orthogonal code. If to these n vectors their negatives are also adjoined, the resulting set of $2n$ vectors is a bi-orthogonal code. Finally, we may transform H to a new Hadamard matrix H' in which the first column of H' consists entirely of $+1$'s, by multiplying those rows of H that begin with -1 by the scalar -1. (Because $(\alpha_i \cdot \alpha_j) = 0$ we will still have $(a\alpha_i \cdot b\alpha_j) = 0$ for any scalars a and b.) Then we remove the first column of H' to get n row vectors of length $n - 1$. Call these row vectors $\sigma_1, \sigma_2, \ldots, \sigma_n$. Then $(\sigma_i \cdot \sigma_j) = \frac{-1}{n-1}$ for all $i \neq j$, because σ_i and σ_j differ from α_i and α_j by dropping a coordinate in which α_i and α_j were equal. For $(\alpha_i \cdot \alpha_j)$ we had $\frac{A-D}{A+D} = \frac{0}{n}$, so for $(\sigma_i \cdot \sigma_j)$ we have $\frac{A-D}{A+D} = \frac{-1}{n-1}$. □

The converse of Theorem 2.2 is obviously also true: If any one of the binary $n \times n$ orthogonal code, the $2n \times n$ bi-orthogonal code, or the $n \times (n - 1)$ simplex code exists, then all three exist, as does the Hadamard matrix of order n (which really is the $n \times n$ orthogonal code). Slightly less obvious is

Theorem 2.3 *If there is a Hadamard matrix H of order $2n > 2$, then there is a simplex code with n binary codewords, each of length $2n - 2$.*

Proof. Since $HH^T = 2nI$, $\frac{1}{\sqrt{2n}}H$ is an orthogonal matrix (in the usual sense of matrix theory), so the columns of H (as well as the rows) are mutually orthogonal. As in Theorem 2.2, we first transform H to H', where all the rows of H' begin with $+1$, and then we drop this "all $+1$'s" column to get H''. Because the second column of H' is orthogonal to the first column of H', it consists of equally many $+1$'s and -1's. This second column of H' is the first column of H''. Thus H'' has n rows beginning with $+1$ and n rows beginning with -1. Keep only the n rows beginning with $+1$'s, and from these drop the initial $+1$, leaving a set S of n vectors of length $2n - 2$. Since both of the dropped positions were agreements between any two of the corresponding rows of H', for any two distinct vectors σ_i and σ_j in S we have $(\sigma_i \cdot \sigma_j) = \frac{A-D}{A+D} = \frac{-2}{2n-2} = \frac{-1}{n-1}$, the simplex bound. □

Here are several general constructions for Hadamard matrices. For each of these, the Hadamard property is easily verified.

1. If $A = (a_{ij})$ is an $n \times n$ matrix and $B = (b_{lk})$ is an $m \times m$ matrix, then the Kronecker product $A * B$ is defined as

$$A * B = \begin{bmatrix} a_{11}B & a_{12}B & \cdots & a_{1n}B \\ a_{21}B & a_{22}B & \cdots & a_{2n}B \\ \vdots & & & \vdots \\ a_{n1}B & a_{n2}B & \cdots & a_{nn}B \end{bmatrix}$$

regarded as an $mn \times mn$ matrix. If A and B are both Hadamard matrices, then so too are $A * B$ and $B * A$.
2. Starting with $A = \begin{bmatrix} +1 & +1 \\ +1 & -1 \end{bmatrix}$ and repeatedly taking the Kronecker product, one obtains Hadamard matrices of all orders 2^k, $k \geq 1$.
3. If B is a Hadamard matrix of order n, then $A * B$ is a Hadamard matrix of order $2n$, where $A = \begin{bmatrix} +1 & +1 \\ +1 & -1 \end{bmatrix}$.
4. Suppose A, B, C, and D are $t \times t$ symmetric circulant matrices of $+1$'s and -1's with $A^2 + B^2 + C^2 + D^2 = 4tI$. (The matrix $M = (m_{ij})$ is a symmetric circulant matrix iff

$$M = \begin{pmatrix} m_{11} & m_{12} & m_{13} & \cdots & m_{13} & m_{12} \\ m_{12} & m_{11} & m_{12} & \cdots & m_{14} & m_{13} \\ m_{13} & m_{12} & m_{11} & \cdots & m_{15} & m_{14} \\ \vdots & \vdots & \cdots & \ddots & \vdots & \vdots \\ m_{13} & m_{14} & m_{15} & \cdots & m_{11} & m_{12} \\ m_{12} & m_{13} & m_{14} & \cdots & m_{12} & m_{11} \end{pmatrix}.$$

$$\text{Then } H = \begin{pmatrix} A & B & C & D \\ -B & A & -D & C \\ -C & D & A & -B \\ -D & -C & B & A \end{pmatrix} \text{ is a Hadamard matrix of order } 4t.$$

This result is Williamson's theorem, and the construction given for H is called Williamson's construction (Williamson 1944).

Williamson's construction can be used to obtain Hadamard matrices of orders 12 ($t = 3$) and 20 ($t = 5$), as well as many larger orders. The first example of a Hadamard matrix of order 92 was found using Williamson's construction with $t = 23$ (Baumert, Golomb, and Hall 1962).

2.5 Cyclic Hadamard matrices

For $n > 1$, the only known example of a Hadamard matrix that is also a circulant matrix has order $n = 4$:

$$H = \begin{pmatrix} -1 & +1 & +1 & +1 \\ +1 & -1 & +1 & +1 \\ +1 & +1 & -1 & +1 \\ +1 & +1 & +1 & -1 \end{pmatrix}.$$

There is a significant literature of partial results that no such example exists with $n > 4$. (See, for example, Turyn (1968) and Jungnickel and Pott (1999).)

However, there are many examples of $n \times n$ Hadamard matrices H that consist of an $(n - 1) \times (n - 1)$ circulant matrix with a border added (top-most row and left-most column of H) consisting entirely of $+1$'s. Some examples of these are

$$H_4 = \begin{bmatrix} +1 & +1 & +1 & +1 \\ +1 & +1 & -1 & -1 \\ +1 & -1 & +1 & -1 \\ +1 & -1 & -1 & +1 \end{bmatrix},$$

$$H_8 = \begin{bmatrix} + & + & + & + & + & + & + & + \\ + & - & - & - & + & - & + & + \\ + & + & - & - & - & + & - & + \\ + & + & + & - & - & - & + & - \\ + & - & + & + & - & - & - & + \\ + & + & - & + & + & - & - & - \\ + & - & + & - & + & + & - & - \\ + & - & - & + & - & + & + & - \end{bmatrix}$$

and

$$H_{12} = \begin{bmatrix}
+ & + & + & + & + & + & + & + & + & + & + & + \\
+ & - & + & - & + & + & + & - & - & - & + & - \\
+ & - & - & + & - & + & + & + & - & - & - & + \\
+ & + & - & - & + & - & + & + & + & - & - & - \\
+ & - & + & - & - & + & - & + & + & + & - & - \\
+ & - & - & + & - & - & + & - & + & + & + & - \\
+ & - & - & - & + & - & - & + & - & + & + & + \\
+ & + & - & - & - & + & - & - & + & - & + & + \\
+ & + & + & - & - & - & + & - & - & + & - & + \\
+ & + & + & + & - & - & - & + & - & - & + & - \\
+ & - & + & + & + & - & - & - & + & - & - & + \\
+ & + & - & + & + & + & - & - & - & + & - & -
\end{bmatrix}.$$

Examples of this type are sometimes called cyclic Hadamard matrices and are in one-to-one correspondence with Paley-Hadamard difference sets. Therefore, these matrices are also called cyclic Paley-Hadamard matrices. (For an extensive treatment of these, see Baumert (1971) and Jungnickel and Pott (1999).) All known examples of cyclic Paley-Hadamard matrices of order $n = 4t$ have $n - 1$ belonging to one of three sequences:

(a) $4t - 1 = 2^k - 1, k \geq 1$.
(b) $4t - 1 = p$, p a prime.
(c) $4t - 1 = p(p + 2)$, where p and $q = p + 2$ form a twin prime.

Examples of type (a) can be obtained for all $k \geq 1$ by taking the top row of the circulant to be an m-sequence, that is, a maximum-length linear shift register sequence of period $2^k - 1$, and replacing the 0's and 1's of the m-sequence by $+1$'s and -1's, respectively. (For the theory of m-sequences, see Golomb (1982). We will present this in Chapter 5 for a general setting.) Additional examples of type (a), for composite values of $k > 5$, are obtained by the Gordon-Mills-Welch (GMW) construction (see Gordon, Mills, and Welch (1962), which we will present in Chapter 8); and more constructions for type (a) are from multi-trace-terms, hyperovals, and the Kasami power functions, which will be introduced in Chapter 9.

Examples of type (b) can be obtained for all primes $p = 4t - 1$ by the Legendre sequence construction, taking the top row of the circulant to be

$$-1, \left(\frac{1}{p}\right), \left(\frac{2}{p}\right), \dots, \left(\frac{p-1}{p}\right),$$

where $\left(\frac{a}{p}\right)$ is the Legendre symbol, or quadratic character, modulo p, defined for $1 \le a \le p - 1$ by

$$\left(\frac{a}{p}\right) = \begin{cases} +1 & \text{if } x^2 \equiv a \,(\text{mod}\, p) \text{ for some } x, \\ -1 & \text{otherwise.} \end{cases}$$

These examples are also called Paley difference sets. Additional examples of type (b) are obtained by Hall's sextic residue sequence construction when $p = 4t - 1 = 4a^2 + 27$ (Hall 1986), described as follows. Let u be a primitive element of $\mathbb{Z}_p$, and let G consist of all sextic residues modulo p, that is,

$$G = \{a \in \mathbb{Z}_p^* \mid x^6 \equiv a \,(\text{mod}\, p), x \in \mathbb{Z}_p\}$$

or equivalently,

$$G = \left\{ u^{6i} \mid 0 \le i < \frac{p-1}{6} \right\}$$

for the primitive element u. We denote by $S = \{s_0, s_1, \ldots, s_{p-1}\}$ the Hall's sextic residue sequence of period p whose elements are given by $s_0 = 1$ and

$$s_i = \begin{cases} 0 & \text{if } i \in G \cup u^3 G \cup u^{i_0} G \\ 1 & \text{otherwise,} \end{cases}$$

where $u^{i_0} G$ is the coset containing 3. (Note the coset $u^j G = \{u^j h \mid h \in G\}$.) Thus the top row of the circulant can be taken as

$$(-1)^{s_i}, i = 0, 1, \ldots, p - 1.$$

If $a \equiv 0 \,(\text{mod}\, p)$, the Legendre symbol $\left(\frac{a}{p}\right)$ is 0. If p and q are distinct odd primes, the Jacobi symbol $\left(\frac{a}{pq}\right)$ is defined to be $\left(\frac{a}{p}\right)\left(\frac{a}{q}\right)$, the product of the Legendre symbols. In the Legendre sequence construction for examples of type (b), we replaced $\left(\frac{0}{p}\right) = 0$ with the value -1. To get the twin prime construction for cyclic Hadamard matrices of type (c) with $4t - 1 = p(p+2) = pq$, we take the top row of the circulant matrix to be a modification of the sequence of Jacobi symbols $\left\{ \left(\frac{0}{pq}\right), \left(\frac{1}{pq}\right), \left(\frac{2}{pq}\right), \ldots, \left(\frac{pq-1}{pq}\right) \right\}$ where we use the Jacobi symbol $\left(\frac{a}{pq}\right)$ whenever this is nonzero; we replace it by $+1$ for $a \in \{0, q, 2q, \ldots, (p-1)q\}$, and we replace it by -1 for $a \in \{p, 2p, 3p, \ldots, (q-1)p\}$.

Sequence lengths of types (b) and (c) are obviously disjoint sets, because no prime is a product of twin primes. The only overlap of lengths of types (a) and (c) occurs with $2^4 - 1 = 3 \cdot 5$, and here the matrix examples are in fact the same. Parametrically, the overlaps of the sequence lengths of type (a) and (b) are precisely the Mersenne primes, $p = 4t - 1 = 2^k - 1$. However, the matrix examples are the same only for $p = 2^2 - 1 = 3$ and $p = 2^3 - 1 = 7$, if we are considering m-sequences and Legendre sequences. At $p = 2^5 - 1 = 31$, the m-sequence construction gives the same result as Hall's sextic residue sequence construction. At $p = 2^7 - 1 = 127$, the m-sequence, the Legendre sequence, and the sextic residue sequence constructions all give inequivalent examples. Moreover, at $4t - 1 = 127$, Baumert (1971), by complete search, found three more inequivalent examples (thus six inequivalent constructions altogether) that did not form part of any known families. At $v = 2^8 - 1 = 255$, Cheng (1983) found four inequivalent cyclic Paley-Hadamard difference sets, of which two were previously unknown. The searches at $2^9 - 1$ (by Dreier and Smith 1991) and at $2^{10} - 1$ (by Gaal and Golomb 1999) have also been completed, finding five and ten inequivalent examples, respectively. However, several new families of constructions have also recently been discovered. Collectively, they explain all the examples that have been found by exhaustive search at $v = 2^n - 1$, for all $n \leq 10$. The set of all known constructions to date is described in Chapter 9. Examples of type (c) were first described by Stanton and Sprott (1958).

3

Finite Fields

Finite fields are used in most of the known constructions of pseudorandom sequences and analysis of periods, correlations, and linear spans of linear feedback shift register (LFSR) sequences and nonlinear generated sequences. They are also important in many cryptographic primitive algorithms, such as the Diffie-Hellman key exchange, the Digital Signature Standard (DSS), the El Gamal public-key encryption, elliptic curve public-key cryptography, and LFSR (or Torus) based public-key cryptography. Finite fields and shift register sequences are also used in algebraic error-correcting codes, in code-division multiple-access (CDMA) communications, and in many other applications beyond the scope of this book. This chapter gives a description of these fields and some properties that are frequently used in sequence design and cryptography. Section 3.1 introduces definitions of algebraic structures of groups, rings and fields, and polynomials. Section 3.2 shows the construction of the finite field $GF(p^n)$. Section 3.3 presents the basic theory of finite fields. Section 3.4 discusses minimal polynomials. Section 3.5 introduces subfields, trace functions, bases, and computation of the minimal polynomials over intermediate subfields. Computation of a power of a trace function is shown in Section 3.6. And, the last section presents some counting numbers related to finite fields.

3.1 Algebraic structures

In this section, we give the definitions of the algebraic structures of groups, rings and fields, polynomials, and some concepts that will be needed for the study of finite fields in the later sections.

We use the following notations for the sets of numbers: $\mathbb{N}$, the set of natural numbers (positive integers); $\mathbb{Z}$, the set of integers; $\mathbb{Q}$, the set of rational numbers; $\mathbb{R}$, the set of real numbers; and $\mathbb{C}$, the set of complex numbers.

3.1.1 Groups

Binary operation

Let S be a set and let $S \times S$ denote the set of all ordered pairs (s, t) with $s \in S$ and $t \in S$. Then a mapping from $S \times S$ into S is called a (binary) operation on S. Under this definition we require that the image of $(s, t) \in S \times S$ must be in S. This is the closure property of an operation.

Algebraic structure

A set S together with one or more operations on S is called an algebraic structure.

Definition 3.1 *A group is a set G together with a binary operation $*$ on G such that the following three properties hold:*

(i) $*$ is associative; that is, for any $a, b, c \in G$,

$$a * (b * c) = (a * b) * c.$$

(ii) There is an identity (or unit) element e in G such that for all $a \in G$,

$$a * e = e * a = a.$$

(iii) For each $a \in G$, there exists an inverse element $a^{-1} \in G$ such that

$$a * a^{-1} = a^{-1} * a = e.$$

Sometimes, we denote the group as a triple $(G, *, e)$. If the group also satisfies
(iv) For all $a, b \in G$,

$$a * b = b * a,$$

then the group is called Abelian *or* commutative.

Note. From the definition, the identity element e of G is unique, and the inverse element of any element $a \in G$ is also unique.

For simplicity, we will frequently use the notation of ordinary multiplication to designate the operation in the group, writing simply ab instead of $a * b$. But it must be emphasized that by doing so we do not assume that the operation actually is ordinary multiplication. If G is an Abelian group, we also write $a + b$ instead of $a * b$ and $-a$ instead of a^{-1}, that is, using additive notation.

The associative law guarantees that expressions such as $a_1 a_2 \cdots a_n$ with $a_j \in G, 1 \le j \le n$, are unambiguous, since no matter how we insert parentheses, the expression will always represent the same element of G. To indicate the n-fold composition of an element $a \in G$ with itself, where $n \in N$, we will write

$$a^n = aa \cdots a, \ (n \text{ factors } a)$$

if using multiplicative notation, and we call a^n the nth power of a. If using additive notation for the operation $*$ on G, we write

$$na = a + a + \cdots + a \ (n \text{ summands } a)$$

and sometimes it is called n times a.

Following customary notation, we have the following rules:

Multiplicative Notation		Additive Notation	
a^{-n}	$= (a^{-1})^n$	$(-n)a$	$= n(-a)$
$a^n a^m$	$= a^{n+m}$	$na + ma$	$= (n+m)a$
$(a^n)^m$	$= a^{nm}$	$m(na)$	$= (mn)a$

For $n = 0 \in \mathbb{Z}$, we adopt $a^0 = e$ by convention in multiplicative notation and $0a = 0$ in additive notation, where the last zero represents the identity element of G.

Example 3.1 The following number sets together with ordinary addition and multiplication are groups: $(\mathbb{Z}, +, 0)$, $(\mathbb{R}, +, 0)$, and $(\mathbb{R}, \cdot, 1)$. We denote by $\mathbb{Z}_n$ the set of the remainders of all integers on division by n where n is a positive integer, that is, $\mathbb{Z}_n = \{0, 1, \ldots, n-1\}$, and $\mathbb{Z}_n^*$ is the set of nonzero elements in $\mathbb{Z}_n$ that are coprime to n. Let $a + b$ and ab be the ordinary sum and product of a and b reduced modulo n, respectively. Then we have

(a) $(\mathbb{Z}_2, +, 0)$, $(\mathbb{Z}_6, +, 0)$, and $(\mathbb{Z}_5, +, 0)$ are groups with respect to addition.
(b) $(\mathbb{Z}_5^*, \cdot, 1)$ forms a group with respect to multiplication.

In general, we have the following results.

Proposition 3.1 *With the notation above,*

(a) $(\mathbb{Z}_n, +, 0)$ *forms a group for any positive integer n, which is called the additive group of integers modulo n.*
(b) $(\mathbb{Z}_p^*, \cdot, 1)$ *forms a group for any prime p, which is called the multiplicative group of integers modulo p.*

Before we give a proof for this proposition, we list the following basic fact on integers whose proof can be found in any book on number theory, say Hardy and Wright (1980), Ireland and Rosen (1991).

Fact 3.1 *Let p be a prime number. For any integer a : $0 < a < p$, a and p are coprime. In other words, the greatest common divisor of a and p, denoted by gcd(a, p), is equal to 1. Moreover, there exist two integers u and v such that $au + pv = 1$ where $0 < u < p$.*

Proof of Proposition 3.1. Let $a \pmod{n}$ denote the remainder of a when divided by n. If the result of an addition or multiplication of ordinary is to be reduced $\pmod{n}$, the same answer is obtained if some integers which appear are reduced $\pmod{n}$ during intermediate steps of the calculation.

(a) For $a, b \in \mathbb{Z}_n$, $a + b$ is the remainder on division by n of the ordinary sum of a and b. So, $a + b \in \mathbb{Z}_n$. (i) For $a, b, c \in \mathbb{Z}_n$, now consider a, b, c as integers. Then $a + (b + c) = (a + b) + c$, as does $a + (b + c) \equiv (a + b) + c \pmod{n}$. (ii) For any $a \in \mathbb{Z}_n$, $a + 0 = 0 + a = a$, so 0 is the identity element in $\mathbb{Z}_n$. (iii) For each $a \in \mathbb{Z}_n$, there exists a positive integer $b \in \mathbb{Z}_n$ such that $a + b = n$. Note that $n \equiv 0 \pmod{n}$. So, $a + b = b + a = 0$. Thus b is the inverse element of a in $\mathbb{Z}_n$.

(b) For $a, b \in \mathbb{Z}_p$, ab is the remainder on division by p of the ordinary product of a and b. So, $ab \in \mathbb{Z}_p^*$. (i) For $a, b, c \in \mathbb{Z}_p^*$, now consider a, b, c as integers. Then $a(bc) = (ab)c$, as does $a(bc) \equiv (ab)c \pmod{p}$, which means that $\mathbb{Z}_p^*$ satisfies the associative law. (ii) It is obvious that 1 is the identity element in $\mathbb{Z}_p^*$. (iii) For any $a \in \mathbb{Z}_p^*$, since p is a prime, according to Fact 1, there exist two integers u and v such that

$$au + pv = 1,$$

where $0 < u < p$. Note that $au + pv \equiv au \pmod{p}$. Therefore we get $au \equiv 1 \pmod{p} \Longrightarrow a^{-1} = u \in \mathbb{Z}_p^*$. Thus u is the inverse of a in $\mathbb{Z}_p^*$. $\square$

Definition 3.2 *A multiplicative group (resp. additive group) G is said to be cyclic if there is an element $a \in G$ such that for any $b \in G$ there is some integer i with $b = a^i$ (resp. $b = ia$). Such an element a is called a* generator *of the cyclic group, and we write $G = <a>$.*

Example 3.2 The following groups are cyclic.

(a) For $(\mathbb{Z}, +, 0)$, the additive group of integers, both 1 and -1 are generators.
(b) For $(\mathbb{Z}_6, +, 0)$, the additive group of integers modulo 6, 1 and 5 are generators.
(c) For $(\mathbb{Z}_3^*, \cdot, 1)$, the multiplicative group of integers modulo 3, 2 is the generator.

(d) For $(\mathbb{Z}_5^*, \cdot, 1)$, the multiplicative group of integers modulo 5, 2 and 3 are generators, that is,

$$\mathbb{Z}_5^* = <2> = \{2^0 = 1, 2^1 = 2, 2^2 = 4, 2^3 \equiv 3 \ (\mathrm{mod}\ 5)\}$$
$$(\text{where } 2^4 \equiv 1 \ (\mathrm{mod}\ 5))$$
$$= <3> = \{3^0 = 1, 3^1 = 3, 3^2 \equiv 4, 3^3 \equiv 2 \ (\mathrm{mod}\ 5)\}$$
$$(\text{where } 3^4 \equiv 1 \ (\mathrm{mod}\ 5)).$$

Definition 3.3 *A group is called* finite *(resp.* infinite*) if it contains finitely (resp. infinitely) many elements. The number of elements in a finite group is called the* order *of the group G . We will write* $|G|$ *for the order of the finite group G.*

3.1.2 Rings and fields

In most of the number systems used in elementary arithmetic there are two distinct binary operations: addition and multiplication. Examples are provided by the integers, the rational numbers, and the real numbers. We now define a type of algebraic structure known as a ring that shares some of the basic properties of these number systems.

Definition 3.4 *A ring* $(R, +, \cdot)$ *is a set R, together with two binary operations, denoted by* $+$ *and* $\cdot$*, such that*

(i) R is an Abelian group with respect to $+$*.*
(ii) $\cdot$ *is associative, that is,* $(a \cdot b) \cdot c = a \cdot (b \cdot c)$ *for all* $a, b, c \in R$*.*
(iii) The distributive laws hold; that is, for all $a, b, c \in R$ *we have*

$$a \cdot (b + c) = (a \cdot b) + (a \cdot c) \ \text{and} \ (b + c) \cdot a = (b \cdot a) + (c \cdot a).$$

Example 3.3 The following are examples of rings from number systems.

(a) $(\mathbb{Z}, +, \cdot)$, $(\mathbb{Q}, +, \cdot)$, $(\mathbb{R}, +, \cdot)$, and $(\mathbb{C}, +, \cdot)$ are rings.
(b) $(\mathbb{Z}_4, +, \cdot)$ forms a ring of four elements. (This is the algebraic structure that underlies $\mathbb{Z}_4$ codes.)
(c) $(\mathbb{Z}_n, +, \cdot)$ forms a ring, called the residue class ring modulo n.

Let $(F, +, \cdot)$ be a ring, and let $F^* = \{a \in R \,|\, a \neq 0\}$, the set of nonzero elements of F.

Definition 3.5 *A field is a ring* $(F, +, \cdot)$ *such that* F^* *together with multiplication* $\cdot$ *forms a commutative group.*

According to the definition, a field is a set F on which two binary operations, called addition and multiplication, are defined and that contains two distinct elements 0 and 1 (we denote the multiplicative identity e by 1) with $0 \neq 1$. Furthermore, $(F, +, 0)$ is an Abelian group with respect to addition having 0 as the identity element, and $(F^*, \cdot, 1)$ forms an Abelian group with respect to multiplication having 1 as the identity element. The two operations of addition and multiplication are linked by the distributive law $a(b + c) = ab + ac$. The second distributive law $(b + c)a = ba + ca$ follows automatically from the commutativity of multiplication. The element 0 is called the zero element and 1 is called the multiplicative identity element or simply the identity.

Definition 3.6 *A finite field is a field that contains a finite number of elements. This number is called the* order *of the field.*

Finite fields are also called Galois fields after their discoverer, Evariste Galois (1811–1832).

Example 3.4 The following structures are fields or finite fields.

(a) $(\mathbb{Q}, +, \cdot)$, $(\mathbb{R}, +, \cdot)$, and $(\mathbb{C}, +, \cdot)$ are fields.
(b) $(\mathbb{Z}_2, +, \cdot)$ forms a finite field of order 2. The elements of this field are 0 and 1, and the operation tables are shown as follows:

$+$	0	1		$\cdot$	0	1
0	0	1		0	0	0
1	1	0		1	0	1

The elements 0 and 1 are called binary elements.

(c) $(\mathbb{Z}_5, +, \cdot)$ forms a finite field of order 5. The elements of this field are 0, 1, 2, 3, and 4, and the operation tables are shown as follows:

$+$	0	1	2	3	4		$\cdot$	0	1	2	3	4
0	0	1	2	3	4		0	0	0	0	0	0
1	1	2	3	4	0		1	0	1	2	3	4
2	2	3	4	0	1		2	0	2	4	1	3
3	3	4	0	1	2		3	0	3	1	4	2
4	4	0	1	2	3		4	0	4	3	2	1

In general, we have the following result.

Proposition 3.2 $(\mathbb{Z}_p, +, \cdot)$ *is a field if p is a prime.*

Proof. From Proposition 1, both $(\mathbb{Z}_p, +, 0)$ and $(\mathbb{Z}_p^*, \cdot, 1)$ are Abelian groups. Because integers satisfy the distributive law, then $a(b + c) = ab + ac$ for any

$a, b, c \in \mathbb{Z}_p$, so that their remainders are equal. Hence $\mathbb{Z}_p$ satisfies the distributive law. According to the definition of fields, $(\mathbb{Z}_p, +, \cdot)$ is a field. $\square$

Note. The converse of the proposition is also true; that is, if $(\mathbb{Z}_n, +, \cdot)$ is a field where $n > 1$ is an positive integer, then n must be a prime.

We denote $(\mathbb{Z}_p, +, \cdot)$ simply by $\mathbb{Z}_p$, or $GF(p)$, which is called the residue class field modulo p. These are the first examples of finite fields that we encounter.

3.1.3 Polynomials

Let R be an arbitrary ring. A polynomial over R is an expression of the form

$$f(x) = \sum_{i=0}^{n} a_i x^i = a_0 + a_1 x + \cdots + a_n x^n,$$

where n is a nonnegative integer, the coefficients $a_i, 0 \le i \le n$, are elements of R, and x is a symbol not belonging to R, called an interdeterminate over R. We adopt the convention that a term $a_i x^i$ with $a_i = 0$ need not be written down. In particular, the polynomial $f(x)$ above may then also be given in the equivalent form $f(x) = a_0 + a_1 x + \cdots a_n x^n + 0x^{n+1} + \cdots + 0x^{n+h}$, where h is any positive integer. When comparing two polynomials $f(x)$ and $g(x)$ over R, it is therefore possible to assume that they both involve the same powers of x.

(a) The polynomials

$$f(x) = \sum_{i=0}^{n} a_i x^i \text{ and } g(x) = \sum_{i=0}^{n} b_i x^i$$

over R are considered equal if and only if $a_i = b_i$ for $0 \le i \le n$.
(b) The sum of $f(x)$ and $g(x)$ is defined by

$$f(x) + g(x) = \sum_{i=0}^{n} (a_i + b_i) x^i.$$

(c) To define the product of two polynomials over R, let

$$f(x) = \sum_{i=0}^{n} a_i x^i \text{ and } g(x) = \sum_{i=0}^{m} b_j x^j$$

and set

$$f(x)g(x) = \sum_{k=0}^{n+m} c_k x^k, \text{ where } c_k = \sum_{0 \le i < k} a_i b_{k-i}.$$

It is easily seen that with these operations the set of polynomials over R forms a ring.

Definition 3.7 *The ring formed by the polynomials over R with the above operations is called the polynomial ring over R and denoted by $R[x]$; that is,*

$$R[x] = \left\{ \sum_{i=0}^{n} a_i x^i \mid a_i \in R, n \geq 0 \right\}.$$

The zero element of $R[x]$ is the polynomial for which all coefficients are 0. This polynomial is called the zero polynomial and denoted by 0.

Definition 3.8 *Let $\sum_{i=0}^{n} a_i x^i$ be a polynomial over R that is not the zero polynomial, so that we can suppose $a_n \neq 0$. Then a_n is called the* leading coefficient *of $f(x)$ and a_0 the* constant term, *whereas n is called the* degree *of $f(x)$, denoted by $n = \deg(f(x)) = \deg(f)$. By convention, we set $\deg(0) = -\infty$. Polynomials of degree ≤ 0 are called* constant polynomials. *If the leading coefficient of $f(x)$ is 1, then $f(x)$ is called a* monic polynomial.

In the following, we consider polynomials over fields. Let F denote a field (not necessarily finite).

Theorem 3.1 *Let $f, g \in F[x]$. Then*

$$\deg(f + g) \leq \max\{\deg(f), \deg(g)\},$$
$$\deg(fg) = \deg(f) + \deg(g).$$

A proof of this theorem can be easily carried out by computing the leading coefficient of the sum and the product of two polynomials, which is omitted here.

Divisability

The polynomial $g \in F[x]$ divides the polynomial $f \in F[x]$ if there exists a polynomial $h \in F[x]$ such that $f = gh$. We also say that g is a divisor of f, f is a multiple of g, or f is divisible by g.

Henceforth, we will use the notation $g \mid f$ if g is a divisor of f, and $g \nmid f$ if g is not a divisor of f or f is not divisible by g where both f and g are polynomials in $F[x]$ or integers in $\mathbb{Z}$.

Theorem 3.2 (Division Algorithm) *Let $g \neq 0$ be a polynomial in $F[x]$. Then for any $f \in F[x]$ there exist polynomials $q, r \in F[x]$ such that*

$$f = qg + r, \text{ where } \deg(r) < \deg(g).$$

Definition 3.9 *Let $f, g \in F[x]$, not both of which are 0. If $d \in F[x]$ satisfying the following conditions: (i) d divides f and g and (ii) any polynomial $c \in F[x]$*

dividing both f and g divides d, then d is called the greatest common divisor *of two polynomials f and g, denoted by* $d = \gcd(f, g)$. *If* $\gcd(f, g) = 1$, *then the two polynomials f and g are said to be* relatively prime.

Theorem 3.3 *Let* $d = \gcd(f, g)$ *with* $f, g, d \in F[x]$. *Then d can be expressed in the form*

$$d(x) = u(x)f(x) + g(x)v(x) \text{ with } u(x), v(x) \in F[x].$$

Proof. The greatest common divisor d of two polynomials $f, g \in F[x]$ can be computed by the Euclidean algorithm as follows. Without loss of generality, we may suppose that $g \neq 0$ and $g \nmid f$. We then repeatedly use the division algorithm in the following manner:

$$
\begin{array}{llll}
f = q_1 g + r_1, & \deg(r_1) < \deg(g), & r_1 \neq 0 \\
g = q_2 r_1 + r_2, & \deg(r_2) < \deg(r_1), & r_2 \neq 0 \\
r_1 = q_3 r_2 + r_3, & \deg(r_3) < \deg(r_2), & r_3 \neq 0 \\
\quad\vdots \\
r_{s-2} = q_s r_{s-1} + r_s, & \deg(r_s) < \deg(r_{s-1}), & r_{s-1} \neq 0 \\
r_{s-1} = q_{s+1} r_s.
\end{array}
$$

Here $q_1, \ldots, q_{s+1}$ and $r_1, \ldots, r_s$ are polynomials in $F[x]$. Since $\deg(g)$ is finite, the procedure must stop after finitely many steps. If the last nonzero remainder r_s has the leading coefficient b, then $d = b^{-1}r_s$. (This step is to make the leading coefficient of d equal to 1.) The polynomials u and v can be found by working backward from the above identities and substituting the $r_s, r_{s-1}, \ldots, r_1$ into $d = b^{-1}r_s$. $\qquad\square$

Definition 3.10 *A polynomial* $p \in F[x]$ *is called* irreducible *over F if p has positive degree and* $p = bc$ *with* $b, c \in F[x]$ *implies that either b or c is a constant polynomial. Otherwise, p is called* reducible *over F.*

Irreducible polynomials are of fundamental importance for the structure of the ring $F[x]$ and the structure of linear feedback shift register sequences, because polynomials in $F[x]$ can be written as products of irreducible polynomials in an essentially unique manner. We list this result in the following theorem without proof. The proof can be found in Redei (1959) or Lidl and Niederreiter (1983).

Theorem 3.4 (UNIQUE FACTORIZATION IN $F[x]$) *Any polynomial* $f \in F[x]$ *of positive degree can be written in the form*

$$f = a p_1^{e_1} p_2^{e_2} \cdots p_k^{e_k},$$

where $a \in F$, $p_1, \ldots, p_k$ are distinct monic irreducible polynomials in $F[x]$, and $e_1, \ldots, e_k$ are positive integers. Moreover, this factorization is unique apart from the order in which the factors occur.

Definition 3.11 *An element $b \in F$ is called a* root *(or* zero*) of the polynomial $f \in F[x]$ if $f(b) = 0$.*

The following concept regarding polynomials over F corresponds to the period of the corresponding sequences.

Definition 3.12 *For a polynomial $f(x)$ over F, the* period *(or* order*) of $f(x)$ is the least positive integer t such that $f(x) \mid (x^t - 1)$, denoted by* per(f) $= t$.

For example, $f(x) = x^3 + x + 1$ over $\mathbb{Z}_2$ has period 7 by noticing that $x^7 + 1 = (x + 1)(x^3 + x + 1)(x^3 + x^2 + 1)$. But if $f(x) = x^5 + x^2$ over $\mathbb{Z}_2$, then $f(x)$ cannot be a factor of $x^t + 1$ for any positive integer t; that is, the period of f does not exist. However, we have the following result whose proof is omitted.

Fact 3.2 *For $f(x) \in F[x]$ where F is finite, if $f(0) \neq 0$, then the period of $f(x)$ exists.*

3.2 Construction of GF(p^n)

In this section, we show the construction for the finite field $GF(p^n)$. In Section 3.1, we have already seen the finite field $GF(p) = \mathbb{Z}_p$ of order p where p is a prime. The elements of $GF(p)$ are $\{0, 1, \ldots, p - 1\}$, and the addition $+$ and and multiplication $\cdot$ are carried out modulo p. We need the following fact (see van der Waerden (1953)) to construct the finite field $GF(p^n)$.

Fact 3.3 *For every prime p and every degree $n > 1$, there is at least one irreducible polynomial of degree n over $\mathbb{Z}_p$.*

Let n be a positive integer. To construct the finite field $GF(p^n)$ of order p^n, we choose $f(x)$ to be an irreducible polynomial over $GF(p)$ of degree n. Let us agree that α is a formal symbol that satisfies $f(\alpha) = 0$. Let

$$GF(p^n) = \{a_0 + a_1\alpha + \cdots + a_{n-1}\alpha^{n-1} \mid a_i \in GF(p)\}.$$

We define two operations: $+$ and $\cdot$ on $GF(p^n)$ as follows. For $g(\alpha), h(\alpha) \in GF(p^n)$,

$$g(\alpha) = \sum_{i=0}^{n-1} a_i\alpha^i \text{ and } h(\alpha) = \sum_{i=0}^{n-1} b_i\alpha^i,$$

Addition: $g(\alpha) + h(\alpha) = \sum_{i=0}^{n-1}(a_i + b_i)\alpha^i \in GF(p^n)$

Multiplication: $g(\alpha) \cdot h(\alpha) = r(\alpha)$

where $r(\alpha)$ is computed as follows:

(i) Multiply $g(\alpha)$ and $h(\alpha)$ according to the multiplication of polynomials; that is,

$$g(\alpha)h(\alpha) = \sum_{i=0}^{n-1} a_i\alpha^i \sum_{j=0}^{n-1} b_j\alpha^j$$

$$= \sum_{k=0}^{2n} c_k\alpha^k = c(\alpha),$$

where

$$c_k = \sum_{i+j=k} a_i b_j.$$

(ii) Applying the division algorithm to $c(\alpha)$ and $f(\alpha)$, we can get two polynomials $q(\alpha)$ and $r(\alpha)$ such that

$$c(\alpha) = q(\alpha)f(\alpha) + r(\alpha) \text{ with } \deg(r(\alpha)) < n.$$

Since α satisfies $f(\alpha) = 0$, we have $c(\alpha) = r(\alpha) \in GF(p^n)$. In other words, $r(x)$ is the remainder of the product $g(x)h(x)$ divided by $f(x)$.

Theorem 3.5 *The set $GF(p^n)$ together with the two operations defined above forms a finite field, and the order of this field is p^n.*

Proof. Using the same argument as we did for the polynomial ring $GF(p)[x]$, it is clear that $(GF(p^n), +, \cdot)$, where two operations $+$ and $\cdot$ are defined above, forms a commutative ring. So, we only need to prove that for each $g \neq 0 \in GF(p^n)$, there exists $g^{-1} \in GF(p^n)$ such that $gg^{-1} = 1$. Since $f(x)$ is irreducible over $GF(p)$ of degree n and $\deg(g(x)) \leq n - 1$, then $g(x)$ is relatively prime to $f(x)$; that is, $\gcd(f(x), g(x)) = 1$. According to Theorem 3.3, there exist two polynomials $u(x), v(x) \in GF(p)[x]$ such that

$$g(x)u(x) + f(x)v(x) = 1.$$

Substituting α into the above identity,

$$g(\alpha)u(\alpha) + f(\alpha)v(\alpha) = 1.$$

Since $f(\alpha) = 0$, we get $g(\alpha)u(\alpha) = 1$. Note that if $\deg(u) \geq n$, applying the division algorithm to $u(x)$ and $f(x)$, we then have $u(x) = q_1(x)f(x) + r_1(x)$ where $\deg(r_1) < n$. By substituting α, it follows that $u(\alpha) = r_1(\alpha) \in GF(p^n)$.

So, we can suppose that $\deg(u) \leq n - 1$. Hence we have $g(\alpha)u(\alpha) = 1$ where $u(\alpha) \in GF(p^n) \Longrightarrow g^{-1} = u(\alpha)$. $\qquad\square$

The polynomial $f(x)$ is called a defining polynomial of $GF(p^n)$ and α is called a defining element of $GF(p^n)$ over $GF(p)$. The finite field $GF(p^n)$ is also called a Galois field. From the construction of $GF(p^n)$, $f(\alpha) = 0$. Thus α is a root of $f(x)$ in $GF(p^n)$. We also say that $GF(p^n)$ is obtained from $GF(p)$ by adjoining to $GF(p)$ a zero of $f(x)$ or $GF(p^n)$ is a finite extension of $GF(p)$.

Example 3.5 Let $p = 2$ and $f(x) = x^3 + x + 1$. Then $f(x)$ is irreducible over $GF(2)$. Let α be a root of $f(x)$; that is, $f(\alpha) = 0$. The finite field $GF(2^3)$ is defined by

$$GF(2^3) = \{a_0 + a_1\alpha + a_2\alpha^2 \mid a_i \in GF(2)\}.$$

$GF(2^3)$, defined by $f(x) = x^3 + x + 1$ and $f(\alpha) = 0$

as a 3-tuple	as a polynomial	as a power of α
$000 =$	0	$= 0$
$100 =$	1	$= 1$
$010 =$	α	$= \alpha$
$001 =$	α^2	$= \alpha^2$
$110 =$	$1 + \alpha$	$= \alpha^3$
$011 =$	$\alpha + \alpha^2$	$= \alpha^4$
$111 =$	$1 + \alpha + \alpha^2$	$= \alpha^5$
$101 =$	$1 + \alpha^2$	$= \alpha^6$
	$\alpha^7 = 1$	

Note that $GF(2^3)^* = <\alpha>$; that is, the nonzero elements of $GF(2^3)$ form a cyclic group of order 7 with generator α, where $\alpha^7 = 1$.

Example 3.6 Let $p = 2$ and $f(x) = x^4 + x + 1$. Then $f(x)$ is irreducible over $GF(2)$. Let α be a root of $f(x)$; that is, $f(\alpha) = 0$. The finite field $GF(2^4)$ is defined by

$$GF(2^4) = \{a_0 + a_1\alpha + a_2\alpha^2 + a_3\alpha^3 \mid a_i \in GF(2)\}.$$

(a) A table of $GF(2^4)$ is given below.

$GF(2^4)$, defined by $f(x) = x^4 + x + 1$ and $f(\alpha) = 0$

$$
\begin{array}{rcl}
0\ 0\ 0\ 0 & = & 0 = \alpha^{\infty} \\
1\ 0\ 0\ 0 & = & 1 = \alpha^{0} \\
0\ 1\ 0\ 0 & = & \alpha \\
0\ 0\ 1\ 0 & = & \alpha^{2} \\
0\ 0\ 0\ 1 & = & \alpha^{3} \\
1\ 1\ 0\ 0 & = & \alpha^{4} \\
0\ 1\ 1\ 0 & = & \alpha^{5} \\
0\ 0\ 1\ 1 & = & \alpha^{6} \\
1\ 1\ 0\ 1 & = & \alpha^{7} \\
1\ 0\ 1\ 0 & = & \alpha^{8} \\
0\ 1\ 0\ 1 & = & \alpha^{9} \\
1\ 1\ 1\ 0 & = & \alpha^{10} \\
0\ 1\ 1\ 1 & = & \alpha^{11} \\
1\ 1\ 1\ 1 & = & \alpha^{12} \\
1\ 0\ 1\ 1 & = & \alpha^{13} \\
1\ 0\ 0\ 1 & = & \alpha^{14}
\end{array}
$$

$$(\alpha^{15} = 1)$$

(b) To add and multiply $1 + \alpha$ and $\alpha + \alpha^{3}$,

$$(1 + \alpha) + (\alpha + \alpha^{3}) = 1 + \alpha^{3}.$$

We may use the definition of multiplication in $GF(p^{n})$ to multiply these two elements. But here we will introduce a simpler way to do that. Note that $1 + \alpha = \alpha^{4}$ and $\alpha + \alpha^{3} = \alpha^{9}$. Hence

$$(1 + \alpha) \cdot (\alpha + \alpha^{3}) = \alpha^{4}\alpha^{9} = \alpha^{4+9} = \alpha^{13} = 1 + \alpha^{2} + \alpha^{3}.$$

So, the polynomial representation is efficient for performing addition, whereas the exponential representation is best for multiplication. Note that we also have $GF(2^{4})^{*} = <\alpha>$; that is, $GF(2^{4})^{*}$ is a cyclic group of order 15 with generator α, where $\alpha^{15} = 1$.

3.3 The basic theory of finite fields

3.3.1 The characteristic of a finite field

Definition 3.13 *If F is a finite field and there exists a positive integer m such that $m\beta = 0$ for every $\beta \in F$, then the least such positive integer m is called the characteristic of F and F is said to have characteristic m.*

Theorem 3.6 *Let F be a finite field. Then the characteristic of F is a prime.*

Proof. Let $|F| = q$. F contains the identity element 1. Since F is finite, the elements 1, $1 + 1 = 2, 1 + 1 + 1 = 3, \ldots$ cannot be all distinct. Therefore, there is the smallest number p such that $p = 1 + 1 + \cdots + 1$ (p times) $= 0$. This p must be a prime number (for if $rs = 0$ then $r = 0$ or $s = 0$). □

Let F be a field. A subset K of F that is itself a field under the operations of F will be called a subfield of F. F is called an extension (field) of K. If $K \neq F$, we say that K is a proper subfield of F. So, $GF(p^n)$ has characteristic p and contains $GF(p)$ as a subfield.

3.3.2 Structures of finite fields

Theorem 3.7 *Let F be a finite field and $|F| = q$ with characteristic p. Then $F = GF(p)$ if $q = p$ or F is a vector space of dimension n over $GF(p)$ if $q > p$; that is, $q = p^n$.*

Proof. If $q = p$, because $GF(p)$ is a subfield of F, we get $F = GF(p)$. Supposing $q > p$, we choose a maximal set of elements of F that are linearly independent over $GF(p)$, say $\alpha_0, \alpha_1, \ldots, \alpha_{n-1}$. Then F contains all the elements

$$a_0\alpha_0 + a_1\alpha_1 + \cdots + a_{n-1}\alpha_{n-1}, \quad a_i \in GF(p),$$

and no others. Thus F is a vector space of dimension n over $GF(p)$ and contains $q = p^n$ elements. □

Theorem 3.7 shows that if F is a finite field of order q and p is the characteristic of F, then $q = p$ or $q = p^n$ ($n \geq 1$). Two finite fields F and G are said to be isomorphic if there is a one-to-one mapping from F onto G that preserves addition and multiplication. All finite fields of order p^n are isomorphic (we omit the proof here). So, we only have two different types of finite fields: one is $GF(p)$, the residue class field modulo p, and the other is $GF(p^n)$, the extension field obtained by adjoining a zero of an irreducible polynomial of degree n over $GF(p)$ to $GF(p)$. Sometimes we denote F with order q by $\mathbb{F}_q$. For a finite field F we denote by F^* the multiplicative group of the nonzero elements of F.

Definition 3.14 *Let G be a group. For $\alpha \in G$, if r is the smallest positive integer such that $\alpha^r = 1$, then r is called the* order *of α, denoted by* $\text{ord}(\alpha) = r$.

In the following, we will establish the cyclic structure of the multiplicative group of a finite field. For doing this, we need the following fact about orders of the elements in a commutative group.

Fact 3.4 *Let $\alpha, \beta \in F^*$ with $\mathrm{ord}(\alpha) = r$ and $\mathrm{ord}(\beta) = s$. Then*

(i) $\mathrm{ord}(\alpha^m) = r/\gcd(r, m)$. Moreover $\mathrm{ord}(\alpha^m) = \mathrm{ord}(\alpha)$ if and only if $\gcd(m, r) = 1$.

(ii) If $\gcd(r, s) = 1$, then $\mathrm{ord}(\alpha\beta) = rs$.

Note that the above fact need not be true for noncommutative groups.

Theorem 3.8 *For every finite field F the multiplicative group F^* (nonzero elements of F) is cyclic.*

Proof. Let $|F| = q$. We need to prove that there is an element in F^* that has order $q - 1$. We first show the following assertion.

If we choose $\alpha \in F^*$ such that the order r of α is the maximum one, then the order s of any element $\beta \in F^*$ divides r; that is, $s|r$.

From Fact 3.4(ii) and the assumption about r, we know that $\gcd(r, s) \neq 1$. For any common prime divisor of s and r, we may write $r = p_1^d a$ and $s = p_1^e b$ where a and b are not divisible by p_1 and $d, e \geq 1$. According to Fact 3.4(i), $\mathrm{ord}(\alpha^{p_1^d}) = a$, $\mathrm{ord}(\beta^b) = p_1^e$, and $\mathrm{ord}(\alpha^{p_1^d}\beta^b) = p_1^e a$. Hence $e \leq d$ or else r would not be maximum. Thus every prime power that is a divisor of s is also a divisor of r, and so $s|r$. Next we prove that $r = q - 1$. It is clear $r \leq q - 1$. Let $g(x) = \prod_{\beta \in F^*}(x - \beta)$. Then $\deg(g) = q - 1$. From the above assertion, every $\beta \in F^*$ satisfies the equation $x^r - 1 = 0$. Consequently, $x^r - 1$ is divisible by $g(x) \Longrightarrow q - 1 = \deg(g) \leq \deg(x^r - 1) = r$. But $r \leq q - 1$; hence, $r = q - 1$. Thus $F^* = <\alpha>$.   □

Definition 3.15 *A generator of the cyclic group $GF(p^n)^*$ is called a* primitive element *of $GF(p^n)$. An irreducible polynomial over $GF(p)$ having a primitive element in $GF(p^n)$ as a zero is called a* primitive polynomial *over $GF(p)$.*

Note that not all irreducible polynomials are primitive. For example, $x^4 + x^3 + x^2 + x + 1$ is irreducible over $GF(2)$, so it can be used to generate the field $GF(2^4)$. However, it is not a primitive polynomial. According to the definition, if F is a finite field of order p^n, an element α of F is a primitive element if it has order $p^n - 1$.

Corollary 3.1 *Every finite field contains a primitive element.*

Proof. Take α to be a generator of the cyclic group F^*.   □

The following corollary follows directly from the proof of Theorem 3.8.

Corollary 3.2 (FERMAT'S THEOREM) *Every element β of a finite field of order p^n satisfies the identity*

$$\beta^{p^n} = \beta,$$

or equivalently, it is a root of the equation

$$x^{p^n} = x.$$

Thus

$$x^{p^n} - x = \prod_{\beta \in F}(x - \beta).$$

Lemma 3.1 *In any finite field of characteristic p,*

$$(x + y)^p = x^p + y^p.$$

Proof. Use the binomial expansion,

$$(x + y)^p = \sum_{k=0}^{P} \binom{p}{k} x^{p-k} y^k,$$

where

$$\binom{p}{0} = \binom{p}{p} = 1.$$

Now if $1 \leq k \leq p - 1$, then $\gcd(k!, p) = 1$. Thus

$$\binom{p}{k} = \frac{p(p-1)\cdots(p-k+1)}{k!}$$

$$= p\frac{(p-1)\cdots(p-k+1)}{k!} \equiv 0 \;(\text{mod } p). \qquad \square$$

The following result is easily proved by mathematical induction.

Corollary 3.3 *In any finite filed of characteristic p,*

$$(x + y)^{p^m} = x^{p^m} + y^{p^m}$$

for any $m \geq 1$.

3.3.3 Representations of elements

According to Theorem 3.7, $GF(p^n)$ is a vector space of dimension n over $GF(p)$. Any set of n linearly independent elements can be used as a basis for this vector space. There are two important bases for $GF(p^n)$. One is the polynomial basis $\{1, \alpha, \ldots, \alpha^{n-1}\}$, which is used to construct $GF(p^n)$ from an irreducible polynomial $f(x)$ over $GF(p)$ of degree n with $f(\alpha) = 0$ (see Section 3.1). Another is called a normal basis $\{\alpha, \alpha^p, \ldots, \alpha^{p^{n-1}}\}$ if these are linearly independent over $GF(p)$. Normal bases for $GF(p^n)$ always exist (see Lidl and Neiderreiter (1983)). We also know that $GF(p^n)^*$ is a cyclic group. Let α be a primitive element of $GF(p^n)$ and $\{\alpha_0, \alpha_1, \ldots, \alpha_{n-1}\}$ be a basis of $GF(p^n)$ over $GF(p)$. Then we can write $GF(p^n)$ as follows:

$$GF(p^n) = \{a_0\alpha_0 + a_1\alpha_1 + \cdots + a_{n-1}\alpha_{n-1} \mid a_i \in GF(p)\}$$

(Vector Representation)

$$= \{\alpha^i \mid 0 \leq i \leq p^n - 2 \text{ or } i = \infty\}$$

(Exponential Representation)

where we denote $0 = \alpha^\infty$.

Example 3.7 For the finite field $GF(2^3)$, defined by $f(x) = x^3 + x + 1$, we have the following representations.

Polynomial Basis			Normal Basis			$\alpha^7 = 1$ Exp.
1	α	α^2	α^3	α^6	α^5	
0	0	0	0	0	0	$0 = \alpha^\infty$
1	0	0	1	1	1	$1 = \alpha^0$
0	1	0	0	1	1	α
0	0	1	1	0	1	α^2
1	1	0	1	0	0	α^3
0	1	1	1	1	0	α^4
1	1	1	0	0	1	α^5
1	0	1	0	1	0	α^6

3.3.4 Computation of $GF(p^n)$

Let α be a primitive element of $GF(p^n)$. For $0 \neq \beta \in GF(p^n)$, then $\beta = \alpha^i$ for some $i \geq 0$. The nonnegative integer i is called a logarithm of β (to the base α). For small finite fields, we have the following look-up table methods for performing computations in $GF(p^n)$, which are much more efficient than using the method from the definition of the multiplication.

Method 1

Using the log and antilog tables, the elements of $GF(p^n)$ are obtained from the vector representation.

Addition: Using a vector representation: For $\beta = b_0\alpha_0 + b_1\alpha_1 + \cdots + b_{n-1}\alpha_{n-1}$ and $\gamma = c_0\alpha_0 + c_1\alpha_1 + \cdots + c_{n-1}\alpha_{n-1}$,

$$\beta + \gamma = (b_0 + c_0)\alpha_0 + (b_1 + c_1)\alpha_1 + \cdots + (b_{n-1} + c_{n-1})\alpha_{n-1}$$

where $b_i + c_i$ is performed in $GF(p)$, that is, modulo p.

Multiplication: First, by looking up in the log table, obtain

$$\beta = \alpha^r \text{ and } \gamma = \alpha^s;$$

then compute

$$\beta\gamma = \alpha^{r+s},$$

where $r + s$ is reduced modulo $p^n - 1$; finally, by looking up in the antilog table, the vector representation of α^{r+s} is retrieved.

Method 2

Using the trinomial table (or Zech's logarithm), the elements of $GF(p^n)$ are presented by the exponential representation. For $0 < t < p^n - 1$, there exists a unique $\tau(t)$ such that

$$1 + \alpha^t = \alpha^{\tau(t)}.$$

$\tau(t)$ is called Zech's logarithm of α^t. The table containing $\tau(t)$ for $0 < t < p^n - 1$ is called the trinomial table of $GF(p^n)$ (or add-one table). Then to add and multiply α^t and α^s:

Addition: $\quad\quad \alpha^t + \alpha^s = \alpha^t(1 + \alpha^{s-t}) = \alpha^t \cdot \alpha^{\tau(s-t)}$
$\quad\quad\quad\quad\quad\quad = \alpha^{t+\tau(s-t)}$ (using the trinomial table)

Multiplication: $\quad \alpha^t\alpha^s = \alpha^{t+s}$.

Example 3.8 (a) For the finite field $GF(2^4)$, given by Example 3.6, we have the following trinomial table:

t	1	2	3	4	5	6	7	8	9	10	11	12	13	14
$\tau(t)$	4	8	14	1	10	13	9	2	7	5	12	11	6	3

For multiplying α^3 times α^5, it is easy because it just involves addition of the exponents; that is, $\alpha^3 \cdot \alpha^5 = \alpha^{3+5} = \alpha^8$. For computing $\alpha^3 + \alpha^5$, we use the above trinomial table:

$$\alpha^3 + \alpha^5 = \alpha^3(1 + \alpha^2) = \alpha^{3+\tau(2)} = \alpha^{3+8} = \alpha^{11}.$$

Table 3.1. $GF(2^5)$ defined by $\alpha^5 + \alpha^3 + 1 = 0$ and its trinomial table

α^t	Vector representation	$\tau(t)$	α^t	Vector representation	$\tau(t)$
0	1 0 0 0 0	∞	16	0 0 1 1 0	7
1	0 1 0 0 0	14	17	0 0 0 1 1	18
2	0 0 1 0 0	28	18	1 0 0 1 1	17
3	0 0 0 1 0	5	19	1 1 0 1 1	8
4	0 0 0 0 1	25	20	1 1 1 1 1	12
5	1 0 0 1 0	3	21	1 1 1 0 1	27
6	0 1 0 0 1	10	22	1 1 1 0 0	15
7	1 0 1 1 0	16	23	0 1 1 1 0	11
8	0 1 0 1 1	19	24	0 0 1 1 1	9
9	1 0 1 1 1	24	25	1 0 0 0 1	4
10	1 1 0 0 1	6	26	1 1 0 1 0	29
11	1 1 1 1 0	23	27	0 1 1 0 1	21
12	0 1 1 1 1	20	28	1 0 1 0 0	25
13	1 0 1 0 1	30	29	0 1 0 1 0	28
14	1 1 0 0 0	1	30	0 0 1 0 1	13
15	0 1 1 0 0	22			

(b) Let $GF(2^5)$ be defined by the primitive polynomial $f(x) = x^5 + x^3 + 1$, and let α be a root of $f(x)$. Then α is a primitive element of $GF(2^5)$. In Table 3.1, the first column (also the fourth column) lists the exponent i in α^i, the second column (and the fifth column) lists the coefficients of α^i under the basis $(1, \alpha, \alpha^2, \alpha^3, \alpha^4)$, and the third column (also the sixth column) lists the trinomial table, that is, the values of $\tau(i)$, $i = 1, \ldots, 30$, where $1 + \alpha^i = \alpha^{\tau(i)}$. Therefore,

$$\alpha^8 + \alpha^{27} = \alpha^8(1 + \alpha^{19}) = \alpha^{8+\tau(19)} = \alpha^{8+8} = \alpha^{16},$$

$$(\tau(19) = 8 \text{ by look-up table})$$

$$\alpha^8 \cdot \alpha^{27} = \alpha^{8+27} = \alpha^4.$$

The second method only needs to store the trinomial table, which contains the values of $\tau(t)$ for $1 \le t < p^n - 1$ (see the above example). This is much more efficient than the first method (which needs to store two tables) for performing computations in small finite fields, say $p^n \le 2^{25}$. In general, for coding applications, the sizes of finite fields are small that fall in this range. However, for cryptographical applications, the fields are usually large, with sizes at least having $\log(p^n) \ge 1024$. Thus the look-up table methods are not applicable to this type of application. So, one has to perform multiplication by the

Euclidean algorithm (see the definition of the multiplication in a finite field in Section 3.2).

3.4 Minimal polynomials

Fermat's theorem (Corollary 3.2) implies that every element α of $GF(q)$ where q is a prime or a power of a prime, say $q = p^n$, satisfies the equation

$$x^q - x = 0. \tag{3.1}$$

This polynomial has all its coefficients from the prime field $GF(p)$ and is monic. However, α may satisfy an equation with a lower degree than in Eq. (3.1).

Definition 3.16 *Let α be an element in $GF(p^n)$. The minimal polynomial of α over $GF(p)$ is defined as the lowest degree monic polynomial $m(x) \in GF(p)[x]$ such that $m(\alpha) = 0$.*

Note. The minimal polynomial of any element in $GF(p^n)$ is unique.

Example 3.9 In $GF(2^4)$, defined by $\alpha^4 + \alpha + 1 = 0$, the minimal polynomials, having coefficients equal to 0 or 1, are listed in the following table.

Element	Minimal polynomial
0	x
1	$x + 1$
α	$x^4 + x + 1$
$\alpha^{-1} = \alpha^{14}$	$x^4 + x^3 + 1$
α^3	$x^4 + x^3 + x^2 + x + 1$
α^5	$x^2 + x + 1$

We will show a method for finding minimal polynomials later.

3.4.1 Properties of minimal polynomials

Assume that $m(x)$ is the minimal polynomial of $\alpha \in GF(p^n)$.

Property 3.1 (IRREDUCIBILITY) $m(x)$ *is irreducible.*

Proof. If $m(x) = g(x)h(x)$ where the degrees of both $g(x)$ and $h(x)$ are greater than zero, then $m(\alpha) = g(\alpha)h(\alpha) = 0$. Thus either $g(\alpha) = 0$ or $h(\alpha) = 0$, which contradicts the fact that $m(x)$ is the lowest degree polynomial with α as a root. $\qquad\square$

Property 3.2 (DIVISIBILITY) *For any $f(x) \in GF(p)[x]$, if $f(\alpha) = 0$, then $m(x)|f(x)$.*

Proof. Applying the division algorithm to divide $f(x)$ by $m(x)$, there exist $q(x), r(x) \in GF(p)[x]$ such that

$$f(x) = q(x)m(x) + r(x), \quad \deg(r(x)) < \deg(m(x)).$$

Substituting $x = \alpha$, this becomes

$$0 = 0 + r(\alpha).$$

Hence, $r(x)$ is a polynomial of lower degree than $m(x)$, but has α as a root. This is a contradiction unless $r(x) = 0$, and then $m(x)|f(x)$.   □

Property 3.3

$$m(x) \,|\, (x^{p^n} - x).$$

Proof. From Corollary 3.2, $\alpha \in GF(p^n) \Longrightarrow \alpha^{p^n} - \alpha = 0$. Applying Property 3.2, $m(x)|(x^{p^n} - x)$.   □

Property 3.4 $\deg(m(x)) \leq n$.

Proof. $GF(p^n)$ is a vector space of dimension n over $GF(p)$. Therefore any $n + 1$ elements, such as $1, \alpha, \ldots, \alpha^n$, are linearly dependent; that is, there exist coefficients $a_i \in GF(p)$, not all zero, such that

$$\sum_{i=0}^{n} a_i \alpha^i = 0.$$

Thus $\sum_{i=0}^{n} a_i x^i \in GF(p)[x]$ is a polynomial of degree $\leq n$ having α as a root. Therefore $\deg(m(x)) \leq n$.   □

Property 3.5 *The minimal polynomial of a primitive element of $GF(p^n)$ has degree n.*

Proof. Let α be a primitive element of $GF(p^n)$, with minimal polynomial $m(x)$ of degree d. As in Theorem 3.5 we may use $m(x)$ to generate a field F of order p^d. Since F contains α, and hence all of $GF(p^n)$, then $d \geq n$. By Property 3.4, it follows that $d = n$.   □

3.4.2 Conjugates and cyclotomic cosets

Property 3.6 *α and α^p have the same minimal polynomial. In particular, in $GF(2^n)$, α and α^2 have the same minimal polynomial.*

Proof. Let $m_\alpha(x) = \sum a_i x^i$ and $m_{\alpha^p}(x) = \sum b_i x^i$, a_i, $b_i \in GF(p)$, be the minimal polynomials of α and α^p over $GF(p)$. Note that $a_i = a_i^p$. Using Lemma 3.1,

$$m_\alpha(\alpha^p) = \sum a_i(\alpha^p)^i = \sum a_i^p(\alpha^i)^p = \sum (a_i\alpha^i)^p$$
$$= \left(\sum a_i\alpha^i\right)^p = m_\alpha(\alpha)^p = 0.$$

According to Property 3.2, $m_{\alpha^p}(x) \mid m_\alpha(x)$. From Property 3.1, $m_\alpha(x)$ is irreducible, so $m_\alpha(x) = m_{\alpha^p}(x)$. □

Definition 3.17 *Let $\alpha \in GF(p^n)$. Then the elements $\alpha, \alpha^p, \ldots, \alpha^{p^{n-1}}$ are called* conjugates *of α with respect to $GF(p)$.*

From Property 3.6 all conjugates of α have the same minimal polynomial.

Definition 3.18 *A* (cyclotomic) coset C_s *modulo $p^n - 1$ (with respect to p) is defined to be*

$$C_s = \{s, sp, \ldots, sp^{n_s-1}\},$$

where n_s is the smallest positive integer such that $s \equiv sp^{n_s}$ (mod $p^n - 1$). The subscript s is chosen as the smallest integer in C_s, and s is called the coset leader *of C_s.*

Note. In group theory, all cosets of a subgroup H of a group G have the same number of elements. The *cyclotomic cosets* are generalized cosets of a cyclic subgroup of the multiplicative group $\mathbb{Z}_m^*$, extended to the entire ring $\mathbb{Z}_m$, and have sizes that divide the order of the cyclic subgroup.

We introduce the following notation related to the cyclotomic cosets.

$$\Gamma_p(n) = \{\text{all coset leaders in } \mathbb{Z}_{p^n-1}\}, \text{ and}$$
$$\Omega_p(n) = \{C_s | s \in \Gamma_p(n)\}.$$

Then $|\Gamma_p(n)| = |\Omega_p(n)|$. Note that the operation of multiplying the elements in $\mathbb{Z}_{p^n-1}$ by p results in a partition of $\mathbb{Z}_{p^n-1}$; that is,

$$\mathbb{Z}_{p^n-1} = \cup_{s \in \Gamma_p(n)} C_s.$$

Example 3.10 We compute the cyclotomic cosets modulo $2^n - 1$ with respect to 2 for $n = 4, 5, 6$, and 7.

(a) For $n = 4$ and $p = 2$, the cyclotomic cosets modulo 15 are:

$$C_0 = \{0\}$$
$$C_1 = \{1, 2, 4, 8\}$$
$$C_3 = \{3, 6, 12, 9\}$$

$$C_5 = \{5, 10\}$$
$$C_7 = \{7, 14, 13, 11\}.$$

Moreover, we have

$$\mathbb{Z}_{15} = C_0 \cup C_1 \cup C_3 \cup C_5 \cup C_7$$

and $\Gamma_2(4) = \{0, 1, 3, 5, 7\}$.

(b) For $n = 5$ and $p = 2$, the cyclotomic cosets modulo 31 are given as follows:

$$C_0 = \{0\}$$
$$C_1 = \{1, 2, 4, 8, 16\}$$
$$C_3 = \{3, 6, 12, 24, 17\}$$
$$C_5 = \{5, 10, 20, 9, 18\}$$
$$C_7 = \{7, 14, 28, 25, 19\}$$
$$C_{11} = \{11, 22, 13, 26, 21\}$$
$$C_{15} = \{15, 30, 29, 27, 23\}.$$

The set $\Gamma_2(5)$, consisting of all coset leaders modulo 31, is given by

$$\Gamma_2(5) = \{0, 1, 3, 5, 7, 11, 15\}.$$

(c) In the following, the cosets modulo 63 and 127, respectively, have the coset leaders listed in the first columns.

Cyclotomic cosets modulo 63

0					
1	2	4	8	16	32
3	6	12	24	48	33
5	10	20	40	17	34
7	14	28	56	49	35
9	18	36			
11	22	44	25	50	37
13	26	52	41	19	38
15	30	60	57	51	39
21	42				
23	46	29	58	53	43
27	54	45			
31	62	61	59	55	47

Cyclotomic cosets modulo 127

0						
1	2	4	8	16	32	64
3	6	12	24	48	96	65
5	10	20	40	80	33	66
7	14	28	56	112	97	67
9	18	36	72	17	34	68
11	22	44	88	49	98	69
13	26	52	104	81	35	70
15	30	60	120	113	99	71
19	38	76	25	50	100	73
21	42	84	41	82	37	74
23	46	92	57	114	101	75
27	54	108	89	51	102	77
29	58	116	105	83	39	78
31	62	124	121	115	103	79
43	86	45	90	53	106	85
47	94	61	122	117	107	87
55	110	93	59	118	109	91
63	126	125	123	119	111	95

For $\alpha \in GF(p^n)$, all α^j's where j runs through a cyclotomic coset have the same minimal polynomial. This leads to the following method for finding minimal polynomials.

3.4.3 Finding minimal polynomials

Algorithm 3.1 An Algorithm for Finding Minimal Polynomials

Input: $f(x)$, a primitive polynomial over $GF(p)$ of degree n.

Output: All irreducible polynomials over $GF(p)$ whose degrees divides n

procedure_MP(f):

1. Generate the finite field $GF(p^n)$ by $f(\alpha) = 0$.
2. Compute all coset leaders mod $(p^n - 1)$: $\Gamma_p(n)$.
3. For each $s \in \Gamma_p(n)$, compute

$$m_s(x) = \prod_{i \in C_s} (x - \alpha^i)$$

4. Return all $m_s(x)$ for $s \in \Gamma_p(n)$.

The polynomial $m_s(x)$, computed in Algorithm 3.1, is the minimal polynomial of α^s and its conjugates. According to Fact 3.4, if $\gcd(s, p^n - 1) = 1$, then $\text{ord}(\alpha^s) = \text{ord}(\alpha) = p^n - 1$. Thus α^s is a primitive element of $GF(p^n)$. Hence $m_s(x)$ is a primitive polynomial over $GF(p)$ of degree n. We list the following formula without proof. The readers may write it out for themselves.

$$\boxed{\; x^{p^n} - x = \text{product of all monic polynomials,} \atop \text{irreducible over } GF(p), \text{ whose} \atop \text{degrees divide } n. \;}$$

So, Algorithm 3.1 gives all of such irreducible polynomials. Thus we have established the following property:

Property 3.7

$$x^{p^n} - x = x \prod_{s \in \Gamma_p(n)} m_s(x),$$

where the size of a coset C_s is equal to the degree of the minimal polynomial of α^s over $GF(p)$ and $\deg(m_s(x))|n$ for every $s \in \Gamma_p(n)$.

Example 3.11 (a) Let $n = 4$, $p = 2$, and $GF(2^4)$ be defined by $\alpha^4 + \alpha + 1 = 0$. The minimal polynomials of the elements of $GF(2^4)$ are listed in Table 3.2.

Table 3.2. *Minimal polynomials of the elements of $GF(2^4)$*

Coset	Element	Minimal polynomial
	0	x
$C_0 = \{0\}$	1	$m_0(x) = x + 1$
$C_1 = \{1, 2, 4, 8\}$	$\alpha, \alpha^2, \alpha^4, \alpha^8$	$m_1(x) = x^4 + x + 1$
$C_3 = \{3, 6, 12, 9\}$	$\alpha^3, \alpha^6, \alpha^{12}, \alpha^9$	$m_3(x) = x^4 + x^3 + x^2 + x + 1$
$C_5 = \{5, 10\}$	α^5, α^{10}	$m_5(x) = x^2 + x + 1$
$C_7 = \{7, 14, 13, 11\}$	$\alpha^7, \alpha^{14}, \alpha^{13}, \alpha^{11}$	$m_7(x) = x^4 + x^3 + 1$

Furthermore,

$$x^{2^4} + x = x m_0(x) m_1(x) m_3(x) m_5(x) m_7(x)$$
$$= x(x + 1)(x^2 + x + 1)(x^4 + x + 1)(x^4 + x^3 + 1) \cdot$$
$$(x^4 + x^3 + x^2 + x + 1).$$

(b) Let $n = 5$, $p = 2$, and $GF(2^5)$ be defined by $\alpha^5 + \alpha^3 + 1 = 0$ from Example 3.8. The minimal polynomials of the elements of $GF(2^5)$ are listed in Table 3.3.

Table 3.3. *Minimal polynomials of the elements of* $GF(2^5)$

Coset	Element	Minimal polynomial
	0	x
$C_0 = \{0\}$	1	$m_0(x) = x + 1$
$C_1 = \{1, 2, 4, 8, 16\}$	$\alpha, \alpha^2, \alpha^4, \alpha^8, \alpha^{16}$	$m_1(x) = x^5 + x^3 + 1$
$C_3 = \{3, 6, 12, 24, 17\}$	$\alpha^3, \alpha^6, \alpha^{12}, \alpha^{24}, \alpha^{17}$	$m_3(x) = x^5 + x^3 + x^2 + x + 1$
$C_5 = \{5, 10, 20, 9, 18\}$	$\alpha^5, \alpha^{10}, \alpha^{20}, \alpha^9, \alpha^{18}$	$m_5(x) = x^5 + x^4 + x^3 + x + 1$
$C_7 = \{7, 14, 28, 25, 19\}$	$\alpha^7, \alpha^{14}, \alpha^{28}, \alpha^{25}, \alpha^{19}$	$m_7(x) = x^5 + x^4 + x^3 + x^2 + 1$
$C_{11} = \{11, 22, 13, 26, 21\}$	$\alpha^{11}, \alpha^{22}, \alpha^{13}, \alpha^{26}, \alpha^{21}$	$m_{11}(x) = x^5 + x^4 + x^2 + x + 1$
$C_{15} = \{15, 30, 29, 27, 23\}$	$\alpha^{15}, \alpha^{30}, \alpha^{29}, \alpha^{27}, \alpha^{23}$	$m_{15}(x) = x^5 + x^2 + 1$

Therefore

$$
\begin{aligned}
x^{2^5} + x &= x m_0(x) m_1(x) m_3(x) m_5(x) m_7(x) m_{11}(x) m_{15}(x) \\
&= x(x + 1)(x^5 + x^3 + 1)(x^5 + x^3 + x^2 + x + 1) \cdot \\
&\quad (x^5 + x^4 + x^3 + x^2 + 1)(x^5 + x^4 + x^3 + x + 1) \cdot \\
&\quad (x^5 + x^4 + x^2 + x + 1)(x^5 + x^2 + 1).
\end{aligned}
$$

When we replace p by q that is a power of p, say $q = p^h$, then all results on finite fields we obtained so far are true. Thus, Algorithm 3.1 can be used for computing all irreducible polynomials over $GF(q)$ of degrees dividing n. It is worth pointing out that Algorithm 3.1 used the natural bijection between the number of the irreducible polynomials over $GF(q)$ of degree $n > 1$ and the coset leaders modulo $q^n - 1$ with size n. Golomb (1969) discovered a very important result about this type of the bijection. Specifically, he showed that the number of irreducible polynomials over $GF(q)$ of degree $n > 1$ is equal to the number of primitive necklaces of length n in q colors. For details of this work, the reader is referred to the paper by Golomb (1969).

Algorithm 3.1 computes all irreducible polynomials over $G(q)$ of degrees dividing n from a given primitive polynomial over $GF(q)$ of degree n. However, the intermediate computation in the algorithm involves aritihemetic in the finite field $GF(q^n)$. For $q = 2$, in the same work of Golomb mentioned above, he discussed a very efficient algorithm, the so called "rational algorithm" for "cubic transformation" which can find many primitive polynomials over $GF(2)$ of degree n for the case that $2^n - 1$ is prime. In the following, we introduce this algorithm and the cubic transformation of polynomials.

Let $f(x)$ be an irreducible polynomial over $GF(2)$ of degree n and α be a root of $f(x)$ in the extension field $GF(2^n)$.

Cubic Transformation (Marsh 1957):

$$M : f(x) \longrightarrow f^*(x) = f(x^{1/3})f(\omega x^{1/3})f(\omega^2 x^{1/3})$$

where $\omega^3 = 1$, a primitive third root of unity. $f^*(x)$ *is the minimal polynomial of* α^3 *over* $GF(2)$.

From this result, it is easily seen that for odd degree n, $2^n - 1$ is not divisible by 3, and the transformation M preserves not only irreducibility, but also primitivity of the roots. In a variety of cases, iteration of M enables one to generate *all* primitive polynomials of degree n from a given one. (These degrees include $n = 3, 5, 7, 13, 17$ and 19.)

The procedure given below is "rational" in the sense that ω and ω^2, which do not appear in the final result $f^*(x)$, do not occur in the intermediate computations either.

Algorithm 3.1′ THE RATIONAL ALGORITHM FOR FINDING MINIMAL POLYNO-MIALS OF CUBES OF ROOTS (GOLOMB 1969)

Input: $f(x)$, *an irreducible polynomial over* $GF(p)$ *of degree* n.
Output: $f^*(x)$, *the minimal polynomial of* α^3 *over* $GF(2)$ *where* α *is a root of* $f(x)$ *in* $GF(2^n)$.

procedure_RA(f):

Divide the exponents of the terms in $f(x)$ *into three classes, A, B, and C, according to the residue class of the exponent modulo 3. We produce the set of exponents for* $f^*(x)$ *from those for* $f(x)$ *by the following four steps:*

1. *Copy the exponents of* $f(x)$.
2. *Adjoin all numbers* $(2u_1 + u_2)/3$ *where* u_1 *and* u_2 *are distinct exponents of* $f(x)$ *in the same residue class modulo 3.*
3. *Adjoin all numbers* $(a + b + c)/3$ *where* $a \in A$, $b \in B$, *and* $c \in C$.
4. *Discard any exponent for* $f^*(x)$ *which is produced an* even *number of times by Steps 1, 2 and 3.*

Return $f^*(x)$ *whose exponents are those which are produced by Steps 1, 2 and 3 and odd number of times.*

Note that if any of the three categories A, B, C is empty, then Step 3 is vacuous. If a category has fewer than two members, it does not contribute to Step 2.

Example (a). We take $f(x) = x^5 + x^3 + 1$, the same as in case (b) in Example 3.11. Then the categories are

A	B	C
0, 3		5

To produce $f^*(x)$, which is $m_3(x)$ in Table 3.3, we follow the four steps in Algorithm 3.1':

	A	B	C	
Step 1	0, 3		5	copy
Step 2		1	2	$\frac{2\times0+3}{3}, \frac{2\times3+0}{3}$
Step 3				vacuous
Step 4 (mod 2 sum)	0, 3	1	2, 5	

Thus $f^*(x)$ has the exponents 0, 1, 2, 3, 5, and $f^*(x) = x^5 + x^3 + x^2 + x + 1 = m_3(x)$. Note that we have

$$3^0 = 1 \in C_1,$$
$$3^1 = 3 \in C_3,$$
$$3^2 = 9 \in C_5,$$
$$3^3 = 27 \in C_{15},$$
$$3^4 (\text{mod } 31) = 19 \in C_7,$$
$$3^5 (\text{mod } 31) = 26 \in C_{11}.$$

(See Example 3.10 for the cosets modulo 31.) Thus, by iterating the above process, that is, computing f^{**}, f^{***}, and so on, we can obtain all six primitive polynomials over $GF(2)$ of degree 5, For example, the process for finding $f^{**}(x)$ from $f^*(x)$ is given as follows.

	A	B	C	
Step 1	0, 3	1	2, 5	copy
Step 2	3	1, 4	2	$\frac{2\times0+3}{3}, \frac{2\times3+0}{3}, \frac{2\times2+5}{3}, \frac{2\times5+2}{3}$
Step 3	3	1	2, 2	$\frac{0+1+2}{3}, \frac{0+1+5}{3}, \frac{3+1+2}{3}, \frac{3+1+5}{3}$
Step 4 (mod 2 sum)	0, 3	1, 4	5	

Thus $f^{**}(x)$ has the exponents 0, 1, 3, 4, 5, and $f^{**}(x) = x^5 + x^4 + x^3 + x + 1$, which is equal to $m_5(x)$ in Table 3.3.

(b). Let $f(x) = x^7 + x + 1$, a primitive polynomial over $GF(2)$ of degree 7, and let α be a root of $f(x)$ in $GF(2^7)$. Then applying Algorithm 3.1' the minimal polynomial of α^3 can be computed as follows.

	A	B	C	
Step 1	0	1, 7		copy
Step 2	3		5	$\frac{2\times1+7}{3}$, $\frac{2\times7+1}{3}$
Step 3				vacuous
Step 4 (mod 2 sum)	0, 3	1	2, 5	

Thus $f^*(x)$ has the exponents 0, 1, 3, 5, 7, and $f^*(x) = x^7 + x^5 + x^3 + x + 1$. Again, according to Example 3.10, we have the following correspondence between the powers 3^i modulo 127 and the coset leaders modulo 127 where the corresponding number r in the third row of the following table is the coset leader modulo 127 for which $(3^i \bmod 127) \in C_r$.

i	0	1	2	3	4	5	6	7	8	9	10	11	12	13	14	15	16	17
3^i	1	3	9	27	81	116	94	28	84	125	121	109	73	92	22	66	71	86
r	1	3	9	27	13	29	47	7	21	63	31	55	19	23	11	5	15	43

By iterating the process, we can obtain all primitive polynomials over $GF(2)$ of degree 7 without involving any arithmetic in $GF(2^7)$.

(c). Let $f(x) = x^9 + x^4 + 1$, a primitive polynomial over $GF(2)$ of degree 9, and let α be a root of $f(x)$ in $GF(2^9)$. Then following four steps in Algorithm 3.1', we can comput the minimal polynomial of α^3 as follows.

	A	B	C	
Step 1	0, 9	4		copy
Step 2	3, 6			$\frac{2\times0+9}{3}$, $\frac{2\times9+0}{3}$
Step 3				vacuous
Step 4 (mod 2 sum)	0, 3, 6, 9	4		

Thus $f^*(x)$ has the exponents 0, 3, 4, 6, 9 and $f^*(x) = x^9 + x^6 + x^4 + x^3 + 1$. Note that $f^*(x) = \Pi_{i=0}^8(x + \alpha^{3 \cdot 2^i})$. So, it is tedious to evaluate this by hand. However, the above process only requires one to calculate sums of integers, dividing by 3, and some simple sorting operations, which can be done by hand very quickly.

The tables of the minimal polynomials of the elements in $GF(2^n)$ for $n = 6, 7, 8, 9$, and 10 are provided in Appendix C. Next we introduce the concept of reciprocal polynomials, which is frequently used in sequence analysis.

3.4.4 Reciprocal polynomials

Definition 3.19 *Let* $f(x) = x^n + a_{n-1}x^{n-1} + \cdots + a_1x + a_0, a_i \in GF(p^n)$ *and* $a_0 \neq 0$. *The* reciprocal polynomial *of* $f(x)$ *is defined as* $\frac{x^n}{a_0} f(x^{-1})$, *denoted by* $f^{-1}(x)$; *that is,* $f^{-1}(x) = \frac{x^n}{a_0} f(x^{-1})$. *In particular, if* $p = 2$, *then the reciprocal polynomial of* $f(x)$ *is given by*

$$f^{-1}(x) = x^n + a_1 x^{n-1} + \cdots + a_{n-1}x + 1;$$

that is, $f^{-1}(x)$ *is obtained by reversing the order of the coefficients of* $f(x)$.

From the definition, the following result is immediate.

Property 3.8 *For* $\alpha \in GF(p^n)$, *the minimal polynomials of* α *and* α^{-1} *are a pair of reciprocal polynomials.*

Example 3.12 From Example 3.11, we compute the pairs of reciprocal polynomials shown in Table 3.4.

Table 3.4. *Pairs of reciprocal polynomials*

n	$f(x)$	$f^{-1}(x)$
4	$m_1(x) = x^4 + x + 1$	$m_1^{-1}(x) = m_7(x) = x^4 + x^3 + 1$
	$m_3(x) = x^4 + x^3 + x^2 + x + 1$	$m_3^{-1}(x) = m_3(x) = x^4 + x^3 + x^2 + x + 1$
	$m_5(x) = x^2 + x + 1$	$m_5^{-1}(x) = m_5(x) = x^2 + x + 1$
5	$m_1(x) = x^5 + x^3 + 1$	$m_1^{-1}(x) = m_{15}(x) = x^5 + x^2 + 1$
	$m_3(x) = x^5 + x^3 + x^2 + x + 1$	$m_3^{-1}(x) = m_7(x) = x^5 + x^4 + x^3 + x^2 + 1$
	$m_5(x) = x^5 + x^4 + x^3 + x + 1$	$m_5^{-1}(x) = m_{11}(x) = x^5 + x^4 + x^2 + x + 1$

Notice that for $n = 4$, the reciprocal polynomial of $x^4 + x^3 + x^2 + x + 1$ is itself.

3.4.5 Periods of the minimal polynomials

For $\alpha \in GF(p^n)$ with order r, from the definition of the order, r is the smallest integer satisfying $\alpha^r = 1$. Let $m_\alpha(x)$ be the minimal polynomial of α. Then

$$m_\alpha(x) \mid (x^r - 1) \tag{3.2}$$

and r is the smallest integer such that Eq. (3.2) is true. According to the definition of the period of polynomials, r is the period of $m_\alpha(x)$. Thus we have established the following assertion.

Theorem 3.9 *For any $0 \neq \alpha \in GF(p^n)$, the period of the minimal polynomial of α is equal to the order of α; that is,*

$$\mathrm{per}(m_\alpha) = \mathrm{ord}(\alpha).$$

For example, for $n = 4$, from Table 3.2, the minimal polynomial of α is $x^4 + x + 1$, which has period 15, and the order of α is 15 since it is a primitive element.

3.5 Trace functions

3.5.1 Subfields

We list the following result without proof.

Fact 3.5 *Suppose that F is a finite extension field of $GF(p)$ that contains all the zeros of $x^{p^n} - x$. Then these zeros form a finite field of order p^n.*

Theorem 3.10 *Let F be the finite field with $q = p^n$ elements, where p is prime.*

(a) *$F = GF(p^n)$ contains a subfield $GF(p^m)$ if and only if m is a positive divisor of n.*
(b) *If $\alpha \in GF(p^n)$ then $\alpha \in GF(p^m)$ if and only if $\alpha^{p^m} = \alpha$.*

To prove Theorem 3.10, we need the following lemma.

Lemma 3.2 *If a, s, t are integers with $a \geq 2, s, t \geq 1$, then*

$$(a^s - 1) \,|\, (a^t - 1) \iff s \,|\, t.$$

(Recall that the vertical bar means "divides.")

Proof. We write $t = qs + r$, where $0 \leq r < s$. Then

$$\frac{a^t - 1}{a^s - 1} = a^r \cdot \frac{a^{qs} - 1}{a^s - 1} + \frac{a^r - 1}{a^s - 1}.$$

Since $a^{qs} - 1 = (a^s - 1)(a^{(q-1)s} + \cdots + a^s + 1)$, $a^{qs} - 1$ is always divisible by $a^s - 1$. The last term is less than 1, and so it is an integer if and only if $r = 0$. □

Proof of Theorem 3.10. (a) If $m|n$, from Fact 3.5, then $GF(p^n)$ contains a subfield $GF(p^m)$. Conversely, if α is a primitive element of $GF(p^m)$, then the order of α is $p^n - 1$. But the order of any element of $GF(p^n)^*$ divides $p^n - 1$. Therefore $(p^m - 1)|(p^n - 1)$, and $m|n$ by Lemma 3.2. (b) is immediate from Corollary 3.2. □

Example 3.13 The subfields of $GF(2^{18})$ can be determined by listing all positive divisors of 18. The containment relations between these various subfields are displayed in the following diagram.

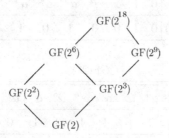

According to Theorem 3.10, the containment relations are equivalent to divisability relations among the positive divisors of 18. Subfields are the basic tools for the constructions of (cascaded, generalized) GMW sequences.

Note. All results discussed in this chapter related to $GF(p)$ or $GF(p^n)$ are true when the prime p is replaced by a power of a prime, say $q = p^h$. From now on, we will discuss the properties of finite fields using the general notation $GF(q)$.

3.5.2 Trace functions

Definition 3.20 *Let q be a prime or a power of a prime. For $\alpha \in F = GF(q^n)$ and $K = GF(q)$, the trace function $Tr_{F/K}(x)$, $x \in F$, is defined by*

$$Tr_{F/K}(x) = x + x^q + \cdots + x^{q^{n-1}}, \quad x \in F.$$

If $\alpha \in F$, $Tr_{F/K}(\alpha)$ is called the trace of α over K, simply denoted as $Tr(\alpha)$ if the context is clear.

In the following, we simply write $(Tr_{F/K}(x))^q = Tr_{F/K}(x)^q$. Note that

$$Tr_{F/K}(x)^q = \sum_{i=0}^{n-1} x^{q^{i+1}} = Tr_{F/K}(x).$$

From Theorem 3.10(b), $Tr_{F/K}(x) \in K$. Hence $Tr_{F/K}(x)$ is a mapping from F to K. If $q = 2$, then $Tr_{F/K}(x)$ is either 0 or 1; that is,

$$Tr_{F/K}(x) = x + x^2 + \cdots + x^{2^{n-1}} \in GF(2) \text{ for all } x \in GF(2^n).$$

Example 3.14 Let $GF(2^3)$ and $GF(2^4)$ be defined by $\alpha^3 + \alpha + 1 = 0$ and $\alpha^4 + \alpha + 1 = 0$, respectively (see Examples 3.5 and 3.6 for the tables of these finite fields). Here α denotes primitive elements in both fields, respectively. We compute the trace functions of the elements of $GF(2^3)$ and $GF(2^4)$ as follows.

$$GF(2^3)$$

x	0	1	α	α^2	α^3	α^4	α^5	α^6
$Tr(x)$	0	1	0	0	1	0	1	1

$$GF(2^4)$$

x	0	1	α	α^2	α^3	α^4	α^5	α^6	α^7	α^8	α^9	α^{10}	α^{11}	α^{12}	α^{13}	α^{14}
$Tr(x)$	0	0	0	0	1	0	0	1	1	0	1	0	1	1	1	1

Theorem 3.11 *Let $F = GF(q^n)$ and $K = GF(q)$. Then the trace function $Tr_{F/K}$ satisfies the following properties:*

(a) $Tr_{F/K}(x + y) = Tr_{F/K}(x) + Tr_{F/K}(y)$ *for all $x, y \in F$.*
(b) $Tr_{F/K}(cx) = cTr_{F/K}(x)$ *for all $c \in K, x \in F$.*
(c) $Tr_{F/K}$ *is a linear transformation from F onto K (i.e., a linear functional), where both F and K are viewed as vector spaces over K.*
(d) $Tr_{F/K}(c) = nc$ *for all $c \in K$.*
(e) $Tr_{F/K}(x^q) = Tr_{F/K}(x)$ *for all $x \in F$ and $Tr_{F/K}(x^p) = Tr_{F/K}(x)^{p^i}$ for $q = p^h$ and any positive integer i.*
(f) $Tr_{F/K}(yx^q) = Tr_{F/K}(y^{q^{n-1}}x)$ *for all $x, y \in F$.*

Proof. (a) For $x, y \in F$ we use Corollary 3.3 to get

$$Tr_{F/K}(x + y) = (x + y) + (x + y)^q + \cdots + (x + y)^{q^{n-1}}$$
$$= x + y + x^q + y^q + \cdots + x^{q^{n-1}} + y^{q^{n-1}}$$
$$= Tr_{F/K}(x) + Tr_{F/K}(y).$$

(b) For $c \in K$ we have $c^{q^j} = c$ for all $j \geq 0$ by Theorem 3.10(b). Therefore

$$Tr_{F/K}(cx) = cx + c^q x^q + \cdots + c^{q^{n-1}} x^{q^{n-1}}$$
$$= cTr_{F/K}(x).$$

(c) The properties (a) and (b), together with the fact that $Tr_{F/K}(x) \in K$ for all $x \in F$, show that $Tr_{F/K}(x)$ is a linear transformation from F into K. To prove that this mapping is onto, it suffices then to show existence of an $\alpha \in F$ with $Tr_{F/K}(\alpha) \neq 0$. Now $Tr_{F/K}(\alpha) = 0$ if and only if α is a root of the polynomial $x^{q^{n-1}} + \cdots + x^q + x$ in F. Since this polynomial can have at most q^{n-1} roots in F and F has q^n elements, the result follows.

(d) This follows immediately from the definitions of the trace function and Theorem 3.10(b).

(e) For $x \in F$ we have $x^{q^n} = x$ by Corollary 3.2, and so $Tr_{F/K}(x^q) = x^q + x^{q^2} + \cdots + x^{q^n} = Tr_{F/K}(x)$. Thus, the first assertion is established. For the second assertion, we only write the proof for $i = 1$. (For $i > 1$, the proof is similar to that for $i = 1$.) Applying Lemma 3.1 in Section 3.3, we have

$$Tr_{F/K}(x^p) = x^p + x^{qp} + \cdots + x^{q^{n-1}p}$$
$$= Tr_{F/K}(x^p) = \left(x + x^q + \cdots + x^{q^{n-1}}\right)^p$$
$$= Tr_{F/K}(x)^p.$$

(f) Again, using Corollary 3.2, and then the result (e), it follows that

$$Tr_{F/K}(yx^q) = Tr_{F/K}\left(y^{q^n}x^q\right) = Tr_{F/K}\left(y^{q^{n-1}}x\right)^q = Tr_{F/K}\left(y^{q^{n-1}}x\right). \qquad \square$$

Theorem 3.12 (TRANSITIVITY OF TRACE) *Let K be a finite field, F a finite extension of K, and E a finite extension of F; that is, $K \subset F \subset E$. Then*

$$Tr_{E/K}(x) = Tr_{F/K}(Tr_{E/F}(x)) \text{ for all } x \in E.$$

In other words, the trace function from E to K is a composition of the trace function from E to F and the trace function from F to K.

Proof. We can suppose that $K = GF(q)$, $F = GF(q^n)$, and $E = GF(q^{nm})$. Then for $x \in E$ we have

$$Tr_{F/K}(Tr_{E/F}(x)) = \sum_{i=0}^{n-1} Tr_{E/F}(x)^{q^i} = \sum_{i=0}^{n-1} \left(\sum_{j=0}^{m-1} x^{q^{nj}}\right)^{q^i}$$
$$= \sum_{i=0}^{n-1}\sum_{j=0}^{m-1} x^{q^{nj+i}} = \sum_{k=0}^{nm-1} x^{q^k} = Tr_{E/K}(x). \qquad \square$$

3.5.3 Norms

Another function from a finite field to one of its subfields is obtained by forming the product of the conjugates of an element of the field with respect to the subfield.

Definition 3.21 For $\alpha \in F = GF(q^n)$ and $K = GF(q)$, the norm $N_{F/K}(\alpha)$ of α over K is defined by

$$N_{F/K}(\alpha) = \alpha \cdot \alpha^q \cdot \cdots \cdot \alpha^{q^{n-1}} = \alpha^{(q^n-1)/(q-1)}.$$

Note that $N_{F/K}(\alpha) \in K$. Thus $N_{F/K}(x)$ is a mapping from F to K.

3.5.4 Dual bases

Definition 3.22 Let $K = GF(q)$ and $F = GF(q^n)$. Then two bases $\{\alpha_1, \ldots, \alpha_n\}$ and $\{\beta_1, \ldots, \beta_n\}$ of F over K are said to be dual (or complementary) bases if for $1 \leq i, j \leq n$:

$$Tr_{F/K}(\alpha_i \beta_j) = \begin{cases} 0 & \text{for } i \neq j \\ 1 & \text{for } i = j. \end{cases}$$

Theorem 3.13 Let $K = GF(q)$ and $F = GF(q^n)$, and let two bases $\{\alpha_1, \ldots, \alpha_n\}$ and $\{\beta_1, \ldots, \beta_n\}$ of F over K be dual bases. If

$$x = x_1\alpha_1 + x_2\alpha_2 + \cdots + x_n\alpha_n \in F, \text{ where } x_j \in K, \tag{3.3}$$

then the coefficients x_j under the basis $\{\alpha_i\}$ are given by

$$x_j = Tr(\beta_j x), 1 \leq j \leq n.$$

Proof. Multiplying β_j times the two sides of the identity (3.3), and then applying the definition of the dual bases, the result follows. $\square$

Example 3.15 Let $GF(2^3) = \{\alpha^3 + \alpha + 1 = 0\}$. Then $\{1, \alpha, \alpha^2\}$ and $\{1, \alpha^2, \alpha\}$ are a pair of the dual bases over $GF(2)$. For α^6, we write

$$\alpha^6 = x_0 + x_1\alpha + x_2\alpha^2, x_i \in GF(2),$$

and then we compute the x_i's by Theorem 3.13,

$$x_0 = Tr(1\alpha^6) = 1, \quad x_1 = Tr(\alpha^2\alpha^6) = Tr(\alpha) = 0,$$
$$x_2 = Tr(\alpha\alpha^6) = Tr(1) = 1.$$

Thus, $\alpha^6 = 1 + \alpha^2$, the same result as we obtained before.

3.5.5 Minimal polynomials over intermediate subfields

Usually, it is easier to write a program for finding primitive polynomials over $GF(q)$ when q is a prime than in the case when q is a power of a prime. In the following theorem, we provide a method for computing primitive polynomials over $GF(q^m)$ in terms of known primitive polynomials over $GF(q)$.

Theorem 3.14 *Let $f(x)$ be a primitive polynomial over $GF(q)$ of degree n and let α be a root of $f(x)$ in the extension field $GF(q^n)$. If $m|n$ and $1 \leq m < n$, let $Q = q^m$ and $l = n/m$. Then the minimal polynomial $g(x)$ of α over $GF(q^m)$ has degree l and is given by*

$$g(x) = (x - \alpha)(x - \alpha^Q) \cdots (x - \alpha^{Q^{l-1}}) \qquad (3.4)$$

$$= x^l + \sum_{j=0}^{l-1} c_j x^j, c_j \in GF(Q).$$

Furthermore, $f(x)$ can be factored into m primitive polynomials over $\mathbb{F}_Q$ of degree l as follows:

$$f(x) = \prod_{i=0}^{m-1} \sigma_i(g(x)), \qquad (3.5)$$

where σ_i is the Frobenius map that raises the coefficients of $g(x)$ to the power q^i and keeps the primitivity of $g(x)$; that is,

$$\sigma_i(g(x)) = \sum_{j=0}^{l-1} c_j^{q^i} x^j,$$

which is primitive over $GF(Q)$.

Proof. Since α is a primitive element in $GF(q^n)$, the minimal polynomial of α over $GF(Q)$ has degree l. Note that $g(\alpha) = 0$; that is, α is a root of $g(x)$. Thus we only need to show that $g(x)$ is a polynomial over $GF(Q)$; that is, the coefficients of $g(x)$ belong to $GF(Q)$. From the construction of $g(x)$ in Eq. (3.4), we know that c_j is a symmetric polynomial in $\alpha^{Q^k}, k = 0, 1, \ldots, l-1$. Thus $c_j^Q = c_j \implies g(x) \in GF(Q)[x]$. Thus, $g(x)$ is the minimal polynomial of α over $GF(Q)$. For the second assertion, it is easy to see that $\sigma_i(g(x)) \in GF(Q)[x]$ because $g(x) \in GF(Q)[x]$. Note that α^{q^i} is a root of $\sigma_i(g(x))$. Thus $\sigma_i(g(x))$ is the minimal polynomial of α^{q^i} over $GF(Q)$. Notice that $\{\alpha^{q^i Q^j} = \alpha^{q^{i+mj}} | 0 \leq i < m; 0 \leq j < l\}$ constitutes all roots of $f(x)$. It follows that Eq. (3.5) is true. $\qquad \square$

From Theorem 3.14, we have the following algorithm to compute primitive polynomials over $GF(q^m)$ in terms of the primitive polynomials over $GF(q)$.

Algorithm 3.2 An Algorithm for Finding Primitive Polynomials over $GF(q^m)$

Input: $f(x)$, *a primitive polynomial over $GF(q)$ of degree n; $1 < m < n$, a proper divisor of n.*

Output: *A primitive polynomial $g(x)$ over $GF(q^m)$ of degree $l = n/m > 1$.*

procedure_MP(f, g):

1. *Generate $GF(q^n)$ from $f(x)$, and set α to be a root of $f(x)$ in $GF(q^n)$.*
2. *Set $\beta = \alpha^d$ where $d = (q^n - 1)/(q^m - 1)$. Then β is a primitive element in $GF(q^m)$. Compute the minimal polynomial of β, say $h(x)$,*

$$h(x) = \prod_{i=0}^{m-1} \left(x - \beta^{q^i}\right).$$

   ($h(x)$ is a primitive polynomial over $GF(q)$ of degree m.) We now use $h(x)$ as a defining polynomial for $GF(q^m)$.
3. *Compute $g(x)$, the minimal polynomial of α over $GF(q^m)$:*

$$g(x) = \prod_{i=0}^{l-1} \left(x - \alpha^{q^{mi}}\right), l = n/m.$$

   $g(x)$ is a primitive polynomial over $GF(q^m)$ of degree l.
4. *Return $g(x)$.*

Example 3.16 Find a primitive polynomial over $GF(2^3)$ of degree 2. We will apply Algorithm 3.2 to this question where the computation is done using the software package Maple.

1. Select $f(x) = x^6 + x + 1$, a primitive polynomial over $GF(2)$ of degree 6, and set α a root of $f(x)$ in $GF(2^6)$ where $GF(2^6)$ is defined by $f(x)$.
2. For $m = 3$, set $d = (2^6 - 1)/(2^3 - 1) = 9$ and $\beta = \alpha^9$; compute

$$h(x) = (x - \beta)(x - \beta^2)(x - \beta^4) = x^3 + x^2 + 1.$$

3. In $GF(2^3)$, defined by $\beta^3 + \beta^2 + 1 = 0$, compute

$$g(x) = (x - \alpha)(x - \alpha^8) = x^2 + (\alpha + \alpha^8)x + \alpha^9$$
$$= x^2 + \alpha^{27}x + \alpha^9 = x^2 + \beta^3 x + \beta,$$

   which is a primitive polynomial over $GF(2^3)$ of degree 2.

Note. This algorithm has important applications for implementation of GMW sequences and interleaved sequences. In Appendix B of this chapter, we compute a table consisting of primitive polynomials over $GF(2^m)$ of degree l for $lm \leq 32$ by Algorithm 3.2.

3.6 Powers of trace functions

In this section, we will describe how to compute a power of a trace function and give a formula for the exponents in the expansion. This expansion has important

applications for computing linear spans of GMW sequences, generalized GMW
sequences, bent function sequences, geometric sequences, and so on.

Let p be a prime, $q = p^r$, and let $f(x)$ be a function from $GF(q^n)$ to
$GF(q)$. By the Lagrange interpolation formula and $x^{q^n} = x$ for $x \in GF(q^n)$,
we can write $f(x)$ in a polynomial form $f(x) = \sum_{i=0}^{q^n-1} c_i x^i$, $c_i \in GF(q^n)$ (we
will discuss this representation in detail in Chapter 6). The weight of $f(x)$,
denoted as $w(f)$, is defined as the number of nonzero coefficients of $f(x)$;
that is,

$$w(f) = |\{i \mid c_i \neq 0, 0 \le i < q^n\}|. \tag{3.6}$$

Any number in $\mathbb{Z}_{p^t}$ can be written as a number in the p-ary number system.
In other words, for $x \in \mathbb{Z}_{p^t}$, we can write it as $x = \sum_{i=0}^{t-1} x_i p^i$, $0 \le x_i < p$.
The Hamming weight of x, as a p-ary number, is defined as the number of
nonzero coefficients of x with respect to the base $\{1, p, \ldots, p^{t-1}\}$; that is,
$w(x) = |\{i \mid x_i \neq 0, 0 \le i < t\}|$. For $y \in \mathbb{Z}_{p^t}$, $y = \sum_{i=0}^{t-1} y_i p^i$, $0 \le y_i < p$, we
have the following property on the uniqueness of the p-ary number system.

Property 3.9 (UNIQUENESS OF p-ARY NUMBERS) *With the above notation,*

$$x = y \iff x_i = y_i, i = 0, 1, \ldots, t - 1.$$

For $m|n$, we have a field tower:

$$GF(q) \subset GF(q^m) \subset GF(q^n).$$

We simply denote by $Tr_m^n(x)$ the trace function $Tr_{GF(q^n)/GF(q^m)}(x) = x +
x^Q + \cdots + x^{Q^{l-1}}$ where $n = ml$ and $Q = q^m$. The objective of this section
is to obtain the prototype of exponents of x in the expansion of a power of
the trace function, that is, the exponents of x in the expansion of $(Tr_m^n(x))^s$,
shortened as $Tr_m^n(x)^s$, where $1 < s < q^m$.

3.6.1 Binary case

We first investigate the simple case: $q = 2$.

Theorem 3.15 (POWER OF TRACE: BINARY CASE) *With the above notation,*
for $q = 2$ and $1 < s < 2^m - 1$, we write $s = 2^{i_1} + \cdots + 2^{i_k}$, a binary number.
Then

(a)

$$Tr_m^n(x)^s = \left(x + x^Q + \cdots + x^{Q^{l-1}}\right)^s \quad (here \ Q = 2^m)$$

$$= \sum_{t \in \mathbb{Z}_l^k} x^{\tau_s(t)}, \tag{3.7}$$

where $\mathbf{t} = (t_1, t_2, \ldots, t_k), 0 \le t_j < l,$ *or equivalently,* $\mathbf{t} \in \mathbb{Z}_l^k = \{(t_1, \ldots, t_k) \mid t_k \in \mathbb{Z}_l\}$ *and*

$$\tau_s(\mathbf{t}) = 2^{i_1 + mt_1} + \cdots + 2^{i_k + mt_k}. \tag{3.8}$$

(b) $\tau_s(\mathbf{t}) < 2^n - 1,$ *for all* $\mathbf{t} \in \mathbb{Z}_l^k.$

(c) $\tau_s(\mathbf{t})$ *is a one-to-one map from* $\mathbb{Z}_l^k$ *to* $\mathbb{Z}_{2^m - 1}.$ *In other words, we have*

$$\tau_s(\mathbf{t}) \ne \tau_s(\mathbf{t}') \iff \mathbf{t} \ne \mathbf{t}' \in \mathbb{Z}_l^k.$$

(d) $\tau_s(\mathbf{t}) \equiv s \pmod{2^m - 1}$ *for all* $\mathbf{t} \in \mathbb{Z}_l^k.$

(e) $w(Tr_m^n(x)^s) = l^k.$

Proof. (a) Expanding $Tr_m^n(x)^s$, then

$$Tr_m^n(x)^s = \left(x + x^Q + \cdots + x^{Q^{l-1}}\right)^s$$

$$= \prod_{j=1}^{k} \left(x^{2^{i_j}} + x^{2^{i_j}Q} + \cdots + x^{2^{i_j}Q^{l-1}}\right).$$

Since $Q = 2^m$, we have

$$Tr_m^n(x)^s = \prod_{j=1}^{k} \left(x^{2^{i_j}} + x^{2^{i_j + m}} + \cdots + x^{2^{i_j + (l-1)m}}\right). \tag{3.9}$$

The exponent of x in the expansion of Eq. (3.9) is a sum of k elements where each is taken from a different row of the following matrix:

$$A = \begin{bmatrix} 1 & 2^m & 2^{2m} & \cdots & 2^{(l-1)m} \\ 2 & 2^{1+m} & 2^{1+2m} & \cdots & 2^{1+(l-1)m} \\ \vdots & & & & \\ 2^{m-1} & 2^{m-1+m} & 2^{m-1+2m} & \cdots & 2^{m-1+(l-1)m} \end{bmatrix}_{m \times l}.$$

In other words, any exponent of x in the expansion (3.9) can be represented as

$$\tau_s(\mathbf{t}) = \quad 2^{i_1 + mt_1} \quad + \quad 2^{i_2 + mt_2} \quad + \cdots + \quad 2^{i_k + mt_k}$$

$$\downarrow \qquad\qquad \downarrow \qquad \cdots \qquad \downarrow$$

$$\text{taken from row } i_1 \quad \text{from row } i_2 \quad \cdots \quad \text{taken from row } i_k$$

where $t_1 \in \mathbb{Z}_l, \ldots, t_k \in \mathbb{Z}_l$; that is, $\mathbf{t} = (t_1, \ldots, t_k) \in \mathbb{Z}_l^k.$ Thus assertion 1 is established.

(b) Note that the numbers in matrix A are the base of binary numbers in $\mathbb{Z}_{2^n - 1}$. Since $s < 2^m - 1$, then $k < m$. Thus the binary number $\tau_s(\mathbf{t})$ has the Hamming weight $k < m < n$. So

$$\tau_s(\mathbf{t}) < 2^n - 1, \quad \text{for all } \mathbf{t} \in \mathbb{Z}_l^k.$$

(c) From Property 3.9, any integer in $\mathbb{Z}_{2^n}$ has a unique binary representation. Together with assertion 2, it follows that

$$\tau_s(\mathbf{t}) \neq \tau_s(\mathbf{t}'), \quad \text{for all } \mathbf{t} \neq \mathbf{t}' \in \mathbb{Z}_l^k.$$

(d) Note that $2^{jm} \equiv 1 \pmod{2^m - 1}$. From Eq. (3.8), we have

$$\tau_s(\mathbf{t}) \equiv 2^{i_1} + \cdots + 2^{i_k} \equiv s \pmod{2^m - 1}, \quad \text{for all } \mathbf{t} \in \mathbb{Z}_l^k.$$

(e) This is immediate from assertion (c) because there are l^k elements in $\mathbb{Z}_l^k$. □

Corollary 3.4 *With the notation in Theorem 3.15, if $s' \neq s$ with $1 < s$, $s' < 2^m - 1$, then*

$$\tau_s(\mathbf{t}) \neq \tau_{s'}(\mathbf{t}')$$

for every pair $\mathbf{t}, \mathbf{t}' \in \mathbb{Z}_l^k$.

Proof. If $\tau_s(\mathbf{t}) = \tau_{s'}(\mathbf{t}')$ for some pair $\mathbf{t}, \mathbf{t}' \in \mathbb{Z}_l^k$, then they are equal modulo $2^m - 1$. In other words, we have

$$\tau_s(\mathbf{t}) \equiv \tau_{s'}(\mathbf{t}') \pmod{2^m - 1}.$$

From Theorem 3.15(d), $\tau_s(\mathbf{t}) \equiv s \pmod{2^m - 1}$ and $\tau_{s'}(\mathbf{t}') \equiv s' \pmod{2^m - 1}$. Consequently

$$s \equiv s' \pmod{2^m - 1},$$

which is a contradiction to $s \neq s'$ with $1 < s, s' < 2^m - 1$. □

Example 3.17 Let $n = ml$.

(a) For $n = 6$, $m = 3$, and $l = 2$, we choose $s = 3 = 1 + 2 \Longrightarrow k = 2$ and $(i_1, i_2) = (0, 1)$. In this case, we have the following field tower:

$$GF(2) \subset GF(2^3) \subset GF(2^6)$$

and $\mathbf{t} = (t_1, t_2) \in \mathbb{Z}_2^2$. We will use the formula (3.7) to expand $Tr_3^6(x)^3$.

$\mathbf{t} = (t_1, t_2) \in \mathbb{Z}_2^2$	$\tau_3(t_1, t_2) = 2^{3t_1} + 2^{1+3t_2}$	(mod 7)
$(0, 0)$	$1 + 2 = 3$	3
$(0, 1)$	$1 + 2^4 = 17$	3
$(1, 0)$	$2^3 + 2 = 10$	3
$(1, 1)$	$2^3 + 2^4 = 24$	3

Then

$$Tr_3^6(x)^3 = x^3 + x^{17} + x^{10} + x^{24}.$$

(b) When $n = 8, m = 4, l = 2, s = 7 = 1 + 2 + 2^2 \implies k = 3, (i_1, i_2, i_3) = (0, 1, 2)$, and $(t_1, t_2, t_3) \in \mathbb{Z}_2^3$, we have

$\mathbf{t} = (t_1, t_2, t_3) \in \mathbb{Z}_2^3$	$\tau_3(t_1, t_2, t_3) = 2^{4t_1} + 2^{1+4t_2} + 2^{2+4t_3}$	(mod 15)
$(0, 0, 0)$	$1 + 2 + 2^2 = 7$	7
$(0, 0, 1)$	$1 + 2 + 2^6 = 67$	7
$(0, 1, 0)$	$1 + 2^5 + 2^2 = 37$	7
$(0, 1, 1)$	$1 + 2^5 + 2^6 = 97$	7
$(1, 0, 0)$	$2^4 + 2 + 2^2 = 22$	7
$(1, 0, 1)$	$2^4 + 2 + 2^6 = 82$	7
$(1, 1, 0)$	$2^4 + 2^5 + 2^2 = 52$	7
$(1, 1, 1)$	$2^4 + 2^5 + 2^6 = 112$	7

Therefore,

$$Tr_4^8(x)^7 = x^7 + x^{67} + x^{37} + x^{97} + x^{22} + x^{82} + x^{52} + x^{112}.$$

(c) When $n = 9, m = 3, l = 3, s = 3 = 1 + 2 \implies k = 2, (i_1, i_2) = (0, 1)$, and $(t_1, t_2) \in \mathbb{Z}_3^2$, we have

$\mathbf{t} = (t_1, t_2) \in \mathbb{Z}_2^3$	$\tau_3(t_1, t_2) = 2^{3t_1} + 2^{1+3t_2}$	(mod 7)
$(0, 0)$	$1 + 2 = 3$	3
$(0, 1)$	$1 + 2^4 = 17$	3
$(0, 2)$	$1 + 2^7 = 129$	3
$(1, 0)$	$2^3 + 2 = 10$	3
$(1, 1)$	$2^3 + 2^4 = 24$	3
$(1, 2)$	$2^3 + 2^7 = 136$	3
$(2, 0)$	$2^6 + 2 = 66$	3
$(2, 1)$	$2^6 + 2^4 = 80$	3
$(2, 2)$	$2^6 + 2^7 = 192$	3

Thus, we get

$$Tr_3^9(x)^3 = x^3 + x^{17} + x^{129} + x^{10} + x^{24} + x^{136} + x^{66} + x^{80} + x^{192}.$$

3.6.2 Nonbinary case

We are now in a position to discuss a power of the trace function for the general case of $q > 2$. First we introduce the multinomial coefficients.

Definition 3.23 Let x_i be interdeterminates, $i = 1, \ldots, l$ and $s > 0$.

$$(x_1 + x_2 + \cdots + x_l)^s = \sum_{i_1, \ldots, i_l} \binom{s}{i_1, \ldots, i_l} x_1^{i_1} \cdots x_l^{i_l},$$

where the summation is taken over all nonnegative integer-valued vectors
$(i_1, \ldots, i_l)$ *such that* $i_1 + \cdots + i_l = s$ *and*

$$\binom{s}{i_1 \cdots i_l} = \frac{s!}{i_1! \cdots i_l!},$$

which is called a multinomial coefficient *and represents the number of possible divisions of s distinct objects into k distinct groups of respective sizes* $i_1, i_2, \ldots, i_l$. *(By convention,* $0! = 1$.)

Recall $Q = q^m$ and $q = p^r$. Let $0 < s < p$. Applying the multinomial formula to $Tr_m^n(x)^s$ with $x_i = x^{Q^i}$, we have

Lemma 3.3

$$Tr_m^n(x)^s = \left(x + x^Q + \cdots + x^{Q^{l-1}}\right)^s = \sum \binom{s}{i_1 \cdots i_l} x^{i_1 + i_2 Q + \cdots + i_l Q^{l-1}},$$

where the summation is taken over all nonnegative integer-valued vectors
$(i_1, \ldots, i_l)$ *such that* $i_1 + \cdots + i_l = s$. *Furthermore, for such a vector,*

$$\binom{s}{i_1 \cdots i_l} \neq 0,$$

and the number of monomials in the expansion, or equivalently, $w(Tr_m^n(x)^s)$, *is given by*

$$w(Tr_m^n(x)^s) = \binom{l + s - 1}{l - 1},$$

where $l = n/m$.

Theorem 3.16 (POWER OF TRACE: NONBINARY CASE) *Let* $s < q^m - 1$, *and* $s = \sum_{i=0}^{rm-1} s_i p^i$, $0 \leq s_i < p$, *a p-ary number.*

(a)

$$Tr_m^n(x)^s = \left(x + x^Q + \cdots + x^{Q^{l-1}}\right)^s$$
$$= \sum_B d_B x^{tr(AB)}, \tag{3.10}$$

where the summation is taken over all matrices B defined as follows:

$$B = \begin{bmatrix} b_{0,0} & b_{0,1} & \cdots & b_{0,rm-1} \\ b_{1,0} & b_{1,1} & \cdots & b_{1,rm-1} \\ \vdots & & & \\ b_{l-1,0} & b_{l-1,1} & \cdots & b_{l-1,rm-1} \end{bmatrix}_{l \times rm}$$

in which the jth column of B is a nonnegative integer-valued vector such that

$$b_{0,j} + b_{1,j} + \cdots + b_{l-1,j} = s_j, \; j = 0, 1, \ldots, rm - 1; \qquad (3.11)$$

the matrix A is given by

$$A = \begin{bmatrix} 1 & p^{rm} & \cdots & p^{rm(l-1)} \\ p & p^{1+rm} & \cdots & p^{1+rm(l-1)} \\ \vdots & & & \\ p^{rm-1} & p^{rm-1+rm} & \cdots & p^{rm-1+rm(l-1)} \end{bmatrix}_{rm \times l} ,$$

the notation $tr(AB)$ denotes the trace of the $rm \times rm$ matrix AB, that is, if we write $AB = (u_{ij})_{rm \times rm}$, then

$$tr(AB) = \sum_{i=0}^{rm-1} u_{i,i};$$

and the coefficient d_B is given by

$$d_B = \prod_{j=0}^{rm-1} \binom{s_j}{b_{0,j}, b_{1,j}, \ldots, b_{l-1,j}} \not\equiv 0 \; (\text{mod } p). \qquad (3.12)$$

(b) *Let $M(s)$ be the set of all such $l \times rm$ matrices B with Eq. (3.11); that is,*

$$M(s) = \{B = (b_{ij})_{l \times rm} \mid 0 \le b_{ij} \in \mathbb{Z} \text{ with Eq. (3.11)}\}.$$

Then

(i) *$tr(AB) < q^n - 1$ for any $B \in M(s)$.*
(ii) *$tr(AB) \ne tr(AC)$ if $B \ne C \in M(s)$.*
(iii) *$tr(AB) \equiv s \; (\text{mod } q^m - 1)$ for every $B \in M(s)$.*

(c) *The number of the monomials in Eq. (3.10) is given by*

$$w(Tr_m^n(x)^s) = |M(s)| = \prod_{i=0}^{rm-1} \binom{l + s_i - 1}{l - 1}.$$

Proof. (a) For $s = \sum_{i=0}^{rm-1} s_i p^i$, $0 \le s_i < p$, notice that $Q = q^m = p^{rm}$, and we have

$$Tr_m^n(x)^s = (x + x^Q + \cdots + x^{Q^{l-1}})^s = \prod_{i=0}^{rm-1} P_i, \qquad (3.13)$$

where

$$P_i = \left(x^{p^i} + x^{p^{i+rm}} + \cdots + x^{p^{i+rm(l-1)}} \right)^{s_i}.$$

According to Lemma 3.3, for a fixed i, an exponent of x in P_i is the inner product of the ith row $A_i = (p^i, p^{i+rm}, \ldots, p^{i+rm(l-1)})$ of A and the ith column $B_i = (b_{0,i}, b_{1,i}, \ldots, b_{l-1,i})^T$ of B; that is,

$$u_{i,i} = (A_i \cdot B_i) = b_{0,i}p^i + b_{1,i}p^{i+rm} + \cdots + b_{l-1,i}p^{i+rm(l-1)}, \quad (3.14)$$

where B_i is a nonnegative integer-valued solution of Eq. (3.11). Thus, every exponent of x in the expansion of (3.13) is the sum of $u_{i,i}$, $i = 0, 1, \ldots, m-1$, which is $tr(AB)$. From Lemma 3.3, each factor in d_B is not congruent to zero modulo p. Thus $d_B \not\equiv 0 \pmod{p}$ for any $B \in M(s)$. Therefore the assertion 1 is true.

(b) Note that the entries in matrix A are the base of the p-ary number system in $\mathbb{Z}_{q^n-1}$ and

$$0 \le b_{i,j} \le s_j < p \Longrightarrow tr(AB) < q^n - 1 = p^{rml} - 1.$$

From the uniqueness of the p-ary number system, $tr(AB) \neq tr(AC)$ if and only if $B \neq C, C \in M(s)$. From Eq. (3.14), note that $q^{mj} = p^{rmj} \equiv 1 \pmod{q^m - 1}$, and then

$$u_{i,i} \equiv b_{0,i}p^i + b_{1,i}p^i + \cdots + b_{l,i}p^i \equiv p^i \sum_{j=0}^{l-1} b_{j,i} \equiv s_i p^i \pmod{q^m - 1}.$$

So, it follows that

$$tr(AB) \equiv \sum_{i=1}^{rm-1} s_i p^i \equiv s \pmod{q^m - 1},$$

which completes the proof of assertion 2.

(c) From assertion 1, we have $d_B \not\equiv 0 \pmod{p}$ for any $B \in M(s)$. Since there are $\binom{l+s_i-1}{l-1}$ ways to choose the ith column of B, there are

$$|M(s)| = \prod_{i=0}^{rm-1} \binom{l+s_i-1}{l-1}$$

ways to form B satisfying Eq. (3.11). Thus the number of monomials in Eq. (3.13) is given by this formula. $\qquad\square$

Corollary 3.5 *With the notation in Theorem 3.16, if $s \neq t < q^m - 1$, then*

$$tr(AB) \neq tr(AC),$$

where $B \in M(s)$ and $C \in M(t)$.

Proof. From Theorem 3.16,

$$tr(AB) \equiv s \ (\text{mod } q^m - 1) \quad \text{and} \quad tr(AC) \equiv t \ (\text{mod } q^m - 1).$$

If $tr(AB) = tr(AC)$, then $tr(AB) \equiv tr(AC)(\text{mod } q^m - 1)$, which implies that $s \equiv t \ (\text{mod } q^m - 1)$. □

For Theorems 3.15 and 3.16, we considered the case $1 < s < q^m - 1$. In the following, we look at the case $s = q^m - 1$. Note that

$$q^m - 1 = \begin{cases} (p-1) + (p-1)p + \cdots + (p-1)p^{rm-1}, & p > 2 \\ 1 + 2 + \cdots + 2^{rm-1}, & p = 2. \end{cases}$$

For the case $p = 2$, we have

$$Tr_m^n(x)^{2^{rm}-1} = \sum_{\mathbf{t} \in \mathbb{Z}_l^m} x^{\tau_s(\mathbf{t})}$$

and

$$w\left(Tr_m^n(x)^{2^{rm}-1}\right) = \left|\left\{\tau_{2^{rm}-1}(\mathbf{t}) \mid \mathbf{t} \in \mathbb{Z}_l^{rm}\right\}\right| = l^{rm}.$$

Because the Hamming weight of $\tau_{2^{rm}-1}(\mathbf{t})$ is equal to rm, $\tau_{2^{rm}-1}(\mathbf{t}) \neq \tau_i(\mathbf{t}')$ for any i with $1 \leq i < 2^{rm} - 1$. For the case of $p > 2$, we have

$$Tr_m^n(x)^{q^m-1} = Tr_m^n(x)^{p^{rm}-1} = \sum_{B \in M(p^{rm}-1)} d_B x^{Tr(AB)}, \quad \text{and}$$

$$w\left(Tr_m^n(x)^{q^m-1}\right) = w\left(Tr_m^n(x)^{p^{rm}-1}\right) = |M(p^{rm} - 1)| = \binom{l+p-2}{l-1}^{rm},$$

in which $tr(AB) \neq tr(AC)$ for $B \in M(p^{rm} - 1)$ and $C \in M(i)$ where $1 < i < q^m - 1$.

3.6.3 Composition of trace functions and arbitrary functions

Let $g(x)$ be a function from $GF(q^m)$ to $GF(q)$. We can write

$$g(x) = \sum_{i=0}^{q^m-1} c_i x^i, \quad c_i \in GF(q^m).$$

From the above discussion, we have established the following theorem on expansion of the composition of $g(x)$ and $Tr_m^n(x)$, denoted by $g \circ Tr_m^n(x)$, a function from $GF(q^n)$ to $GF(q)$.

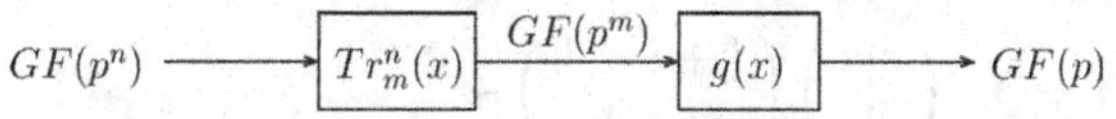

$$GF(p^n) \longrightarrow \boxed{Tr_m^n(x)} \xrightarrow{GF(p^m)} \boxed{g(x)} \longrightarrow GF(p)$$

Figure 3.1. Diagram of composition of g and Tr_m^n.

Theorem 3.17 *Let*

$$f(x) = g \circ Tr_m^n(x).$$

For $p = 2$ and $q = 2^r$,

$$f(x) = \sum_{c_i \neq 0} c_i \sum_{\mathbf{t} \in \mathbb{Z}_l^{w(i)}} x^{\tau_{w(i)}(\mathbf{t})}$$

and

$$w(f) = \sum_{c_i \neq 0} l^{w(i)}. \tag{3.15}$$

For $p > 2$,

$$f(x) = \sum_{c_i \neq 0} c_i \sum_{B \in M(i)} d_B x^{tr(AB)}$$

and the number of nonzero monomials in the expansion is given by

$$w(f) = \sum_{c_i \neq 0} |M(i)|. \tag{3.16}$$

These functions are illustrated in Figure 3.1.

Remark 3.1 The results shown in Theorems 3.15–3.17 have important applications for computation of linear spans of GMW sequences and many other sequences. In fact, this value is just the number of nonzero coefficients of f represented relative to the basis $(1, x, \ldots, x^{q^n-1})$. In Chapter 6, we will show that $f(x)$ corresponds to a sequence over $GF(q)$ and $w(f)$ is equal to the linear span of the sequence.

Remark 3.2 (MAXIMUM WEIGHT) Note that the function g has maximum weight when all coefficients c_i are nonzero. In this case, f has maximum weight that is derived as follows.

$$w(f) = \sum_{s=1}^{q^m-1} |M(s)| + 1$$

$$= \sum_{s=1}^{q^m-1} \prod_{i=1}^{rm-1} \binom{l+s_i-1}{l-1} + 1 \quad \left(s = \sum_{i=0}^{rm-1} s_i p^i, 0 \leq s_i < p \right)$$

$$= 1 + \sum_{i=1}^{p-1} \binom{rm}{1} \binom{l+i-1}{l-1} +$$

$$\binom{rm}{2} \sum_{0 < i,j < p} \binom{l+i-1}{l-1} \binom{l+j-1}{l-1} + \cdots$$

$$+ \binom{rm}{t} \sum_{0 < i_1, \dots, i_t < p} \binom{l+i_1-1}{l-1} \cdots \binom{l+i_t-1}{l-1}$$

$$+ \cdots + \binom{rm}{rm} \binom{l+p-2}{l-1}^{rm}.$$

In particular, if $p = 2$, then the above formula becomes

$$w(f) = (l+1)^{rm}.$$

Therefore, f has maximum weight if and only if g has maximum weight.

Example 3.18 Let $n = 4$, $m = l = 2$, $q = p = 3$, and $s = 5$. Then $s = 2 + 3 \implies s_0 = 2$ and $s_1 = 1$. Directly expanding $Tr_2^4(x)^5$, we have

$$\begin{aligned} Tr_2^4(x)^5 &= (x + x^9)^5 \\ &= (x + x^9)^{2+3} \\ &= (x + x^9)^2 (x^3 + x^{27}) \\ &= (x^2 + 2x^{10} + x^{18})(x^3 + x^{27}) \\ &= x^5 + x^{29} + 2x^{13} + 2x^{37} + x^{21} + x^{45}. \end{aligned}$$

Using Theorem 3.16,

$$Tr_2^4(x)^5 = \sum_{B \in M(5)} d_B x^{tr(AB)},$$

where

$$A = \begin{bmatrix} 1 & 3^2 \\ 3 & 3^3 \end{bmatrix} \quad \text{and} \quad B = \begin{bmatrix} b_{0,0} & b_{0,1} \\ b_{1,0} & b_{1,1} \end{bmatrix},$$

where $(b_{0,0}, b_{1,0})$ is a nonnegative integer-valued solution of $b_{0,0} + b_{1,0} = 2$ and $(b_{0,1}, b_{1,1})$ is a nonnegative integer-valued solution of $b_{0,1} + b_{1,1} = 1$. Thus we have $(b_{0,0}, b_{1,0}) \in \{(2, 0), (0, 2), (1, 1)\}$ and $(b_{0,1}, b_{1,1}) \in \{(0, 1), (1, 0)\}$. We list all the corresponding matrices B (here the brackets for matrices are omitted), d_B, and $tr(AB)$ values in the following table.

B	d_B	$tr(AB)$	(mod 8)
2 0 0 1	1	29	5
2 1 0 0	1	5	5
0 0 2 1	1	45	5
0 1 2 0	1	21	5
1 1 1 0	2	13	5
1 0 1 1	2	37	5

Therefore, we have

$$Tr_2^4(x)^5 = x^{29} + x^5 + x^{45} + x^{21} + 2x^{13} + 2x^{37}.$$

For small values of n, q, and s, one may directly expand a power of a trace function, which seems simpler. However, for large n, the exponents in the expansion of a power of a trace function can be easily calculated by computer using the representation of the exponents in Theorems 3.15 and 3.16. These representations also provide insight into the properties of those exponents in certain algebraic forms.

Note. Equation (3.16) also includes the binary case in Theorem 3.15. For the binary case, each column of the matrix B has only one nonzero entry (which is one) and the other entries in the column are zero.

3.7 The numbers of irreducible polynomials and coset leaders

In the section, we will give the formulae for determining the number of irreducible or primitive polynomials over $\mathbb{F}_q$ of degree n. Let $I_q(n)$ be the number of irreducible polynomials over $\mathbb{F}_q$ of degree n. Then

$$I_q(n) = \frac{1}{n} \sum_{m|n} \mu(m) q^{n/m},$$

where $\mu(x)$ is the Möbius function defined by

$$\mu(x) = \begin{cases} 1 & \text{if } x = 1 \\ (-1)^r & \text{if } x \text{ is the product of } r \text{ distinct primes} \\ 0 & \text{otherwise.} \end{cases}$$

This result can be found in many books, say Golomb (1967). As an example, we have

$$I_2(2) = \frac{1}{2}(\mu(1)2^2 + \mu(2)2) = \frac{1}{2}(4 - 2) = 1$$

$$I_2(3) = \frac{1}{3}(\mu(1)2^3 + \mu(3)2) = \frac{1}{3}(8 - 2) = 2$$

$$I_2(4) = \frac{1}{4}(\mu(1)2^4 + \mu(2)2^2) = \frac{1}{4}(16 - 4) = 3$$

$$I_2(5) = \frac{1}{5}(\mu(1)2^5 + \mu(5)2) = \frac{1}{5}(32 - 2) = 6$$

$$I_2(6) = \frac{1}{6}(\mu(1)2^6 + \mu(2)2^3 + \mu(3)2^2 + \mu(6)2)$$

$$= \frac{1}{6}(64 - 8 - 4 + 2) = 9.$$

The number of primitive polynomials over $\mathbb{F}_q$ of degree n, $P_q(n)$, is given by

$$P_q(n) = \frac{\phi(q^n - 1)}{n},$$

where $\phi(x)$ is the the Euler phi function that denotes the number of integers in the range from 1 to x that are coprime to x. From Property 3.7, $I_q(m)$ for $m > 1$ is equal to the number of the cosets modulo $q^n - 1$ with respect to q that have size m. Therefore, we have

$$|\Gamma_q(n)| = \sum_{m|n} I_q(m) - 1,$$

because the polynomial x of degree 1 corresponds to no coset. In other words, the number of coset leaders modulo $q^n - 1$ with respect to q is equal to the sum of the number of irreducible polynomials over $GF(q)$ of degree m, where m runs through all divisors of n. Thus, we have

$$\sum_{m|n} m I_q(m) = q^n. \tag{3.17}$$

In the following table, we present the values of $I_2(n)$, $\Gamma_2(n)$, and $P_2(n)$ for $1 \le n \le 31$.

Values of $I_2(n)$, $\Gamma_2(n)$, and $P_2(n)$ for $1 \le n \le 31$

n	$2^n - 1$	$I_2(n)$	$\Gamma_2(n)$	$P_2(n)$
1	1	2	1	1
2	3	1	2	1
3	7	2	3	2
4	15	3	5	2
5	31	6	7	6
6	63	9	13	6
7	127	18	19	18
8	255	30	35	16
9	511	56	59	48
10	1023	99	107	60
11	2047	186	187	176
12	4095	335	351	144
13	8191	630	631	630
14	16383	1161	1181	756
15	32767	2182	2191	1800
16	65535	4080	4115	2048
17	131,071	7710	7711	7710
18	262,143	14532	14601	7776
19	524,287	27594	27595	27594
20	1,048,575	52377	52487	24000
21	2,097,151	99858	99879	84672
22	4,194,303	190,557	190,745	120,032
23	8,388,607	364,722	364,723	356,960
24	16,777,215	698,870	699,251	276,480
25	33,554,431	1,342,176	1,342,183	1,296,000
26	67,108,863	2,580,795	2,581,427	1,719,900
27	134,217,727	4,971,008	4,971,067	4,202,496
28	268,435,455	9,586,395	9,587,579	4,741,632
29	536,870,911	18,512,790	18,512,791	18,407,808
30	1,073,741,823	35,790,267	35,792,567	17,820,000
31	2,147,483,647	69,273,666	69,273,667	69,273,666

Remark 3.3 We provide some tables in Appendix B that contain primitive polynomials over $GF(2)$ of degree up to 32, primitive polynomials over $GF(p)$ of degree n where $2 < p \le 127$ and $p^n \le 2^{32}$, and primitive polynomials over $GF(2^m)$ of degree l where $m \in \{2, 3, \ldots, 8\}$ and $l \cdot m \le 32$.

Note

For more detailed treatments about the relationships among irreducible polynomials, cyclotomic cosets, and combinatorial necklaces, see Golomb (1969). Algorithm 3.2 for finding the minimal polynomial of cubes of the roots of a given irreducible polynomial also appeared in that paper. For further references on finite fields, see Lidl and Niederreiter (1983), McWilliams and Sloane (1977), and McEliece (1986).

Appendix A: A Maple Program for Step 3 in Algorithm 3.1

Comments: In this Maple program, we set the parameters for computing the minimal polynomial of α^3 in $GF(2^4)$ defined by the primitive polynomial $f(x) = x^4 + x + 1$ where α is a root of $f(x)$. So, for computing the minimal polynomials for different elements in different fields, one should reset the parameters.

```
writeto('output-file'):
n:=4;
p:=2;
q:=p^n-1;
alias(alpha=RootOf(z^4+z+1)):
C:=[3, 6, 12, 9];   #coset with coset leader 3 modulo 15
m:=1:   # a variable that holding the minimal polynomial
for d in C do
    m:=Expand(m*(x+alpha^d)) mod p
od:
print(m);
quit;
```

Appendix B: Primitive polynomials

In Table 3.5, we list specific primitive polynomials over $GF(2)$ of every degree up to 32. (These were taken from Menezes, van Corschot, and Vanstone (1996).) In the second column of Table 3.5, we represent a primitive polynomial $f(x) = x^n + c_{n-1}x^{n-1} + \cdots + c_1x + c_0$ as a vector $(c_0, c_1, \ldots, c_{n-1})$. For example, for $n = 4$, the primitive polynomial is $f(x) = x^4 + x + 1$. Table 3.6 contains primitive polynomials over $GF(p)$ of degree n where p is a prime with $2 < p \leq 127$ and $p^n < 2^{32}$. (These were computed by Amr Youssef.) The primitive polynomials over $GF(2^m)$ of degree n where $m \in \{2, 3, \ldots, 8\}$ and $n = ml \leq 32$ are listed in Table 3.7. (These were computed using Theorem 3.14.) The data listed in Table 3.7 were taken from the course projects of the graduate course Sequence Design and Cryptography, Spring 2002, University of Waterloo. The notation used in Table 3.7 is as follows:

Table 3.5. *Specific primitive polynomials over*
GF(2) of degree n: $1 \leq n \leq 32$

n	$(c_0, c_1, \ldots, c_{n-1})$
1	1
2	11
3	110
4	1100
5	10010
6	110000
7	1100000
8	10111000
9	100010000
10	1001000000
11	10100000000
12	110010100000
13	1101100000000
14	11010100000000
15	110000000000000
16	1011010000000000
17	10010000000000000
18	111001000000000000
19	1110010000000000000
20	10010000000000000000
21	101000000000000000000
22	1100000000000000000000
23	10000100000000000000000
24	110110000000000000000000
25	1001000000000000000000000
26	11100010000000000000000000
27	111001000000000000000000000
28	1001000000000000000000000000
29	10100000000000000000000000000
30	110010100000000000000000000000
31	1001000000000000000000000000000
32	11000000000000000000000000011000

- $f(x)$ is a primitive polynomial over $GF(2)$ of degree n, taken from Table 3.5, and α is a root of $f(x)$ in the extension $GF(2^n)$.
- n is a composite number and $n = ml$ where both m and l are proper factors of n.
- $\beta = \alpha^d$ where $d = \frac{2^n-1}{2^m-1}$ is a primitive element of $GF(2^m)$ and $h(x)$ is the minimal polynomial of β that is used as the defining polynomial of $GF(2^m)$.
- $g(x) = \prod_{i=0}^{l-1}(x - \alpha^{Q^i})$, where $Q = 2^m$ is the minimal polynomial of α over $GF(2^m)$ of degree l, which is primitive over $GF(2^m)$ of degree l.

Table 3.6. *Primitive polynomials over GF(p) of degree n with $2 \leq p$ (a prime) ≤ 127 and $p^n < 2^{32}$*

p	$f(x)$	p	$f(x)$	p	$f(x)$
3	$x^2 + x + 2$	13	$x^2 + x + 2$	43	$x^2 + x + 3$
3	$x^3 + 2x^2 + 1$	13	$x^3 + x^2 + 2$	43	$x^3 + x^9$
3	$x^4 + x^3 + 2$	13	$x^4 + x^3 + x^2 + 6$	43	$x^4 + x + 20$
3	$x^5 + x^4 + x^2 + 1$	13	$x^5 + x^4 + x^3 + 6$	43	$x^5 + x^4 + 9$
3	$x^6 + x^5 + 2$	13	$x^6 + x^5 + x^3 + 6$		
3	$x^7 + x^6 + x^4 + 1$	13	$x^7 + x^4 + 2$	47	$x^2 + x + 13$
3	$x^8 + x^5 + 2$	13	$x^8 + x^7 + x^6 + 11$	47	$x^3 + x^2 + 2$
3	$x^9 + x^7 + x^5 + 1$			47	$x^4 + x^3 + 5$
3	$x^{10} + x^9 + x^7 + 2$	17	$x^2 + x + 3$	47	$x^5 + x^4 + 6$
3	$x^{11} + x^{10} + x^4 + 1$	17	$x^3 + x^2 + 7$		
3	$x^{12} + x^{11} + x^7 + 2$	17	$x^4 + x^3 + 5$	53	$x^2 + x + 5$
3	$x^{13} + x^{12} + x^6 + 1$	17	$x^5 + x^4 + 5$	53	$x^3 + x^2 + 2$
3	$x^{14} + x^{13} + 2$	17	$x^6 + x^5 + 3$	53	$x^4 + x^3 + 2$
3	$x^{15} + x^{14} + x^4 + 1$	17	$x^7 + x^6 + 7$	53	$x^5 + x^4 + 12$
3	$x^{16} + x^9 + 2$				
3	$x^{17} + x^{16} + x^8 + 1$	19	$x^2 + x + 2$	59	$x^2 + x + 2$
3	$x^{18} + x^{17} + x^5 + 2$	19	$x^3 + x^2 + 6$	59	$x^3 + x^2 + 9$
		19	$x^4 + x^3 + 2$	59	$x^4 + x^3 + 18$
5	$x^3 + x^2 + 2$	19	$x^5 + x^4 + 5$	59	$x^5 + x^4 + 4$
5	$x^4 + x^3 + x + 3$	19	$x^6 + x^5 + 15$		
5	$x^5 + x^2 + 2$	19	$x^7 + x^6 + 5$	61	$x^2 + x + 2$
5	$x^6 + x^5 + 2$			61	$x^3 + x^2 + 6$
5	$x^7 + x^6 + 2$	23	$x^2 + x + 7$	61	$x^4 + x^3 + 17$
5	$x^8 + x^5 + x^3 + 3$	23	$x^3 + x^2 + 6$	61	$x^5 + x^4 + 55$
5	$x^9 + x^7 + x^6 + 3$	23	$x^4 + x^3 + 20$		
5	$x^{10} + x^9 + x^7 + 3$	23	$x^5 + x^4 + 6$	67	$x^2 + x + 12$
5	$x^{11} + x^{10} + 2$	23	$x^6 + x^5 + 7$	67	$x^3 + x^2 + 6$
5	$x^{12} + x^7 + x^4 + 3$			67	$x^4 + x^3 + 12$
		29	$x^2 + x + 3$		
7	$x^2 + x + 3$	29	$x^3 + x^2 + 3$	71	$x^2 + x + 11$
7	$x^3 + x^2 + x + 2$	29	$x^4 + x^3 + 2$	71	$x^3 + x^2 + 8$
7	$x^4 + x^3 + x^2 + 3$	29	$x^5 + x^4 + 2$	71	$x^4 + x^3 + 13$
7	$x^5 + x^4 + 4$	29	$x^6 + x^5 + 11$		
7	$x^6 + x^5 + x^4 + 3$			73	$x^2 + x + 11$
7	$x^7 + x^5 + 4$	31	$x^2 + x + 12$	73	$x^3 + x^2 + 5$
7	$x^8 + x^7 + 3$	31	$x^3 + x^2 + 9$	73	$x^4 + x^3 + 33$
7	$x^9 + x^8 + x^3 + 2$	31	$x^4 + x^3 + 13$		
7	$x^{10} + x^9 + x^8 + 3$	31	$x^5 + x^4 + 10$	79	$x^2 + x + 3$
		31	$x^6 + x^5 + 12$	79	$x^3 + x^2 + 2$
11	$x^2 + x + 7$			79	$x^4 + x^3 + 7$
11	$x^3 + x^2 + 3$	37	$x^2 + x + 5$		
11	$x^4 + x^3 + 8$	37	$x^3 + x^2 + 17$	83	$x^2 + x + 2$
11	$x^5 + x^4 + x^3 + 3$	37	$x^4 + x^3 + 22$	83	$x^3 + x^2 + 11$
11	$x^6 + x^5 + x + 7$	37	$x^5 + x^4 + 2$	83	$x^4 + x^3 + 24$
11	$x^7 + x^6 + 4$				
11	$x^8 + x^7 + x^6 + 7$	41	$x^2 + x + 12$	89	$x^2 + x + 6$
		41	$x^3 + x^2 + 11$	89	$x^3 + x^2 + 6$
		41	$x^4 + x^3 + 26$	89	$x^4 + x^3 + 14$
		41	$x^5 + x^4 + 11$		
				97	$x^2 + x + 5$
				97	$x^3 + x^2 + 5$
				97	$x^4 + x^3 + 15$
101	$x^2 + x + 3$	107	$x^2 + x + 5$	113	$x^2 + x + 10$
101	$x^3 + x^2 + 27$	107	$x^3 + x^2 + 10$	113	$x^3 + x^2 + 10$
101	$x^4 + x^3 + 11$	107	$x^4 + x^3 + 8$	113	$x^4 + x^3 + 6$
103	$x^2 + x + 5$	109	$x^2 + x + 6$	127	$x^2 + x + 3$
103	$x^3 + x^2 + 7$	109	$x^3 + x^2 + 6$	127	$x^3 + x^2 + 15$
103	$x^4 + x^3 + 44$	109	$x^4 + x^3 + 24$	127	$x^4 + x^3 + 3$

Table 3.7. *Primitive polynomials over* $GF(2^m)$ *of degree* l *with* $n = ml \leq 32$

l	$h(x)$	$g(x)$
		$GF(2^2)$
2	$x^2 + x + 1$	$x^2 + x + \beta$
3	$x^2 + x + 1$	$x^3 + x^2 + \beta^2 x + \beta$
4	$x^2 + x + 1$	$x^4 + x^3 + \beta x^2 + \beta x + \beta$
5	$x^2 + x + 1$	$x^5 + x^4 + \beta^2 x^3 + \beta x^2 + \beta$
6	$x^2 + x + 1$	$x^6 + x^2 + x + \beta$
7	$x^2 + x + 1$	$x^7 + x^6 + \beta x^5 + \beta^2 x^4 + \beta^2 x^2 + x + \beta$
8	$x^2 + x + 1$	$x^8 + x^7 + \beta^2 x^6 + \beta^2 x^5 + x^4 + x^3 + x^2 + \beta$
		$GF(2^3)$
2	$x^3 + x^2 + 1$	$x^2 + \beta^3 x + \beta$
3	$x^3 + x + 1$	$x^3 + \beta x^2 + \beta^5 x + \beta$
4	$x^3 + x + 1$	$x^4 + \beta x^3 + \beta^3 x^2 + x + \beta$
5	$x^3 + x + 1$	$x^5 + \beta^2 x^3 + \beta^2 x^2 + x + \beta$
6	$x^3 + x^2 + 1$	$x^6 + \beta^3 x^5 + x^4 + x^3 + \beta^3 x^2 + \beta^2 x + \beta$
7	$x^3 + x^2 + 1$	$x^7 + \beta^3 x^6 + \beta^2 x^5 + x^4 + \beta^3 x^3 + \beta^4 x + \beta$
8	$x^3 + x^2 + 1$	$x^8 + \beta^3 x^6 + \beta^5 x^5 + \beta x^4 + x^3 + \beta x^2 + \beta x + \beta$
		$GF(2^4)$
2	$x^4 + x + 1$	$x^2 + \beta^2 x + \beta$
3	$x^4 + x + 1$	$x^3 + \beta^{10} x^2 + \beta^{13} x + \beta$
4	$x^4 + x + 1$	$x^4 + \beta^4 x^3 + \beta x^2 + \beta^6 x + \beta$
5	$x^4 + x^3 + 1$	$x^5 + x^4 + \beta^5 x^3 + \beta^{14} x^2 + x + \beta$
6	$x^4 + x^3 + 1$	$x^6 + \beta^{10} x^5 + \beta^4 x^4 + \beta^{14} x^3 + \beta^{11} x^2 + \beta^9 x + \beta$
		$GF(2^5)$
2	$x^5 + x^4 + x^3 + x^2 + 1$	$x^2 + \beta^{23} x + \beta$
3	$x^5 + x^3 + x^2 + x + 1$	$x^3 + \beta^{20} x + \beta$
4	$x^5 + x^2 + 1$	$x^4 + \beta^2 x^3 + \beta x^2 + \beta^8 x + \beta$
5	$x^5 + x^3 + x^2 + x + 1$	$x^5 + \beta^8 x^4 + \beta^{30} x^3 + \beta^{21} x^2 + \beta^{19} x + \beta$
		$GF(2^6)$
2	$x^6 + x^5 + 1$	$x^2 + \beta^{30} x + \beta$
3	$x^6 + x^5 + x^4 + x + 1$	$x^3 + \beta^{56} x^2 + x + \beta$
4	$x^6 + x^5 + x^4 + x + 1$	$x^4 + \beta^{54} x^3 + \beta^{34} x^2 + \beta^{54} x + \beta$
		$GF(2^7)$
2	$x^7 + x^6 + 1$	$x^2 + \beta^{91} x + \beta$
3	$x^7 + x^5 + x^3 + x + 1$	$x^3 + \beta^{115} x^2 + \beta^{66} x + \beta$
4	$x^7 + x^6 + x^5 + x^2 + 1$	$x^4 + \beta^{66} x^3 + \beta^{92} x^2 + \beta^{113} x + \beta$
		$GF(2^8)$
2	$x^8 + x^4 + x^3 + x^2 + 1$	$x^2 + \beta^{50} x + \beta$
3	$x^8 + x^7 + x^3 + x^2 + 1$	$x^3 + \beta^{84} x^2 + \beta^{36} x + \beta$
4	$x^8 + x^6 + x^5 + x^4 + 1$	$x^4 + \beta^{57} x^3 + \beta^{28} x^2 + \beta^{212} x + \beta$

Table 3.8. *Minimal polynomials of elements in* $\mathbb{F}_{2^5}$ *and* $\mathbb{F}_{2^6}$

n = 5		n = 6	
Coset leader i	f_{α^i}	Coset leader i	f_{α^i}
1	100101	1	1100001
3	111101	3	1110101
5	110111	5	1110011
7	101111	7	1001001
11	111011	9	1011
15	101001	11	1011011
		13	1101101
		15	1010111
		21	111
		23	1100111
		27	1101
		31	1000011

Table 3.9. *Minimal polynomials of elements in* $\mathbb{F}_{2^7}$

Coset leader i	f_{α^i}	Coset leader i	f_{α^i}
1	11000001	21	11010011
3	11010101	23	11100101
5	11110001	27	11101111
7	10111111	29	10001001
9	10011101	31	10101011
11	10010001	43	11001011
13	10100111	47	10001111
15	11111101	55	10111001
19	11110111	63	10000011

Table 3.10. *Minimal polynomials of elements in* $\mathbb{F}_{2^8}$

Coset leader i	f_{α^i}	Coset leader i	f_{α^i}	Coset leader i	f_{α^i}
1	101110001	23	110001101	53	111000011
3	111011101	25	110110001	55	100011011
5	110011111	27	111111001	59	101100101
7	100101101	29	101100011	61	111100111
9	101111011	31	101101001	63	101110111
11	111001111	37	111110101	85	111
13	110101001	39	100111111	87	110001011
15	111010111	43	110000111	91	101011111
17	11001	45	100111001	95	111110011
19	101001101	47	100101011	111	110111101
21	110100011	51	11111	119	10011
				127	100011101

Table 3.11. *Minimal polynomials of elements in* $\mathbb{F}_{2^9}$

Coset leader i	f_{α^i}	Coset leader i	f_{α^i}	Coset leader i	f_{α^i}
1	1000100001	41	1100111011	93	1001111101
3	1001101001	43	1101001111	95	1001110111
5	1000110011	45	1011111001	103	1111010101
7	1001100101	47	1101101011	107	1100010101
9	1100100011	51	1010101111	109	1111111011
11	1011010001	53	1010100101	111	1111100011
13	1110111001	55	1011110101	117	1111001011
15	1000011011	57	1010111101	119	1100000001
17	1101101101	59	1111001101	123	1110000101
19	1010000111	61	1111101001	125	1000101101
21	1110100001	63	1010011001	127	1001011001
23	1001011111	73	1101	171	1010110111
25	1100011111	75	1101111111	175	1000010111
27	1111000111	77	1001001011	183	1101110011
29	1101011011	79	1110001111	187	1001101111
31	1101100001	83	1010100011	191	1100110001
35	1000000011	85	1110110101	219	1011
37	1111011001	87	1010010101	223	1100010011
39	1011001111	91	1101001001	239	1011011011
				255	1000010001

Appendix C: Minimal polynomials

Minimal polynomials have important applications in the implementation of many binary sequences with good correlation, such as 3-term sequences, 5-term sequences, most of the signal sets with low cross-correlation, and the boolean functions transformed from trace representations of sequences. For these applications, we need to compute the minimal polynomials of elements in finite fields. In this appendix, we list the minimal polynomials of elements in $\mathbb{F}_{2^n}$ for $5 \leq n \leq 10$. For example, for $\alpha^i \in \mathbb{F}_{2^n}$, the corresponding minimal polynomial of α^i is equal to the polynomial

$$f_{\alpha^i}(x) = \prod_{j=0}^{s-1} \left(x + \alpha^{i2^j}\right) = c_0 + c_1 x + \cdots + c_{s-1} x^{s-1} + x^s, \quad c_i \in \mathbb{F}_2$$

where s is the smallest number such that $i2^s \equiv i \pmod{2^n - 1}$. We represent this as a vector $(c_0, c_1, \ldots, c_{s-1}, c_s)$ where $s|n$. For example, for $n = 6$ in Table 3.8, the minimal polynomial of α^3 is given as $f_{\alpha^3}(x) = 1 + x + x^2 + x^4 + x^6 \leftrightarrow 1110101$, and the minimal polynomial of α^9 is given as $f_{\alpha^9}(x) = 1 + x^2 + x^3 \leftrightarrow 1011$.

Table 3.12. *Minimal polynomials of elements in* $\mathbb{F}_{2^{10}}$

Coset leader i	f_{α^i}	Coset leader i	f_{α^i}	Coset leader i	f_{α^i}
1	10010000001	73	11100010111	175	10110001111
3	11110000001	75	11111000101	179	10010001011
5	10110000101	77	11010000101	181	11011011111
7	10011111111	79	11100111001	183	11110000111
9	11110101001	83	11001001111	187	10100001011
11	10101100001	85	11100011101	189	11000110001
13	11110110001	87	10010011001	191	10001101111
15	11010101101	89	11101011001	205	11010001001
17	10110010111	91	10101101011	207	10101100111
19	11011111101	93	11111111111	213	11011001101
21	11010111111	95	10100110001	215	10000101101
23	11011000001	101	10110100001	219	10011101101
25	11000100101	103	10111110111	221	10011010111
27	11011110111	105	11100001111	223	10100100011
29	10001100101	107	10011110011	235	11110010011
31	11000100011	109	11100100001	237	10100011111
35	11001000011	111	11001010001	239	11101010101
37	11000110111	115	11101111101	245	10100011001
39	11100010001	117	10011001001	247	11000010011
41	10100111101	119	11011010011	251	10111111011
43	10011000101	121	11100101011	253	10000110101
45	10001100011	123	10001000111	255	10000001111
47	11111110011	125	10000011011	343	10111000111
49	10101010111	127	11111111001	347	11000010101
51	11100110101	147	10110111001	351	11111101011
53	11110001101	149	10101000011	367	10111100101
55	11010100111	151	10000100111	375	10111000001
57	10001010011	155	10010101001	379	11101100011
59	10011100111	157	11010110101	383	10100001101
61	11001111111	159	11101111011	439	11101000111
63	10110101011	167	11001111001	447	10010101111
69	10000011101	171	10110011011	479	11101001101
71	11001011011	173	11111011011	511	10000001001
33	101111	99	111011	165	100101
231	110111	363	101001	495	111101
341	111				

Exercises for Chapter 3

1. Give a proof for Corollary 3.3 in Section 3.2. In other words, prove that in any finite field of characteristic p,

$$(x + y)^{p^m} = x^{p^m} + y^{p^m}$$

for any $m \geq 1$.

2. Let $GF(2^6)$ be defined by the primitive polynomial $f(x) = x^6 + x + 1$ and let α be a root of $f(x)$. Compute: $\alpha^9 + \alpha^{23}$, $\alpha^9 \cdot \alpha^{23}$, and $1 + \alpha^7$.

3. Let p be a prime number. Prove that

$$x^{p^n} - x = \text{product of all monic polynomials,}$$
$$\text{irreducible over } GF(p), \text{ whose}$$
$$\text{degree divides } n.$$

4. The cyclotomic coset containing s consists of

$$C_s = \{s, sp, sp^2, \ldots, sp^{n_s-1}\},$$

where n_s is the smallest positive integer such that $p^{n_s} s \equiv s \pmod{p^n - 1}$. Prove that $n_s \mid n$.

5. Let α be a primitive element in $GF(p^n)$. Prove that $m_s(x)$ defined in the algorithm in Section 3.4 is a polynomial over $GF(p)$. In other words, if

$$m_s(x) = \prod_{i \in C_s}(x - \alpha^i),$$

where C_s is the cyclotomic coset modulo $p^n - 1$ containing s as the coset leader, then the coefficients of $m_s(x)$ belong to $GF(p)$.

6. Let $GF(2^5)$ be defined by the primitive polynomial $f(x) = x^5 + x^3 + 1$ and let α be a root of $f(x)$.
   (a) Compute the cyclotomic cosets modulo 31 that contain 5 and 13, respectively.
   (b) Compute the minimal polynomials of α^5 and α^{13}.
   (c) What are orders of α^5 and α^{13}? What are periods of these two polynomials?
   (d) Are these two polynomials a pair of reciprocal polynomials?

7. Let $GF(2^6)$ be defined by the primitive polynomial $f(x) = x^6 + x + 1$ and let α be a root of $f(x)$. Compute $Tr(1)$, $Tr(\alpha)$, $Tr(\alpha^2)$, $Tr(\alpha^3)$, $Tr(\alpha^4)$, $Tr(\alpha^6)$.

8. Let $GF(2^3)$ be defined by the primitive polynomial $g(x) = x^3 + x^2 + 1$ and α be a root of $g(x)$. Let $f(x) = x^2 + \alpha^3 x + \alpha$. Then $f(x)$ is a primitive polynomial over $GF(2^3)$ of degree 2. We can construct the finite field $GF(2^6)$ in an alternative way by constructing $GF(q^2)$ where $q = 8$ by $f(x)$. Let β be a root of $f(x)$. Then $GF(q^2)$ can be considered as a vector space over $GF(q)$ of dimension 2.
   (a) Write the first six elements

$$1, \beta, \beta^2, \beta^3, \beta^4, \beta^5$$

   in the vector representation forms under the basis $\{1, \beta\}$. For example,

$$\beta^2 = \alpha^3 \beta + \alpha,$$

so the vector representation of β^2 is (α, α^3); and

$$\begin{aligned}
\beta^3 = \beta \cdot \beta^2 &= \beta(\alpha^3 \beta + \alpha) \\
&= \alpha^3 \beta^2 + \alpha \beta \\
&= \alpha^3(\alpha^3 \beta + \alpha) + \alpha \beta \\
&= \alpha^6 \beta + \alpha^4 + \alpha \beta = \alpha^2 \beta + \alpha^4
\end{aligned}$$

so, the vector representation of β^3 is (α^2, α^4).

(b) What is the minimal polynomial of β over $GF(2)$?

4

Feedback Shift Register Sequences

Feedback shift register (FSR) sequences have been widely used as synchronization, masking, or scrambling codes and for white noise signals in communication systems, signal sets in CDMA communications, key stream generators in stream cipher cryptosystems, random number generators in many cryptographic primitive algorithms, and testing vectors in hardware design. Golomb's popular book *Shift Register Sequences*, first published in 1967 and revised in 1982 is a pioneering book that discusses this type of sequences. In this chapter, we introduce this topic and discuss the synthesis and the analysis of periodicity of linear feedback shift register sequences. We give different (though equivalent) definitions and representations for LFSR sequences and point out which are most suitable for either implementation or analysis. This chapter contains seven sections, which are organized as follows. In Section 4.1, we give a general description for feedback shift registers at the gate level for the binary case and as a finite field configuration for the q-ary case. In Sections 4.2–4.4, we introduce the definition of LFSR sequences from the point of view of polynomial rings and discuss their characteristic polynomials, minimal polynomials, and periods. Then, we show the decomposition of LFSR sequences. We provide the matrix representation of LFSR sequences in Section 4.5 as another historic approach and discuss their trace representation for the irreducible case in detail in Section 4.6, which is a more modern approach. (The general case will be treated in Chapter 6.) LFSRs with primitive minimal polynomials are basic building blocks for nonlinear generators. The trace representation of LFSR sequences is a powerful tool for the analysis of unpredictability or randomness of pseudorandom sequences and for the design of pseudorandom sequences with desired properties. In Section 4.7, we present the generating function method for studying LFSR sequences.

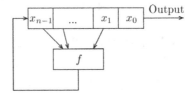

Figure 4.1. A block diagram for an FSR.

4.1 Feedback shift registers

In this section, we give a definition and some of the basic terms for feedback shift register sequences. We denote $F = GF(2) = \{0, 1\}$ and

$$F^n = \{(a_0, a_1, \ldots, a_{n-1}) \mid a_i \in F\},$$

a vector space over F of dimension n. A function with n binary inputs and one binary output is called a boolean function of n variables, that is, $f : F^n \to F$, which can be represented as follows:

$$f(x_0, x_1, \ldots, x_{n-1}) = \sum c_{i_1 i_2 \ldots i_t} x_{i_1} x_{i_2} \ldots x_{i_t}, \; c_{i_1 i_2 \ldots i_t} \in F, \tag{4.1}$$

where the sum runs through all subsets $\{i_1, \ldots, i_t\}$ of $\{0, 1, \ldots, n-1\}$. This shows that there are 2^{2^n} different boolean functions of n variables.

4.1.1 Basic concepts and examples

An n-stage shift register is a circuit consisting of n consecutive 2-state storage units (flip-flops) regulated by a single clock. At each clock pulse, the state (1 or 0) of each memory stage is shifted to the next stage in line. A shift register is converted into a code generator by including a feedback loop, which computes a new term for the left-most stage, based on the n previous terms. In Figure 4.1, we see a diagram of a feedback shift register.

Each of the squares is a 2-state storage unit. The n binary storage elements are called the stages of the shift register, and their contents (regarded as either a binary number or a binary vector, n bits in length) is called a state of the shift register. $(a_0, a_1, \ldots, a_{n-1}) \in F^n$ is called an initial state of the shift register. The feedback function $f(x_0, x_1, \ldots, x_{n-1})$ is a boolean function of n variables, defined in Eq. (4.1). At every clock pulse, there is a transition from one state to the next. To obtain a new value for stage n, we compute $f(x_0, x_1, \ldots, x_{n-1})$ of all the present terms in the shift register and use this in stage n. For example, the next state of the shift register in Figure 4.1 becomes

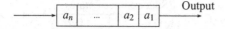

where

$$a_n = f(a_0, a_1, \ldots, a_{n-1}).$$

After the consecutive clock pulses, a feedback shift register outputs a sequence:

$$a_0, a_1, \ldots, a_n, \ldots. \tag{4.2}$$

The sequence satisfies the following recursive relation:

$$a_{k+n} = f(a_k, a_{k+1}, \ldots, a_{k+n-1}), k = 0, 1, \ldots. \tag{4.3}$$

Any n consecutive terms of the sequence in (4.2)

$$a_k, a_{k+1}, \ldots, a_{k+n-1}$$

represents a state of the shift register in Figure 4.1. A state (or vector) diagram is a diagram that is drawn based on the successors of each of the states. The output sequence is called a feedback shift register sequence. If the feedback function $f(x_0, x_1, \ldots, x_{n-1})$ is a linear function, then the output sequence is called a LFSR sequence. Otherwise, it is called a nonlinear feedback shift register (NLFSR) sequence. Sometimes, we also say that **a** is generated by an LFSR (or NLFSR). Here linear means that the feedback function computes the modulo 2 sum of a subset of the stages of the shift register.

Example 4.1 In Figure 4.2, we see a 3-stage shift register with a (nonlinear) feedback function $f(x_0, x_1, x_2) = x_0 x_1$. From this, we can compute the next-state function, as shown in the following table, a successor table, for each of the eight states of the shift register. (Note. The shifting direction in the table is reversed from Figure 4.2.)

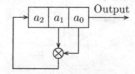

Figure 4.2. A 3-stage nonlinear feedback shift register.

Succession of states

$x_0 x_1 x_2$	$x_0 x_1 x_2$
Current state	Next state
000	000
001	010
010	100
011	110
100	000
101	010
110	101
111	111

The state diagram of the eight states is shown in Figure 4.3. From the state diagram, we can directly observe the autonomous behavior of the device. For example, the initial state 100 leads to the output sequence 1000..., and this state diagram is shown in Figure 4.4.

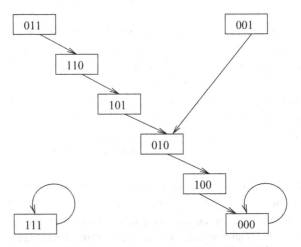

Figure 4.3. The state diagram of the FSR in Figure 4.2.

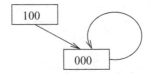

Figure 4.4. The state diagram with the initial state $x_0 x_1 x_2 = 100$.

Table 4.1. *Truth table of*
$$f = x_0 + x_1$$

$x_0 x_1 x_2$	$f = x_0 + x_1$
000	0
001	0
010	1
011	1
100	1
101	1
110	0
111	0

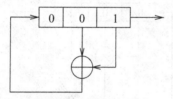

Figure 4.5. A 3-stage LFSR for Example 4.2.

Example 4.2 A 3-stage LFSR is shown in Figure 4.5 with the linear feedback function $f(x_0, x_1, x_2) = x_0 + x_1$. The truth table of this feedback function is given in Table 4.1, and the state diagram of the corresponding LFSR is shown in Figure 4.6. The output sequence with the initial state 100 is

$$10010111001011\ldots,$$

which is seen to repeat periodically with a period of 7.

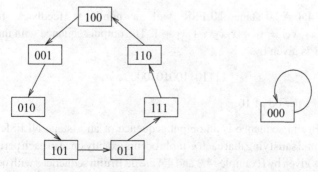

Figure 4.6. State diagram of the LFSR in Figure 4.5.

Table 4.2. *Truth table of*
$$f = x_0 + x_1x_2 + x_2 + 1$$

$x_0x_1x_2$	$x_0 + x_1x_2 + x_2 + 1$
000	1
001	0
010	1
011	1
100	0
101	1
110	0
111	0

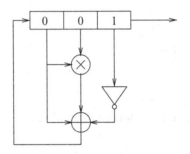

Figure 4.7. The 3-stage NLFSR for Example 4.3.

Example 4.3 A 3-stage LFSR with a nonlinear feedback function $f(x_0, x_1, x_2) = x_0 + x_1x_2 + x_2 + 1$, is shown in Figure 4.7. The truth table of this boolean function is given in Table 4.2. From the truth table, the state diagram of the NLFSR is easily obtained, as shown in Figure 4.8. Therefore, the output sequence with the initial state 100 is

$$1000101110001011\ldots.$$

Example 4.4 A 4-stage NLFSR with a nonlinear feedback function $f(x_0, x_1, x_2, x_3) = x_0 + x_1x_2x_3 + x_1 + 1$. The output sequence with the initial state 1111 is given by

$$1111011001010000\ldots,$$

which has a period of 16.

A de Bruijn sequence is an output sequence of an n-stage NLFSR having period 2^n and satisfying that each n-tuple occurs exactly once in each period. The sequences given by Examples 4.3 and 4.4 are de Bruijn sequences with periods 8 and 16, respectively. For a general discussion of nonlinear feedback shift register

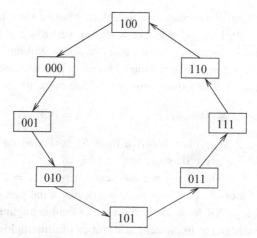

Figure 4.8. State diagram of Figure 4.7

sequences including the number of de Bruijn sequences and constructions for several subsets of de Bruijn sequences, see Chapter 6 in *Shift Register Sequences* (Golomb 1967).

We will see that the periods of LFSRs are completely determined by their feedback functions in a mathematically predictable way. However, for NLFSRs, there are only a few results on the period problem in the literature.

Remark 4.1 The de Bruijn sequence in Example 4.3 can be obtained by inserting an extra 0 into the run of two consecutive zeros of the LFSR sequence in Example 4.2. But the feedback functions of these two sequences are completely different. A sequence generated by an n-stage LFSR with period $2^n - 1$ is called a maximal length sequence, or m-sequence for short (we will formally define it later). From any m-sequence of period $2^n - 1$, we can obtain a de Bruijn sequence by inserting an extra 0 into the run of $n - 1$ consecutive zeros of the m-sequence. However, this type of de Bruijn sequence is not secure for use in stream cipher cryptosystems.

4.1.2 Finite field configuration for q-ary feedback shift register sequences

Let $F = GF(q)$ where q is a prime or a power of a prime (so $|F| = q$). Referring back to Figure 4.1, each stage is replaced by a q-state storage unit and the feedback function is replaced by a function from F^n to F, that is,

$$f(x_0, x_1, \ldots, x_{n-1}) = \sum c_{i_0, i_1, \ldots, i_{n-1}} x_0^{i_0} x_1^{i_1} \ldots x_{n-1}^{i_{n-1}}, \ c_{i_0, i_1, \ldots, i_{n-1}} \in F, \quad (4.4)$$

which is a polynomial function in n interdeterminates. Here the exponents $i_0, i_1, \ldots, i_{n-1} \in \{0, 1, \ldots, q - 1\}$, because $x^q = x$ for any $x \in F$. The output sequence of the shift register is called a q-ary feedback shift register sequence. In other words, we define an abstract model for q-ary FSR sequences as follows. Let $\mathbf{a} = \{a_i\}_{i \geq 0}$, $a_i \in F$ be a q-ary sequence whose elements are given by

$$a_{n+k} = f(a_k, a_{k+1}, \ldots, a_{k+n-1}), k = 0, 1, \ldots, \tag{4.5}$$

where $f(x_0, x_1, \ldots, x_{n-1})$ is a function from F^n to F defined by Eq. (4.4). Then $\mathbf{a}$ is called a q-ary FSR sequence, and $(a_0, a_1, \ldots, a_{n-1})$ is still called an initial state of $\mathbf{a}$. The significant difference between $q = 2$ and $q > 2$ is that Figure 4.1 for binary FSR sequences is a device at the gate level, but Figure 4.1 for a q-ary FSR ($q > 2$) is only a finite field configuration. The latter needs more circuits to implement arithmetics of finite fields for feedback function computation.

For whichever case $q = 2$ or $q > 2$, the design of LFSR sequences with desired properties requires us to understand the functionality of three components of an LFSR: the initial state, the feedback function, and the output sequences. In other words, we want to understand how the behavior of the output sequences is completely determined by initial states and feedback functions. From now on, we treat the cases $q = 2$ and $q > 2$ together when we discuss their algebraic properties. From Eq. (4.4), the following result is immediate.

Property 4.1 *There are q^{q^n} different functions from F^n to F.*

4.1.3 Periodic Property

Definition 4.1 *The sequence $a_0, a_1, \ldots$ is denoted as $\mathbf{a}$ or $\{a_i\}$. If $a_i \in F$, then we say that $\mathbf{a}$ is a q-ary sequence or a sequence over F. If there exist integers $r > 0$ and $u \geq 0$ such that*

$$a_{i+r} = a_i \text{ for all } i \geq u, \tag{4.6}$$

then the sequence is said to be ultimately periodic *with parameters (r, u), and r is called a* period *of the sequence. The smallest number r satisfying Eq. (4.6) is called a* (least) period *of the sequence. If $u = 0$, then the sequence is said to be* periodic. *When the context is clear, we simply say the period of $\mathbf{a}$ instead of the least period of $\mathbf{a}$.*

For example, the output sequence $00011011011\ldots$ of a 4-stage LFSR with the feedback function $f(x_0, x_1, x_2, x_3) = x_2 + x_3$ and the initial state $a_0 a_1 a_2 a_3 = 0001$ is an ultimately periodic sequence, where $u = 2$ and the period r is 3. The output sequence of the feedback shift register sequence in Figure 4.5 is a periodic sequence with period 7.

Theorem 4.1 *Any q-ary feedback shift register sequence is ultimately periodic with period $r \leq q^n$ where n is the number of the stages. In particular, if $q = 2$, then $r \leq 2^n$.*

Proof. In a q-ary feedback shift register with n stages, there are q^n possible states. Each state uniquely determines its successor. Hence, the first time a previous state is repeated, a period for the sequence is established. (If state S at time t_1 is the same as state S' at time t_2, then the states at times $t_1 + 1$ and $t_2 + 1$ are the same, as are the states at times $t_1 + 2$ and $t_2 + 2$, etc.) Thus the maximum possible period is q^n, the number of the different states. □

4.1.4 Linear feedback shift register sequences

If the feedback function $f(x_0, x_1, \ldots, x_{n-1})$ is a linear function, that is, if it can be expressed as

$$f(x_0, x_1, \ldots, x_{n-1}) = c_0 x_0 + c_1 x_1 + \ldots + c_{n-1} x_{n-1}, c_i \in F,$$

then the recursive relation shown in Eq. (4.3) becomes the following linear recursive relation:

$$a_{k+n} = \sum_{i=0}^{n-1} c_i a_{k+i}, k = 0, 1, \ldots. \tag{4.7}$$

Thus, an LFSR sequence is also called a linear recursive sequence (or linear recurring sequence) over F in the literature, where F could be $GF(2)$ or any finite field $GF(q)$. Note that there are only q^n different n-stage LFSRs. In particular, for $q = 2$, we only have 2^n different n-stage binary LFSRs.

Theorem 4.2 *Let $\mathbf{a}$ be a sequence generated by an n-stage LFSR over F. Then the period of $\mathbf{a}$ is less than or equal to $q^n - 1$. In particular, if $q = 2$, the period of any binary n-stage LFSR sequence is less than or equal to $2^n - 1$.*

Proof. Note that the successor state of $00 \cdots 0$ (n times 0) of an n-stage LFSR is again $00 \cdots 0$. Using the same argument as in the proof of Theorem 4.1, we see that the period of $\mathbf{a}$ is $\leq q^n - 1$, since the state $00 \cdots 0$ cannot be part of any other period. □

In the rest of this chapter, we will restrict ourselves to LFSR sequences. It is worth pointing out that it is difficult to generate nonlinear sequences with the desired properties using an NLFSR directly. Most of the methods for generating nonlinear sequences make use of one or several LFSRs together with some control operations.

4.2 Definition of LFSR sequences in terms of polynomial rings

To characterize the periodicity of LFSR sequences, we introduce another equivalent definition of LFSR sequences over F in terms of the polynomial ring $F[x]$. We use the notation $F = GF(q)$. The basic mathematical tools of this section are from linear algebra.

4.2.1 The left shift operator

Let $V(F)$ be a set consisting of all infinite sequences whose elements are taken from F; that is,

$$V(F) = \{\mathbf{a} = (a_0, a_1, \ldots) \mid a_i \in F\}. \tag{4.8}$$

Let

$$\mathbf{a} = (a_0, a_1, a_2, \ldots),$$
$$\mathbf{b} = (b_0, b_1, b_2, \ldots)$$

be two sequences in $V(F)$ and let $c \in F$. We define addition and scalar multiplication on $V(F)$ as follows:

$$\mathbf{a} + \mathbf{b} = (a_0 + b_0, a_1 + b_1, a_2 + b_2, \ldots),$$
$$c\mathbf{a} = (ca_0, ca_1, ca_2, \ldots).$$

It is easy to verify that $V(F)$ is a linear space (i.e., a vector space) over F under these two operations. We also denote the zero sequence $0 = (0, 0, 0, \ldots)$. (0 represents an element in $V(F)$ or in F depending on the context. Sometimes, we also use $\mathbf{0}$ for the zero sequence.) Thus, an LFSR sequence is a sequence

$$\mathbf{a} = (a_0, a_1, a_2, \ldots)$$

in $V(F)$ whose elements satisfy the linear recursive relation

$$a_{n+k} = \sum_{i=0}^{n-1} c_i a_{k+i}, k = 0, 1, \ldots. \tag{4.9}$$

For $\mathbf{a} = (a_0, a_1, a_2, \ldots) \in V(F)$, we define a (left) shift operator L as follows:

$$L\mathbf{a} = (a_1, a_2, a_3, \ldots).$$

Note that L is a linear transformation of $V(F)$. Generally, for any positive integer i, we have

$$L^i \mathbf{a} = (a_i, a_{i+1}, a_{i+2}, \ldots).$$

By convention, we write $L^0 \mathbf{a} = I\mathbf{a} = \mathbf{a}$, where I is the identity transformation on $V(F)$. By using the left shift operation L, the formula (4.9) can be written as

$$L^n \mathbf{a} = \sum_{i=0}^{n-1} c_i L^i \mathbf{a},$$

or equivalently,

$$\left(L^n - \sum_{i=0}^{n-1} c_i L^i \right) \mathbf{a} = 0. \qquad (4.10)$$

We write

$$f(x) = x^n - (c_{n-1} x^{n-1} + \cdots + c_0),$$
$$f(L) = L^n - (c_{n-1} L^{n-1} + \cdots + c_0 I), \text{ and } f(L)\mathbf{a} = 0.$$

From Eq. (4.10), the definition of LFSR sequences (or linear recursive sequences) is equivalent to the following definition.

Definition 4.2 *For any infinite sequence* $\mathbf{a}$ *in* $V(F)$, *if there exists a nonzero monic polynomial* $f(x) \in F[x]$ *such that*

$$f(L)\mathbf{a} = 0,$$

then $\mathbf{a}$ *is called a* linear recursive sequence, *or equivalently, an* LFSR sequence. *The polynomial* $f(x)$ *is called the* characteristic polynomial *of* $\mathbf{a}$ *over* F. *The reciprocal polynomial of* $f(x)$ *is called the* feedback polynomial *of* $\mathbf{a}$.

Note that the definition of reciprocal polynomials has been introduced in Section 3 of Chapter 3. For any nonzero polynomial $f(x) \in F[x]$, we use $G(f)$ to represent the set consisting of all sequences in $V(F)$ with

$$f(L)\mathbf{a} = 0.$$

Since $f(L)$ is also a linear transformation, $G(f)$ is a subspace of $V(F)$.

Note. By convention, the constant polynomial 1 is the characteristic polynomial of the zero sequence $00 \ldots$.

Theorem 4.3 *Let* $f(x) \in F[x]$ *be a monic polynomial of degree n. Then* $G(f)$ *is a linear space of dimension n. Hence it contains* q^n *different sequences. In particular, if* $q = 2$, $G(f)$ *contains* 2^n *different binary sequences.*

Proof. For a sequence

$$\mathbf{a} = (a_0, a_1, \ldots, a_{n-1}, a_n, \ldots) \in G(f),$$

since $\deg(f) = n$, once the first n terms $(a_0, a_1, \ldots, a_{n-1})$ (or equivalently, an initial state) are given, the other terms of $\mathbf{a}$ can be determined by the formula (4.9) starting from a_n. There are q^n different ways to choose an n-tuple $(a_0, a_1, \ldots, a_{n-1})$ in F^n. Therefore $|G(f)| = q^n$. $\qquad\square$

Note that the sequences in $V(F)$ may or may not be periodic. Definition 4.2 of LFSR sequences makes it easier to determine periodicity of the sequences. We will discuss this in the next section. We conclude this section with a summary of three definitions that we already encountered plus some examples. Note that these three definitions are equivalent. Thus when we say that a sequence $\mathbf{a} = (a_0, a_1, a_2, \ldots)$ is generated by an n-stage LFSR, we mean any one of the following three equivalent definitions.

1. $\mathbf{a} = (a_0, a_1, a_2, \ldots)$ is an output sequence of an LFSR with the linear feedback function

$$f(x_0, x_1, \ldots, x_{n-1}) = \sum_{i=0}^{n-1} c_i x_i, \ c_i \in F$$

and an initial state

$$(a_0, a_1, \ldots, a_{n-1}),$$

so that the elements of $\mathbf{a}$ satisfy the following recursive relation:

$$a_{k+n} = \sum_{i=0}^{n-1} c_i a_{k+i}, \ k = 0, 1, \ldots. \tag{4.11}$$

2. $\mathbf{a}$ is a linear recursive sequence that satisfies the above recursive relation (4.11).
3. There exists a monic polynomial $f(x) = x^n - \sum_{i=0}^{n-1} c_i x^i \in F[x]$ of degree n such that

$$f(L)\mathbf{a} = 0,$$

or equivalently, $\mathbf{a} \in G(f)$.

The polynomial $f(x)$ is called the characteristic polynomial of $\mathbf{a}$ for each of these definitions and the reciprocal polynomial of $f(x)$ is called the feedback polynomial of $\mathbf{a}$.

Example 4.5 A sequence $\mathbf{a} = (000100110101111)$ of period 15 is generated by an LFSR with the feedback function $f(x_0, x_1, x_2, x_3) = x_0 + x_1$, and the LFSR implementation is shown in Figure 4.9. Thus, $\mathbf{a}$ satisfies the

Table 4.3. *All sequences in* $G(x^4 + x + 1)$

Initial state	Sequence
0000	0000...
0001	000100110101111
0010	001001101011110
0011	001101011110001
0100	010011010111100
0101	010111100010011
0110	011010111100010
0111	011110001001101
1000	100010011010111
1001	100110101111000
1010	101011110001001
1011	101111000100110
1100	110001001101011
1101	110101111000100
1110	111000100110101
1111	111100010011010

linear recursive relation

$$a_{4+k} = a_k + a_{1+k}, k = 0, 1, \ldots$$

and its characteristic polynomial is given by

$$f(x) = x^4 + x + 1.$$

Equivalently, **a** is a linear recursive sequence that satisfies the following linear recursive relation:

$$a_{4+k} = a_k + a_{1+k}, k = 0, 1, \ldots,$$

or equivalently, $\mathbf{a} \in G(f)$; that is,

$$f(L)\mathbf{a} = (L^4 + L + I)\mathbf{a} = 0.$$

In particular, there are $2^4 = 16$ different sequences in $G(f)$, because there are 16 ways to choose a 4-tuple (a_0, a_1, a_2, a_3). We list all of these sequences in Table 4.3.

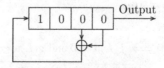

Figure 4.9. A 4-stage LFSR.

4.3 Minimal polynomials and periods

In the previous section, we associated LFSR sequences over F with polynomials over F. This enables us to use the extensive theory of periods of polynomials to investigate the periods of LFSR sequences. In this section, we discuss the minimal polynomials and the periods of LFSR sequences in terms of the polynomial ring $F[x]$.

4.3.1 Minimal polynomials of LFSR sequences

Let $\mathbf{a}$ be an LFSR sequence. According to the definition, there is a nonzero monic polynomial $f(x)$ such that

$$f(L)\mathbf{a} = 0. \tag{4.12}$$

In fact, for the fixed sequence $\mathbf{a}$, there are many polynomials for which Eq. (4.12) is satisfied. For example, given the LFSR sequence,

$$\mathbf{a} = 011011\ldots;$$

then the polynomial $f(x) = x^2 + x + 1$ satisfies the property $f(L)\mathbf{a} = 0$. But the polynomial $x^3 + 1$ also has this property. To find relations among these polynomials, for the LFSR sequence $\mathbf{a}$, we define

$$A(\mathbf{a}) = \{f(x) \in F[x] \mid f(L)\mathbf{a} = 0\}.$$

In other words, $A(\mathbf{a})$ is the set consisting of all polynomials satisfying the condition

$$f(L)\mathbf{a} = 0.$$

According to the definition of characteristic polynomials of a sequence, $A(\mathbf{a})$ consists of all characteristic polynomials of $\mathbf{a}$.

Theorem 4.4 *Let $\mathbf{a}$ be an LFSR sequence and $A(\mathbf{a})$ be defined as above. Then $A(\mathbf{a})$ satisfies the following properties:*

(a) The zero polynomial belongs to $A(\mathbf{a})$.
(b) If $f(x), g(x) \in A(\mathbf{a})$, then $f(x) \pm g(x) \in A(\mathbf{a})$.
(c) If $f(x) \in A(\mathbf{a})$ and $h(x) \in F[x]$, then $h(x)f(x) \in A(\mathbf{a})$.

Proof. (a) $0\mathbf{a} = 0 \Longrightarrow 0 \in A(\mathbf{a})$.
(b)

$$f(x), g(x) \in A(\mathbf{a}) \Longrightarrow f(L)\mathbf{a} = 0 \text{ and } g(L)\mathbf{a} = 0$$
$$\Longrightarrow (f(L) \pm g(L))\mathbf{a} = f(L)\mathbf{a} \pm g(L)\mathbf{a} = 0$$
$$\Longrightarrow f(x) \pm g(x) \in A(\mathbf{a}).$$

(c)

$$f(x) \in A(\mathbf{a}) \Longrightarrow f(L)\mathbf{a} = 0$$
$$\Longrightarrow (h(L)f(L))\mathbf{a} = h(L)(f(L)\mathbf{a}) = h(L)0 = 0. \qquad \square$$

Note. Since $A(\mathbf{a})$ is closed with respect to all of these operations, $A(\mathbf{a})$ is not merely a linear space, but also an algebra.

Definition 4.3 *A monic polynomial of the lowest degree in $A(\mathbf{a})$ is called a* minimal polynomial *of* **a** *over F.*

In other words, the minimal polynomial of a sequence represents the LFSR of shortest length that can generate the sequence.

Remark 4.2 According to the definition of LFSR sequences, the constant polynomial 1 is the minimal polynomial of the zero sequence $00\ldots$, and the polynomial $x - 1$ is the minimal polynomial of any constant sequence $(c, c, \ldots), 0 \neq c \in F$.

Theorem 4.5 *Let* $\mathbf{a} \in V(F)$ *and $m(x)$ be the minimal polynomial of* **a**. *Then the minimal polynomial of* **a** *is unique and satisfies the following two properties:*

(a) $m(L)\mathbf{a} = 0$.
(b) For $f(x) \in F[x]$, $f(L)\mathbf{a} = 0$ if and only if $m(x)|f(x)$; that is, $m(x)$ divides $f(x)$.

Proof. We first establish the validity of these two assertions on minimal polynomials of sequences, and then we show their uniqueness. If $\mathbf{a} = 00\ldots$, then it is clear that the results are true. Now we suppose that **a** is a nonzero sequence. Because $m(x) \in A(\mathbf{a})$, then $m(L)\mathbf{a} = 0$ is satisfied automatically, which gives (a). For the assertion (b), if $m(x)|f(x)$, we can write $f(x) = m(x)g(x)$. Since $m(x)$ is the minimal polynomial of **a**, then $m(x) \in A(\mathbf{a})$. By Proposition 4.4(c), $f(x) \in A(\mathbf{a})$. Next, we show that if $f(x) \in A(\mathbf{a})$ then $m(x)|f(x)$. Applying the division algorithm to $f(x)$ and $m(x)$, there exist $q(x), r(x) \in F[x]$ such that

$$f(x) = q(x)m(x) + r(x), 0 \leq \deg(r(x)) < \deg(m(x)).$$

Again using Proposition 4.4(c), it follows that $q(L)m(L)\mathbf{a} = 0$. Hence

$$0 = f(L)\mathbf{a} = (q(L)m(L) + r(L))\mathbf{a} = q(L)m(L)\mathbf{a} + r(L)\mathbf{a} = r(L)\mathbf{a}.$$

So if $r(x) \neq 0$, then $r(L)\mathbf{a} = 0 \Longrightarrow r(x) \in A(\mathbf{a})$. But $\deg(r(x)) < \deg(m(x))$, which contradicts $m(x)$ is a polynomial in $A(\mathbf{a})$ with lowest degree. Therefore $r(x) = 0 \Longrightarrow f(x) = q(x)m(x) \Longrightarrow m(x)|f(x)$. If there is another polynomial $m_1(x)$ in $A(\mathbf{a})$ that is also a minimal polynomial of **a**, then according to (b), we have both $m(x)|m_1(x)$ and $m_1(x)|m(x)$. Since the polynomials in $A(\mathbf{a})$ are monic, $m(x) = m_1(x)$. $\qquad\square$

Note that for any $\mathbf{a} \in G(f)$, $f(x)$ need not be the minimal polynomial of $\mathbf{a}$. However, we have the following result.

Corollary 4.1 *If $f(x) \neq 0$, $\mathbf{a} \in G(f)$, then the minimal polynomial of $\mathbf{a}$, say $m(x)$, divides $f(x)$.*

Proof. According to the definition of $G(f)$, $\mathbf{a} \in G(f) \Longrightarrow f(L)\mathbf{a} = 0$. Applying Theorem 4.5(b), $m(x)|f(x)$. $\qquad\square$

Example 4.6 Let $F = GF(2)$ and $f(x) = x^5 + x^4 + 1$. Then the following sequences belong to $G(f)$:

$$\begin{aligned}
\mathbf{a} &= 100101110010\ldots & &\text{of period 7} \\
\mathbf{b} &= 01101101\ldots & &\text{of period 3} \\
\mathbf{c} &= 111110101001100010000\ldots & &\text{of period 21}
\end{aligned}$$

which have the minimal polynomials $f_1(x) = x^3 + x + 1$, $f_2(x) = x^2 + x + 1$, and $f(x)$, respectively. Note that $f(x) = f_1(x)f_2(x)$. Thus the minimal polynomials of $\mathbf{a}$ and $\mathbf{b}$ are divisors of $f(x)$.

Corollary 4.2 *With the same notation as in Corollary 4.1, if $f(x)$ is irreducible, then $f(x)$ is the minimal polynomial of any nonzero sequence in $G(f)$.*

Proof. Note that $f(x)$ has only 1 and itself as its factors, and the sequence having 1 as its minimal polynomial is the zero sequence. The result follows immediately. $\qquad\square$

Example 4.7 Let $f(x)$ be as given in Example 4.5. Since the characteristic polynomial $f(x)$ is irreducible, then all the 15 nonzero sequences have $f(x)$ as their minimal polynomial. (In fact, they are simply cyclic shifts of each other.)

According to the definition of minimal polynomials, the degree of the minimal polynomial of $\mathbf{a}$ is equal to the length of the shortest LFSR that can generate $\mathbf{a}$. This is a very important security parameter for measuring unpredictability of pseudorandom sequences used as key stream sequences in stream cipher cryptosystems. The degree of the minimal polynomial of a sequence is called the linear span (or linear complexity) of the sequence. We will give a formal definition below.

4.3.2 Periodicity

For any periodic sequence, we have the following result.

Theorem 4.6 *If* **a** *is an ultimately periodic sequence with parameters* (u, r), *then the minimal polynomial of* **a** *is* $m(x) = x^u m_1(x)$ *with* $m_1(0) \neq 0$ *and* $m_1(x)|(x^r - 1)$. *Hence, it can be generated by an LFSR.*

Proof. Note that

$$a_{k+r} = a_k, k = u, u + 1, \ldots$$

$\Longrightarrow (L^r - 1)L^u(\mathbf{a}) = 0 \Longrightarrow m(x)|x^u(x^r - 1)$. We write $m(x) = x^u m_1(x)$. Then $m_1(0) \neq 0$ and $m_1(x)$ divides $x^r - 1$. Therefore, **a** can be generated by an LFSR with the characteristic polynomial $m(x)$. $\quad\square$

The following corollary follows immediately from Theorem 4.6.

Corollary 4.3 *If* **a** *is periodic with period* r, *then its minimal polynomial* $m(x)$ *divides* $x^r - 1$.

From the proof of Theorem 4.6, if **a** is ultimately periodic, then the minimal polynomial of **a** can be written as $m(x) = x^u m_1(x)$ with $m_1(0) \neq 0$.

Definition 4.4 *Let* **a** *be an ultimately periodic sequence over* F. *Then the degree of the minimal polynomial of* **a** *is called the* linear span *or* linear complexity *of* **a**.

In other words, the linear span of a periodic sequence is the length of the shortest LFSR that can generate the sequence.

Note. This definition can be extended to any finite segment of a sequence.

Lemma 4.1 *Let* r *be the least period of* **a**. *If* l *is a period of* **a**, *then* $r|l$.

Proof. l is a period of **a** $\Longrightarrow L^l \mathbf{a} = \mathbf{a}$; that is, $a_i = a_{i+l}, \forall i \geq 0$. On the other hand, because r is the least period of **a**, we also have $L^r \mathbf{a} = \mathbf{a}$. Applying the division algorithm for integers to l and r, there exist two integers q and t such that

$$l = qr + t, 0 \leq t < r.$$

Note that $L^{qr}\mathbf{a} = \mathbf{a}$. Thus $L^l \mathbf{a} = L^{qr+t}\mathbf{a} = L^t(L^{qr}\mathbf{a}) = L^t \mathbf{a} = \mathbf{a}$. Because r is the smallest number having this property, $t = 0$. Hence, $r|l$. $\quad\square$

Let us denote by per(s) a period of a sequence s or of a polynomial s.

Theorem 4.7 *Let* **a** *be an LFSR sequence with minimal polynomial* $m(x)$.

(a) If $m(0) \neq 0$, *then* **a** *is periodic. In this case,*

$$\text{per}(\mathbf{a}) = \text{per}(m(x)).$$

In other words,

> *the period of an LFSR sequence is equal to*
>
> *the period of the minimal polynomial of the sequence*

(b) The converse of the first assertion is also true; that is, if **a** *is periodic, then*
    $m(0) \neq 0$.

Proof. Let **a** be an ultimately periodic sequence over F with parameters (u, r)
and $m(x)$ be its minimal polynomial over F. According to Definition 4.1, **a** is
periodic if and only if $u = 0$. From Theorem 4.6, $m(x) = x^u m_1(x)$ with $m_1(0) \neq$
0 and $m_1(x)|(x^r - 1)$. Thus **a** is periodic if and only if $m(x) = m_1(x)$. According
to the definitions of periods of sequences and polynomials, it is immediate that
$\text{per}(\mathbf{a}) = \text{per}(m(x))$. The assertion (a) is now established. Conversely, if **a** is
periodic, then $u = 0$ and $x^r - 1 \in A(\mathbf{a})$. According to Theorem 4.5, we have
$m(x)|(x^r - 1) \Longrightarrow m(0) \neq 0$, which gives the assertion (b). $\square$

Thus far, we have obtained a criterion for determining whether an ultimately
periodic sequence is periodic in terms of the evaluation of its minimal polynomial (or any characteristic polynomial of the sequence) at 0. Next, we will
show the relationships among the period of the sequence, the period of its minimal polynomial, and the order of a root of the minimal polynomial when the
minimal polynomial is irreducible.

Theorem 4.8 *Let* **a** *be an LFSR sequence with minimal polynomial $m(x)$. Assume that $m(x)$ is an irreducible polynomial over $F = GF(q)$ of degree n. Let
α be a root of $m(x)$ in the extension field $GF(q^n)$. Then*

$$\boxed{\text{per}(\mathbf{a}) = \text{per}(m(x)) = \text{ord}(\alpha)}$$

In other words, the period of the sequence **a***, the period of the minimal polynomial of* **a***, and the order of a root of the minimal polynomial of* **a** *are equal.*

Proof. Note that $m(x)$ is the minimal polynomial of α. From Theorem 3.9 in
Section 3.5 of Chapter 3, we have $\text{per}(m(x)) = \text{ord}(\alpha)$. According to Theorem
4.7, the assertion is established. $\square$

This is an important discovery in the history of sequence design and analysis
(Golomb 1955). The period of an LFSR sequence is equal to the period of its
minimal polynomial. If the minimal polynomial is irreducible, then the period
of the sequence is equal to the order of a root (all roots have the same order) of
the minimal polynomial in the extension field. The following diagram illustrates
this relationship.

$$\text{period of } \mathbf{a} = a_0, a_1, \dots$$

$$\Updownarrow$$

$$\text{period of the minimal polynomial } m(x) \text{ of } \mathbf{a}$$

$$\Updownarrow$$

$$\text{for } m(x) \text{ being irreducible, order of } \alpha, \text{ a root of } m(x)$$

$$\text{per}(\mathbf{a}) = \text{ord}(\alpha) = \text{per}(m(x))$$

Example 4.8 Let $\mathbf{a} = (1000010101110110001111100110100)$, generated by $f(x) = x^5 + x^3 + 1$. The period of $f(x)$ is 31. So $\text{per}(\mathbf{a}) = 31$. Furthermore, $f(x)$ is primitive over $GF(2)$. Let α be a root of $f(x)$. Then α is a primitive element in $GF(2^5)$. Thus, the order of α is 31. Therefore, we have

$$\text{per}(\mathbf{a}) = \text{per}(f(x)) = \text{ord}(\alpha) = 31$$

We define the following sets:

- S, the set consisting of all periodic LFSR sequences over $GF(q)$, and S_0, the subset of S in which the minimal polynomials of the sequences are irreducible over $GF(q)$;
- P, the set of all polynomials over $GF(q)$ with nonzero constant terms, and P_0, the subset of P that are the irreducible polynomials over $GF(q)$; and
- F, the set consisting of all finite fields.

By combining the results of Theorems 4.7 and 4.8, we have the following one-to-one correspondences among these sets:

$$S \leftrightarrow P$$
$$S_0 \leftrightarrow P_0 \leftrightarrow F$$

These relations are also illustrated in Figure 4.10.

Note. It is worth pointing out that all irreducible polynomials of degree n over $GF(q)$ generate finite fields with order q^n, and all these fields are isomorphic, namely $GF(q^n)$. The above one-to-one correspondence $P_0 \leftrightarrow F$ represents the different computations in $GF(q^n)$ invoked by different irreducible polynomials.

For the rest of this chapter, we will restrict ourselves to periodic sequences.

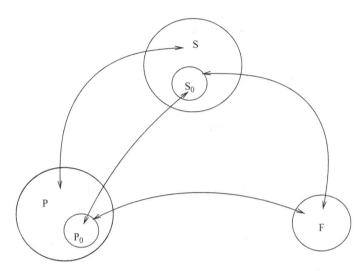

Figure 4.10. One-to-one correspondences among LFSR sequences, polynomials, and finite fields.

4.3.3 Structure of $G(f)$ for f irreducible

Definition 4.5 *Two periodic sequences* $\mathbf{a} = \{a_i\}$ *and* $\mathbf{b} = \{b_i\}$ *are called* (cyclically) shift equivalent *if there exists an integer k such that*

$$a_i = b_{i+k}, \quad \forall i \geq 0. \tag{4.13}$$

In this case, we write $\mathbf{a} = L^k(\mathbf{b})$, *or simply* $\mathbf{a} \sim \mathbf{b}$. *Otherwise, they are called* (cyclically) shift distinct.

Note that $\sim$ is an equivalence relation on $V(F)$. A set in which all sequences are shift equivalent is called a *shift-equivalent class*. One shift-equivalent class of $G(f)$ corresponds to one cycle of states in the state diagram of the LFSR with $f(x)$.

Theorem 4.9 *Let $f(x)$ be an irreducible polynomial over $GF(q)$ of degree n. Then the number of shift-equivalent classes of nonzero LFSR sequences in $G(f)$ is given by*

$$(q^n - 1)/\mathrm{per}(f).$$

Proof. Let $0 \neq \mathbf{a} \in G(f)$. According to Proposition 1(b), $f(L)L^k\mathbf{a} = 0 \Longrightarrow$ $L^k\mathbf{a} \in G(f)$. Since f is irreducible, f is the minimal polynomial of $\mathbf{a}$. Using Theorem 4.8, $\mathrm{per}(\mathbf{a}) = \mathrm{per}(f(x)) = r$. So, $L^r\mathbf{a} = \mathbf{a}$. Let G_1 denote the set consisting of $\mathbf{a}$ and all its shifts; that is, $G_1 = \{\mathbf{a}, L\mathbf{a}, \ldots, L^{r-1}\mathbf{a}\}$. Then every

Table 4.4. *Shift-equivalent classes of $G(f)$*

G_1	G_2	G_3
00011	01010	11110
00110	10100	11101
01100	01001	11011
11000	10010	10111
10001	00101	01111

sequence in G_1 has period r, and all the sequences in G_1 are shift equivalent. Next we take $\mathbf{b} \in G(f)$ which does not belong to G_1. By performing the shift operator, we obtain $G_2 = \{L^i\mathbf{b}|0 \leq i \leq r - 1\}$, which is a shift-equivalent class in $G(f)$ with the same cardinal number r as that of G_1. Continuing the process in this manner, we get $(q^n - 1)/r$ shift-equivalent classes. $\qquad\square$

In the language of the state diagram, this theorem shows that for an LFSR with an irreducible polynomial, in its state diagram there are $(q^n - 1)/\mathrm{per}(f)$ cycles with length $\mathrm{per}(f)$ and one cycle of length 1, that is, the zero sequence. For example, let $q = 2$. Then $f(x) = x^4 + x^3 + x^2 + x + 1 \in GF(2)[x]$ is irreducible over $GF(2)$. In $G(f)$, there are three shift-equivalent classes, $G_i, i = 1, 2$, and 3, in which each sequence has period 5, as shown in Table 4.4. Thus, we have

$$G(f) = \{0\} \cup G_1 \cup G_2 \cup G_3.$$

The state diagram of this LFSR is shown in Figure 4.11.

As a consequence of Theorem 4.9, we have the following assertion.

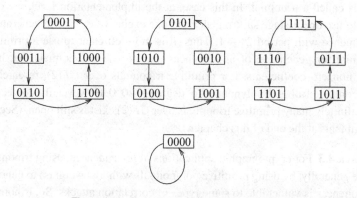

Figure 4.11. State diagram of the 4-stage LFSR.

Corollary 4.4 *With the notation of Theorem 4.9, if $f(x)$ is primitive, then any nonzero sequence* **a** *in $G(f)$ has period $q^n - 1$ and*

$$G(f) = \{L^i \mathbf{a} \mid 0 \le i \le q^n - 2\} \cup \{0\}.$$

In particular, if $q = 2$, then any nonzero sequence **a** *in $G(f)$ has period $2^n - 1$ and*

$$G(f) = \{L^i \mathbf{a} \mid 0 \le i \le 2^n - 2\} \cup \{0\}.$$

Definition 4.6 *A q-ary sequence generated by an n-stage LFSR is called a* maximal length *sequence if it has period $q^n - 1$ (an m-sequence for short). In particular, if $q = 2$, a binary m-sequence is a sequence generated by an n-stage LFSR with period $2^n - 1$.*

Example 4.9 For $q = 2$, (a) $1001011\ldots$ is an m-sequence of period 7 with the minimal polynomial $f(x) = x^3 + x + 1$. (b) The sequences in Example 4.5 of Section 4.2 are m-sequences of period 15 with the minimal polynomial $f(x) = x^4 + x + 1$. (c) The sequence in Example 4.8 is an m-sequence of period 31 with the minimal polynomial $f(x) = x^5 + x^3 + 1$.

According to Corollary 4.4, in order to generate an m-sequence of period $q^n - 1$ over F by an LFSR, we only need to select a primitive polynomial over F of degree n as the characteristic polynomial of this LFSR. In particular, for $q = 2$, we may sometimes choose primitive polynomials having the following form:

$$f(x) = x^n + x^k + 1.$$

This is called a trinomial. In this case, at the implementation level, we only need to use one n-stage shift register and one exclusive-or gate to generate an m-sequence with period $2^n - 1$. Thus, it is more efficient to use a primitive trinomial for generation of m-sequences than to use polynomials that have more nonzero coefficients. For primitive trinomials over $GF(2)$, researchers have computed all such polynomials of degree ≤ 2000. However, the conjecture that infinitely many primitive trinomials over $GF(2)$ exist is still open. (See the assignments at the end of this chapter.)

Remark 4.3 For cryptographic applications of m-sequences, using trinomials (more generally, by using primitive polynomials with low weights) to generate m-sequences is vulnerable to some types of correlation attacks. So, trinomials are not recommended for this type of applications.

4.4 Decomposition of LFSR sequences

In this section, we will analyze the structures of $G(f)$ when $f(x)$ is a product of distinct irreducible polynomials over $\mathbb{F}_q$. The method used here can be easily generalized to the case for which $f(x)$ satisfies $f(0) \neq 0$. Before we give the main result of this section, we establish the following lemma.

Lemma 4.2 *For any nonzero monic polynomials* $f(x), g(x) \in F[x]$,

(a) $G(f) \subset G(g)$ *if and only if* $f(x) \mid g(x)$.
(b) $G(f) \cap G(g) = G(d)$ *where* $d = \gcd(f, g)$.
(c) $G(f) + G(g) = G(h)$ *where* $h = \mathrm{lcm}[f, g]$, *the least common multiple of* f *and* g.

Proof. (a) Assume that $f(x) \mid g(x)$. For any $\mathbf{a} \in G(f)$, $f(L)\mathbf{a} = 0$. $f(x) \mid g(x)$ $\implies g(x) = t(x)f(x) \implies g(L)\mathbf{a} = t(L)f(L)\mathbf{a} = 0 \implies \mathbf{a} \in G(g) \implies G(f) \subset G(g)$. Conversely, we choose $\mathbf{a} \in G(f)$ such that the minimal polynomial of $\mathbf{a}$ is $f(x)$. According to Theorem 4.5, $g(L)\mathbf{a} = 0 \implies f(x) \mid g(x)$.

(b) From (a), $G(d) \subset G(f)$, $G(d) \subset G(g) \implies G(d) \subset G(f) \cap G(g)$. Assume that $\mathbf{a} \in G(f) \cap G(g)$; that is,

$$f(L)\mathbf{a} = 0 \quad \text{and} \quad g(L)\mathbf{a} = 0.$$

Since $d(x) = \gcd(f(x), g(x))$, there exist two polynomials $v(x)$ and $u(x)$ such that

$$d(x) = u(x)f(x) + v(x)g(x).$$

Therefore $d(L)\mathbf{a} = u(L)f(L)\mathbf{a} + v(L)g(L)\mathbf{a} = 0 \implies G(f) \cap G(g) \subset G(d)$. Together with $G(d) \subset G(f) \cap G(g)$, we get $G(f) \cap G(g) = G(d)$.

(c) From (a), $G(f) \subset G(h)$, $G(g) \subset G(h)$. Thus

$$G(f) + G(g) \subset G(h).$$

Since the sets on both sides of the above inclusion are linear spaces, we only need to prove their dimensions are equal. Let $\dim(V)$ denote the dimension of the linear space V. Notice that $\mathrm{lcm}[f, g]\gcd(f, g) = fg$. According to the dimension formula, we have

$$\begin{aligned}
\dim(G(f) + G(g)) &= \dim(G(f)) + \dim(G(g)) - \dim(G(f) \cap G(g)) \\
&= \dim(G(f)) + \dim(G(g)) - \dim(G(d)) \\
&= \deg(f) + \deg(g) - \deg(d) \\
&= \deg(h) \\
&= \dim(G(h)).
\end{aligned}$$

$\square$

Theorem 4.10 *Let $f(x) = f_1(x) \cdots f_s(x)$ where the f_i are distinct irreducible polynomials over F, $s > 0$. Then $G(f)$ can be decomposed as a direct sum of subspaces $G(f_i)$, $1 \leq i \leq s$; that is,*

$$G(f) = G(f_1) \oplus G(f_2) \oplus \cdots \oplus G(f_s).$$

Proof. We will use induction to prove this result. If $s = 1$, the result is true. Assume that the result is true for $s = k - 1$. For $s = k$, let $f(x) = f_1(x) \cdots f_k(x)$ and $g(x) = f_2(x) \cdots f_k(x)$. Then $f(x) = f_1(x)g(x)$ and $\gcd(f_1(x), g(x)) = 1$. According to Lemma 4.2(b) and (c),

$$G(f) = G(f_1) + G(g),$$
$$G(f_1) \cap G(g) = \{0\}.$$

Thus we have the decomposition of a direct sum:

$$G(f) = G(f_1) \oplus G(g). \tag{4.14}$$

Applying the induction hypothesis to $G(g)$, we get

$$G(g) = G(f_2) \oplus \cdots \oplus G(f_s).$$

Substituting this into Eq. (4.14), the result follows for $s = k$. Therefore the result is true for every $s > 0$. $\qquad \square$

We state the following result without proof.

Fact 4.1 *Let $f(x)$ be the same as in Theorem 4.10. Then the period of $f(x)$ is equal to $\mathrm{lcm}[\mathrm{per}(f_1), \ldots, \mathrm{per}(f_s)]$, the least common multiple of the periods of the $f_i(x)$'s.*

From Fact 4.1 and Theorem 4.10, the following corollary is immediate.

Corollary 4.5 *Let $f(x)$ be the same as in Theorem 4.10, and let $\mathbf{a} \in G(f)$ whose minimal polynomial is $f(x)$.*

(a) $\mathbf{a}$ *can be decomposed as*

$$\mathbf{a} = \mathbf{a}_1 + \cdots + \mathbf{a}_s, 0 \neq \mathbf{a}_i \in G(f_i), i = 1, \ldots, s.$$

(b) The period of $\mathbf{a}$ is equal to $\mathrm{lcm}[\mathrm{per}(\mathbf{a}_1), \ldots, \mathrm{per}(\mathbf{a}_s)]$ where $\mathrm{per}(\mathbf{a}_i) = \mathrm{per}(f_i)$.

(c) The linear span of $\mathbf{a}$ is given by $\sum_{i=1}^{s} \deg(f_i)$.

Example 4.10 Let $F = GF(2)$, and let $f_1(x) = x^4 + x + 1$, $f_2(x) = x^2 + x + 1$, and $f(x) = f_1(x)f_2(x) = x^6 + x^5 + x^4 + x^3 + 1$. Since $f_1(x)$ and

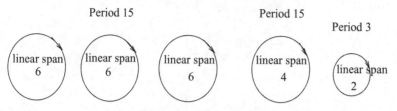

Figure 4.12. Cycle structure of $G(f)$.

$f_2(x)$ are coprime, according to Theorem 4.10 we have

$$G(f) = G(f_1) \oplus G(f_2) = G(x^4 + x + 1) \oplus G(x^2 + x + 1).$$

We take two nonzero sequences, say **a** from $G(x^4 + x + 1)$ and **b** from $G(x^2 + x + 1)$. Notice that both $f_1(x)$ and $f_2(x)$ are primitive. Thus, any nonzero binary sequence in $G(f)$ can be written as

$$cL^i\mathbf{a} + dL^j\mathbf{b}, \ 0 \le i \le 14, \ 0 \le j \le 2, \ c, d \in GF(2).$$

The number of all the nonzero sequences in $G(f)$ is $2^6 - 1 = 63$, which can be classified according to their linear spans as follows:

- three shift-distinct classes of period 15 with linear span 6: $\mathbf{a} + L^i\mathbf{b}$, $0 \le i \le 2$;
- one shift-distinct class of period 15 with linear span 4: **a**, an m-sequence of period 15; and
- one shift-distinct class of period 3 with linear span 2: **b**, an m-sequence of period 3.

The state cycles for these sequences in $G(f)$ are illustrated in Figure 4.12. For example, we have the following decomposition:

$$
\begin{array}{r}
\mathbf{a} = 100010011010111 \\
+ \, \mathbf{b} = 011011011011011 \\
\hline
\mathbf{c} = 111001000001100
\end{array}
$$

where $\mathbf{c} = \mathbf{a} + \mathbf{b}$ can be implemented by two LFSRs. (See Figures 4.13 and 4.14, which are equivalent.)

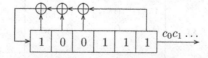

Figure 4.13. A 6-stage LFSR.

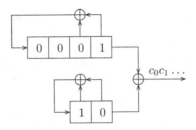

Figure 4.14. Decomposition of the LFSR in Figure 4.10.

From Theorem 4.10, we understand that any LFSR sequence can be decomposed into a sum of LSFRs with irreducible characteristic polynomials (which may or may not be primitive polynomials). This implies that LFSRs with primitive or irreducible characteristic polynomials can be used as basic blocks for building more complicated nonlinear pseudorandom sequence generators.

4.5 The matrix representation

In Section 4.1, we saw that each state of an n-stage LFSR is a vector in the n-dimensional space F^n. The shift register is then a linear operator that changes the current state into its successor vector according to the feedback. In other words, the transformation of each nonzero sequence in $G(f)$, from state

$$(a_k, a_{k+1}, \ldots, a_{k+n-1})$$

to its successor

$$(a_{k+1}, a_{k+2}, \ldots, a_{k+n}),$$

can be regarded as a linear operator on F^n. It is a familiar fact that a linear operator, operating on an n-dimensional vector space F^n, is conveniently studied when it is represented by an $n \times n$ matrix. We know that

$$a_{n+k} = c_0 a_k + c_1 a_{k+1} + \cdots + c_{n-1} a_{k+n-1}, \quad k \geq 0.$$

Hence, a shift register matrix takes the form

$$M = \begin{bmatrix} 0 & 0 & 0 & \cdots & 0 & c_0 \\ 1 & 0 & 0 & \cdots & 0 & c_1 \\ 0 & 1 & 0 & \cdots & 0 & c_1 \\ & & \vdots & & & \\ 0 & 0 & 0 & \cdots & 1 & c_{n-1} \end{bmatrix}$$

and

$$(a_{k+1}, a_{k+2}, \ldots, a_{k+n}) = (a_k, a_{k+1}, \ldots, a_{k+n-1})M$$
$$= (a_{k-1}, a_k, \ldots, a_{k-1+n-1})M^2$$
$$= \ldots$$
$$= (a_0, a_1, \ldots, a_{n-1})M^{k+1}.$$

We called the matrix M a state transform matrix of the LFSR. Note that $\det(M) = (-1)^n c_0$. Thus M is invertible if and only if $c_0 \neq 0$.

As an example, the state transform matrix of the 3-stage LFSR in Example 4.2 in Section 4.1 is given by

$$M = \begin{bmatrix} 0 & 0 & 1 \\ 1 & 0 & 1 \\ 0 & 1 & 0 \end{bmatrix}.$$

In particular,

$$(001) = (100) \begin{bmatrix} 0 & 0 & 1 \\ 1 & 0 & 1 \\ 0 & 1 & 0 \end{bmatrix}$$

$$(010) = (001) \begin{bmatrix} 0 & 0 & 1 \\ 1 & 0 & 1 \\ 0 & 1 & 0 \end{bmatrix}.$$

The characteristic equation of the matrix M is defined by

$$f(x) = \det|xI - M| = \begin{vmatrix} x & 0 & 0 & \cdots & 0 & -c_0 \\ -1 & x & 0 & \cdots & 0 & -c_1 \\ 0 & -1 & x & \cdots & 0 & -c_2 \\ & & & \vdots & & \\ 0 & 0 & 0 & \cdots & -1 & x - c_{n-1} \end{vmatrix}$$

$$= x^n - c_{n-1}x^{n-1} - \cdots - c_1 x - c_0$$

where I is the identity matrix. Notice that the characteristic polynomial $f(x)$ is the same as the characteristic polynomial of the LFSR. Under this representation, determining the periodicity of the LFSR is equivalent, except for degenerate cases, to finding the smallest positive integer r such that $M^r = I$. (Note. r is also called the order of the matrix M in the group consisting of all invertible matrices whose entries are taken from F.)

A well-known theorem of matrix theory (the Cayley-Hamilton Theorem) asserts that every matrix formally satisfies its characteristic equation: thus $f(M) = 0$, where 0 here is the matrix of all zeros. If $f(x)$ divides $x^r - 1$, then M is a root of $x^r - I = 0$. In other words, if $f(x)$ divides $x^r - 1$, then

$M^r = I$. Conversely, if $f(x)$ is irreducible, it divides every polynomial that has the root M in common with it and will divide $x^r - 1$ if $M^r = I$. Roughly, this is a matrix theory proof of Theorem 4.7. The rest of the theory of LFSR sequences can be done entirely by matrix theory and is frequently done so in the literature (see Golomb 1967).

4.6 Trace representation of LFSRs

In this section, we present trace representations for nonzero sequences in $G(f)$ when $f(x)$ is irreducible. We will discuss trace representations for NLFSR sequences with period N where N is a factor of $q^n - 1$ in Chapter 6 in terms of the Fourier transforms of periodic sequences. Starting with this section, we will simply denote the finite field $GF(Q) = \mathbb{F}_Q$ and $Tr_{\mathbb{F}_{q^n}/F_q}(x) = Tr(x) = x + x^q + \cdots + x^{q^{n-1}}$, the trace function from $\mathbb{F}_{q^n}$ to $\mathbb{F}_q$, if the context is clear. (For more properties of the trace functions, see Section 5 of Chapter 3.)

4.6.1 Trace representation

Let

$$f(x) = x^n + c_{n-1}x^{n-1} + \cdots + c_1 x + c_0$$

be an irreducible polynomial over $\mathbb{F}_{q^n}$ of degree n, and let α be a root of $f(x)$; that is,

$$\alpha^n + c_{n-1}\alpha^{n-1} + \cdots + c_1\alpha + c_0 = 0. \tag{4.15}$$

Then we can construct a finite field $\mathbb{F}_{q^n}$ with $f(x)$ as a defining polynomial.

Lemma 4.3 *Let* $\mathbf{a} = \{a_i\}$ *whose elements are given by*

$$a_i = Tr(\beta\alpha^i), \ i \geq 0, \ \beta \in \mathbb{F}_{q^n}. \tag{4.16}$$

Then $\mathbf{a} \in G(f)$.

Proof. If $\beta = 0$, then $\mathbf{a}$ is a zero sequence, so $\mathbf{a} \in G(f)$. If $\beta \neq 0$, then

$$
\begin{aligned}
&a_{k+n} + c_{n-1}a_{k+n-1} + \cdots + c_1 a_{k+1} + c_0 a_k \\
&= Tr(\beta\alpha^{k+n}) + c_{n-1}Tr(\beta\alpha^{k+n-1}) + \cdots + c_1 Tr(\beta\alpha^{k+1}) + c_0 Tr(\beta\alpha^k) \\
&\quad \text{(by Eq. (4.16))} \\
&= Tr(\beta\alpha^k(\alpha^n + c_{n-1}\alpha^{n-1} + \cdots + c_1\alpha + c_0))
\end{aligned}
$$

$$= Tr(\beta\alpha^k 0) = 0, \; \forall k \geq 0 \; \text{(by Eq. (4.15))}$$
$$\implies f(L)\mathbf{a} = 0$$
$$\implies \mathbf{a} \in G(f). \qquad \Box$$

Theorem 4.11 *With the above notation, for any sequence* $\mathbf{a} = \{a_i\} \in G(f)$, *there exists some* $\beta \in \mathbb{F}_{q^n}$ *such that*

$$a_i = Tr(\beta\alpha^i), \; \forall i \geq 0. \qquad (4.17)$$

Proof. Because there are q^n different sequences in $G(f)$, from Lemma 4.3, we have $\mathbf{a} = \{a_i\}$, with $a_i = Tr(\beta\alpha^i)$, belonging to $G(f)$ for any $\beta \in \mathbb{F}_{q^n}$. So, we only need to prove that two sequences related to two different elements in $\mathbb{F}_{q^n}$ are different. Because the trace function is a linear function from $\mathbb{F}_{q^n}$ to $\mathbb{F}_q$, this reduces to proving that if $\beta \neq 0$, then $\mathbf{a} \neq 0$. If not, we have $\mathbf{a} = 0$. We may write down the first n terms as follows:

$$0 = Tr(\beta) = \beta + \beta^q + \cdots + \beta^{q^{n-1}}$$
$$0 = Tr(\beta\alpha) = \beta\alpha + \beta^q\alpha^q + \cdots + \beta^{q^{n-1}}\alpha^{q^{n-1}}$$
$$\vdots$$
$$0 = Tr(\beta\alpha^{n-1}) = \beta\alpha^{n-1} + \beta^q\alpha^{(n-1)q} + \cdots + \beta^{q^{n-1}}\alpha^{(n-1)q^{n-1}}.$$

This is a system of n linear equations in n unknowns $\beta, \beta^q, \ldots, \beta^{q^{n-1}}$. We rewrite the above n linear equations in the following matrix form

$$\begin{bmatrix} 1 & 1 & 1 & \cdots & 1 \\ \alpha & \alpha^q & \alpha^{q^2} & \cdots & \alpha^{q^{n-1}} \\ \alpha^2 & \alpha^{2q} & \alpha^{2q^2} & \cdots & \alpha^{2q^{n-1}} \\ & & \vdots & & \\ \alpha^{n-1} & \alpha^{(n-1)q} & \alpha^{(n-1)q^2} & \cdots & \alpha^{(n-1)q^{n-1}} \end{bmatrix} \begin{bmatrix} \beta \\ \beta^q \\ \beta^{q^2} \\ \vdots \\ \beta^{q^{n-1}} \end{bmatrix} = \begin{bmatrix} 0 \\ 0 \\ 0 \\ \vdots \\ 0 \end{bmatrix}.$$

Since $f(x)$ is irreducible,

$$\alpha, \alpha^q, \alpha^{q^2}, \ldots, \alpha^{q^{n-1}}$$

are all the n roots of $f(x)$, which are distinct. Therefore, the above matrix is a Vandemonde matrix. Hence, it is invertible. Thus $\beta = 0$, which is a contradiction to the assumption. $\qquad \Box$

The formula (4.17) is called the trace representation for LFSR sequences with irreducible characteristic (or feedback) polynomials.

From Theorem 4.11, the following corollary is immediate.

Corollary 4.6 *Let* **a** *be a sequence over* $\mathbb{F}_q$. *Then* **a** *is an m-sequence with period* $q^n - 1$ *if and only if the elements of* **a** *can be represented by*

$$a_i = Tr(\beta\alpha^i), \ \forall i \geq 0, \ 0 \neq \beta \in \mathbb{F}_{q^n}$$

where α *is a primitive element in* $\mathbb{F}_{q^n}$.

Example 4.11 Let $q = 2$.

(a) Let $f(x) = x^4 + x^3 + x^2 + x + 1$, which defines $\mathbb{F}_{2^4}$, and let α be a root of $f(x)$ in $\mathbb{F}_{2^4}$; that is, $\alpha^4 + \alpha^3 + \alpha^2 + \alpha + 1$. Then for $\mathbf{a} = (10001) \in G(f)$, the trace representation of **a** is given by

$$a_i = Tr(\beta\alpha^i), i = 0, 1, \ldots,$$

where $\beta = 1 + \alpha$.

(b) Let $f(x) = x^3 + x + 1$ and $\mathbb{F}_{2^3}$ be defined by $f(\alpha) = 0$. Then $\mathbf{a} = (1001011) \in G(f) \leftrightarrow$ the trace representation $a_i = Tr(\alpha^i)$, $i = 0, 1, \ldots$.

(c) Let $f(x) = x^4 + x + 1$ and $\mathbb{F}_{2^4}$ be defined by $f(\alpha) = 0$. Then

$$\mathbf{a} = (100110101111000) \in G(f) \leftrightarrow$$

the trace representation $a_i = Tr(\alpha^3\alpha^i), i = 0, 1, \ldots$.

(See Example 3.14 in Chapter 3 for the computation of $Tr(x)$ in $\mathbb{F}_{2^3}$ and $\mathbb{F}_{2^4}$.)

4.6.2 S-Decimation

Definition 4.7 *Let* $f(x)$ *be an irreducible polynomial over* $\mathbb{F}_q$ *of degree n and s be a positive integer. For* $\mathbf{a} \in G(f)$, *if the elements of a sequence* $\mathbf{b} = \{b_i\}$ *are defined by*

$$b_i = a_{si} \ \forall i \geq 0,$$

then **b** *is called an s-decimation sequence of* **a**, *denoted by* $\mathbf{b} = \mathbf{a}^{(s)}$.

For example, let $\mathbf{a} = (1001011)$. For $\mathbf{b} = \mathbf{a}^{(3)}$; that is, $b_i = a_{3i}, i = 0, 1, \ldots$, we have $\mathbf{b} = (1110100)$.

Theorem 4.12 *Let* $f(x)$ *be an irreducible polynomial over* $\mathbb{F}_q$ *of degree n and let s be a positive integer. Let* α *be a root of* $f(x)$ *in the extension* $\mathbb{F}_{q^n}$. *For* $0 \neq \mathbf{a} \in G(f)$, *if the s-decimation* $\mathbf{a}^{(s)} \neq 0$, *then the minimal polynomial of* $\mathbf{a}^{(s)}$ *is equal to the minimal polynomial of* α^s *over* $\mathbb{F}_q$.

Proof. Because $\mathbf{a} \in G(f)$, there exists $\beta \in \mathbb{F}_{q^n}$ such that $a_i = Tr(\beta\alpha^i), \forall i \geq 0$.

Therefore

$$\mathbf{a}^{(s)} = (Tr(\beta), Tr(\beta\alpha^s), Tr(\beta\alpha^{2s}), \ldots).$$

Let $\gamma = \alpha^s$. Then

$$\mathbf{a}^{(s)} = (Tr(\beta), Tr(\beta\gamma), Tr(\beta\gamma^2), \ldots).$$

Let $g(x)$ be the minimal polynomial of γ. From Lemma 4.3, $\mathbf{a}^{(s)} \in G(g)$. Because $g(x)$ is irreducible, if $\mathbf{a}^{(s)} \neq 0$, then $g(x)$ is the minimal polynomial of $\mathbf{a}^{(s)}$. $\qquad\square$

According to Theorem 4.8, $\mathrm{per}(\mathbf{a}) = \mathrm{per}(f) = \mathrm{ord}(\alpha)$. Then the period of s-decimation $\mathbf{a}^{(s)}$ is equal to $\mathrm{ord}(\alpha^s) = \mathrm{ord}(\alpha)/(s, \mathrm{ord}(\alpha))$. Hence

$$\mathrm{per}(\mathbf{a}^{(s)}) = \mathrm{per}(\mathbf{a})/(s, \mathrm{per}(\mathbf{a})).$$

Corollary 4.7 *With the above notation, if $f(x)$ is primitive, then $0 \neq \mathbf{a} \in G(f)$ is an m-sequence. In this case, an s-decimation of $\mathbf{a}$ is also an m-sequence if and only if $\gcd(s, q^n - 1) = 1$. Moreover, the number of shift-distinct m-sequences over $\mathbb{F}_q$ of period $q^n - 1$ is given by*

$$\phi(q^n - 1)/n,$$

which is equal to the number of the primitive polynomials over $\mathbb{F}_q$ of degree n. (Here $\phi(\cdot)$ is Euler's phi-function.)

Example 4.12 Let $\mathbb{F}_{2^4}$ be defined by $f(x) = x^4 + x + 1$, and let α be a root of $f(x)$ in $\mathbb{F}_{2^4}$. Let $f_{\alpha^s}(x)$ be the minimal polynomial of α^s over $\mathbb{F}_2$. We select $\mathbf{a} \in G(f)$ as follows:

$$\begin{aligned} \mathbf{a} &= (100010011010111) \\ &= \left(Tr(\alpha^{14}), Tr(\alpha^{14}\alpha), Tr(\alpha^{14}\alpha^2), \ldots, Tr(\alpha^{14}\alpha^{14})\right). \end{aligned}$$

All possible distinct decimated sequences of $\mathbf{a}$ (up to shift equivalence) and their corresponding minimal polynomials and periods are listed as follows.

All decimations of **a** up to shift equivalence

s	$\mathbf{a}^{(s)}$	Period $\mathbf{a}^{(s)}$	Minimal polynomial $f_{\alpha^s}(x)$	Period $f_{\alpha^s}(x)$	Order α^s
1	100010011010111	15	$x^4 + x + 1$	15	15
3	10001	5	$x^4 + x^3 + x^2 + x + 1$	5	5
5	101	3	$x^2 + x + 1$	3	3
7	111010110010001	15	$x^4 + x^3 + 1$	15	15

4.7 Generating functions of LFSRs

In this section, we will introduce another method to represent a periodic LFSR sequence, called a generating function of the LFSR sequence. This method is frequently used in studying the combinatorial structure of objects. Assume that $\mathbf{a} = \{a_i\}$ is a periodic sequence over $\mathbb{F}_q$, generated by $f(x) = x^n + \sum_i c_i x^i$ with $c_0 \neq 0$, $c_i \in \mathbb{F}_q$. We associate the sequence $\mathbf{a}$ with a polynomial $a(x) = \sum_{i=0}^{\infty} a_i x^i$. Then we have

$$a_{n+k} + \sum_i c_i a_{k+i} = 0, k = 0, 1, \ldots. \tag{4.18}$$

Next we rewrite the linear recursive relation given by Eq. (4.18) into polynomial form. We denote the product of $a(x)$ and $f^{-1}(x)$, the reciprocal of $f(x)$, by $d(x) = \sum_{i=0}^{\infty} d_i x^i$; that is,

$$a(x) f^{-1}(x) = \sum_{i=0}^{\infty} d_i x^i. \tag{4.19}$$

From Eq. (4.18), we have

$$d_j = a_j + \sum_i c_i a_{j-n+i} = 0, j = n, n+1, \ldots.$$

Thus the right-hand side of Eq. (4.19) becomes

$$\sum_{i=0}^{\infty} d_i x^i = \sum_{i=0}^{n-1} d_i x^i, \tag{4.20}$$

where

$$d_0 = c_0 a_0$$
$$d_1 = c_0 a_1 + c_1 a_0$$
$$\vdots$$
$$d_i = \sum_{k+j=i} a_k a_j$$
$$\vdots$$
$$d_{n-1} = \sum_{k+j=n-1} a_k a_j.$$

In other words, the d_i's can be determined by the initial state $(a_0, a_1, \ldots, a_{n-1})$ of the LFSR and the coefficients of the characteristic polynomial. We may write the above relation in the following matrix form.

$$
\begin{bmatrix} d_0 \\ d_1 \\ d_2 \\ \vdots \\ d_{n-1} \end{bmatrix} = \begin{bmatrix} c_0 & 0 & 0 & \cdots & 0 \\ c_1 & c_0 & 0 & \cdots & 0 \\ c_2 & c_1 & c_0 & \cdots & 0 \\ \vdots & & & & \\ c_{n-1} & c_{n-2} & c_{n-3} & \cdots & c_0 \end{bmatrix} \begin{bmatrix} a_0 \\ a_1 \\ a_2 \\ \vdots \\ a_{n-1} \end{bmatrix} \qquad (4.21)
$$

Consequently, Eqs. (4.19) and (4.20) yield

$$
a(x) = \frac{d(x)}{f^{-1}(x)}, \qquad (4.22)
$$

where $\deg(d(x)) \leq n - 1$. Usually, $\frac{d(x)}{f^{-1}(x)}$ is called a generating function of the sequence **a**. If an initial state of the LFSR is given, then $d(x)$ can be determined by Eq. (4.21). Let the matrix in Eq. (4.21) be A. Then A is invertible. Thus, if $d(x)$ is any polynomial over $GF(q)$ of degree less than n, then the corresponding initial state of **a** can be determined by

$$
\begin{bmatrix} a_0 \\ a_1 \\ \vdots \\ a_{n-1} \end{bmatrix} = A^{-1} \begin{bmatrix} d_0 \\ d_1 \\ \vdots \\ d_{n-1} \end{bmatrix}.
$$

Therefore, in terms of the generating function method, the set $G(f)$ that consists of all sequences generated by $f(x)$ is given by

$$
G(f) = \left\{ \frac{d(x)}{f^{-1}(x)} \text{ such that } d(x) \in GF(q)[x] \text{ and } \deg(d) < n \right\}.
$$

Example 4.13 Let $q = 2$, $f(x) = x^4 + x + 1$, and $d(x) = x^3 + 1$. Then $f^{-1}(x)$, the reciprocal polynomial of $f(x)$, is given by $f^{-1}(x) = x^4 + x^3 + 1$. The fraction

$$
\frac{d(x)}{f^{-1}(x)} = \frac{x^3 + 1}{x^4 + x^3 + 1} = 1 + x^4 + x^7 + x^8 + x^{10} + x^{12} + x^{13} + x^{14} + \cdots
$$

gives a sequence 100010011010111 with an initial state 1000 and period 15.

Note

In the literature, the maximal length sequences or m-sequences are also called pseudonoise (PN) sequences when $q = 2$. Golomb studied these sequences in his popular book (Golomb 1967), from which our discussion of binary LFSR sequences is taken. For general q, a power of a prime, most of the results on m-sequences appeared in Zierler's work (1959). Using the set $G(f)$ to discuss the structure of LFSR sequences is also due to Zierler. For this approach, see also Shishung Ding's book, which was published in 1982 in Chinese, McEliece's book (1986), and Lidl and Niederreiter's book (1983, Chapter 8). Another early reference for q-ary m-sequences was Selmer's book (1966). For correlation attacks on pseudorandom sequence generators for which m-sequences are generated by primitive polynomials with low weights, see Meier and Staffelbach (1988).

Exercises for Chapter 4

1. Given a 3-stage shift register with the boolean feedback function $f(x_0, x_1, x_2) = x_0 + x_1 x_2$:
   (a) Draw the state diagram of the FSR.
   (b) If the initial state is set as $(a_0, a_1, a_2) = (011)$, determine the output sequence and the period of the sequence.
2. Let $f(x_0, x_1, \ldots, x_{n-1})$ be a boolean function in n variables that is employed as the feedback function of a shift register. Prove that the cycles in the state diagram have no branch points if and only if the feedback function can be decomposed into

$$f(x_0, x_1, \ldots, x_{n-1}) = x_0 + g(x_1, \ldots, x_{n-1}),$$

   where $g(x_1, \ldots, x_{n-1})$ is a boolean function in $n - 1$ variables. (Hint: The cycles in the state diagram have no branch points if and only if two distinct state vectors have distinct successors. If $(a_0, a_1, \ldots, a_{n-1})$ and $(b_0, b_1, \ldots, b_{n-1})$ differ in any component other than the first, then their successors $(a_1, \ldots, a_n)$ and $(b_1, \ldots, b_n)$ are still distinct. So, one only needs to consider whether $(a_0, a_1, \ldots, a_{n-1})$ and $(a_0 + 1, b_1, \ldots, b_{n-1})$ have distinct successors.)
3. Design an LFSR over $GF(2)$ for implementation of the linear recurrence relation

$$a_{5+k} = a_{3+k} + a_k, k = 0, 1, \ldots.$$

   Determine the characteristic polynomial $f(x)$ of the sequence and the number of sequences in $G(f)$. Write the first 50 bits of the output sequence with a nonzero initial state.
4. Design an LFSR over $GF(2)$ for implementation of the linear recurrence relation

$$a_{6+k} = a_{1+k} + a_k, k = 0, 1, \ldots.$$

   Determine the characteristic polynomial $f(x)$ of the sequence and the number of sequences in $G(f)$.
5. Design an LFSR over $GF(2)$ for implementation of the linear recurrence relation

$$a_{7+k} = a_{1+k} + a_k, k = 0, 1, \ldots.$$

   Determine the characteristic polynomial $f(x)$ of the sequence and the number of sequences in $G(f)$.
6. Construct two different (shift-distinct) de Bruijn sequences with period 16.

7. Let $f(x) = x^5 + x^4 + 1$ over $\mathbb{F}_2$ be the characteristic polynomial of a 5-stage LFSR.
    (a) Write the first 50 bits of the output sequence with the initial state 00101 and determine the period and the minimal polynomial of the sequence. (Hint: $f(x) = (x^3 + x + 1)(x^2 + x + 1)$.)
    (b) Write the first 50 bits of the output sequence with the initial state 01000 and determine the period and the minimal polynomial of the sequence.
    (c) Determine the number of sequences in $G(f)$ and draw the state diagram.
8. Design an LFSR over $GF(2)$ that generates a binary m-sequence with period 1023.
9. Determine the number of LFSRs over $GF(2)$ that generate a binary m-sequence with period $2^8 - 1 = 255$.
10. Let α be a primitive element of $\mathbb{F}_{2^n}$. Let $\mathbf{a} = \{a_i\}$ be a binary m-sequence of degree n of period $2^n - 1$ whose elements are given by $a_i = Tr(\beta \alpha^i)$, $\forall i \geq 0$ where $Tr(x) = x + x^2 + \cdots x^{2^{n-1}}$, the trace function from $\mathbb{F}_{2^n}$ to $\mathbb{F}_2$, $\beta \in F$, $\forall i$. Prove that $\mathbf{a}$ has the following property: $a_{2i} = a_i$, $\forall i \geq 0$, if and only if $\beta = 1$. (This property is also referred to as constant-on-cosets.)
11. Find the initial state of an m-sequence that is generated by the primitive polynomial $f(x) = x^7 + x + 1$ that satisfies the property of being constant-on-cosets.
12. An m-sequence of period 31 with the minimal polynomial $f(x) = x^5 + x^3 + 1$ is given by:

$$\mathbf{a} = (1000010101110110001111100110100).$$

Determine its 7-decimation sequence $\mathbf{a}^{(7)}$ and compute the minimal polynomial of this sequence.
13. Let $\mathbf{a}$ be an m-sequence over $GF(2)$ of period 511. Determine the periods and linear spans of the following decimation sequences:

$$\mathbf{a}^{(r)}, \text{ where } r = 2, 3, 5, 14, 146.$$

(It is not necessary to generate the m-sequences.)
14. An m-sequence $\mathbf{a}$ of period 127 with the minimal polynomial $f(x) = x^7 + x + 1$ is given by:

```
1 0 0 0 0 0 0 1 0 0 0 0 0 1 1 0 0 0 0 1 0 1 0 0 0 1 1 1 1 0
0 1 0 0 0 1 0 1 1 0 0 1 1 1 0 1 0 1 0 0 1 1 1 1 1 0 1 0 0 0
0 1 1 1 0 0 0 1 0 0 1 0 0 1 1 0 1 1 0 1 0 1 1 0 1 1 1 1 0 1
1 0 0 0 1 1 0 1 0 0 1 0 1 1 1 0 1 1 1 0 0 1 1 0 0 1 0 1 0 1
0 1 1 1 1 1 1
```

    (a) The set Γ consisting of all the coset leaders modulo 127 is given by

$$\Gamma = \{1, 3, 5, 7, 9, 11, 13, 15, 19, 21, 23, 27, 29, 31, 43, 47, 55, 63\}.$$

    Find the individual terms a_i, $\forall i \in \Gamma$.
    (b) Verify that the sequence $\mathbf{a}$ is constant-on-cosets.

The following three unsolved problems and conjectures related to shift register sequences are proposed by S.W. Golomb.

15. (**Golomb's Conjecture**) There exist infinitely many n such that $f(x) = x^n + x^k + 1$ is a primitive polynomial over $GF(2)$, where $1 \leq k < n$ and k may differ for different n.

    Notes regarding Golomb's conjecture:

    (a) It is easy to show that there are infinitely many irreducible trinomials over $GF(2)$. For example, $x^{2 \cdot 3^k} + x^{3^k} + 1$ is irreducible for every $k = 0, 1, 2, \ldots$, with period 3^{k+1}, but it is primitive only for $k = 0$.

    (b) Can you prove that there are infinitely many primitive polynomials over $GF(2)$ which have no more than t terms, for any fixed integer t? This would be a new result.

    (c) It seems to be true for every degree $n \geq 5$ that there are primitive 5-term polynomials (pentanomials) of degree n over $GF(2)$. This would be a far stronger result than (b) above.

16. It is known that an n-stage shift register with the boolean feedback function $f(x_0, x_1, \ldots, x_{n-1})$ produces pure cycles (without branches) if and only if we can write $f(x_0, x_1, \ldots, x_{n-1}) = x_0 + g(x_1, \ldots, x_{n-1})$ (see Problem 2). It is also known that for $n > 2$, the number of (pure) cycles, for this case, is even (odd) if and only if the number of ones in the truth table for $g(x_1, \ldots, x_{n-1})$ is even (odd). Are there other general, qualitative results about the cycles of a nonlinear shift register that can be similarly and simply stated in terms of the boolean functions $f(x_0, x_1, \ldots, x_{n-1})$ or $g(x_1, \ldots, x_{n-1})$?

17. If $f(x_0, x_1, \ldots, x_{n-1}) = x_0 + g(x_1, \ldots, x_{n-1})$, and $g(0, \ldots, 0) = 0$, then the "all zero state" forms a pure cycle by itself. What further conditions on g will guarantee that the remaining $2^n - 1$ states lie on a single cycle of the shift register? (It is not necessary to find conditions for all $2^{2^{n-1}-n}$ such sequences. It would be very interesting to find conditions for even a small family of nonlinear sequences of period $2^n - 1$.)

5

Randomness Measurements and m-Sequences

Randomness of a sequence refers to the unpredictablity of the sequence. Any deterministically generated sequence used in practical applications is not truly random. The best that can be done here is to single out certain properties as being associated with randomness and to accept any sequence that has these properties as random or more properly, a pseudorandom sequence. In this chapter, we will discuss the randomness of sequences whose elements are taken from a finite field. In Section 5.1, we present Golomb's three randomness postulates for binary sequences, namely the balance property, the run property, and the (ideal) two-level autocorrelation property, and the extension of these randomness postulates to nonbinary sequences. M-sequences over a finite field possess many extraordinary randomness properties except for having the lowest possible linear span, which has stimulated researchers to seek nonlinear sequences with similarly such favorable properties for years. In Section 5.2, we show that m-sequences satisfy Golomb's three randomness postulates. In Section 5.3, we introduce the interleaved structures of m-sequences and the subfield decomposition of m-sequences. In Sections 5.4–5.6, we present the shift-and-add property, constant-on-cosets property, and 2-tuple balance property of m-sequences, respectively. The last section is devoted to the classification of binary sequences of period $2^n - 1$.

5.1 Golomb's randomness postulates and randomness criteria

We discussed some general properties of auto- and crosscorrelation in Chapter 1 for sequences whose elements are taken from the real number field or the complex number field. In this chapter, we will relate those concepts to periodic sequences whose elements are taken from a finite field $\mathbb{F}_q$. Their correlation will be defined in terms of character functions, which map the elements of sequences over $\mathbb{F}_q$ into the complex field.

Let **a** be a sequence of period N over $\mathbb{F}_q$. As usual, we also write $\mathbf{a} = (a_0, a_1, \ldots, a_{N-1})$, an element in a vector space of dimension N over $\mathbb{F}_q$.

5.1.1 Golomb's three randomness postulates for binary sequences

Historical note

These three randomness postulates first appeared in Golomb's June 1955, Glenn L. Martin Co. report on *Sequences with Randomness Properties*, which was the basis for Chapter 3 of his 1967 book *Shift Register Sequences*.

Run

For a binary sequence **a** with period N, k consecutive zeroes (or ones) preceded by one (or zero) and followed by one (or zero) is called a run of zeroes (or ones) of length k.

Autocorrelation

The autocorrelation function of **a**, denoted by $C_{\mathbf{a}}(\tau)$, is defined as

$$C_{\mathbf{a}}(\tau) = \sum_{i=0}^{N-1} (-1)^{a_{i+\tau}+a_i}; \tag{5.1}$$

that is, $C_{\mathbf{a}}(\tau)$ is the (unnormalized) dot product of two vectors

$$((-1)^{a_0}, (-1)^{a_1}, \ldots, (-1)^{a_{N-1}}) \quad \text{and} \quad ((-1)^{a_\tau}, (-1)^{a_{\tau+1}}, \ldots, (-1)^{a_{\tau+N-1}})$$

defined in Chapter 1. Here the indices are reduced modulo N. We also denote this correlation by $C(\tau)$ if the context is clear. Here τ can be considered as a phase shift value of the sequence **a**. $C(\tau)$ measures the amount of similarity between the sequence and its phase shift. This is always highest for $\tau = 0$, because $C(0) = \sum_{i=0}^{N-1}(-1)^{a_i+a_i} = N$.

Golomb proposed the following three randomness postulates to measure apparant randomness of binary periodic sequences (Golomb 1967).

R-1 In every period, the number of zeroes is nearly equal to the number of ones. (More precisely, the disparity is not to exceed 1; i.e., $\left| \sum_{i=0}^{N-1}(-1)^{a_i} \right| \leq 1$.)

R-2 In every period, half the runs have length 1, one fourth have length 2, one eighth have length 3, and so on, as long as the number of runs so indicated exceeds 1. Moreover, for each of these lengths, there are equally many runs of 0's and of 1's.

R-3 The autocorrelation function $C(\tau)$ is two-valued, given by

$$C(\tau) = \begin{cases} N & \text{if } \tau \equiv 0 \ (\text{mod } N) \\ K & \text{if } \tau \not\equiv 0 \ (\text{mod } N), \end{cases} \tag{5.2}$$

where K is a constant. If $K = -1$ for N odd and $K = 0$ for N even, then we say that the sequence has the (ideal) 2-level autocorrelation function.

If $N = 2^n - 1$, we have the following refined postulates.

R-1 In every period, 0's occur $2^{n-1} - 1$ (or 2^{n-1}) times and 1's occur 2^{n-1} (or $2^{n-1} - 1$) times. This property is referred to as the balance property.

R-2 In every period, runs of 0's (or 1's) of length $k : 1 \le k \le n - 2$ occur 2^{n-2-k} times. A run of 0's of length $n - 1$ occurs once and a run of 1's of length n occurs once. This is referred to as the run property.

R-3 The autocorrelation function $C(\tau)$ is two-valued and is given by

$$C(\tau) = \begin{cases} 2^n - 1 & \text{if } \tau \equiv 0 \ (\text{mod } 2^n - 1) \\ -1 & \text{if } \tau \not\equiv 0 \ (\text{mod } 2^n - 1). \end{cases} \tag{5.3}$$

Example 5.1 We consider the following two m-sequences.

(a) Let $\mathbf{a} = (1110010)$ with the minimal polynomial $f(x) = x^3 + x + 1$. There are four 1's and three 0's, which is adequate to satisfy R-1. Of the four runs, half have length 1 and one fourth have length 2. The autocorrelation is 7 in phase and -1 out of phase. Thus, this sequence satisfies R-1, R-2, and R-3.

(b) Let $\mathbf{a} = (000100110101111)$ with the minimal polynomial $f(x) = x^4 + x + 1$. It can be easy verified that this sequence also satisfies R-1, R-2, and R-3.

Binary sequences with 2-level autocorrelation have many applications in communications such as radar distance ranging, sonar distance ranging, hardware testing, coding theory, and cryptography. A binary sequence of period $2^n - 1$ with 2-level autocorrelation corresponds to a cyclic Hadamard difference set with parameters $(2^n - 1, 2^{n-1} - 1, 2^{n-2} - 1)$, which will be shown in Chapter 7. M-sequences are the first class of binary sequences of period $2^n - 1$ with 2-level autocorrelation for any positive integer n. M-sequences correspond to the Singer Hadamard difference sets that were discovered in 1938 by Singer (1938). Golomb found m-sequences from the approach of LFSR sequences in 1954. In recent years, there has been a breakthrough for the constructions of such sequences, after the Gordon, Mills, and Welch construction of 1960. In the next section, we show that m-sequences have the 2-level autocorrelation property as well as the other randomness properties. We will introduce GMW sequences and their generalizations in Chapter 8 and those discovered after 1997 in Chapter 9. Except for the constructions for binary 2-level autocorrelation sequences of period $2^n - 1$, from Chapter 2, we have three more classes of binary 2-level autocorrelation sequences of period N. These are the Legendre sequences for $N = p = 4t - 1$ when p is a prime, the Hall's sextic residue sequences if p also satisfies the condition that $p = 4t - 1 = 4u^2 + 27$, and the twin prime sequences for $N = p(p + 2)$ where p and $p + 2$ are both prime. These are all the known construction for binary 2-level autocorrelation sequences to date (except when the period is $2^n - 1$).

In the following, we recall the construction for the Legendre sequences from Chapter 2. Let p be an odd prime number and $p \equiv 3$ (mod 4). The Legendre symbol (i/p) was defined in Section 2.5 of Chapter 2; that is, for $i \not\equiv 0$ (mod p), by

$$\left(\frac{i}{p}\right) = \begin{cases} 1 & \text{if there is an integer } x \text{ for which } x^2 \equiv i \ (\text{mod } p) \\ -1 & \text{otherwise.} \end{cases}$$

A Legendre sequence (or quadratic residue sequence), denoted by $\mathbf{a} = \{a_i\}$, is defined as

$$a_i = \begin{cases} 0 & \Longleftrightarrow \left(\frac{i}{p}\right) = 1 \\ 1 & \Longleftrightarrow \left(\frac{i}{p}\right) = -1 \end{cases} \tag{5.4}$$

and $a_0 = 1$. O. Perron (1952) observed that the sequence $\mathbf{a}$ has the property R-3 for p a prime of the form $4n - 1$. (Note: Interchanging 1's and 0's is an inessential distinction, in that it does not affect the autocorrelation; and two sequences that differ only in this regard will not usually be considered different; see Chapter 2.) Generally, Legendre sequences satisfy R-1 and R-3, but not R-2.

Example 5.2 Let $p = 11$. Then

$$\left(\tfrac{1}{p}\right) = 1, \quad \left(\tfrac{2}{p}\right) = -1,$$

$$\left(\tfrac{3}{p}\right) = 1, \quad \left(\tfrac{4}{p}\right) = 1, \quad \left(\tfrac{5}{p}\right) = 1,$$

$$\left(\tfrac{6}{p}\right) = -1, \quad \left(\tfrac{7}{p}\right) = -1, \quad \left(\tfrac{8}{p}\right) = -1,$$

$$\left(\tfrac{9}{p}\right) = 1, \quad \left(\tfrac{10}{p}\right) = -1.$$

So, **a** has period 11 and

$$\mathbf{a} = (10100011101).$$

In one period, there are six 1's and five 0's, so that it satisfies R-1. It can be checked that it satisfies R-3. There are six runs, including three runs of 0's and three runs of 1's, among which there are two runs of 0's of length 1, one run of 1's of length 1, no run of 0's of length 2, one run of 1's of length 2, and so on. So R-2 is violated by two runs of 0's of length 1 balanced by only one run of 1's of length 1.

Example 5.3 Let $p = 31$. Then the Legendre sequence of period 31 is given by

$$\mathbf{a} = (1001001000011101010001111011011).$$

In one period, there are sixteen 1's and fifteen 0's, so that it satisfies R-1. It can be checked that it satisfies R-3. In one period, there are sixteen runs, including eight runs of 0's and eight runs of 1's; and four runs of 1's of length 1 and four runs of 0's of length 1, one run of 1's of length 2, two runs of 0's of length 2, and so on. So R-2 is violated by two runs of 0's of length 2 balanced by only one run of 1's of length 2.

5.1.2 Extending Golomb's three randomness postulates to nonbinary sequences

We now generalize Golomb's three randomness postulates to sequences whose elements are taken from the finite field $\mathbb{F}_q$ where $q = p^r$, where p is a prime and r is a positive integer. In particular, for $q = 2$, the following three postulates reduce to those presented above. To proceed, we need some preparation for sequences over $\mathbb{F}_q$. We assume that **a** is a sequence over $\mathbb{F}_q$ of period N.

Run

For $\eta, \lambda, \xi \in \mathbb{F}_q$ and $\eta \neq \lambda$ and $\xi \neq \lambda$, if $\eta, \underbrace{\lambda, \ldots, \lambda}_{k}, \xi$ appears in the sequence **a**, then we say that $\underbrace{\lambda, \ldots, \lambda}_{k}$ is a run of λ's of length k.

Additive character

Let $\omega = e^{2\pi i/p}$ be a primitive pth root of unity. The *(canonical) additive character* of $\mathbb{F}_{p^r}$ is defined by

$$\chi(x) = e^{2\pi i Tr(x)/p}, \quad x \in \mathbb{F}_{p^r}, \tag{5.5}$$

Table 5.1. $\mathbb{F}_{3^3}$ *defined by the polynomial* $f(x) = x^3 + 2x^2 + 1$

1	α	α^2	α^i	1	α	α^2	α^i
1	0	0	0	0	1	0	1
0	0	1	2	2	0	1	3
2	2	1	4	2	2	0	5
0	2	2	6	1	0	1	7
2	1	1	8	2	2	2	9
1	2	1	10	2	1	0	11
0	2	1	12	2	0	0	13
0	2	0	14	0	0	2	15
1	0	2	16	1	1	2	17
1	1	0	18	0	1	1	19
2	0	2	20	1	2	2	21
1	1	1	22	2	1	2	23
1	2	0	24	0	1	2	25

where $Tr(x)$ is the trace function from $\mathbb{F}_{p^r}$ to $\mathbb{F}_p$, given by

$$Tr(x) = Tr_{\mathbb{F}_{p^r}/\mathbb{F}_p}(x) = x + x^p + \cdots + x^{p^{r-1}}, x \in \mathbb{F}_{p^r}.$$

In the following, we use an example to illustrate this concept.

Example 5.4 Let $f(x) = x^3 + 2x^2 + 1$, a primitive polynomial over $\mathbb{F}_3$, and let α be a root of $f(x)$ in the extension field $\mathbb{F}_{3^3}$. We wish to compute the additive character $\chi(x)$ for $x = \alpha + 2\alpha^2 \in \mathbb{F}_{3^3}$. First we construct the table for the finite field $\mathbb{F}_{3^3}$ in Table 5.1. Let $\omega = e^{2\pi i/3}$. Using Table 5.1, note that $\alpha + 2\alpha^2 = \alpha^{25}$. We have

$$Tr(\alpha + 2\alpha^2) = Tr(\alpha^{25}) = \alpha^{25} + \alpha^{23} + \alpha^{17}$$
$$= (\alpha + 2\alpha^2) + (2 + \alpha + 2\alpha^2) + (1 + \alpha + 2\alpha^2) = 0,$$

where $Tr(x)$ is the trace function from $\mathbb{F}_{3^3}$ to $\mathbb{F}_3$. So,

$$\chi(\alpha + 2\alpha^2) = \omega^{Tr(\alpha + 2\alpha^2)} = \omega^0 = 1.$$

The following notation for the trace functions related to intermediate fields was introduced in Chapter 3. Here we reproduce it for the reader's convenience. Let $F = \mathbb{F}_q$, $K = \mathbb{F}_{q^m}$, and $E = \mathbb{F}_{q^n}$ where $m|n$. We denote the trace function from E to K by $Tr_m^n(x) = x + x^Q + \cdots + x^{Q^{l-1}}$ where $Q = q^m$ and $l = n/m$. In other words, the dimensions of E and K, viewed as linear spaces over F, appear as the superscript and subscript of Tr, respectively.

Autocorrelation of nonbinary sequences

In Chapter 1, we introduced the definition of autocorrelation of sequences whose entries are taken from the real number field or the complex number field. For a sequence $\mathbf{a}$ over $\mathbb{F}_q$ of period N, we define its autocorrelation through the additive character of $\mathbb{F}_q$, which is a mapping from the finite field, $\mathbb{F}_q$, into the complex number field, $\mathbb{C}$; that is, the autocorrelation of $\mathbf{a}$ is defined by

$$C_{\mathbf{a}}(\tau) = \sum_{i=0}^{N-1} \chi(a_{i+\tau})\chi^*(a_i), \quad \tau = 0, 1, \ldots \tag{5.6}$$

where $\chi^*(x) = (\chi(x))^*$, the complex conjugate of $\chi(x)$. In other words, the autocorrelation of $\mathbf{a}$ at τ is equal to the Hermitian dot product $(\chi(L^\tau \mathbf{a}) \cdot \chi(\mathbf{a}))$ where

$$\chi(L^\tau \mathbf{a}) = (\chi(a_\tau), \chi(a_{\tau+1}), \ldots, \chi(a_{\tau+N-1})), \quad \tau = 0, 1, \ldots$$

where the indices are reduced modulo N. We are now in a position to generalize Golomb's three randomness postulates to q-ary sequences.

R-1 Let N_λ be the number of $\lambda \in \mathbb{F}_q$ that occur in one period of $\mathbf{a}$; that is,

$$N_\lambda = |\{j | a_j = \lambda, 0 \le j < N\}|.$$

For each $\lambda \ne \beta$, $|N_\lambda - N_\beta| \le 1$. In particular, for $N = q^n - 1$, in every period, each nonzero element in $\mathbb{F}_q$ occurs q^{n-1} times, and the zero element occurs $q^{n-1} - 1$ times. (A more general concept is as follows: There exists a constant c in $\mathbb{F}_q$ such that $b_i = a_i + c$ for which $\mathbf{b} = \{b_i\}$ satisfies the above distribution.) This property, which states that elements occur as equally often as possible within one period, is referred to as the balance property of nonbinary sequences.

R-2 If $N \ne q^n - 1$, we set $n = \lfloor \log_q N \rfloor$ where $\lfloor x \rfloor$ represents the largest integer that is less than or equal to x. In every period,
  (a) for $1 \le k \le n - 2$, the runs of every element in F of length k occur $(q - 1)^2 q^{n-k-2}$ times;
  (b) the runs of every nonzero element of length $n - 1$ occur $q - 2$ times;
  (c) the runs of the zero element of length $n - 1$ occur $q - 1$ times; and
  (d) the run of every nonzero element of length n occurs once.
This is referred to as the run property of nonbinary sequences.

R-3 The autocorrelation function $C_{\mathbf{a}}(\tau)$ is two-valued and is given by

$$C_{\mathbf{a}}(\tau) = \begin{cases} N & \text{if } \tau \equiv 0 \pmod{N} \\ -1 & \text{if } \tau \not\equiv 0 \pmod{N}. \end{cases} \tag{5.7}$$

In this case, the sequence $\mathbf{a}$ is said to have an (ideal) 2-level autocorrelation function.

5.1.3 Component sequences

It is worth pointing out that if $r > 1$ for $q = p^r$, then an autocorrelation function of a periodic sequence over $\mathbb{F}_q$ can be defined in another way, which will be introduced below. Let $\{\alpha_0, \alpha_1, \ldots, \alpha_{r-1}\}$ and $\{\beta_0, \beta_1, \ldots, \beta_{r-1}\}$ be a pair of dual bases of $\mathbb{F}_{p^r}$ over $\mathbb{F}_p$. According to Theorem 3.13 in Chapter 3, a_i, the ith element of **a**, can be represented as

$$a_i = \sum_{j=0}^{r-1} Tr(\beta_j a_i)\alpha_j, \ \forall i \geq 0, \tag{5.8}$$

where $Tr(x)$ is the trace function from $\mathbb{F}_{p^r}$ to $\mathbb{F}_p$. Let

$$\mathbf{a}_j = \{a_{ij}\}_{i \geq 0}, \tag{5.9}$$

where

$$a_{ij} = Tr(\beta_j a_i) \in \mathbb{F}_p. \tag{5.10}$$

So, $\mathbf{a}_j$ is a sequence over $\mathbb{F}_p$. We call $\mathbf{a}_j$, $0 \leq j \leq r - 1$, the jth component sequence of **a** with respect to $\mathbb{F}_p$.

Proposition 5.1 *If* **a** *is an m-sequence over* $\mathbb{F}_q$ $(q = p^r)$, *then all component sequences with respect to* $\mathbb{F}_p$ *are m-sequences over* $\mathbb{F}_p$ *of degree* nr.

Proof. Let $f(x)$ be the minimal polynomial of **a** over $\mathbb{F}_q$ of degree n and α be a root of $f(x)$ in the extension $\mathbb{F}_{q^n}$. Then **a** has the following trace representation:

$$a_i = Tr_{\mathbb{F}_{q^n}/F_q}(\gamma\alpha^i), i = 0, 1, \ldots, \gamma \in \mathbb{F}_{q^n}. \tag{5.11}$$

Consistent with the definition of the component sequences, the ith element in the jth component sequence of **a** is given by

$$\begin{aligned}
a_{ij} &= Tr(\beta_j a_i) \ (\text{here } Tr(x) = Tr_{\mathbb{F}_q/\mathbb{F}_p}(x)) \\
&= Tr_{\mathbb{F}_q/F_p}(\beta_j Tr_{\mathbb{F}_{q^n}/F_q}(\gamma\alpha^i)) \qquad (\text{by (5.5.11)}) \\
&= Tr_{\mathbb{F}_{q^n}/\mathbb{F}_p}(\beta_j\gamma\alpha^i), \beta_j, \gamma \in \mathbb{F}_{q^n},
\end{aligned}$$

where the last identity follows from the transitivity of the trace function in Theorem 3.12 in Chapter 3. Since α is a primitive element of $\mathbb{F}_{q^n} = \mathbb{F}_{p^{nr}}$, from Lemma 4.3 in Chapter 4, $\mathbf{a}_j$ is an m-sequence over $\mathbb{F}_p$ of degree nr. □

Another way to define an autocorrelation function of **a** is to define it as the sum of the autocorrelations of its component sequences; that is, we define

$$D_{\mathbf{a}}(\tau) = \sum_{j=0}^{r-1} C_{\mathbf{a}_j}(\tau), \ \forall\tau \tag{5.12}$$

which is called a component autocorrelation function of **a**. Note that these two definitions are the same when $r = 1$; that is, we have $C_{\mathbf{a}}(\tau) = D_{\mathbf{a}}(\tau), \forall \tau \geq 0$. Under this definition, the randomness property R-3 can be restated as follows.

R-3′ The component autocorrelation function $D_{\mathbf{a}}(\tau)$ is two-valued, given by

$$D_{\mathbf{a}}(\tau) = \begin{cases} rN & \text{if } \tau \equiv 0 \ (\text{mod } N) \\ -r & \text{if } \tau \not\equiv 0 \ (\text{mod } N). \end{cases} \tag{5.13}$$

A crosscorrelation function between two periodic sequences $\mathbf{a} = \{a_i\}$, of period s, and $\mathbf{b} = \{b_i\}$, of period t, over $\mathbb{F}_q$ can be defined as

$$C_{\mathbf{a},\mathbf{b}}(\tau) = \sum_{i=0}^{N-1} \chi(a_{i+\tau})\chi^*(b_i), \quad \tau = 0, 1, \ldots, \tag{5.14}$$

where $N = \text{lcm}[s, t]$. Alternatively, it can be defined through the crosscorrelations of their components:

$$D_{\mathbf{a},\mathbf{b}}(\tau) = \sum_{j=0}^{r-1} C_{\mathbf{a}_j, \mathbf{b}_j}(\tau). \tag{5.15}$$

In particular, the crosscorrelation of **a** and **b** defined by Eq. (5.14) becomes the following formulae without involving the trace function for the case that q is a prime.

$$C_{\mathbf{a},\mathbf{b}}(\tau) = \sum_{i=0}^{N-1} (-1)^{a_{i+\tau}+b_i}, \quad \tau = 0, 1, \ldots, \text{ for } q = 2; \tag{5.16}$$

$$C_{\mathbf{a},\mathbf{b}}(\tau) = \sum_{i=0}^{N-1} \omega^{a_{i+\tau}-b_i}, \quad \tau = 0, 1, \ldots, \text{ for } q = p > 2. \tag{5.17}$$

When $\mathbf{a} = \mathbf{b}$, the above two identities become the autocorrelation function for $q = 2$ and $q = p$, respectively.

Remark 5.1 In some applications to communication systems, the main concern is how to obtain correlation between signals in order to distinguish them. The ways to compute (or define) correlation are not essential. However, for a sequence whose elements are taken from a finite field $\mathbb{F}_q$ where $q = p^r$ with $r > 1$, both definitions of the autocorrelation function, given by Eqs. (5.6) and (5.13), and the crosscorrelation, defined by Eqs. (5.14) and (5.15), do not completely characterize correlation of the sequences. In fact, the correlation defined in terms of the character function only uses one component sequence, and the second definition of the correlation is defined as the sum of the correlation functions of their component sequences. Up to now, no adequate definitions for correlations of sequences over $\mathbb{F}_{p^r}$ where $r > 1$ have appeared in the literature.

We propose some problems related to this question in the exercises for this chapter.

In the next section, we prove that any m-sequence over $\mathbb{F}_q$ satisfies the extended form of Golomb's three randomness postulates. In particular, if $q = 2$, any binary m-sequence satisfies Golomb's three randomness postulates. We conclude this section by presenting a set of randomness criteria which serve as design principles for sequence design for applications in communications and cryptography.

(Unconditional) Randomness Criteria

For a sequence over $\mathbb{F}_q$ of period N,

1. *Period requirement:* long period.
2. *Statistical properties:* the balance property R-1, run property R-2, and ideal k-tuple distribution where $1 \le k \le n = \lfloor \log_q N \rfloor$. (We say a sequence satisfies the ideal k-tuple distribution if for $1 \le k \le n$, each of the k-tuples $d_0, d_1, \ldots, d_{k-1}$, $d_i \in \mathbb{F}_q$, occurs nearly equally many times in one period. A detailed discussion of this property will be given in the next section.)
3. *Correlation:*
   (a) Two-level autocorrelation R-3.
   (b) Low-valued crosscorrelation: Let S be a set consisting of finite sequences with period N. If for any two sequences in S,

$$0 \le |C_{\mathbf{a},\mathbf{b}}(\tau)| \le c\sqrt{N} \tag{5.18}$$

   where $\tau \not\equiv 0 \pmod{N}$ when $\mathbf{a} = \mathbf{b}$, and $c > 0$ is a constant, then we say that S has a low crosscorrelation. Note that (5.18) is a variation of the Welch bound (1974).
4. *Linear span:* large ratio of linear span to period,

$$\rho(\mathbf{a}) = \frac{LS(\mathbf{a})}{N} > \delta,$$

where $\delta > 0$ is a constant for N large. Sometimes, we call the ratio $\rho(\mathbf{a})$ the normalized linear span of $\mathbf{a}$. Note that $0 < \rho(\mathbf{a}) \le 1$ for a fixed N.

The criterion for linear span is related to the application of the so-called Berlekamp-Massey algorithm for computing the linear spans of sequences. If the linear span of a sequence is n, then from any known $2n$ consecutive terms of the sequence, applying the algorithm, one can find the linear recursive relation that generates the sequence. This simple fact of linear algebra, already used by Golomb in 1954, scarcely needs to be named for anyone. Berlekamp and Massey have each done far more important things for which they deserve to be remembered. For cryptographical applications of sequences, a large normalized linear span reflects a certain unpredictability of the sequence. However, it is not easy to construct sequences having all these randomness properties. Roughly speaking, the pseudorandom sequences employed in communication systems utilize their correlation properties. On the other hand, in cryptographic applications, large normalized linear span is a necessary requirement. Thus, sequence design is devoted to seeking sequences with good correlation, large normalized linear span, and efficient implementations in hardware and software, with different trade-offs between these features depending on the application.

5.2 Randomness properties of m-sequences

First we give a generalization of the balance property R-1 and introduce the affine operator on sequences. Let **a** be a sequence over $\mathbb{F}_q$ of period $q^n - 1$.

R-4 In every period of **a**, each nonzero n-tuple $(\lambda_1, \lambda_2, \ldots, \lambda_n) \in \mathbb{F}_q^n$ occurs exactly once.

Consider all k with $1 \leq k \leq n$. In every period of **a**, if each nonzero k-tuple $(\lambda_1, \lambda_2, \ldots, \lambda_k) \in \mathbb{F}_q^k$ occurs in **a** q^{n-k} times and the zero k-tuple $(0, \ldots, 0)$ occurs $q^{n-k} - 1$ times, then we say that the sequence satisfies the ideal k-tuple distribution. Precisely, let $\mathbf{a} = \{a_i\}$ be a sequence over $\mathbb{F}_q$ with period $q^n - 1$, and let

$$R(k) = \{(a_i, a_{i+1}, \ldots, a_{i+k-1}) \mid 0 \leq i < q^n - 1\}.$$

For $1 \leq k \leq n$, **a** satisfies the (ideal) k-tuple distribution if every nonzero k-tuple $(\lambda_1, \lambda_2, \ldots, \lambda_k) \in \mathbb{F}_q^k$ occurs q^{n-k} times in $R(k)$ and the zero k-tuple occurs $q^{n-k} - 1$ times in $R(k)$. Sometimes, R-4 is also called the span n property of the sequence. More specifically, the span n property requires that no k-tuple, for any $k \geq n$, occurs more than once.

Let $\mathbf{c} = (c, c, \ldots,), c \in \mathbb{F}_q$ be a constant vector and have the same length as the **a** vector when we write **a** as a vector. According to the definition, if there exists some $c \in \mathbb{F}_q$ such that a k-tuple $\underbrace{(c, c, \ldots, c)}_{k}$ occurs $q^{n-k} - 1$ times in one period of **a** and each of the other k-tuples occurs q^{n-k} times, then $\mathbf{a} - \mathbf{c}$ has the ideal k-tuple distribution. We refer to the operation of subtracting a constant vector from a sequence as an affine operation on the sequence. When we talk about the ideal k-tuple distribution of **a**, we always mean that one of those sequences obtained from applying the affine operator has this property, and we choose this one as **a**. Note that the affine operation does not change the autocorrelation of the sequence.

Proposition 5.2 *If* **a** *satisfies the randomness property* R-4, *then* **a** *has the ideal k-tuple distribution for every k with* $1 \leq k \leq n$.

Proof. Because each nonzero n-tuple appears once in every period, there are q^{n-k} ways to extend a k-tuple $(\lambda_1, \lambda_2, \ldots, \lambda_k)$ to an n-tuple:

$$(\lambda_1, \lambda_2, \ldots, \lambda_k, \lambda_{k+1}, \ldots, \lambda_n)$$

except for $(\lambda_1, \lambda_2, \ldots, \lambda_k) = (0, \ldots, 0)$. For the latter, there are $q^{n-k} - 1$ ways to extend it, where $(\lambda_{k+1}, \ldots, \lambda_n) = (0, \ldots, 0)$ is excluded. $\square$

Property 5.1 *Every m-sequence satisfies the randomness postulate* R-4.

Proof. Let **a** be an m-sequence over $\mathbb{F}_q$ of period $q^n - 1$. We need to show that each nonzero n-tuple appears exactly once in one period of **a**. It is equivalent to prove that $(a_i, a_{i+1}, \ldots, a_{i+n-1}) \neq (a_j, a_{j+1}, \ldots, a_{j+n-1})$ for all $0 \leq i \neq j \leq q^n - 2$. Let $f(x)$ be the minimal polynomial of **a** over $\mathbb{F}_q$. According to Corollary 4.6 in Chapter 4, all sequences in $G(f)$, the set consisting of all sequences generated by $f(x)$, are shift equivalent. In other words, $G(f) = \{L^i \mathbf{a} \mid 0 \leq i \leq q^n - 2\} \cup \{0\}$. Moreover, $G(f)$ is a linear space over $\mathbb{F}_q$ of dimension n. So, there are q^n different sequences in $G(f)$. Because different initial states $(a_i, a_{i+1}, \ldots, a_{i+n-1})$ correspond to different sequences $L^i \mathbf{a}$ in $G(f)$, each nonzero n-tuple appears in one period of **a** exactly once. $\square$

From Property 5.1 and Proposition 5.2, the next result follows immediately.

Property 5.2 *Every m-sequence over* $\mathbb{F}_q$ *of period* $q^n - 1$ *has the ideal k-tuple distribution for every k with* $1 \leq k \leq n$.

Property 5.3 *Every m-sequence satisfies the randomness property* R-1; *that is, every m-sequence is balanced.*

Proof. In Property 5.2, by setting $k = 1$, the result follows. $\square$

Combining this property with the trace representation of m-sequences, we have established the following corollary.

Corollary 5.1 *For any* $c \in \mathbb{F}_q$, *the equation*

$$Tr(\beta x) = c, \beta \neq 0 \in \mathbb{F}_{q^n},$$

has q^{n-1} *solutions in* $\mathbb{F}_{q^n}$ *where* $Tr(x)$ *is the trace function from* $\mathbb{F}_{q^n}$ *to* $\mathbb{F}_q$.

Property 5.4 *Every m-sequence satisfies the randomness property* R-2, *the run property.*

Proof. Let **a** be an m-sequence over $\mathbb{F}_q$ of period $q^n - 1$. We know it satisfies R-4.

1. For $1 \leq k \leq n - 2$, any run of $\lambda, \ldots, \lambda$ of length k is contained within an n-tuple of **a**. Without loss of generality, we may assume that the run takes the following form in an n-tuple:

$$(\eta_0, \underbrace{\lambda, \ldots, \lambda}_{k}, \eta_{k+1}, \underbrace{\eta_{k+2}, \ldots, \eta_{n-1}}_{n-k-2}),$$

where $\lambda, \eta_i \in \mathbb{F}_q$, $\eta_0 \neq \lambda$, and $\eta_{k+1} \neq \lambda$. Notice that each n-tuple appears in **a** only once per period, and there are $q - 1$ ways to choose each of η_0 and η_{k+1} and q^{n-k-2} ways to choose an $(n - 2 - k)$-tuple $\eta_{k+2}, \ldots, \eta_{n-1}$. Hence $\underbrace{\lambda, \ldots, \lambda}_{k}$ occurs $(q - 1)^2 q^{n-k-2}$ times in one period.

2. For $\lambda \neq 0$ we consider

$$\eta, \underbrace{\lambda, \ldots, \lambda}_{n-1}, \eta \neq \lambda$$

as a run of λ's of length $n - 1$. There are $q - 1$ ways to select η. Therefore, there are $q - 1$ successors for the above states. Among these $q - 1$ successors, one is given by $\underbrace{\lambda, \ldots, \lambda}_{n}$, which is not a run of λ's of length $n - 1$. Hence, there are $q - 2$ η's such that $\underbrace{\lambda, \ldots, \lambda}_{n-1}$ is a run of length $n - 1$.

Therefore, the assertion is established.

3. There are $q - 1$ ways to select η such that

$$\eta, \underbrace{0, \ldots, 0}_{n-1}$$

is an run of 0's of length $n - 1$. The reason is that for each $\eta \neq 0$, the successor of the above state is as follows:

$$\underbrace{0, \ldots, 0}_{n-1}, \theta,$$

where $\theta \neq 0$, because the constant term in any primitive polynomial is not zero.

4. For each $\lambda \neq 0$,

$$\underbrace{\lambda, \ldots, \lambda}_{n}$$

occurs once, because each nonzero n-tuple occurs once in one period of **a**. $\qquad \square$

Property 5.5 *Any m-sequence* $\mathbb{F}_q$ *satisfies the randomness property R-3. In other words, any m-sequence over* $\mathbb{F}_q$ *has the (ideal) 2-level autocorrelation function.*

This result can be proved using various approaches. Here, we introduce a proof that can be easily extended to GMW sequences and their generalizations. First, we need some results on exponential sums to proceed to the proof.

Basic Properties of Exponential Sums

Lemma 5.1 *Let* $\chi(x)$, $x \in \mathbb{F}_{p^r}$, *be defined by Eq. (5.5). For* $\beta \in \mathbb{F}_{p^r}$,

$$\sum_{x \in \mathbb{F}_{p^r}} \chi(\beta x) = \sum_{x \in \mathbb{F}_{p^r}} \omega^{Tr(\beta x)} = \begin{cases} 0 & \text{if } \beta \neq 0 \\ p^r & \text{if } \beta = 0, \end{cases}$$

where $Tr(x)$ *is the trace function from* $\mathbb{F}_{p^r}$ *to* $\mathbb{F}_p$. *In particular, if* $p = 2$, *then*

$$\sum_{x \in \mathbb{F}_{2^r}} (-1)^{Tr(\beta x)} = \begin{cases} 0 & \text{if } \beta \neq 0 \\ 2^r & \text{if } \beta = 0. \end{cases}$$

Proof. Since ω is a primitive pth root of unity, ω is a root of the equation $x^p - 1 = 0$. So,

$$\sum_{x \in \mathbb{F}_p} \omega^x = \sum_{i=0}^{p-1} \omega^i = 0. \tag{5.19}$$

According to Corollary 5.1, if $\beta \neq 0$, then $Tr(\beta x) = y$ has p^{r-1} solutions in $\mathbb{F}_{p^r}$ for every $y \in \mathbb{F}_p$. Thus, for $\beta \neq 0$, using Eq. (5.19),

$$\sum_{x \in \mathbb{F}_{p^r}} \chi(\beta x) = \sum_{x \in \mathbb{F}_{p^r}} \omega^{Tr(\beta x)}$$

$$= p^{r-1} \sum_{y \in \mathbb{F}_p} \omega^y$$

$$= 0.$$

Note that $\chi(0) = 1$. So, if $\beta = 0$, it is clear that

$$\sum_{x \in \mathbb{F}_{p^r}} \chi(\beta x) = \sum_{x \in \mathbb{F}_{p^r}} \chi(0) = p^r. \qquad \square$$

Lemma 5.2 *Let* $\chi(x)$, $x \in \mathbb{F}_{p^r}$, *be defined by Eq. (5.5). Then*

$$\chi(x)\chi^*(y) = \chi(x - y), \forall x, y \in \mathbb{F}_{p^r}.$$

Proof. According to the definition of $\chi(x)$ in Eq. (5.5), for any $x, y \in \mathbb{F}_{p^r}$,

$$\chi(x)\chi^*(y) = \omega^{Tr(x)}\omega^{-Tr(y)}$$

$$= \omega^{Tr(x)-Tr(y)}$$

$$= \omega^{Tr(x-y)}$$

$$= \chi(x - y). \qquad \square$$

Proof of Property 5.5. Let $E = \mathbb{F}_{q^n}$, $F = \mathbb{F}_q$ with $q = p^r$, $K = \mathbb{F}_p$, and $\omega = e^{2\pi i/p}$. We will use the transitivity of the trace functions, as shown in Theorem 3.12 in Chapter 3. Note that the shift operator does not change the autocorrelation of the sequence. Thus, we may assume that $\beta = 1$ in the trace representation of **a**; that is, $a_i = Tr_{E/F}(\alpha^i), \forall i \geq 0 \implies a_{i+\tau} = Tr_{E/F}(\alpha^{i+\tau}) = Tr_{E/F}(\alpha^\tau \alpha^i)$. Since $\tau \not\equiv 0 \pmod{q^n - 1} \implies 1 - \alpha^\tau \neq 0$, we have

$$C(\tau) = \sum_{i=0}^{q^n-2} \chi(a_{i+\tau})\chi^*(a_i) = \sum_{i=0}^{q^n-2} \chi(a_{i+\tau} - a_i)$$

$$= \sum_{i=0}^{q^n-2} \chi(Tr_{E/F}(\alpha^\tau \alpha^i) - Tr_{E/F}(\alpha^i))$$

$$= \sum_{i=0}^{q^n-2} \chi(Tr_{E/F}((\alpha^\tau - 1)\alpha^i))$$

$$= \sum_{i=0}^{q^n-2} \chi(Tr_{E/F}(y\alpha^i)), \; y = \alpha^\tau - 1 \neq 0,$$

$$= \sum_{x \in E} \omega^{Tr_{F/K}(Tr_{E/F}(yx))} - 1$$

$$= \sum_{x \in E} \omega^{Tr_{E/K}(yx)} - 1$$

$$= \sum_{x \in E} \chi(yx) - 1.$$

According to Lemma 5.1,

$$\sum_{x \in E} \chi(yx) = \begin{cases} 0 & \text{if } y \neq 0 \\ p^{nr} & \text{if } y = 0. \end{cases}$$

Therefore $C(\tau) = -1$ if $\tau \not\equiv 0 \pmod{q^n - 1}$, and $C(\tau) = p^{nr} - 1 = q^n - 1$ if $\tau \equiv 0 \pmod{q^n - 1}$. $\qquad\square$

Note. If $q = 2$, then the above proof can be simplified as follows:

$$C(\tau) = \sum_{i=0}^{2^n-2} (-1)^{a_{i+\tau}+a_i}$$

$$= \sum_{i=0}^{2^n-2} (-1)^{Tr((1+\alpha^\tau)\alpha^i)}$$

$$= \sum_{x \in \mathbb{F}_{2^n}} (-1)^{Tr(yx)} - 1, \quad \text{by setting } y = 1 + \alpha^\tau, x = \alpha^i.$$

Applying Lemma 5.1 to the case $p = 2$, the result follows immediately. (In fact, for the binary case, there is a much simpler proof without involving the trace representation of m-sequences; see Section 5.6.)

Corollary 5.2 *Every m-sequence over $\mathbb{F}_q$ $(q = p^r)$ has the randomness property R-3$'$.*

Proof. According to Proposition 5.1 in Section 5.1, the component sequences of $\mathbf{a}$ are m-sequences over $\mathbb{F}_p$ of degree nr. Applying Property 5.5, $C_{\mathbf{a}_j}(\tau) = q^n - 1$ if $\tau \equiv 0 \pmod{q^n - 1}$. Otherwise $C_{\mathbf{a}_j}(\tau) = -1$. This establishes the assertion. $\qquad\square$

Example 5.5 For $q = 2, n = 6$, let $f(x) = x^6 + x + 1$, and $\mathbf{a} \in G(f)$, an m-sequence of degree 6, whose elements are listed below. Here we can arrange this sequence into a 7×9 array by placing the first 9 bits in the first row, the second 9 bits in the second row, and so on. In this fashion, after we use up the elements in one period of $\mathbf{a}$, we obtain a 7×9 matrix, in which each column is either an m-sequence of period 7 or a zero sequence. This is called the interleaved array structure of m-sequences, which will be discussed in the next section.

$$
\begin{array}{ccccccccc}
0 & 0 & 0 & 0 & 0 & 1 & 0 & 0 & 0 \\
0 & 1 & 1 & 0 & 0 & 0 & 1 & 0 & 1 \\
0 & 0 & 1 & 1 & 1 & 1 & 0 & 1 & 0 \\
0 & 0 & 1 & 1 & 1 & 0 & 0 & 1 & 0 \\
0 & 1 & 0 & 1 & 1 & 0 & 1 & 1 & 1 \\
0 & 1 & 1 & 0 & 0 & 1 & 1 & 0 & 1 \\
0 & 1 & 0 & 1 & 1 & 1 & 1 & 1 & 1 \\
\end{array}
$$

Let $N(\mathbf{v})$ be the number of times a vector $\mathbf{v}$ occurs as a run in $\mathbf{a}$ in each period. We list the run distribution of the sequence in the following table.

Verifying process		Property satisfied
32 1's	31 0's	R-1
$N(1) = 8,$	$N(0) = 8$	
$N(11) = 4,$	$N(00) = 4$	
$N(111) = 2,$	$N(000) = 2$	R-2
$N(1111) = 1,$	$N(0000) = 1$	
$N(11111) = 0,$	$N(00000) = 1$	
$N(111111) = 1,$	$N(000000) = 0$	
$C(\tau) = \begin{cases} 63 & \tau \equiv 0 \ (\mathrm{mod}\ 63) \\ -1 & \tau \not\equiv 0 \ (\mathrm{mod}\ 63) \end{cases}$		R-3

In the following, we will show some examples of nonbinary m-sequences represented in an array by interleaving elements of sequences in the same fashion as we did for the m-sequence of period 63 in the above example.

Example 5.6 Examples of nonbinary m-sequences. (The reader is encouraged to verify their randomness properties R-1, R-2, and R-3).

Case 1. $q = p > 2$, a prime.

(a) Select $p = 3, n = 3$, and $f(x) = x^3 - x^2 - 2$, primitive over $\mathbb{F}_3$. We write $\mathbf{a}$ as a 2×13 array:

$$0 \ \ 0 \ \ 1 \ \ 1 \ \ 1 \ \ 0 \ \ 2 \ \ 1 \ \ 1 \ \ 2 \ \ 1 \ \ 0 \ \ 1$$
$$0 \ \ 0 \ \ 2 \ \ 2 \ \ 2 \ \ 0 \ \ 1 \ \ 2 \ \ 2 \ \ 1 \ \ 2 \ \ 0 \ \ 2$$

(b) Select $p = 3, n = 4$, and $f(x) = x^4 - x^3 - 2x^2 - 2x - 1$. We write $\mathbf{a}$ as a 2×40 array:

1121202111102210202001021001211220002202
2212101222201120101002012002122110001101

or as an 8×10 array:

$$
\begin{array}{cccccccccc}
1 & 1 & 2 & 1 & 2 & 0 & 2 & 1 & 1 & 1 \\
1 & 0 & 2 & 2 & 1 & 0 & 2 & 0 & 2 & 0 \\
0 & 1 & 0 & 2 & 1 & 0 & 0 & 1 & 2 & 1 \\
1 & 2 & 2 & 0 & 0 & 0 & 2 & 2 & 0 & 2 \\
2 & 2 & 1 & 2 & 1 & 0 & 1 & 2 & 2 & 2 \\
2 & 0 & 1 & 1 & 2 & 0 & 1 & 0 & 1 & 0 \\
0 & 2 & 0 & 1 & 2 & 0 & 0 & 2 & 1 & 2 \\
2 & 1 & 1 & 0 & 0 & 0 & 1 & 1 & 0 & 1 \\
\end{array}
$$

(c) Choose $p = 5$, $n = 3$, and $f(x) = x^3 - 4x^2 - 3$. Then **a** can be arranged into a 4×31 array:

$$
\begin{array}{l}
0014120332224243340432042342201 \\
0032310441112124420241021421103 \\
0041430223331312210123013213304 \\
0023240114443431130314034134402 \\
\end{array}
$$

(d) Select $p = 11$, $n = 2$, and $f(x) = x^2 - 10x - 4$. Then **a** can be represented by a 10×12 array:

$$
\begin{array}{cccccccccccc}
0 & 1 & 10 & 5 & 2 & 7 & 1 & 5 & 10 & 10 & 8 & 10 \\
0 & 7 & 4 & 2 & 3 & 5 & 7 & 2 & 4 & 4 & 1 & 4 \\
0 & 5 & 6 & 3 & 10 & 2 & 5 & 3 & 6 & 6 & 7 & 6 \\
0 & 2 & 9 & 10 & 4 & 3 & 2 & 10 & 9 & 9 & 5 & 9 \\
0 & 3 & 8 & 4 & 6 & 10 & 3 & 4 & 8 & 8 & 2 & 8 \\
0 & 10 & 1 & 6 & 9 & 4 & 10 & 6 & 1 & 1 & 3 & 1 \\
0 & 4 & 7 & 9 & 8 & 6 & 4 & 9 & 7 & 7 & 10 & 7 \\
0 & 6 & 5 & 8 & 1 & 9 & 6 & 8 & 5 & 5 & 4 & 5 \\
0 & 9 & 2 & 1 & 7 & 8 & 9 & 1 & 2 & 2 & 6 & 2 \\
0 & 8 & 3 & 7 & 5 & 1 & 8 & 7 & 3 & 3 & 9 & 3 \\
\end{array}
$$

Case 2. $q = p^r$, p a prime and $r > 1$.

(e) Construct an *m*-sequence over $\mathbb{F}_{2^2}$ of degree 3 with period 63. Let $\mathbb{F}_{2^2}$ be defined by the irreducible polynomial $h(x) = x^2 + x + 1$ and let β be a root of $h(x)$. So,

$$
\mathbb{F}_{2^2} = \{0, 1, \beta, \beta^2\}, \quad \beta^2 + \beta + 1 = 0.
$$

Let $f(x) = x^3 + x^2 + \beta^2 x + \beta$. From Appendix B of Chapter 3, $f(x)$ is a primitive polynomial over $\mathbb{F}_{2^2}$ of degree 3. Let $\mathbf{a} = \{a_i\} \in G(f)$. Then we have

$$
a_{3+k} = a_{2+k} + \beta^2 a_{1+k} + \beta a_k, \quad k = 0, 1, \ldots.
$$

The elements in the first period of **a** with an initial state $(1, 1, 1)$, represented in a 3×21 array, is given below.

$$
\begin{array}{ccccccccccccccccccccc}
1 & 1 & 1 & 0 & 1 & \beta^2 & 0 & 0 & 1 & 1 & \beta & \beta^2 & 0 & 1 & 0 & \beta^2 & 1 & \beta^2 & 1 & 1 & \beta^2 \\
\beta & \beta & \beta & 0 & \beta & 1 & 0 & 0 & \beta & \beta & \beta^2 & 1 & 0 & \beta & 0 & 1 & \beta & 1 & \beta & \beta & 1 \\
\beta^2 & \beta^2 & \beta^2 & 0 & \beta^2 & \beta & 0 & 0 & \beta^2 & \beta^2 & 1 & \beta & 0 & \beta^2 & 0 & \beta & \beta^2 & \beta & \beta^2 & \beta^2 & \beta
\end{array}
$$

(Check for the run property R-2 of nonbinary sequences!)

(f) Construct an *m*-sequence over $\mathbb{F}_{2^3}$ of degree 2 with period 63. Let $\mathbb{F}_{2^3}$ be defined by the irreducible polynomial $h(x) = x^3 + x^2 + 1$ and let β be a root of $h(x)$. So,

$$
\mathbb{F}_{2^3} = \{0, 1, \beta, \beta^2, \beta^3, \beta^4, \beta^5, \beta^6\}, \quad \beta^3 + \beta^2 + 1 = 0.
$$

Let $f(x) = x^2 + \beta^3 x + \beta$. From Appendix B of Chapter 3, $f(x)$ is a primitive polynomial over $\mathbb{F}_{2^3}$ of degree 2. Let $\mathbf{a} = \{a_i\} \in G(f)$. Then

$$
a_{2+k} = \beta^3 a_{1+k} + \beta a_k, \quad k = 0, 1, \ldots.
$$

The elements in the first period of **a** with an initial state $(0, \beta^3)$, represented by a 7×9 array, is given as follows.

$$
\begin{array}{ccccccccc}
0 & \beta^3 & \beta^6 & \beta^5 & \beta^5 & \beta^2 & \beta^3 & \beta^5 & \beta^3 \\
0 & \beta^4 & 1 & \beta^6 & \beta^6 & \beta^3 & \beta^4 & \beta^6 & \beta^4 \\
0 & \beta^5 & \beta & 1 & 1 & \beta^4 & \beta^5 & 1 & \beta^5 \\
0 & \beta^6 & \beta^2 & \beta & \beta & \beta^5 & \beta^6 & \beta & \beta^6 \\
0 & 1 & \beta^3 & \beta^2 & \beta^2 & \beta^6 & 1 & \beta^2 & 1 \\
0 & \beta & \beta^4 & \beta^3 & \beta^3 & 1 & \beta & \beta^3 & \beta \\
0 & \beta^2 & \beta^5 & \beta^4 & \beta^4 & \beta & \beta^2 & \beta^4 & \beta^2
\end{array}
$$

5.3 Interleaved structure of *m*-sequences

In this section, we show the interleaved structure and subfield decomposition of *m*-sequences. (These are the fundamental structures for the construction of GMW sequences and generalized GMW sequences.)

5.3.1 Interleaved sequences

Let $\mathbf{u} = (u_0, u_1, \ldots, u_{st-1})$ be a sequence over $\mathbb{F}_q$ of period st where both s and t are unequal to 1. We can arrange the elements of the sequence **u** into an

$s \times t$ array as follows:

$$
A = \begin{bmatrix}
u_0 & u_1 & \cdots & u_{t-1} \\
u_t & u_{t+1} & \cdots & u_{t+t-1} \\
u_{2t} & u_{2t+1} & \cdots & u_{2t+t-1} \\
\vdots & & & \\
u_{(s-1)t} & u_{(s-1)t+1} & \cdots & u_{(s-1)t+t-1}
\end{bmatrix}
$$

Definition 5.1 *If each column vector of the above array is either a phase shift of a q-ary sequence, say* **a***, of period s or a zero sequence, then we say that* **u** *is an* (s, t) *interleaved sequence over* $\mathbb{F}_q$ *(with respect to* **a***). We also say that A is the array form of* **u** *and call the jth column vector of A, A_j, the jth column sequence of* **u***.*

According to the definition, A_j is a transpose of $L^{e_j}(\mathbf{a})$ or $(0, 0, \ldots, 0)$ where e_j is a nonnegative integer. Here, we omit the transpose notation because we consider it as a sequence. Thus, we may write

$$
A_j = L^{e_j}(\mathbf{a}), 0 \le j < t, e_j \in \mathbb{Z}_{q-1} \cup \{\infty\},
$$

where we set $e_j = \infty$ if $A_j = (0, 0, \ldots, 0)$ by convention. The vector $(e_0, e_1, \ldots, e_{t-1})$ is said to be a shift sequence of **u** (with respect to **a**) and **a** a base sequence of **u**. For given **e** and **a**, the interleaved sequence **u** is uniquely determined; that is,

$$
u_{it+j} = a_{i+e_j}, \ 0 \le i < s, \ 0 \le j < t. \tag{5.20}
$$

We also denote the array form of **u** by $A(\mathbf{a}, \mathbf{e})$ in which the jth column of the matrix is given by the e_jth shift of **a**, $L^{e_j}(\mathbf{a})$.

Example 5.7 Let $\mathbf{e} = (\infty, 3, 6, 5, 5, 2, 3, 5, 3)$ and $\mathbf{a} = (1, 1, 1, 0, 1, 0, 0)$, a binary m-sequence of period 7. From $(\mathbf{a}, \mathbf{e})$, we can construct a 7×9 array A with $A_j = L^{e_j}(\mathbf{a})$:

$$
A(\mathbf{a}, \mathbf{e}) = \begin{bmatrix}
0 & 0 & 0 & 0 & 0 & 1 & 0 & 0 & 0 \\
0 & 1 & 1 & 0 & 0 & 0 & 1 & 0 & 1 \\
0 & 0 & 1 & 1 & 1 & 1 & 0 & 1 & 0 \\
0 & 0 & 1 & 1 & 1 & 0 & 0 & 1 & 0 \\
0 & 1 & 0 & 1 & 1 & 0 & 1 & 1 & 1 \\
0 & 1 & 1 & 0 & 0 & 1 & 1 & 0 & 1 \\
0 & 1 & 0 & 1 & 1 & 1 & 1 & 1 & 1
\end{bmatrix}.
$$

We read it out row by row, from left to right within a row beginning with the top-most row. Then we have the following $(7, 9)$ interleaved sequence of period 63:

u = (000001000011000101001111010001110010010110111011001101010111111).

This sequence is just the *m*-sequence of period 63 given in Example 5.5.

In general, any *m*-sequence can be written as an interleaved sequence, unless the period is a prime number.

5.3.2 Interleaved structure and subfield decomposition of *m*-sequences

Assume that **a** is an *m*-sequence over $\mathbb{F}_q$ of degree n. Let $f(x)$ be the minimal polynomial of **a**, α a root of $f(x)$ in the extension $\mathbb{F}_{q^n}$, $d = \frac{q^n-1}{q-1}$, and $\beta = \alpha^d$. Then β is a primitive element in $\mathbb{F}_q$. Using the trace representation of **a**, for $0 \le k < q^n - 1$, we write $k = id + j$ where $0 \le j < d$, $0 \le i < q - 1$. Then

$$a_k = a_{id+j} = Tr(\eta\alpha^{id+j}) = \alpha^{id}Tr(\eta\alpha^j) = \beta^i a_j;$$

that is,

$$a_{id+j} = \beta^i a_j, \ 0 \le j < d, \ 0 \le i < q - 1. \tag{5.21}$$

We arrange $\{a_i\}$ into a $(q - 1)$ by d array, $A = (a_{ij})$, whose entries are given by $a_{ij} = a_{id+j}$. Then A is the array form of the sequence **a** that has the following structures:

(a) The other row vectors are multiples of the first row. In particular, we have

$$A = \begin{bmatrix} R \\ \beta R \\ \vdots \\ \beta^{q-2}R \end{bmatrix},$$

where $R = (a_0, a_1, \ldots, a_{d-1})$, the first row of A. (This is called the projective property of *m*-sequences.)

(b) Each column vector of the array is either an *m*-sequence of degree 1 or a zero sequence, which depends on whether $Tr(\alpha^j)$ is equal to zero for $0 \le j < d$. Also all nonzero column sequences are shift-equivalent *m*-sequences. Precisely, the $(q - 1) \times d$ array is given by

$$A = \begin{bmatrix} a_0 & a_1 & \cdots & a_j & \cdots & a_{d-1} \\ \beta a_0 & \beta a_1 & \cdots & \beta a_j & \cdots & \beta a_{d-1} \\ \vdots & & & & & \\ \beta^{q-2}a_0 & \beta^{q-2}a_1 & \cdots & \beta^{q-2}a_j & \cdots & \beta^{q-2}a_{d-1} \end{bmatrix}$$
$$= (A_0, A_1, \ldots, A_{d-1}),$$

where A_j is the jth column of A, which is given by

$$A_j = (a_j, \beta a_j, \ldots, \beta^{q-2} a_j)^T, \quad 0 \leq j < d, \tag{5.22}$$

which is either an m-sequence over $\mathbb{F}_q$ of degree 1 or a zero sequence depending on whether or not a_j is zero. (Here $\mathbf{v}^T$ represents the transpose of the vector $\mathbf{v}$.)

(c) Furthermore, the base sequence of $\mathbf{a}$ is $(1, \beta, \ldots, \beta^{q-2})$ and the shift sequence of $\mathbf{a}$ is determined by

$$e_j = \begin{cases} e & \text{if } 0 \neq a_j = \beta^e \\ \infty & \text{if } a_j = 0. \end{cases} \tag{5.23}$$

Theorem 5.1 *With the above notation, there are $\frac{q^{n-1}-1}{q-1}$ zero columns in A.*

Proof. From Property 5.3, $\mathbf{a}$ is balanced. In other words, there are $q^{n-1} - 1$ zeroes in $\mathbf{a} = (a_0, a_1, \ldots, a_{q^n-2})$. Applying the assertion (b), for a fixed j, $A_j = 0$ if and only if $a_j = 0 \Longrightarrow a_{id+j} = \beta^i a_j = 0$ for $0 \leq i < q - 1$ if $a_j = 0$. Thus all zeroes in $\mathbf{a}$ are partitioned into the zero columns of A. Thus the number of all zero columns of A is $\frac{q^{n-1}-1}{q-1}$. $\qquad\square$

If n is a composite number, then we can arrange an m-sequence into an array in various ways according to different factorizations of n.

Theorem 5.2 *If n is a composite number, let m be a proper factor of n and $d = (q^n - 1)/(q^m - 1)$. Then any m-sequence over $\mathbb{F}_q$ of degree n can be arranged into a $(q^m - 1) \times d$ array where each column sequence is either an m-sequence over $\mathbb{F}_q$ of degree m or a zero sequence for which all the m-sequences are shift equivalent. Furthermore, there are*

$$\frac{q^{n-m} - 1}{q^m - 1}$$

zero columns in the array.

Proof. Let $Q = q^m$, let α be a primitive root of $\mathbb{F}_{q^n}$, and let $a_i = Tr_1^n(\gamma \alpha^i)$, a trace representation of $\mathbf{a}$. We arrange the elements of $\mathbf{a}$ into a $(Q - 1) \times d$ array $A = (a_{ij})$ in which $a_{ij} = a_{id+j}$. Using the trace representation of $\mathbf{a}$ and the transitivity property of the trace function, for $k = id + j, 0 \leq j < d$, we have

$$\begin{aligned} a_k = a_{id+j} &= Tr_1^n(\gamma \alpha^{id+j}) \\ &= Tr_1^m\left(\alpha^{id} Tr_m^n(\gamma \alpha^j)\right) \\ &= Tr_1^m(\beta^i b_j), \end{aligned}$$

where $\beta = \alpha^d$ and $b_j = Tr_m^n(\gamma\alpha^j)$, $0 \le j < d$. Note that β is a primitive element of $\mathbb{F}_Q$. Then for $b_j \ne 0$, the jth column of A is given by

$$A_j = \left(Tr_1^m(b_j), Tr_1^m(b_j\beta), \ldots, Tr_1^m(b_j\beta^{Q-2})\right)^T,$$

which is an m-sequence over $\mathbb{F}_q$ of degree m, generated by the minimal polynomial of β over $\mathbb{F}_q$. Therefore, all nonzero column vectors in the array are shift equivalent. If $b_j = 0$, then A_j is a zero sequence. This establishes the first assertion. Note that $A_j = 0$ if and only if $b_j = 0$. Thus the number of zero columns of A is equal to the number of zero b_j's for j from 0 to $d - 1$. On the other hand, $\mathbf{b} = \{b_j\}$ is an m-sequence of $\mathbb{F}_Q$ of degree $l = n/m$. From Theorem 5.1(c), this number is given by $\frac{(Q)^{l-1}-1}{Q-1} = \frac{q^{n-m}-1}{q^m-1}$, which completes the proof. $\square$

Before we provide some examples to illustrate the results in Theorems 5.1–5.2, we extend the trace operator from $F = \mathbb{F}_{q^r}$ to the linear space $V(F)$, the set consisting of all sequences over F, introduced in Section 4.2 of Chapter 4, and to $M_{r,s}(F)$, the set consisting of all $r \times s$ matrices over F. For $\mathbf{u} = \{u_i\}$, $u_i \in F$, and $A = (a_{ij})_{r \times s}$, $a_{ij} \in F$, that is, $A \in M_{r,s}(F)$, we define

$$Tr(\mathbf{u}) = (Tr(u_0), Tr(u_1), \ldots, Tr(u_i), \ldots) \tag{5.24}$$

$$Tr(A) = (Tr(a_{ij}))_{r \times s}, \tag{5.25}$$

where $Tr(x)$ is the trace function from F to $\mathbb{F}_q$ and $Tr(\mathbf{u})$ can be considered to be either a vector of length N, the period of $\mathbf{u}$, or an infinite sequence that depends on the representation of $\mathbf{u}$. From the proof of Theorem 5.2, we have

$$\mathbf{a} = Tr_1^m(\mathbf{b}) \quad \text{and} \quad A = Tr_1^m(B), \tag{5.26}$$

where A and B are $(q^m - 1) \times d$ array forms of $\mathbf{a}$ and $\mathbf{b}$, respectively. Therefore, if $\mathbf{a}$ is an m-sequence over $\mathbb{F}_q$ of degree n, then $\mathbf{a}$ can be constructed by applying the trace function from $\mathbb{F}_{q^m}$ to $\mathbb{F}_q$ to an m-sequence over $\mathbb{F}_{q^m}$ of degree l where $n = ml$.

Remark 5.2 The m-sequences $\mathbf{b}$ and $\mathbf{a}$ have the same shift sequence when both are regarded as $(q^m - 1, d)$ interleaved sequences. The sequence $\mathbf{a}$ is generated by the minimal polynomial of α, a primitive element of $\mathbb{F}_{q^n}$, over $\mathbb{F}_q$, and $\mathbf{b}$ is generated by the minimal polynomial of α over $\mathbb{F}_{q^m}$. The column sequences of $\mathbf{a}$ are either m-sequences over $\mathbb{F}_q$ of degree m or zero sequences, and those of $\mathbf{b}$ are either m-sequences over $\mathbb{F}_{q^m}$ of degree 1 or zero sequences. However, they have the same columns, which are zero sequences, because both are determined by the zero elements of $\mathbf{b}$.

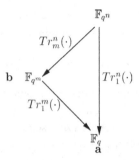

Figure 5.1. Subfield decomposition of m-sequences.

Definition 5.2 *With the notation above, Eq. (5.26) is called a* subfield decomposition *of* **a**, *and* **b** *a* subfield factor sequence *of* **a**.

The subfield decomposition of m-sequences, represented by Eq. (5.26), is a fundamental tool for the construction of the GMW sequences and their generalizations and for the classification of general 2-level autocorrelation sequences as well. Theoretically, it only utilizes the transitivity of the trace function. However, it has an immense effect in practice for the design of sequences with good correlation properties. This will be thoroughly explained in Chapter 8. The following diagram, Figure 5.1, emphasizes this relation between the subfield decomposition and the sequence composition through the trace function composition.

Example 5.8 Let $q = 2$ and $n = 6$. In this case, there are two subfields in $\mathbb{F}_{2^6}$. Let $f(x) = x^6 + x + 1$ be a primitive polynomial over $\mathbb{F}_2$ and α a root of $f(x)$ in $\mathbb{F}_{2^6}$. Let **a** be an m-sequence generated by $f(x)$ with an initial state $(0, 0, 0, 0, 0, 1)$, the same m-sequence given in Example 5.5. According to Theorem 5.2, we have two different subfield decompositions for **a**.

(a) *Case 1: $m = 3$.* In this case, the 7×9 array form of **a**, denoted by A, is shown in Example 5.5. We reproduce it here:

$$A = \begin{bmatrix} 0 & 0 & 0 & 0 & 0 & 1 & 0 & 0 & 0 \\ 0 & 1 & 1 & 0 & 0 & 0 & 1 & 0 & 1 \\ 0 & 0 & 1 & 1 & 1 & 1 & 0 & 1 & 0 \\ 0 & 0 & 1 & 1 & 1 & 0 & 0 & 1 & 0 \\ 0 & 1 & 0 & 1 & 1 & 0 & 1 & 1 & 1 \\ 0 & 1 & 1 & 0 & 0 & 1 & 1 & 0 & 1 \\ 0 & 1 & 0 & 1 & 1 & 1 & 1 & 1 & 1 \end{bmatrix}.$$

We set $d = 63/7 = 9$ and $\beta = \alpha^9$. Then β is a primitive element in $\mathbb{F}_{2^3}$ whose minimal polynomial is $h(x) = x^3 + x^2 + 1$. In this case, the subfield $\mathbb{F}_{2^3}$ sequence $\mathbf{b} = \{b_i\}$ is just the m-sequence given in Example 5.6(f), reproduced as follows:

$$B = \begin{bmatrix} 0 & \beta^3 & \beta^6 & \beta^5 & \beta^5 & \beta^2 & \beta^3 & \beta^5 & \beta^3 \\ 0 & \beta^4 & 1 & \beta^6 & \beta^6 & \beta^3 & \beta^4 & \beta^6 & \beta^4 \\ 0 & \beta^5 & \beta & 1 & 1 & \beta^4 & \beta^5 & 1 & \beta^5 \\ 0 & \beta^6 & \beta^2 & \beta & \beta & \beta^5 & \beta^6 & \beta & \beta^6 \\ 0 & 1 & \beta^3 & \beta^2 & \beta^2 & \beta^6 & 1 & \beta^2 & 1 \\ 0 & \beta & \beta^4 & \beta^3 & \beta^3 & 1 & \beta & \beta^3 & \beta \\ 0 & \beta^2 & \beta^5 & \beta^4 & \beta^4 & \beta & \beta^2 & \beta^4 & \beta^2 \end{bmatrix}.$$

Thus we can verify that

$$\mathbf{a} = Tr_1^3(\mathbf{b}) \text{ and } A = Tr_1^3(B),$$

where $Tr_1^3(x)$ is the trace function from $\mathbb{F}_{2^3}$ to $\mathbb{F}_2$, for which $Tr_1^3(\beta^i) = 1, i \in C_1 = \{1, 2, 4\}$, and $Tr_1^3(\beta^i) = 0, i \in C_3 = \{3, 5, 6\}$. Furthermore, we can verify that both A and B have the same shift sequence, $\mathbf{e} = (\infty, 3, 6, 5, 5, 2, 3, 5, 3)$, which is given by the exponents of the elements at the top row of B relative to the base β.

(b) *Case 2: m = 2*. In this case, $\mathbf{a}$ can be arranged into a 3×21 array as follows:

$$C = \begin{bmatrix} 0 & 0 & 0 & 0 & 0 & 1 & 0 & 0 & 0 & 0 & 1 & 1 & 0 & 0 & 0 & 1 & 0 & 1 & 0 & 0 & 1 \\ 1 & 1 & 1 & 0 & 1 & 0 & 0 & 0 & 1 & 1 & 1 & 0 & 0 & 1 & 0 & 0 & 1 & 0 & 1 & 1 & 0 \\ 1 & 1 & 1 & 0 & 1 & 1 & 0 & 0 & 1 & 1 & 0 & 1 & 0 & 1 & 0 & 1 & 1 & 1 & 1 & 1 & 1 \end{bmatrix}.$$

We set $d = 63/3 = 21$ and $\beta = \alpha^{21}$. Then β is a primitive element in $\mathbb{F}_{2^2}$ whose minimal polynomial over $\mathbb{F}_2$ is given by $h(x) = x^2 + x + 1$. In this case, the subfield $\mathbb{F}_{2^2}$ sequence $\mathbf{d} = \{d_i\}$ is just the sequence given in Example 5.6(e), reproduced here for easy verification of the shift sequence.

$$D = \begin{bmatrix} 1 & 1 & 1 & 0 & 1 & \beta^2 & 0 & 0 & 1 & 1 & \beta & \beta^2 & 0 & 1 & 0 & \beta^2 & 1 & \beta^2 & 1 & 1 & \beta^2 \\ \beta & \beta & \beta & 0 & \beta & 1 & 0 & 0 & \beta & \beta & \beta^2 & 1 & 0 & \beta & 0 & 1 & \beta & 1 & \beta & \beta & 1 \\ \beta^2 & \beta^2 & \beta^2 & 0 & \beta^2 & \beta & 0 & 0 & \beta^2 & \beta^2 & 1 & \beta & 0 & \beta^2 & 0 & \beta & \beta^2 & \beta & \beta^2 & \beta^2 & \beta \end{bmatrix}.$$

Thus, we have

$$\mathbf{a} = Tr_1^2(\mathbf{d}) \text{ and } C = Tr_1^2(D).$$

Both C and D have the same shift sequence given by

$$(0, 0, 0, \infty, 1, 2, \infty, \infty, 0, 0, 1, 2, \infty, 0, \infty, 2, 0, 2, 0, 0, 2),$$

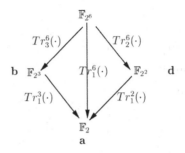

Figure 5.2. Subfield decompositions of a binary m-sequence of period 63.

which are the exponents of the elements at the top row of D relative to the base β. These two different subfield decompositions of **a** are illustrated in Figure 5.2.

5.3.3 Difference property of shift sequences of m-sequences

First, we extend the shift sequence **e** of **a**, an m-sequence over $\mathbb{F}_q$ of degree n, regarded as a $(q^m - 1, d)$ interleaved sequence where $d = (q^n - 1)/(q^m - 1)$ and $m|n$. For $k = id + j, 0 \leq j < d$ with $d \leq k < q^n - 1$, we define

$$e_k = e_{id+j} = \begin{cases} \infty & \text{if } e_j = \infty \\ e_j + i \ (\text{mod } q^m - 1) & \text{otherwise.} \end{cases} \tag{5.27}$$

From the subfield decomposition of **a**, we have $\mathbf{a} = Tr_1^m(\mathbf{b})$. Thus, $e_k = \infty$ if $b_k = 0$ and $e_k = e$ if $a_k = \beta^e$ for $k = 0, 1, \ldots$. Thus $\mathbf{e} = \{e_k\}$ has period $q^n - 1$. The following result is immediate.

Proposition 5.3 *With the above notation, assume that* **v** *is a shift of* **a**; *that is, there exists some* $0 \leq \tau < q^n - 1$ *such that* $\mathbf{v} = L^\tau(\mathbf{a})$. *Then the shift sequence of* **v**, *denoted by* **k**, *is given by*

$$\mathbf{k} = L^\tau(\mathbf{e}).$$

In other words, the shift sequence of a τ-shift of **a** is a τ-shift of its exponent sequence when **a** is regarded as an interleaved sequence.

Example 5.9 Consider the m-sequence over $\mathbb{F}_2$ of period 63 in Example 5.8 when it is regarded as a $(7, 9)$ interleaved sequence. Then its extended shift sequence is given as follows.

$$
\begin{array}{ccccccccc}
\infty & 3 & 6 & 5 & 5 & 2 & 3 & 5 & 3 \\
\infty & 4 & 0 & 6 & 6 & 3 & 4 & 6 & 4 \\
\infty & 5 & 1 & 0 & 0 & 4 & 5 & 0 & 5 \\
\infty & 6 & 2 & 1 & 1 & 5 & 6 & 1 & 6 \\
\infty & 0 & 3 & 2 & 2 & 6 & 0 & 2 & 0 \\
\infty & 1 & 4 & 3 & 3 & 0 & 1 & 3 & 1 \\
\infty & 2 & 5 & 4 & 4 & 1 & 2 & 4 & 2
\end{array}
$$

The differences $e_{j+s} - e_j$ from the shift sequences of m-sequences have an interesting property, which is stated below as a fact.

Fact 5.1 *With the above notation, for $s \not\equiv 0 \pmod{d}$, we define two sets:*

$$T_0(s) = \{e_{j+s} - e_j \pmod{q^m - 1} \mid e_{j+s} \neq \infty \text{ and } e_j \neq \infty, 0 \leq j < d\} \tag{5.28}$$

$$T_\infty(s) = \{j \mid e_{j+s} = \infty \text{ and } e_j = \infty, 0 \leq j < d\}. \tag{5.29}$$

Then each element in $\mathbb{Z}_{q^m-1}$ occurs q^{n-2m} times in $T_0(s)$ and

$$|T_\infty(s)| = (q^{n-2m} - 1)/(q^m - 1)$$

for every s with $s \not\equiv 0 \pmod{d}$.

This property has found several applications for constructions of sonar sequences, Costas arrays, and interleaved signal sets. There are two approaches to proving this property. One is to use the concepts of the Singer cyclic Hadamard difference sets and projective hyperplanes over finite fields. The other is to use an earlier result of Zierler (1959). (Zierler's result will be introduced in Section 5.6.)

For simplicity of presenting these differences, we introduce the following notation. Let

$$T(s) = \{e_{j+s} - e_j \mid 0 \leq j < d\},$$

where the differences are defined as

$$
e_{j+s} - e_j = \begin{cases}
e_{j+s} - e_j \pmod{q^m - 1} & \text{if both } e_{j+s} \text{ and } e_j \text{ are not equal to } \infty \\
* & \text{if one of } e_{j+s} \text{ and } e_j \text{ is equal to } \infty \\
\infty & \text{if both } e_{j+s} \text{ and } e_j \text{ are equal to } \infty.
\end{cases}
$$

Thus, $T_0(s)$ contains the integer numbers in $T(s)$ and $T_\infty(s)$ contains the indices that correspond to the symbol ∞ in $T(s)$.

Example 5.10 Let $q = 2$ and $m = 3$. We compute $T(s)$ for the binary m-sequence of period 63 whose shift sequence **e** is given in Example 5.9. In the following table, for each s, $T_0(s)$ consists of the integer numbers appearing in $T(s)$, and $T_\infty(s)$ is empty, because there is no ∞ in $T(s)$.

s	$T(s)$								
1	*	3	6	0	4	1	2	5	*
2	*	2	6	4	5	3	0	*	1
3	*	2	3	5	0	1	*	6	4
4	*	6	4	0	5	*	1	2	3
5	*	0	6	5	*	2	4	1	3
6	*	2	4	*	6	5	3	1	0
7	*	0	*	6	2	4	3	5	1
8	*	*	5	2	1	4	0	6	3

According to Fact 5.1, each element in $\mathbb{Z}_7$ occurs once in $T_0(s)$ for $s : 1 \le s < 9$. The above table verifies this result.

Example 5.11 Let $n = 9$, $m = 3$, $v = 2^3 - 1 = 7$, and $d = (2^9 - 1)/7 = 73$. Let **u** be an m-sequence of degree 9 generated by the primitive polynomial $c(x) = x^9 + x^4 + 1$ over $\mathbb{F}_2$. Hence, it can be arranged into a $(7, 73)$ interleaved structure, shown below, where the base sequence, denoted by **a**, is given by **a** $= (1001011)$, and the shift sequence **e** is given below for the first $d = 73$ elements. (Note that the other elements in the extended **e** can be computed by (5.27).)

M-sequence of period 511

u $\leftrightarrow$ A(**a**, **e**), **a** $= (1001011)$
100001000110000100111001010101100001101111010011011100100010100001010101
001111110110010010010110111111100100110101001100110000000110001100101000 11
010010111111101000101100011101011001011001111000111110111010000011010101 01
101110110000010110101111101010101000000101001010111100101110111000000 1110
011101001001110101110101000100100001100111000010111101101100110100000 1110
111100001111111110000011111011111000101110011001000000100101001101101010 0011
110011111001101100010101001000111000110110101010111000100110001000100000000

Shift sequence of the m-sequence of period 511

0	1	2	2	4	6	4	4	1	5	5	1	1	2	1	0	2	∞	3	6
3	4	2	0	2	5	4	5	2	5	0	1	4	∞	∞	5	6	1	5	0
6	3	1	5	4	∞	0	6	4	3	3	3	1	∞	3	1	4	2	3	∞
0	2	2	∞	1	5	∞	5	∞	3	3	2	5							

For s with $1 \le s \le 5$, we list the elements in $T(s)$ in the following tables. Thus, $T_0(s)$ consists of all the integer numbers in $T(s)$ and $T_\infty(s)$ consists of those indices that correspond to ∞ in $T(s)$. For example, we have $T_\infty(1) = \{23\}$ since both e_{24} and e_{23} are equal to ∞. We can verify that each element in $\mathbb{Z}_7$ occurs eight times in $T_0(s)$ and ∞ occurs once in $T_\infty(s)$ for the listed $s = 1, 2, 3, 4,$ and 5.

Elements in $T(s)$

$T(1)$

1	1	0	2	2	5	0	4	4	0	3	0	1	6	6	2	*	*	3	4
1	5	5	2	3	6	1	4	3	2	1	3	*	∞	*	1	2	4	2	6
4	5	4	6	*	*	6	5	6	0	0	5	*	*	5	3	5	1	*	*
2	0	*	*	4	*	*	*	*	0	6	3	3							

$T(2)$

2	1	2	4	0	5	4	1	4	3	3	1	0	5	1	*	1	*	0	5
6	3	0	5	2	0	5	0	5	3	4	*	*	*	*	3	6	6	1	3
2	2	3	*	3	*	4	4	6	0	5	*	2	*	1	1	6	*	4	*
2	*	6	*	*	0	∞	5	*	6	2	6	4							

$T(3)$

2	3	4	2	0	2	1	1	0	3	4	0	6	0	*	3	4	*	1	3
4	5	3	4	3	4	1	2	6	6	*	*	1	*	*	0	1	5	5	1
6	1	*	2	2	*	3	4	6	5	*	0	0	*	6	2	*	5	6	*
*	6	3	∞	4	*	*	5	*	2	5	0	5							

$T(4)$

4	5	2	2	4	6	1	4	0	4	3	6	1	*	2	6	1	*	6	1
6	1	2	5	0	0	3	3	2	*	*	4	2	*	*	2	0	2	3	5
5	*	6	1	0	*	3	4	4	*	0	5	3	*	0	*	3	0	6	∞
1	3	*	*	*	5	*	4	*	5	6	1	5							

$T(5)$

6	3	2	6	1	6	4	4	1	3	2	1	*	1	5	3	2	*	4	3
2	0	3	2	3	2	4	6	*	*	5	5	4	*	*	1	4	0	0	4
*	4	5	6	6	*	3	2	*	0	5	1	1	*	*	6	5	0	*	*
5	*	3	∞	2	5	*	0	*	6	0	1	0							

5.4 Trinomial property

In this section, we discuss a characteristic of binary m-sequences that is called the shift-and-add property and introduce the concept of trinomial pairs for any nonlinear binary sequences of period N.

Theorem 5.3 (SHIFT-AND-ADD PROPERTY FOR BINARY SEQUENCES) *If* **a** *is a binary sequence of period N such that either* $L^i\mathbf{a} + L^j\mathbf{a} = L^k\mathbf{a}$ *or* $L^i\mathbf{a} + L^j\mathbf{a} = 0$, *then there exists some n such that* $N = 2^n - 1$ *and* **a** *is an m-sequence of period* $2^n - 1$. *The converse is also true. In other words, if* **a** *is a binary m-sequence of period* $2^n - 1$, *then for each pair* $(i, j) : 0 \le i, j \le 2^n - 1$, *either there exists some* $k : 0 \le k \le 2^n - 2$ *such that*

$$L^i\mathbf{a} + L^j\mathbf{a} = L^k\mathbf{a}$$

or $L^i\mathbf{a} + L^j\mathbf{a} = 0$.

Proof. Let $S = \{L^i\mathbf{a} \mid 0 \le i \le N - 1\} \cup \{0\}$. Then S is closed under addition. Since the only scalars in $\mathbb{F}_2$ are 0 and 1, S is also closed under multiplication by scalars. Therefore, S is a subspace of $V(\mathbb{F}_2)$. Let n be the dimension of S. Then $N + 1 = 2^n \implies N = 2^n - 1$. Let $f(x)$ be the minimal polynomial of **a** over $\mathbb{F}_2$, and let $\deg(f) = m$. Note that $S \subset G(f) \implies n \le m$. On the other hand, $\{\mathbf{a}, L\mathbf{a}, \ldots, L^{m-1}\mathbf{a}\} \subset G(f)$, which are linearly independent over $\mathbb{F}_2$. So, $m \le n$. Thus $m = n$. Because $\operatorname{per}(f) = \operatorname{per}(\mathbf{a}) = 2^n - 1$, $f(x)$ is primitive over $\mathbb{F}_2$. Therefore **a** is an m-sequence of period $2^n - 1$.

Conversely, if $i = j$, then $L^i\mathbf{a} + L^j\mathbf{a} = 0$. We now assume that $i \ne j$. Without loss of generality, we may assume $i < j$. Note that $L^i\mathbf{a} + L^j\mathbf{a} = L^i(\mathbf{a} + L^{j-i}\mathbf{a})$, which is a shift of $\mathbf{a} + L^{j-i}\mathbf{a}$. So, it suffices to show that $\mathbf{a} + L^j\mathbf{a}$ is a shift of **a**. Let $f(x)$ be the minimal polynomial of **a** over $\mathbb{F}_2$. Then $G(f)$ can be written as

$$G(f) = \{L^i\mathbf{a} \mid 0 \le i \le 2^n - 2\} \cup \{0\}.$$

Because $G(f)$ is a linear space, $\mathbf{a} + L^j\mathbf{a} \in G(f)$. Thus, there exists some $k : 0 \le k \le 2^n - 2$ such that $\mathbf{a} + L^j\mathbf{a} = L^k\mathbf{a}$.   □

Theorem 5.4 (SHIFT-AND-ADD PROPERTY FOR NONBINARY m-SEQUENCES) *Let* **a** *be an m-sequence over* $\mathbb{F}_q$ *of period* $q^n - 1$. *Then for each pair* $(i, j) : 0 \le i, j \le q^n - 1$, *either there exists some* $k : 0 \le k \le q^n - 2$ *such that*

$$L^i\mathbf{a} + L^j\mathbf{a} = L^k\mathbf{a}$$

or $L^i\mathbf{a} + L^j\mathbf{a}$ *is the zero sequence.*

This can be shown using a similar process to the one in the second part of the proof of Theorem 5.3, so we omit it.

Note. A sequence over $\mathbb{F}_q$ of period N that satisfies the shift-and-add property when $q > 2$ may not be an m-sequence. (See Gong, Di Porto, and Wolfowicz (1993) for such counterexamples.)

From Theorem 5.3, for $\tau \not\equiv 0 \pmod{2^n - 1}$, $\mathbf{a} + L^{\tau}\mathbf{a}$ is a shift of $\mathbf{a}$, say $L^{\tau'}\mathbf{a}$. Because $\mathbf{a}$ satisfies R-1, so do the shifts of $\mathbf{a}$. Therefore, there are 2^{n-1} 1's occurring in one period of $L^{\tau'}\mathbf{a}$ and $2^{n-1} - 1$ 0's occurring in one period of it. Thus, $C_{\mathbf{a}}(\tau)$, the autocorrelation of $\mathbf{a}$, is equal to -1. This is another proof for binary m-sequences satisfying the randomness postulate R-3.

According to Theorem 5.3, a binary m-sequence $\mathbf{a} = \{a_i\}$ of period $N = 2^n - 1$ is characterized by the shift-and-add property: for each τ, $0 \le \tau \le N$ there is a corresponding τ' such that $\{a_i\} + \{a_{i+\tau}\} = \{a_{i+\tau'}\}$. The N cyclic shifts of $\{a_i\}$ together with the zero vector of length N form an n-dimensional subspace of $V(\mathbb{F}_2)$. In the light of this property of m-sequences, we give the following definition for general binary sequences with period $2^n - 1$.

Definition 5.3 *We say that a binary sequence* $\mathbf{a} = \{a_i\}$ *of period* $N = 2^n - 1$ *has the* trinomial property *if there is (at least) one pair of positive integers,* τ *and* τ', *such that*

$$\{a_i\} + \{a_{i+\tau}\} = \{a_{i+\tau'}\}, \quad or \quad \mathbf{a} + L^{\tau}\mathbf{a} = L^{\tau'}\mathbf{a}, \tag{5.30}$$

where $0 \le \tau, \tau' \le 2^n - 2$. *(Clearly such a sequence has linear span less than or equal to* τ' *if* $\tau < \tau'$ *and corresponds to the trinomial* $x^{\tau'} + x^{\tau} + 1$.*)* (τ, τ') *is called a* trinomial pair *of* $\mathbf{a}$.

For example, for $\mathbf{a} = (011111101110100)$, we have

$$\mathbf{a} = (011111101110100)$$
$$L^5(\mathbf{a}) = (110111010001111)$$
$$\mathbf{a} + L^5(\mathbf{a}) = (101000111111011) = L^{10}(\mathbf{a}).$$

Thus $(5, 10)$ is a trinomial pair of $\mathbf{a}$. However, $\mathbf{a}$ is not an m-sequence of period 15. In fact, the minimal polynomial of $\mathbf{a}$ is given by $(x^4 + x + 1)$ $(x^2 + x + 1) = x^6 + x^5 + x^4 + x^3 + 1$.

If a binary sequence does not possess the trinomial property, then no trinomials are multiples of the minimal polynomial of the sequence. One type of correlation attack is to make use of trinomial pairs of binary sequences. (See Zeng and Huang (1990) and Meier and Staffelbach (1988).) Note that any binary m-sequence of period $2^n - 1$ has (τ, τ') as its trinomial pair for τ taking all integers in $\mathbb{Z}_{2^n-1}$ where τ' is determined by τ. Thus the number of trinomial pairs of a binary sequence is inversely related to the nonlinearity of the sequence. In other words, more trinomial pairs imply less nonlinearity.

We conclude this section by providing the generalized shift-and-add property that characterizes the nonbinary m-sequences.

Generalized shift-and-add property for nonbinary m-sequences

Let $\mathbf{a}$ be a sequence over $\mathbb{F}_q$ of period $N = q^n - 1$. Then $\mathbf{a}$ is an m-sequence if and only if for each pair $(i, j) : 0 \leq i, j \leq q^n - 1$, and $c, d \in \mathbb{F}_q$, either there exists some $k : 0 \leq k \leq q^n - 2$ such that

$$cL^i\mathbf{a} + dL^j\mathbf{a} = L^k\mathbf{a}$$

or $cL^i\mathbf{a} + dL^j\mathbf{a}$ is equal to zero. In other words, $\mathbf{a}$ is an m-sequence if and only if the N cyclic shifts of $\mathbf{a}$ together with the zero sequence form an n-dimensional linear space over $\mathbb{F}_q$. We say that a sequence $\mathbf{a} = \{a_i\}$ over $\mathbb{F}_q$ of period $N = q^n - 1$ has the trinomial property if there is (at least) one pair of positive integers, τ and τ', and some scalers c and d in $\mathbb{F}_q$, such that

$$c\{a_i\} + d\{a_{i+\tau}\} = \{a_{i+\tau'}\} \ \text{ or } \ c\mathbf{a} + dL^\tau\mathbf{a} = L^{\tau'}\mathbf{a}, \tag{5.31}$$

where $0 \leq \tau, \tau' \leq q^n - 2$. Similar to the binary case, such a sequence has linear span less than or equal to τ' if $\tau < \tau'$ and corresponds to the trinomial $x^{\tau'} - dx^\tau - c$. (τ, τ') is called a trinomial pair of $\mathbf{a}$.

5.5 Constant-on-cosets property

In this section, we present the constant-on-coset property of m-sequences.

Theorem 5.5 *Let* $\mathbf{a} = (a_0, \ldots, a_{q^n-2})$ *be an m-sequence over* $\mathbb{F}_q$ *of degree n. Then there exists some* τ, $0 \leq \tau < q^n - 2$, *such that*

$$a_{\tau+q\cdot j} = a_{\tau+j}, \forall j = 0, 1, \ldots.$$

Proof. We use the trace representation of m-sequences to show this property. Let $f(x)$ be the minimal polynomial of $\mathbf{a}$ over $\mathbb{F}_q$, and let α be a root of $f(x)$ in $\mathbb{F}_{q^n}$. Then the elements of $\mathbf{a}$ can be represented by

$$a_i = Tr(\eta\alpha^i), i = 0, 1, \ldots, \ \text{ and } \eta \in \mathbb{F}_{q^n}^*,$$

where $Tr(x)$ is the trace function from $\mathbb{F}_{q^n}$ to $\mathbb{F}_q$. When $\eta = 1$, notice that $Tr(x^q) = Tr(x)$ (Theorem 3.11(e) in Chapter 3). We have

$$a_{qi} = Tr(\alpha^{qi}) = Tr((\alpha^i)^q) = Tr(\alpha_i)^q = a_i, i = 0, 1, \ldots;$$

that is,

$$a_{qi} = a_i, \quad i = 0, 1, \dots .$$

In general, if $\eta = \alpha^{\tau'}$ where $\tau' \neq 0$, we take $\tau \equiv -\tau' \pmod{q^n - 1}$. It follows that

$$a_{\tau+i} = Tr(\alpha^{\tau'} \alpha^{-\tau'} \alpha^i) = Tr(\alpha^i), \quad i = 0, 1, \dots .$$

From the discussion of the first case, we have

$$a_{\tau+qi} = a_{\tau+i}, \quad i = 0, 1, \dots . \qquad \Box$$

Definition 5.4 *Let* **b** *be a sequence over* $\mathbb{F}_q$ *of period* $q^n - 1$. *If there exists some* τ *such that* $b_{\tau+qi} = b_{\tau+i}, i = 0, 1, \dots$, *that is,* $L^{\tau}(\mathbf{b})$ *takes a fixed value on every cyclotomic coset modulo* $q^n - 1$, *then we say that* **b** *satisfies the* constant-on-cosets *property,* τ *is a* characteristic phase *of* **b**, *and* $L^{\tau}(\mathbf{b})$ *is constant-on-cosets.*

From Theorem 5.5, every m-sequence satisfies the constant-on-cosets property. We will extend this property to all 2-level autocorrelation sequences in Chapter 7. Applying Theorem 5.5 and Definition 5.4, the following result is immediate.

Corollary 5.3 *Let* α *be a root of* $f(x)$ *in* $\mathbb{F}_{q^n}$, *and let* **a** *be an* m-sequence over $\mathbb{F}_q$ *of degree* n *whose elements are given by*

$$a_i = Tr(\eta \alpha^i), i = 0, 1, \dots, \text{ and } \eta \in \mathbb{F}_{q^n}^*.$$

Then **a** *is constant-on-cosets if and only if* $\eta \in \mathbb{F}_q^*$. *In particular, if* $q = 2$, *then* **a** *is constant-on-cosets if and only if* $\eta = 1$.

Example 5.12 (a) For $n = 4$ and $p = 2$, the cyclotomic cosets mod 15 are

$$C_0 = \{0\}$$
$$C_1 = \{1, 2, 4, 8\}$$
$$C_3 = \{3, 6, 12, 9\}$$
$$C_5 = \{5, 10\}$$
$$C_7 = \{7, 14, 13, 11\}$$

and the coset leaders are $\Gamma_2(4) = \{0, 1, 3, 5, 7\}$.

Let $\mathbf{a} = (011110101100100)$ be an m-sequence generated by $f(x) = x^4 + x^3 + 1$. We can verify that

$$a_i = \begin{cases} 1 & \text{if } i \in C_1 \cup C_3 \\ 0 & \text{if } i \in C_0 \cup C_5 \cup C_7. \end{cases}$$

Thus $\mathbf{a}$ is constant-on-cosets.

(b) For $n = 5$ and $p = 2$, the cyclotomic cosets modulo 31 are

$$C_0 = \{0\}$$
$$C_1 = \{1, 2, 4, 8, 16\}$$
$$C_3 = \{3, 6, 12, 24, 17\}$$
$$C_5 = \{5, 10, 20, 9, 18\}$$
$$C_7 = \{7, 14, 28, 25, 19\}$$
$$C_{11} = \{11, 22, 13, 26, 21\}$$
$$C_{15} = \{15, 30, 29, 27, 23\}.$$

The set $\Gamma_2(5)$, consisting of all coset leaders modulo 31, is given by

$$\Gamma_2(5) = \{0, 1, 3, 5, 7, 11, 15\}.$$

For $\mathbf{a} = (1001001000011101010001111011011)$, which is a quadratic residue sequence modulo 31, by inspection we have

$$a_i = \begin{cases} 1 & \text{if } i \in C_0 \cup C_3 \cup C_{11} \cup C_{15} \\ 0 & \text{if } i \in C_1 \cup C_5 \cup C_7. \end{cases}$$

Thus $\mathbf{a}$ is constant-on-cosets. For $\mathbf{b} = (1111101110001010110100001100100)$, which is an m-sequence generated by $f(x) = x^5 + x^4 + x^3 + x^2 + 1$, we have

$$b_i = \begin{cases} 1 & \text{if } i \in C_0 \cup C_1 \cup C_3 \cup C_7 \\ 0 & \text{if } i \in C_5 \cup C_{11} \cup C_{15}. \end{cases}$$

Thus $\mathbf{b}$ is constant-on-cosets.

(c) For $n = 6$ and $p = 2$, the cyclotomic cosets modulo 63 are given by

$$C_0 = \{0\}$$
$$C_1 = \{1, 2, 4, 8, 16, 32\}$$
$$C_3 = \{3, 6, 12, 24, 48, 33\}$$
$$C_5 = \{5, 10, 20, 40, 17, 34\}$$
$$C_7 = \{7, 14, 28, 56, 49, 35\}$$
$$C_{11} = \{11, 22, 44, 25, 50, 37\}$$
$$C_{13} = \{13, 26, 52, 41, 19, 38\}$$
$$C_{15} = \{15, 30, 60, 57, 51, 39\}$$

$$C_{23} = \{23, 46, 29, 58, 53, 43\}$$
$$C_{31} = \{31, 62, 61, 59, 55, 47\}$$
$$C_9 = \{9, 18, 36\}$$
$$C_{27} = \{27, 54, 45\}$$
$$C_{21} = \{21, 42\}.$$

For the m-sequence in Example 5.5, we have

	i													
Coset leaders of C_i	0	1	3	5	7	11	13	15	23	31	9	27	21
a_i	0	0	0	1	0	1	0	1	1	1	0	0	1

Thus **a** is constant-on-cosets.

From Corollary 5.3, to generate an m-sequence with the characteristic phase, we should use the initial state $(Tr(1), Tr(\alpha), \ldots, Tr(\alpha^{n-1}))$ for $q = 2$ and $(Tr(\eta), Tr(\eta\alpha), \ldots, Tr(\eta\alpha^{n-1}))$, $\eta \in \mathbb{F}_q^*$ for $q > 2$. This method involves computation in $\mathbb{F}_{q^n}$. In the following, we give a method for finding the characteristic phase of an m-sequence without involving computation in $\mathbb{F}_{q^n}$.

Theorem 5.6 *Let $f(x)$ be a primitive polynomial over $\mathbb{F}_q$ with degree n and* **a** $\in G(f)$; *that is,* **a** *is an m-sequence generated by $f(x)$. Let*

$$\mathbf{b} = \sum_{k=0}^{n-1} \mathbf{a}^{q^k}.$$

Then **b** *is constant-on-cosets if $a_0 \neq 0$.*

Proof. Let α be a root of $f(x)$ in $\mathbb{F}_{q^n}$. Then the trace representation of **a** is given by

$$a_i = Tr(\eta\alpha^i), \quad i = 0, 1, \ldots, \quad \text{and } \eta \in \mathbb{F}_{q^n}^*. \tag{5.32}$$

We denote by $\mathbf{u} = \{u_i\}$ the sequence when $\eta = 1$. Thus, **u** is constant-on-cosets from Corollary 5.3. Note that b_i, the ith element of **b**, is equal to the sum of $a_i, a_{qi}, \ldots, a_{q^{n-1}i}$. Using the trace property that $Tr(xy) = xTr(y)$ for $x \in \mathbb{F}_q$, $y \in \mathbb{F}_{q^n}$ (Theorem 3.11 in Section 3.5 of Chapter 3), we have

$$b_i = \sum_{k=0}^{n-1} a_{iq^k} = \sum_{k=0}^{n-1} Tr(\eta\alpha^{iq^k}) = Tr\left(\eta \sum_{k=0}^{n-1} \alpha^{iq^k}\right)$$
$$= Tr(\eta Tr(\alpha^i)) \overset{(5.32)}{=} Tr(\eta u_i) = u_i Tr(\eta) = u_i a_0.$$

Thus $\mathbf{b} = a_0\mathbf{u}$. If $a_0 \neq 0$, then **b** is constant-on-cosets because multiplying a nonzero constant does not change the constant-on-coset property of **u**. $\square$

Usually, we may choose $(1, 0, \ldots, 0)$ as an initial state of the LFSR $f(x)$ and compute

$$b_i = \sum_{k=0}^{n-1} a_{iq^k}, i = 0, 1, \ldots, n - 1.$$

Using $(b_0, b_1, \ldots, b_{n-1})$ as an initial state of the LFSR f, the generated m-sequence is constant-on-cosets.

5.6 Two-tuple balance property

In this section, we present an important property of m-sequences that plays an essential role in determining autocorrelations of GMW sequences and their generalizations.

Definition 5.5 (Two-Tuple Balance Property) *Let* $\mathbf{a} = \{a_i\}$ *be a sequence over* $\mathbb{F}_q$ *of period* $N = q^n - 1$, *and let* $d = (q^n - 1)/(q - 1)$,

$$T = \{(a_i, a_{i+\tau}) | 0 \le i < q^n - 1\}$$

and

$$N_{\lambda,\mu}(\tau) = |\{i \mid (a_i, a_{i+\tau}) = (\lambda, \mu), 0 \le i < q^n - 1\}|, \lambda, \mu \in \mathbb{F}_q.$$

We say that $\mathbf{a}$ *has the* 2-*tuple balance property if* $\mathbf{a}$ *satisfies the following conditions:*

(a) If $\tau \not\equiv 0 \pmod{d}$, *then*

$$N_{\lambda,\mu}(\tau) = q^{n-2} \;\; \forall(\lambda, \mu) \ne (0, 0)$$
$$N_{0,0}(\tau) = q^{n-2} - 1.$$

(b) If $\tau \equiv 0 \pmod{d}$, *that is,* $\tau = jd$, $j = 0, 1, \ldots$, *then there exists some* $\mu \in \mathbb{F}_q$ *such that*

$$(a_i, a_{i+jd}) = (\lambda, \mu\lambda), \forall i, \mu \ne 1 \; if \; \tau \ne 0$$

and

$$N_{\lambda,\mu\lambda}(jd) = q^{n-1} \;\; \forall \lambda \ne 0$$
$$N_{0,0}(jd) = q^{n-1} - 1.$$

In particular, if $q = 2$, then $d = 2^n - 1$. Thus, $\mathbf{a}$ is 2-tuple balanced if and only if $N_{\lambda,\mu}(\tau) = 2^{n-2}$ when $(\lambda, \mu) \neq (0, 0)$ and $N_{0,0}(\tau) = 2^{n-2} - 1$ for every $\tau \not\equiv 0 \pmod{2^n - 1}$.

Theorem 5.7 (Zierler, 1959) *Every m-sequence over $\mathbb{F}_q$ satisfies the 2-tuple balance property.*

Proof. Let $\mathbf{a}$ be an m-sequence over $\mathbb{F}_q$ of degree n. Without loss of generality, we may assume that the elements of $\mathbf{a}$ are given by

$$a_i = Tr(\alpha^i), i = 0, 1, \ldots,$$

where α is a primitive element of $\mathbb{F}_{q^n}$. Then

$$a_{i+\tau} = Tr(\alpha^\tau \alpha^i), i = 0, 1, \ldots.$$

Therefore, T can be rewritten as

$$T = \{(Tr(x), Tr(\beta x)) \mid x \in \mathbb{F}_{q^n}\}, \text{ where } \beta = \alpha^\tau.$$

For any $(\lambda, \mu) \in \mathbb{F}_q \times \mathbb{F}_q$, we consider the following system of equations:

$$\begin{cases} Tr(x) = \lambda \\ Tr(\beta x) = \mu. \end{cases} \tag{5.33}$$

Note that Eq. (5.33) is a system of two linear equations in n variables $x, x^q, \ldots, x^{q^{n-1}}$. So, the coefficient matrix of Eq. (5.33), denoted by A, is given by

$$A = \begin{bmatrix} 1 & 1 & \cdots 1 \\ \beta & \beta^q & \cdots \beta^{q^{n-1}} \end{bmatrix}.$$

The rank of A, denoted by Rank(A), is equal to either 1 or 2 depending on whether or not β belongs to $\mathbb{F}_q$. In detail, Eq. (5.33) has q^{n-2} solutions in $\mathbb{F}_{q^n}$ if and only if

$$\text{Rank}(A) = 2 \iff \beta \notin \mathbb{F}_q$$
$$\iff \tau \not\equiv 0 \pmod{d}.$$

On the other hand, Eq. (5.33) has q^{n-1} solutions in $\mathbb{F}_{q^n}$ if and only if

$$\text{Rank}(A) = 1 \iff \beta \in \mathbb{F}_q$$
$$\iff \tau \equiv 0 \pmod{d},$$

which establishes the assertion. $\qquad\square$

In the following, we will show that for any q, the 2-tuple balance property is a sufficient condition for a sequence to have a 2-level autocorrelation function.

Proposition 5.4 *Assume that* **a** *is a sequence over* $\mathbb{F}_q$ *with period* $q^n - 1$. *If* **a** *satisfies the 2-tuple balance property, then it has 2-level autocorrelation.*

Proof. Notice that

$$C(\tau) = \sum_{i=0}^{q^n-2} \chi(a_{i+\tau})\chi^*(a_i)$$

$$= \sum_{i=0}^{q^n-2} \chi(a_{i+\tau} - a_i). \tag{5.34}$$

According to Condition 1 in Definition 5.5, if $\tau \not\equiv 0 \pmod{d}$, then Eq. (5.34) becomes

$$C(\tau) = q^{n-2} \sum_{\lambda,\mu \in \mathbb{F}_q} \chi(\mu - \lambda) - 1$$

$$= q^{n-2} \sum_{\mu \in \mathbb{F}_q} \chi(\mu) \sum_{\lambda \in \mathbb{F}_q} \chi(-\lambda) - 1$$

$$= -1.$$

The last identity follows from Lemma 5.1 in Section 5.2. If $\tau \equiv 0 \pmod{d}$, applying Condition 2 in Definition 5.5 to Eq. (5.34), we have

$$C(\tau) = q^{n-1} \sum_{\lambda \in \mathbb{F}_q} \chi(\mu\lambda - \lambda) - 1, 1 \neq \mu \in \mathbb{F}_q$$

$$= q^{n-1} \sum_{\lambda \in \mathbb{F}_q} \chi(\lambda(\mu - 1)) - 1$$

$$= -1.$$

The last identity also follows from Lemma 5.1 because $\mu \neq 1$. $\square$

This is another proof for the assertion that every m-sequence has 2-level autocorrelation. For the binary case, this is the third proof for m-sequences satisfying R-3. For $q = 2$, the 2-tuple balance property characterizes 2-level autocorrelation sequences. In other words, a binary sequence of period $2^n - 1$ has a 2-level autocorrelation function if and only if it satisfies the 2-tuple balance property (see Exercise 5 for its being necessary). However, if $q > 2$, it is an open problem whether any 2-level autocorrelation sequence satisfies the 2-tuple balance property. Gong and Song (2002) recently showed that if a sequence over $\mathbb{F}_q$ with $q > 2$ has 2-level autocorrelation and an interleaved structure, then it satisfies the 2-tuple balance property.

5.7 Classification of binary sequences of period $2^n - 1$

Every binary m-sequence of period $2^n - 1$ has linear span n. Therefore, the normalized linear span is $n/(2^n - 1)$, which converges to zero as $n \to \infty$. Investigations of using m-sequences or LFSRs with primitive characteristic polynomials as basic blocks for the constructions of nonlinear sequences with the randomness properties R-1 to R-4, large periods, and large linear spans have been carried out since at least the 1970s, especially for the design of secure communication systems. However, there are certain constraints on sequences to achieve good randomness properties. Up to now, no sequences have been found that possess all the randomness properties listed at the end of Section 5.1. In the following, we introduce a classification of binary sequences of period $2^n - 1$ that was studied by Golomb (1980).

Let U be the set of all binary sequences of period $N = 2^n - 1$ that contain $(N + 1)/2$ ones and $(N - 1)/2$ zeroes in each period, i.e., U consists of all binary sequences with the randomness property R-1. Let PN be the subset of U consisting of the m-sequences of period N. Some of the sets that are intermediate between U and PN are the following.

- R, the subset of U consisting of those sequences with the run property R-2;
- S, the subset of U consisting of those sequences with the randomness property R-4, the ideal n-tuple distribution, or span n property;
- C, the subset of U consisting of those sequences with the 2-level autocorrelation property R-3;
- M, the subset of U consisting of those sequences that satisfy the constant-on-cosets property. Another characterization of M is that M is that subset of U consisting of those sequences having two as a multiplier. That is, for some cyclic shift $\mathbf{a}' = (a_0', a_1', \ldots, a_{r-1}')$ of $\mathbf{a} = (a_0, a_1, \ldots, a_{r-1})$, we have $a_{2i}' = a_i'$ for all i. Because 2 is a multiplier for all Hadamard difference sets, we have $C \subset M$. (This relation will be discussed in Chapter 7 in detail.)

The hierarchy of inclusions among these sets of sequences is shown in a lattice diagram in Figure 5.3. By enumeration of elements in these sets, we have the following results.

$$|U| = \binom{2^n - 1}{2^{n-1}} / (2^n - 1)$$
$$|PN| = \phi(2^n - 1)/n$$
$$|S| = 2^{2^{n-1} - n}$$

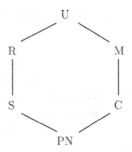

Figure 5.3. Lattice hierarchy for subsets of U.

If $n = p$ is a prime number, we can also determine that

$$|M| = \binom{2t}{t} \quad \text{where } t = \frac{2^{p-1} - 1}{p}.$$

All sequences in S can be obtained from the set B, consisting of all de Bruijn sequences of period 2^n, by deleting one 0 from the run of 0's of length n. (The sequences in S are also called modified de Bruijn sequences.) It was first established in the 1890s that $|B| = 2^{2^{n-1}-n}$ (see Golomb 1982). Chan, Games, and Key (1982) proved that the linear span of any de Bruijn sequence of period 2^n, denoted by L, is bounded by

$$2^{n-1} + n \le L \le 2^n - 1, \tag{5.35}$$

where both the lower bound and the upper bound are achievable (Etzion and Lempel 1984). Therefore any de Bruijn sequence has large linear span and satisfies R-4, so that R-1 and R-2, but not R-3, are satisfied. However, for a modified de Bruijn sequence, that is, a sequence in S, the lower bound of the linear span of the sequence dramatically drops to

$$n \le L \le 2^n - 2. \tag{5.36}$$

No theoretical results on the linear spans of the sequences in S, except for the PN set, have been established. However, experimental results show that the linear span of sequences in S varies in the range from n to $2^n - 2$. In the following, we give some examples to show the linear spans of de Bruijn sequences of period 16 and their corresponding sequences of period 15 in S.

Example 5.13 Examples of de Bruijn sequences with period 16.

(a) We take a de Bruijn sequence of period 16 as follows:

$$\mathbf{a} = (1101011110010000).$$

The sequence **a** has linear span 15, which is the maximum in Eq. (5.35) and the minimal polynomial of the sequence is $\sum_{i=0}^{15} x^i$. By removing one 0 from the run of 0's of length 4, we then get a sequence **b** $\in S$,

$$\mathbf{b} = (110101111001000),$$

which has linear span 14 and the minimal polynomial $\sum_{i=0}^{14} x^i$. This also achieves the maximum linear span for modified de Bruijn sequences of period 15.

(b) The following is another de Bruijn sequence of period 16,

$$(1001011110000110),$$

which has linear span 12 and the minimal polynomial $(x^3 + x^2 + x + 1)^4$. The corresponding sequence in S is

$$(100101111000110),$$

which still has linear span 12 and the minimal polynomial $x^{12} + x^9 + x^6 + x^3 + 1$.

(c) The following de Bruijn sequence

$$\mathbf{a} = (0000100110101111)$$

is obtained by inserting one 0 into the run of 0's of length 3 in the following m-sequence of period 15:

$$\mathbf{b} = (000100110101111).$$

a has linear span 15, but **b** is an m-sequence of degree 4 whose linear span is equal to 4.

Conjecture 5.1 (GOLOMB'S CONJECTURE, 1980)

$$S \cap C = PN.$$

In other words, if a binary sequence satisfies both the ideal n-tuple distribution R-4 and 2-level autocorrelation R-3, then the sequence is an m-sequence.

The importance of this conjecture is that, if true, one cannot design good pseudorandom sequences that have not only 2-level autocorrelation with ideal n-tuple distribution but also large linear span. Two-level autocorrelation with ideal n-tuple distribution must be compromised if large linear span is required because the truth of the conjecture implies that the former two conditions require an m-sequence, which has the smallest linear span.

Up to now, no counterexamples to this conjecture have been found, and there are none for period less than or equal to $2^{10} - 1$. Incidentally, sequences in $R \cap M$ that are not PN occur as early as period $2^5 - 1 = 31$, and sequences in $M \cap S$ occur as early as period $2^7 - 1 = 127$. In fact, sequences in $R \cap C$ (and therefore having the three postulates R-1, R-2, and R-3) that are not PN sequences also occur as early as period $2^7 - 1 = 127$ (Golomb 1980). Thus the only open problem, in the binary case, is whether $S \cap C = PN$. But if we consider sequences over $\mathbb{F}_q$ of period $q^n - 1$ where $q > 2$, then the corresponding conjecture fails. One can easily construct a nonbinary sequence that belongs to $S \cap C$, but not PN, which is presented as follows (see Gong, Di Porto, and Wolfowicz (1993) for details).

Construction of Nonbinary Sequences in $S \cap C$. Let $\mathbf{a} = \{a_i\}$ be an m-sequence over $\mathbb{F}_{p^r}$ of degree n where $r > 1$. Let $\sigma = \sum_{i=0}^{r-1} c_i x^{p^i}$, $c_i \in \mathbb{F}_{p^r}$, be an invertible linear transformation on $\mathbb{F}_{p^r}$. Let $\mathbf{b} = \{b_i\}$ whose elements are given by

$$b_i = \sigma(a_i), i = 0, 1, \ldots.$$

It can be verified that $\mathbf{b}$ satisfies both the 2-level autocorrelation and span n property. But the linear span of $\mathbf{b}$ could be as large as rn. So, it is no longer an m-sequence over $\mathbb{F}_{p^r}$ when the number of nonzero coefficients in $\sigma(x)$ is greater than 1.

Note

For the materials on Golomb's three randomness postulates, the shift-and-add property, and the constant-on-cosets property of binary m-sequences, see Golomb (1982). An approach to define an autocorrelation function for a periodic nonbinary sequence is to use exponential sums. For more treatments on exponential sums, see Chapter 5 in Lidl and Niederreiter (1997), and Chapter 8 in Ireland and Rosen (1991), and for a complete treatment, see Berndt and Evans (1979) and Berndt, Evans, and Williams (1998). The autocorrelation defined by Eq. (5.13), which is the sum of the autocorrelation functions of the component sequences, was introduced by Park and Komo in 1989. The run property for q-ary m-sequences and their component sequences was discussed by Gong in 1986. Interleaved structures of m-sequences were first discussed by Zierler in 1959. It was then further explored by Games (1987). The difference property of the shift sequences of m-sequences was also studied by Games (1987), and Fact 5.1 was taken from that paper. The trinomial pairs of nonlinear sequences were introduced by Golomb and Gong. In 1999, Golomb and Gong derived some necessary and sufficient conditions for finding trinomial pairs for an nonlinear binary sequence of period $2^n - 1$. The 2-tuple balance property of m-sequences over $\mathbb{F}_q$ is due to Zierler (1959). Here we presented a simple proof using the trace function. In Section 5.7, the lower bound on linear spans of de Bruijn sequences was established by Chan, Games, and Key in 1982, and Etzion and Lempel proved that this lower bound was achievable in 1984. Modified de Bruijn sequences were discussed by Mayhew and Golomb in 1990, and the results on sequences in B and in S for $n = 6$ and $n = 7$ were also included in that paper. The classification of binary sequences of period $2^n - 1$ was due to Golomb (1980). The construction of nonbinary sequences satisfying both the span n property and 2-level autocorrelation but not in the PN set was given by Gong, Di Porto, and Wolfowicz (1993).

Exercises for Chapter 5

1. Let $\mathbf{a} = (111101100101000)$ be a binary sequence of period 15.
   (a) Determine if the sequence satisfies the Golomb randomness postulates R-1 and R-2.
   (b) Compute the autocorrelation function $C(\tau)$ for $\tau = 0, 1, \ldots, 6$.
   (c) Is the sequence an m-sequence? Why?
2. Determine the Legendre sequence modulo 17. Verify whether it has the balance property R-1 and the run property R-2, and compute the autocorrelation function $C(\tau)$, for $\tau = 0, 1, \ldots, 16$.
3. Compute the autocorrelation function of the Legendre sequence of period 31, $C(\tau)$, where $\tau = 0, 1, 3, 5, 7, 11, 15$.
4. Determine the run distribution for the m-sequence over $\mathbb{F}_{2^2}$ given in Example 5.6(e) in Section 5.2.
5. Let $\mathbf{a} = \{a_i\}$ be a binary sequence of period $2^n - 1$ and let $\mathbf{a}$ satisfy R-1 with $2^{n-1} - 1$ 0's and 2^{n-1} 1's in every period, and R-3. Let

$$N_{k,j}(\tau) = |\{ i \mid (a_i, a_{i+\tau}) = (k, j),\ 0 \le i \le 2^n - 2\}|,\ k, j \in \{0, 1\}.$$

   Prove that for every τ with $0 < \tau \le 2^n - 2$,

$$N_{0,0}(\tau) = 2^{n-2} - 1 \text{ and } N_{0,1}(\tau) = N_{1,0}(\tau) = N_{1,1}(\tau) = 2^{n-2}.$$

6. Determine the first 12 bits for each component sequence of the m-sequence over $\mathbb{F}_{2^3}$ given in Example 5.6(f) in Section 5.2.
7. Determine the shift sequence of the m-sequence given in Example 5.6(b), regarded as an $(8, 10)$ interleaved sequence with respect to $(1, 1, 0, 1, 2, 2, 0, 2)$, an m-sequence over $\mathbb{F}_3$ of degree 2. Verify the result of Fact 5.1.
8. Let $\mathbf{a} = (1110100)$ be a binary m-sequence of degree 3 and period 7. Compute $\tau(j)$ such that

$$\mathbf{a} + L^j \mathbf{a} = L^{\tau(j)} \mathbf{a},\ 1 \le j \le 6.$$

9. Let $\mathbf{a} = \{a_i\}$ be a binary sequence of period 63 whose elements in one period are given by

```
0 1 1 0 1 0 0 1 1 0 0 0 0 1 1 1 1 0 0 1 0
1 0 0 0 0 1 0 1 0 1 0 1 0 0 1 0 0 1 1 0 1
1 0 0 0 0 0 0 1 0 1 1 0 0 0 1 1 0 0 1 0 0
```

Verify that $(11, 25)$ is a trinomial pair of $\mathbf{a}$. (Hint: Check if $\mathbf{a} + L^{11}\mathbf{a} = L^{25}\mathbf{a}$.)

10. Construct a binary sequence of period 15 that is constant-on-cosets but does not have the two-level correlation property R-3.

11. Construct a binary sequence of period 15 that satisfies R-4.

12. Give the initial state of an m-sequence of period 127 for which the sequence is constant-on-cosets.

13. Verify that the following binary sequence of period 127 has the run property and that $C(9) = 3$.

```
1 1 1 1 1 1 1 0 1 1 1 1 1 0 0 1 1 1 1 0
1 0 1 1 1 0 0 0 0 1 1 0 1 1 1 0 1 0 0 1
1 0 0 0 1 0 1 0 1 1 0 0 0 0 0 1 0 1 1 1
1 0 0 0 1 1 1 0 1 1 0 1 1 0 0 1 0 0 1 0
1 0 0 1 0 0 0 0 1 0 0 1 1 1 0 0 1 0 1 1
0 1 0 0 0 1 0 0 0 1 1 0 0 1 1 0 1 0 1 0
1 0 0 0 0 0 0
```

Hence it does not have the two-level correlation property R-3. (Note: This sequence is constructed from an m-sequence of period 127, generated by the primitive polynomial $f(x) = x^7 + x + 1$, by complementing all bits in the m-sequence except for the first bit. In other words, if we complement all bits of the above sequence except for the first bit, then the resulting sequence is an m-sequence generated by $f(x)$. This comes from a general construction for binary sequences that have the run property, but violate the 2-level autocorrelation property. See Golomb (1980).) Verify whether this sequence is constant-on-cosets.

14. Proposition 5.4 shows that for any nonbinary sequence over $\mathbb{F}_q$ ($q > 2$) of period $q^n - 1$, if it satisfies the 2-tuple balance property, then it also has the 2-level autocorrelation property. However, the converse is not true. Construct a counterexample.

15. Can you construct a nonbinary sequence over $\mathbb{F}_q$ of period $q^n - 1$ that is not an m-sequence over $\mathbb{F}_q$, but has both the 2-tuple balance property and the span n property R-4? Can you give a general construction for this type of sequence? (Hint: Consider applying some operations on m-sequences.)

16. *Research Problem: Investigation of Correlation of Nonbinary Sequences.* For two nonbinary sequences $\mathbf{a} = \{a_i\}$ and $\mathbf{b} = \{b_i\}$, $a_i, b_i \in GF(p^r)$ ($r > 1$) of period N, there are three possible ways to define the cross- and autocorrelation of $\mathbf{a}$ and $\mathbf{b}$. The first two methods are given by Eqs. (5.14) and (5.15). The third one is given as

follows. Let η be a primitive p^rth root of unity.

$$C_{\mathbf{a},\mathbf{b}}(\tau) = \sum_{i=0}^{N-1} \eta^{b_{i+\tau} - a_i}, \tau = 0, 1, \ldots, \tag{5.37}$$

where the elements of $GF(p^r)$ are represented by the p-ary numbers.

(a) Investigate the differences between these three definitions for $r > 1$. Which method do you think is best? Can you construct a class of nonbinary sequences over $GF(p^r)$ $(r > 1)$ with 2-level autocorrelation, according to the third method, given above, that are not m-sequences over $GF(p^r)$?

(b) Assume that $\mathbf{b}$ is an m-sequence over $GF(2^r)$ of degree $2n$ and $\mathbf{a}$ is an m-sequence over $GF(2^r)$ of degree n. What is the crosscorrelation between $\mathbf{a}$ and $\mathbf{b}$ using the third method? (Note. The sequence $\mathbf{b} + L^i(\mathbf{a})$ is a Kasami (small set) sequence over $GF(2^r)$.)

6

Transforms of Sequences and Functions

The discrete Fourier transform, Hadamard transform (or Walsh transform), and convolution transform all play important roles in signal processing and coding practice in several engineering disciplines. In this chapter, we introduce these transforms for periodic sequences over a finite field $\mathbb{F}_q$. The discrete Fourier transform yields the trace representation of a periodic sequence. The number of nonzero Fourier spectral coefficients is equal to the linear span of the sequence. These results are included in Sections 6.1–6.3. The trace representation of a periodic sequence over $\mathbb{F}_q$ is a function from $\mathbb{F}_{q^n}$ to $\mathbb{F}_q$. We discuss this one-to-one correspondence in Section 6.4. Furthermore, the Hadamard transform and convolution transform of sequences can be defined in terms of their trace representations, which are introduced in Sections 6.5. Consequently, the correlation between two sequences is transferred to correlation between their trace representations. This is presented in Section 6.6. In Section 6.7, we show some basic properties of the Hadamard transform and convolution transform. Thus, the Fourier transform serves as a bridge for a connection between sequences and functions. It is worth pointing out that all newly discovered binary sequences with 2-level autocorrelation functions were proved in terms of their Hadamard transforms. Therefore, the contents introduced in this chapter are fundamental tools for the design of sequences with special properties. The last section features the discrete Fourier transform and the Hadamard transform in their matrix representations.

6.1 The (discrete) Fourier transform

Let q be a prime or a power of a prime, and let N be a positive integer for which there exists some integer $n \geq 1$ such that $N \mid (q^n - 1)$. The smallest integer n that satisfies this property is called the order of q modulo N. (Note.

162

$N \mid (q^n - 1) \Rightarrow q^n \equiv 1 \pmod{N}$.) In the rest of this chapter, we mean that n is the order of q modulo N when we say that $N \mid (q^n - 1)$.

Let $\mathbf{a} = \{a_t\}$ be a sequence over $\mathbb{F}_q$ with period $N > 1$ where $N \mid (q^n - 1)$. We also write $\mathbf{a} = \{a_t\} = (a_0, a_1, \ldots, a_{N-1})$ as usual.

Definition 6.1 *Let α be an element in $\mathbb{F}_{q^n}$ of order N. Then the* (discrete) Fourier Transform (DFT) *of $\{a_t\}$ is defined as*

$$A_k = \sum_{t=0}^{N-1} a_t \alpha^{tk}, \quad k = 0, 1, \ldots, N - 1. \tag{6.1}$$

Lemma 6.1 *The inverse formula of Eq. (6.1) is given as follows:*

$$a_t = \frac{1}{N} \sum_{k=0}^{N-1} A_k \alpha^{-kt}, \quad t = 0, 1, \ldots, N - 1. \tag{6.2}$$

To prove Lemma 6.1, we need the following lemma.

Lemma 6.2

$$\sum_{i=0}^{N-1} \alpha^{di} = \begin{cases} 0 & \text{if } d \not\equiv 0 \pmod{N} \\ N & \text{if } d \equiv 0 \pmod{N}. \end{cases}$$

Proof. If $d \equiv 0 \pmod{N}$, then $\sum_{k=0}^{N-1} \alpha^{i \cdot 0} = N$. We now suppose that $d \not\equiv 0$ $\pmod{N}$. Note that for each $\gamma \in \mathbb{F}_{q^n}^*$ with $\gamma \neq 1$, the order of γ, denoted by $\mathrm{ord}(\gamma)$, is greater than 1, that is, $\mathrm{ord}(\gamma) > 1$. Hence

$$\gamma^s = 1 \implies (\gamma - 1)(\gamma^{s-1} + \cdots + \gamma + 1) = 0$$
$$\implies \gamma^{s-1} + \cdots + \gamma + 1 = 0. \tag{6.3}$$

For $d \not\equiv 0 \pmod{N}$, let $\gamma = \alpha^d$ and s be the order of γ. Then $s = \mathrm{ord}(\gamma) > 1$. The order of γ can be determined as follows: $s = \mathrm{ord}(\alpha^d) = \frac{N}{\gcd(d,N)}$ (see Fact 3.4 in Section 3.2 of Chapter 3). We define $v = \gcd(d, N)$. If $v = 1$, then $s = N$. From Eq. (6.3), for the case $s = N$ the result is immediate. If $v > 1$, we have $1 < s < N$. Therefore, for any k with $0 \leq k < N$, we may write

$$k = is + j, 0 \leq i < v, 0 \leq j < s.$$

Again using Eq. (6.3), we have

$$\sum_{j=0}^{s-1} \gamma^j = 0 \text{ and } \gamma^{is} = 1 \ \forall i.$$

Therefore

$$\sum_{k=0}^{N-1} \alpha^{dk} = \sum_{k=0}^{N-1} \gamma^k = \sum_{i=0}^{v-1} \gamma^{is} \sum_{j=0}^{s-1} \gamma^j = v \cdot 0 = 0. \qquad \square$$

Table 6.1. *DFT and its inverse*

DFT:	$A_k = \sum_{t=0}^{N-1} a_t \alpha^{tk}, k = 0, 1, \ldots, N-1$
Inverse DFT:	$a_t = -\sum_{k=0}^{N-1} A_k \alpha^{-kt}, t = 0, 1, \ldots, N-1 (\frac{1}{N} \equiv -1 \bmod p)$
$q = 2$:	$a_t = \sum_{k=0}^{N-1} A_k \alpha^{-kt}, \quad t = 0, 1, \ldots, N-1$

Proof of Lemma 6.1. Multiplying (6.1) by α^{-jk}, and summing over k, we have

$$\sum_{k=0}^{N-1} \alpha^{-jk} A_k = \sum_{k=0}^{N-1} \sum_{t=0}^{N-1} a_t \alpha^{tk} \alpha^{-jk} = \sum_{t=0}^{N-1} a_t \sum_{k=0}^{N-1} \alpha^{(t-j)k}.$$

Applying Lemma 6.2, the term a_t survives only for t such that $t - j \equiv 0$ (mod N). Consequently, the right-hand side in the above identities is equal to Na_j. The result follows.   □

A_k is called a (Fourier) spectrum of the sequence **a**. Note that $\mathbf{A} = \{A_k\}$ is also a sequence with period N, which is called a (Fourier) spectral sequence of **a**. We also use the notation $\mathbf{a} \leftrightarrow \mathbf{A}$ to represent that $\mathbf{A}$ is the Fourier spectral sequence of **a**. In general, $A_k \in \mathbb{F}_{q^n}$. (Note that all indices are reduced modulo N throughout this chapter.)

Suppose that $\mathbb{F}_q$ has characteristic p. Because N is a factor of $q^n - 1$, we have $N \equiv -1$ (mod p). So, $N^{-1} \equiv -1$ (mod p). We feature the DFT of the sequence $\mathbf{a} = \{a_t\}$ and its inverse transformation in Table 6.1.

Example 6.1 Let $q = 2$, $n = 3$, $\mathbb{F}_{2^3}$ be defined by the primitive polynomial $f(x) = x^3 + x + 1$, and α be a root of $f(x)$. Let $\mathbf{a} = (0100010)$ be a binary sequence of period 7. Thus $N = 7$. Hence

$$A_k = \sum_{t=0}^{6} a_t \alpha^{tk} = \alpha^k + \alpha^{5k}, k = 0, 1, \ldots.$$

In detail,

$$A_0 = \sum_{t=0}^{6} a_t = a_1 + a_5 = 0$$

$$A_1 = \sum_{t=0}^{6} a_t \alpha^t = \alpha + \alpha^5 = \alpha^6$$

$$A_2 = \sum_{t=0}^{6} a_t \alpha^{2t} = \alpha^2 + \alpha^3 = \alpha^5$$

$$A_3 = \sum_{t=0}^{6} a_t \alpha^{3t} = \alpha^3 + \alpha = 1$$

$$A_4 = \sum_{t=0}^{6} a_t \alpha^{4t} = \alpha^4 + \alpha^6 = \alpha^3$$

$$A_5 = \sum_{t=0}^{6} a_t \alpha^{5t} = \alpha^5 + \alpha^4 = 1$$

$$A_6 = \sum_{t=0}^{6} a_t \alpha^{6t} = \alpha^6 + \alpha^2 = 1.$$

Thus the spectral sequence of **a** is

$$\mathbf{A} = (0, \alpha^6, \alpha^5, 1, \alpha^3, 1, 1).$$

If **A** is given as above, using the inverse DFT,

$$a_t = \sum_{k=0}^{6} A_k \alpha^{-tk}, t = 0, 1, \ldots,$$

we obtain

$$a_0 = \sum_{k=0}^{6} A_k = \alpha^6 + \alpha^5 + 1 + \alpha^3 + 1 + 1 = 0$$

$$a_1 = \sum_{k=0}^{6} 6 A_k \alpha^{-k} = \alpha^6 \alpha^{-1} + \alpha^5 \alpha^{-2} + \alpha^{-3} + \alpha^3 \alpha^{-4} + \alpha^{-5} + \alpha^{-6}$$

$$= \alpha^5 + \alpha^3 + \alpha^4 + \alpha^6 + \alpha^2 + \alpha = 1$$

$$a_2 = \sum_{k=0}^{6} A_k \alpha^{-2k} = \alpha^6 \alpha^{-2} + \alpha^5 \alpha^{-4} + \alpha^{-6} + \alpha^3 \alpha^{-1} + \alpha^{-3} + \alpha^{-5}$$

$$= \alpha^4 + \alpha + \alpha + \alpha^2 + \alpha^4 + \alpha^2 = 0$$

$$a_3 = \sum_{k=0}^{6} A_k \alpha^{-2k} = \alpha^6 \alpha^{-3} + \alpha^5 \alpha^{-6} + \alpha^{-2} + \alpha^3 \alpha^{-5} + \alpha^{-1} + \alpha^{-4}$$

$$= \alpha^3 + \alpha^6 + \alpha^5 + \alpha^5 + \alpha^6 + \alpha^3 = 0.$$

Similarly, we compute

$$a_4 = 0, \ a_5 = 1, \text{ and } a_6 = 0,$$

which verifies **a** = (0100010).

Note that the Fourier transform yields a one-to-one correspondence between the sequence and its Fourier transform, that is, the Fourier transform of a sequence is uniquely determined by the sequence, and vice versa. We write this result as the following proposition.

Proposition 6.1 *Let* $\mathbf{a} \leftrightarrow \mathbf{A}$ *and* $\mathbf{b} \leftrightarrow \mathbf{B}$. *Then* $\mathbf{a} = \mathbf{b}$ *if and only if* $\mathbf{A} = \mathbf{B}$.

Note. From Corollary 4.3 of Section 4.3 in Chapter 4, the minimal polynomial of $\mathbf{a}$, denoted by $m(x)$, divides $x^N - 1$ when $\mathbf{a}$ has period N. Because $N \mid (q^n - 1)$, we have $(x^N - 1) \mid (x^{q^n-1} - 1)$, which implies $m(x) \mid (x^{q^n-1} - 1)$. Thus $m(x)$ has no multiple root. The converse of this assertion is also true. In other words, if $\mathbf{a}$ is a sequence over $\mathbb{F}_q$, then the minimal polynomial of $\mathbf{a}$ over $\mathbb{F}_q$ has no multiple roots if and only if $\mathbf{a}$ is a periodic sequence with period N for which there exists some positive integer n such that $N \mid (q^n - 1)$. Thus the Fourier transform that we introduced here applies to sequences whose minimal polynomials have no multiple roots. For the case of sequences with minimal polynomials having multiple roots, there is not much known.

6.2 Trace representation

From the inverse Fourier transform of $\mathbf{a}$, we may group nonzero monomial terms in the inverse Fourier transform according to different trace terms. In this way, we may obtain the trace representation of the sequence. In the following, we first generalize the concept of cyclotomic cosets modulo $q^n - 1$, introduced in Chapter 3, to the case of modulo N where N is a divisor of $q^n - 1$.

6.2.1 Cyclotomic cosets modulo N with respect to q

Let $C = \{1, q, \ldots, q^{n-1}\}$. Then C is a subgroup of the multiplicative group of $\mathbb{Z}_N$. In terms of C, we define a relation, say $\sim$, on $\mathbb{Z}_N$ as follows: for any $a, b \in \mathbb{Z}_N$

$$a \sim b \iff a \equiv q^i b \;(\text{mod } N) \text{ for some } i.$$

We can easily verify that $\sim$ is an equivalence relation; that is, $\sim$ satisfies the following conditions. For any $a, b, c \in \mathbb{Z}_N$,

1. *Reflexivity*: $a \sim a$ for all a.
2. *Symmetry*: $a \sim b$ implies $b \sim a$.
3. *Transitivity*: if $a \sim b$ and $b \sim c$, then $a \sim c$.

Thus $\sim$ induces a partition on $\mathbb{Z}_N$. We denote by C_s the equivalence class containing s. It can be represented as

$$C_s = \{s, sq, sq^2, \ldots, sq^{n_s-1}\},$$

where n_s is the smallest positive integer such that

$$q^{n_s} s \equiv s \;(\text{mod } N). \tag{6.4}$$

Definition 6.2 *The set C_s is called a* (cyclotomic) *coset modulo N (with respect to q), the smallest number in C_s a* coset leader *modulo N (with respect to q), and n_s, the size of the coset C_s, the* order *of s with respect to q modulo N.*

The number n_s is important in the trace representation of sequences. In the following proposition, we present some properties related to n_s, whose validity can be easily verified from the definition and basic number theory.

Proposition 6.2 *With the above notation,*

(a) n_s *is the smallest positive integer such that $q^{n_s}s \equiv s \pmod{N}$.*
(b) $n_s = |C_s|$.
(c) n_s *is equal to the degree of the minimal polynomial of α^s over $\mathbb{F}_q$ where α is an element of order N in $\mathbb{F}_{q^n}$.*
(d) n_s *divides n; that is, $n_s \mid n$.*

We assign s, which is the coset leader modulo N, as the subscript of the coset C_s. Let $\Gamma(N)$ be the set that consists of all cyclotomic coset leaders modulo N with respect to q. If $N = q^n - 1$, we also write $\Gamma(N)$ as $\Gamma_q(n)$, which is the same as was introduced in Section 3.4 of Chapter 3 for the case $q = p$. Thus we have a partition of $\mathbb{Z}_N$ as follows:

$$Z_N = \cup_{s \in \Gamma(N)} C_s. \tag{6.5}$$

Example 6.2 Let $q = 2$ and $N = 21$. Since 21 is a factor of $63 = 2^6 - 1$ but not a factor of $31 = 2^5 - 1$, we have $n = 6$. The cyclotomic cosets modulo 21 are as follows:

$$C_0 = \{0\},$$
$$C_1 = \{1, 2, 4, 8, 16, 11\},$$
$$C_3 = \{3, 6, 12\},$$
$$C_5 = \{5, 10, 20, 19, 17, 13\},$$
$$C_7 = \{7, 14\},$$
$$C_9 = \{9, 18, 15\}.$$

Hence $\Gamma(21) = \{0, 1, 3, 5, 7, 9\}$, the set consisting of all coset leaders modulo 21. Furthermore, $\mathbb{Z}_{21}$ can be partitioned as:

$$\mathbb{Z}_{21} = C_0 \cup C_1 \cup C_3 \cup C_5 \cup C_7 \cup C_9.$$

6.2.2 Conjugate property of spectral sequences

From the Fourier transform of the sequence $\mathbf{a}$, it is known that the spectrum A_k belongs to $\mathbb{F}_{q^n}$. However, since the elements of $\mathbf{a}$ are taken from $\mathbb{F}_q$, we have the following restriction on the values of A_k.

Lemma 6.3 *With the above notation, for any k with $1 \leq k \leq N - 1$,*

$$A_{kq^j} = A_k^{q^j}, 0 \leq j < n_k \text{ and } A_0 = \sum_{t=0}^{N-1} a_t,$$

where $n_k = |C_k|$ as determined by Eq. (6.4).

Proof. Note that $a_t \in \mathbb{F}_q$, so $a_t^{q^j} = a_t$. Thus

$$A_{kq^j} = \sum_{t=0}^{N-1} a_t \alpha^{tkq^j} = \sum_{t=0}^{N-1} a_t^{q^j} (\alpha^{tk})^{q^j} = \sum_{t=0}^{N-1} (a_t \alpha^{tk})^{q^j}$$

$$= \left(\sum_{t=0}^{N-1} a_t \alpha^{tk} \right)^{q^j} = A_k^{q^j},$$

where the second to last identity comes from Corollary 3.3 in Section 3.2. ☐

Considering $\mathbf{a} = (0100010)$ and $\mathbf{A} = (0, \alpha^6, \alpha^5, 1, \alpha^3, 1, 1)$ in Example 6.1, we have

$$A_1 = \alpha^6,$$
$$A_1^2 = (\alpha^6)^2 = \alpha^5 = A_2,$$
$$A_1^4 = (\alpha^6)^4 = \alpha^3 = A_4,$$
$$A_3 = 1, A_6 = A_3^2 = 1, A_5 = A_3^4 = 1.$$

Lemma 6.3 also shows that we only need to compute the spectra for those indices that are coset leaders. For the other indices, the spectra can be computed by the conjugate operation.

Example 6.3 Let $q = 2, n = 6$, $\mathbb{F}_{2^6}$ be defined by $\alpha^6 + \alpha + 1 = 0$, and $\mathbb{F}_{2^6}^* = \{\alpha^i | 0 \leq i \leq 62\}$. Let $N = 21$ and let a binary sequence $\mathbf{a}$ of period 21 be defined by

Index t	0	1	2	3	4	5	6	7	8	9	10
a_t	0	1	0	0	0	0	0	0	1	0	1
Index t	11	12	13	14	15	16	17	18	19	20	
a_t	0	0	0	0	0	1	0	0	0	0	

From Example 6.2, we have

$$\Gamma(21) = \{0, 1, 3, 5, 7, 9\},$$

which is the set consisting of all coset leaders modulo 21. Let $\beta = \alpha^3$. Then $\text{ord}(\beta) = 21$. From the DFT,

$$A_k = \sum_{t=0}^{20} a_t \beta^{tk} = \beta^k + \beta^{8k} + \beta^{10k} + \beta^{16k},$$

$$= \alpha^{3k} + \alpha^{24k} + \alpha^{30k} + \alpha^{48k}, \ k = 0, 1, \dots.$$

Note that the computation is performed in $\mathbb{F}_{2^6}$. Hence

$$A_0 = \sum_{t=0}^{20} a_t = 0$$
$$A_1 = \alpha^3 + +\alpha^{24} + \alpha^{30} + \alpha^{48} = \alpha^{47}$$
$$A_3 = \alpha^9 + \alpha^9 + \alpha^{27} + \alpha^{18} = 1$$
$$A_5 = \alpha^{15} + \alpha^{57} + \alpha^{24} + \alpha^{51} = \alpha^{37}$$
$$A_7 = \alpha^{21} + \alpha^{42} + \alpha^{21} + \alpha^{21} = 1$$
$$A_9 = \alpha^{27} + \alpha^{27} + \alpha^{18} + \alpha^{54} = \alpha^9.$$

Applying Lemma 6.3, for $j \in C_1 = \{1, 2, 4, 8, 16, 11\}$,

$$A_1 = \alpha^{47} \implies A_2 = A_1^2 = (\alpha^{47})^2 = \alpha^{31},$$
$$A_4 = A_1^4 = (\alpha^{47})^4 = \alpha^{62}, A_8 = A_1^8 = (\alpha^{47})^8 = \alpha^{61},$$
$$A_{16} = A_1^{16} = (\alpha^{47})^{16} = \alpha^{59}, A_{11} = A_1^{32} = (\alpha^{47})^{11} = \alpha^{55}.$$

For $j \in C_3 = \{3, 6, 12\}$,

$$A_3 = 1 \implies A_6 = A_3^2 = 1, A_{12} = A_3^4 = 1.$$

For $j \in C_5 = \{5, 10, 20, 19, 17, 13\}$,

$$A_5 = \alpha^{37} \implies A_{10} = A_5^2 = \alpha^{11},$$
$$A_{20} = A_5^4 = \alpha^{22}, A_{19} = A_5^8 = \alpha^{44},$$
$$A_{17} = A_5^{16} = \alpha^{25}, A_{13} = A_5^{32} = \alpha^{50}.$$

For $j \in C_7 = \{7, 14\}$,

$$A_7 = 1 \implies A_{14} = A_7^2 = 1.$$

For $j \in C_9 = \{9, 18, 15\}$,

$$A_9 = \alpha^9 \implies A_{18} = A_9^2 = \alpha^{18}, A_{15} = A_9^4 = \alpha^{36}.$$

Thus, we have obtained the spectral sequence $\mathbf{A} = \{A_k\}$, whose elements are listed below. (We list the exponents of α for $\{A_k\}$ and use the notation $0 = \alpha^\infty$ by convention.)

Index k	0	1	2	3	4	5	6	7	8	9	10
A_k	∞	47	31	0	62	37	0	0	61	9	11
Index k	11	12	13	14	15	16	17	18	19	20	
$A_k =$	55	0	50	0	36	59	25	18	44	22	

6.2.3 The formula

Theorem 6.1 (TRACE REPRESENTATION OF SEQUENCES) *The inverse formula of the Fourier transform of the sequence* $\mathbf{a}$ *can be written in the following form:*

$$a_t = - \sum_{j \in \Gamma(N)} Tr_1^{n_j}(A_j \alpha^{-jt}), \quad t = 0, 1, \ldots, N-1, \, A_j \in \mathbb{F}_{q^{n_j}}, \qquad (6.6)$$

where $n_j = |C_j|$, $Tr_1^{n_j}(x)$ *is the trace function from* $\mathbb{F}_{q^{n_j}}$ *to* $\mathbb{F}_q$, *and* A_j *and* $\alpha^{-jt} \in \mathbb{F}_{q^{n_j}}$. *In particular, if* $q = 2$, *that is, if* $\mathbf{a}$ *is a binary sequence of period* N, *we have*

$$a_t = \sum_{j \in \Gamma(N)} Tr_1^{n_j}(A_j \alpha^{-jt}), \quad t = 0, 1, \ldots, N-1, \, A_j \in \mathbb{F}_{2^{n_j}}. \qquad (6.7)$$

Proof. Since $\mathbb{Z}_N$ is a disjoint union of all the different cosets modulo N with respect to q, we group the terms in the inverse DFT of $\mathbf{a}$ according to the different cosets. First, we consider the following partial sum in the inverse DFT of $\mathbf{a}$ for which all indices belong to C_j:

$$S = A_j \alpha^{-tj} + A_{jq} \alpha^{-tjq} + \cdots + A_{jq^{n_j-1}} \alpha^{-tjq^{n_j-1}}. \qquad (6.8)$$

According to Lemma 6.3, $A_{jq^v} = A_j^{q^v}$ for $0 \le v < n_j$. Therefore, S can be written as:

$$S = A_j \alpha^{-tj} + A_j^q (\alpha^{-tj})^q + \cdots + A_j^{q^{n_j-1}} (\alpha^{-tj})^{q^{n_j-1}}$$
$$= Tr_1^{n_j}(A_j \alpha^{-tj}).$$

To have $S \in \mathbb{F}_q$, we need to show that both A_j and α^{-j} belong to $\mathbb{F}_{q^{n_j}}$. From Proposition 6.2, n_j is equal to the degree of the minimal polynomial of α^{-j} over $\mathbb{F}_q$. Thus $\alpha^{-j} \in \mathbb{F}_{q^{n_j}}$. Again, from Proposition 6.2, we have $jq^{n_j} \equiv j$ (mod N). Using Lemma 6.3, $A_j = A_{jq^{n_j}} = A_j^{q^{n_j}} \implies A_j \in \mathbb{F}_{q^{n_j}}$. Thus we have proved that both A_j and α^{-j} belong to $\mathbb{F}_{q^{n_j}}$. Therefore $S \in \mathbb{F}_q$. Hence the formula (6.6) follows. $\qquad \square$

The formulae (6.6) is called the trace representation of the sequence **a**.

Note that the trace representation of a periodic sequence over $\mathbb{F}_q$ is the inverse of the DFT of the sequence by grouping terms as the sums of trace terms.

Example 6.4 Considering the previous two examples, for $\mathbf{a} = (0100010)$ we have $A_1 = \alpha^6$ and $A_3 = 1$ where $\alpha^3 + \alpha + 1 = 0$, which defines $\mathbb{F}_{2^3}$. Applying Theorem 6.1, we have

$$a_t = Tr(\alpha^6 \alpha^{-t}) + Tr(\alpha^{-3t}), t = 0, 1, \ldots$$

where $Tr(x)$ is the trace function from $\mathbb{F}_{2^3}$ to $\mathbb{F}_2$. Note that $\alpha^{-1} = \alpha^6$ and $Tr(\alpha^{2^i}) = Tr(\alpha)$. Alternatively, we may write the trace representation of **a** as follows:

$$a_t = Tr(\alpha^6 \alpha^{6t}) + Tr(\alpha^{4t}) = Tr(\alpha^3 \alpha^{3t}) + Tr(\alpha^t), t = 0, 1, \ldots.$$

For $\mathbf{a} = (0100000010100000010000)$ in Example 6.3, we have

$$a_t = Tr_1^6(\alpha^{47}\beta^{-t}) + Tr_1^3(\beta^{-3t}) + Tr_1^6(\alpha^{37}\beta^{-5t}) + Tr_1^2(\beta^{-7t}) + Tr_1^3(\alpha^9 \beta^{-9t}), \forall t.$$

In the following, we discuss a relationship between characteristic polynomials and the Fourier spectra of sequences.

Theorem 6.2 *With the same notation as in Theorem 6.1, the minimal polynomial of* **a** *is a product* $f_1(x)f_2(x) \cdots f_s(x)$ *where the* $f_j(x)$'s *are distinct irreducible polynomials over* $\mathbb{F}_q$ *and* $f_j(x)$ *is the minimal polynomial of* α^{-r_j} *over* $\mathbb{F}_q$ *for which* $A_{r_j} \neq 0$, *where* $r_j \in \Gamma(N)$, $1 \leq j \leq s$.

Proof. Let $\mathbf{a}_j = \{a_{ij}\}_{i \geq 0}$ where

$$a_{ij} = Tr_1^{n_j}(\beta_j \alpha^{-ir_j}), i = 0, 1, \ldots, \text{ by setting } \beta_j = -A_{r_j}.$$

Then

$$\mathbf{a} = \mathbf{a}_1 + \cdots + \mathbf{a}_s.$$

Therefore, we have $\mathbf{a}_j \in G(f_j)$, the set consisting of all LFSR sequences generated by $f_j(x)$, and $f_j(x)$ is the minimal polynomial of the sequence $\mathbf{a}_j$ over $\mathbb{F}_q$ (see Lemma 4.3 in Section 4.6 of Chapter 4). Let $f(x) = f_1(x) \cdots f_s(x)$. Since $f(x)$ is a multiple of $f_j(x)$ for every j, $f(x)$ is a characteristic polynomial of $\mathbf{a}_j$; that is, $f(L)\mathbf{a}_j = 0$ (see Theorem 4.5 in Section 4.3 of Chapter 4). Hence

$$f(L)\mathbf{a} = f(L)\mathbf{a}_1 + \cdots + f(L)\mathbf{a}_s = 0 \Longrightarrow \mathbf{a} \in G(f).$$

Again using Theorem 4.5, the minimal polynomial of the sequence **a** is a factor of $f(x)$. Note that for each j, $\mathbf{a}_j$ is not the zero sequence, and $G(f)$ is a direct sum of $G(f_j)$. From Theorem 4.10 in Section 4.4 of Chapter 4, $f(x)$ is the minimal polynomial of **a** over $\mathbb{F}_q$. □

Table 6.2. *The second version of the DFT and the inverse*

$$A_k = -\sum_{t=0}^{N-1} a_t \alpha^{-tk}, k \in \Gamma(N)$$

$$a_t = \sum_{k \in \Gamma(N)} Tr_1^{n_k}(A_k \alpha^{kt}), t = 0, 1, \ldots, N-1$$

Remark 6.1 If **a** is a sequence over $\mathbb{F}_q$ of period N where $N \mid (q^n - 1)$, then the minimal polynomial of **a** has no multiple roots in the extension field $\mathbb{F}_{q^n}$. The proof of Theorem 6.2 provides an alternative proof that **a** has the trace representation. In other words, directly using Theorem 4.11 in Section 4.6 of Chapter 4 and Theorem 4.10 (Section 4.4 of Chapter 4), we can prove that **a** has the trace representation. However, this method is not constructive. The method that we introduced in Theorem 6.1 is constructive, because we can compute the spectral sequence from a given sequence, and therefore, the trace representation.

The spectral sequence $\{A_k\}$ of the sequence $\mathbf{a} = \{a_t\}$ depends on a particular element α in $\mathbb{F}_{q^n}$ of order N. For a better picture of the trace representation, we can compute the spectrum of a sequence by using the inverse of α. In this way, we obtain the trace representation shown in Table 6.2 (we also move the negative sign in a_t to A_k). We will refer to the formulae introduced in Table 6.1 as the first version of the DFT.

Example 6.5 Let $q = 2, n = 4, N = 15$, and $\mathbb{F}_{2^4}$ be defined by $\alpha^4 + \alpha + 1 = 0$. Let $\mathbf{a} = (111011000101001)$, a binary sequence of period 15. (Note that this is a modified de Bruijn sequence.) We have $\Gamma_2(4) = \{0, 1, 3, 5, 7\}$, the set consisting of the coset leaders modulo 15 with respect to 2. Using the second version of the DFT, that is, the formulae in Table 6.2, for computing the spectral sequence of **a**:

$$A_0 = \sum_0^{14} a_t = 0,$$

$$A_1 = \sum_0^{14} a_t \alpha^{-t} = 1, \quad A_3 = \sum_0^{14} a_t \alpha^{-3t} = \alpha,$$

$$A_5 = \sum_0^{14} a_t \alpha^{-5t} = 0, \quad A_7 = \sum_0^{14} a_t \alpha^{-7t} = \alpha^6.$$

Thus **a** has the following trace representation:

$$a_t = Tr(\alpha^t) + Tr(\alpha \alpha^{3t}) + Tr(\alpha^6 \alpha^{7t}), t = 0, 1, \ldots$$

where $Tr(x)$ is the trace function from $\mathbb{F}_{2^4}$ to $\mathbb{F}_2$. The minimal polynomials of α, α^3, and α^7 are $f_1(x) = x^4 + x + 1$, $f_3(x) = x^4 + x^3 + x^2 + x + 1$, and

$f_7(x) = x^4 + x^3 + 1$, respectively. According to Theorem 6.2, the minimal polynomial of **a** is given by

$$f_1(x)f_3(x)f_7(x) = x^{12} + x^9 + x^6 + x^3 + 1 = f_3(x^3).$$

6.2.4 A fast algorithm

In the following, we integrate the previous discussion into an algorithm for computing a trace representation of a sequence, which also results in the DFT of the sequence. When it is applied to compute the DFT of the sequence, this algorithm only takes approximately $1/n$ times the computation cost of the process when the definition of the DFT is used directly. We retain the notations $\Gamma(N)$, A_j, and n_j.

Algorithm 6.1 AN ALGORITHM FOR COMPUTING THE TRACE REPRESENTATION

Input: *n, a positive integer;*
           q, a power of a prime;
           $f(x)$, an irreducible polynomial over $\mathbb{F}_q$ of degree n; and
           $\mathbf{a} = (a_0, a_1, \ldots, a_{N-1})$, a sequence over $\mathbb{F}_q$ of period $N \mid (q^n - 1)$.

Output: *The trace representation of **a** as an $s \times 3$ array M*
           in which each row contains (j, n_j, A_j), where $A_j \neq 0$, $j \in \Gamma(N)$.

Procedure_tracerep(**a**)

1. *Generate the finite field $\mathbb{F}_{q^n}$ using $f(x)$ and choose α as an element in $\mathbb{F}_{q^n}$ of order N.*
2. *Compute j, each coset leader modulo N, and n_j, the order of C_j, and set $I = \{(j, n_j) \text{ such that } j \in \Gamma(N)\}$.*
3. *Compute $A_0 = \sum_{t=0}^{N-1} a_t$; if $A_0 \neq 0$ **then return** $(0, A_0, 1)$*
4. **for** $0 \neq j$ in $\Gamma(N)$ **do**
     (1) Using the first version of the DFT, compute $A_j = \sum_{t=0}^{N-1} a_t \alpha^{tj}$; or using the second version of the DFT, compute $A_j = \sum_{t=0}^{N-1} a_t \alpha^{-tj}$
     (2) if $A_j \neq 0$ put (j, n_j, A_j) into the array M //one row of the array M
5. *Return M*

For example, for the parameters in Example 6.3, Algorithm 6.1 outputs the following array (using the first version of the DFT):

$$
\begin{array}{ccc}
j & n_j & A_j
\end{array}
$$

$$
M = \begin{bmatrix}
1 & 6 & \alpha^{47} \\
3 & 3 & 1 \\
5 & 6 & \alpha^{37} \\
7 & 2 & 1 \\
9 & 3 & \alpha^{9}
\end{bmatrix}.
$$

For the parameters in Example 6.5 using the second version of the DFT, Algorithm 6.1 outputs:

$$
\begin{array}{ccc}
j & n_j & A_j
\end{array}
$$

$$
M = \begin{bmatrix}
1 & 4 & 1 \\
3 & 4 & \alpha \\
7 & 4 & \alpha^{6}
\end{bmatrix}.
$$

6.3 Linear spans and spectral sequences

The linear span of a sequence is equal to the degree of the minimal polynomial that generates the sequence. According to Theorem 6.2, this quantity can be determined from the Fourier spectrum of the sequence. In this section, we will present two variations of this method. Note that this is not an efficient way to determine linear spans of sequences. However, it provides some insight into their dynamical behavior. A more efficient algorithm for computing linear spans of sequences is Berlekamp's algorithm (1968) which was further analyzed by Massey (1969).

Let $\mathbf{b}$ be a sequence over $\mathbb{F}_{q^n}$ with period s. The Hamming weight of $\mathbf{b} = \{b_t\}$, denoted by $w(\mathbf{b})$, is defined as

$$
w(\mathbf{b}) = |\{t \mid b_t \neq 0, 0 \leq t < s\}|.
$$

(Recall that $|S|$ denotes the number of elements in the set S.)

Theorem 6.3 (NONZERO SPECTRA METHOD) *Let* $\mathbf{a} = \{a_i\}$ *be a sequence over* $\mathbb{F}_q$ *of period* N *where* N *divides* $q^n - 1$, *and let* $\mathbf{A} = \{A_i\}$ *be its spectral sequence. Then* $LS(\mathbf{a})$, *the linear span of* $\mathbf{a}$, *is given by*

$$
LS(\mathbf{a}) = |\{k \mid A_k \neq 0, 0 \leq k < N\}|. \tag{6.9}
$$

In other words, the linear span of the sequence is equal to the Hamming weight of its spectral sequence; that is,

$$
LS(\mathbf{a}) = w(\mathbf{A}).
$$

Proof. $LS(\mathbf{a})$, the linear span of $\mathbf{a}$, is equal to the degree of the minimal polynomial of $\mathbf{a}$ over $\mathbb{F}_q$. According to Theorem 6.1, the minimal polynomial of $\mathbf{a}$, say $f(x)$, satisfies $f(x) = f_1(x) \cdots f_s(x)$, where f_j is the minimal polynomial of α^{-j}. Thus the degree of $f(x)$ is equal to $n_1 + \cdots + n_s$ where $n_j = \deg(f_j)$. Hence $LS(\mathbf{a}) = n_1 + \cdots + n_s$, which is the number of nonzero elements in $\{A_k\}$. Thus the result follows. $\qquad\square$

Applying Theorem 6.3 to the spectral sequence $\{A_i\}$ given by Eq. (6.1), we have the following dual result of Theorem 6.3.

Theorem 6.4 *The Hamming weight of the sequence* $\mathbf{a}$ *is equal to the linear span of its spectral sequence; that is,*

$$w(\mathbf{a}) = LS(\mathbf{A}).$$

In particular, if the sequence $\mathbf{a}$ *is a binary sequence of period* $2^n - 1$, *then it has the balance property; that is,* $w(\mathbf{a}) = 2^{n-1}$ *if and only if* $LS(\mathbf{A}) = 2^{n-1}$.

We show these two dual results in the following diagram:

	$\overset{DFT}{\longleftrightarrow}$	
$\mathbf{a}$		$\mathbf{A}$
$LS(\mathbf{a})$	$=$	$w(\mathbf{A})$
$w(\mathbf{a})$	$=$	$LS(\mathbf{A})$

Example 6.6 Let $q = 2, n = 5, N = 2^5 - 1 = 31$, and $\mathbb{F}_{2^5}$ be defined by $\alpha^5 + \alpha^3 + 1 = 0$, where α is a primitive element of $\mathbb{F}_{2^5}$, and let

$$\mathbf{a} = (1100010101111110101000010010011).$$

We have

$$\Gamma_2(5) = \{0, 1, 3, 5, 7, 11, 15\},$$

the set consisting of all coset leaders modulo 31. Then the trace representation of the sequence (by the second version of the DFT) is as follows:

$$a_t = Tr(\alpha^{25}\alpha^t) + Tr(\alpha^{5t}) + Tr(\alpha^{7t}), \, t = 0, 1, \ldots.$$

In other words, we have $A_1 = \alpha^{25}$, $A_5 = 1$, $A_7 = 1$, and $A_0 = A_3 = A_{11} = A_{15} = 0$. Since $n_j = 5$ for all $0 \neq j \in \Gamma_2(5)$, the linear span of $\mathbf{a}$ is 15, which is the Hamming weight of the spectral sequence $\mathbf{A} = \{A_k\}$; that is,

$$LS(\mathbf{a}) = w(\mathbf{A}) = 15.$$

Note that the Hamming weight of **a** is equal to 16, so the linear span of the spectral sequence is 16; that is,

$$LS(\mathbf{A}) = w(\mathbf{a}) = 16.$$

Remark 6.2 Theorems 6.3 and 6.4 have significant applications in coding theory. For example, several important bounds on the minimum distance of a code, such as the BCH bound and the Rose bound, are obtained by applying Theorem 6.4.

From the pair $\mathbf{a} = \{a_t\}$ and $\mathbf{A} = \{A_k\}$ of the first version of the Fourier transform (similar to the second version of the DFT), we define two polynomials $a(x)$ and $A(x)$ that are associated as **a** and **A**, as follows:

a	$\overset{DFT}{\longleftrightarrow}$	**A**
$a(x) = \sum_{t=0}^{N-1} a_t x^t$		$A(x) = -\sum_{k=0}^{N-1} A_k x^k$

These two polynomials are called a time domain polynomial and a spectral polynomial of the sequence, respectively. Then we have the following identities (the first version of the DFT):

a	$\overset{DFT}{\longleftrightarrow}$	**A**
$a(\alpha^k)$	$=$	A_k
a_t	$=$	$A(\alpha^{-t})$

From Theorem 6.3, the linear span of **a** is equal to the number of zeros of $a(x)$ in $\mathbb{F}_{q^n}$ subtracted from N. In detail,

$$LS(\mathbf{a}) = w(\mathbf{A})$$
$$= |\{k \mid A_k \neq 0, 0 \leq k < N\}|$$
$$= N - |\{k \mid A_k = 0, 0 \leq k < N\}|$$
$$= N - |\{k \mid a(\alpha^k) = 0, 0 \leq k < N\}|.$$

Note that α^k is a root of $x^N - 1$ for $0 \leq k < N$. Thus, α^k is a root of $a(x)$ if and only if α^k is a root of $\gcd(a(x), x^N - 1)$. Therefore, we have established the following result.

Corollary 6.1 (GCD Method)

$$LS(\mathbf{a}) = N - \deg(\gcd(a(x), x^N - 1))$$

where $\deg(g(x))$ denotes the degree of the polynomial $g(x)$.

Example 6.7 Let $q = 2$, $N = 15 \Rightarrow n = 4$, and let $\mathbf{a} = (111001001110001)$ be a binary sequence with period 15. Then

$$a(x) = 1 + x + x^2 + x^5 + x^8 + x^9 + x^{10} + x^{14}.$$

Because $a(x)$ can be factorized into the following factors:

$$a(x) = (x^4 + x + 1)(x + 1)(x^2 + x + 1)^2(x^5 + x^4 + x^2 + x + 1),$$

this implies

$$\gcd(a(x), x^{15} + 1) = (x^4 + x + 1)(x + 1)(x^2 + x + 1),$$

which implies

$$LS(\mathbf{a}) = 15 - 7 = 8.$$

6.4 One-to-one correspondence between sequences and functions

The trace representation of a sequence of period $N \mid (q^n - 1)$ induces a function from $\mathbb{F}_{q^n}$ to $\mathbb{F}_q$. In this section, we investigate a one-to-one correspondence between sequences with period $N \mid (q^n - 1)$ and functions from $\mathbb{F}_{q^n}$ to $\mathbb{F}_q$. Then, in the next section, we introduce the Hadamard transform and the convolution transform for the sequences in terms of their corresponding functions. In the following discussion, we will first restrict ourselves to the case $q = 2$. Let

- n be a positive integer;
- S_2 be the set of all binary sequences with period $N \mid (2^n - 1)$; and
- $\mathcal{F}_2$ be the set of all (polynomial) functions from $\mathbb{F}_{2^n}$ to $\mathbb{F}_2$.

6.4.1 DFT of functions and their trace representations

Any function f in $\mathcal{F}_2$ (i.e., f is a function from $\mathbb{F}_{2^n}$ to $\mathbb{F}_2$), using Lagrange interpolation and noticing that $x^{2^n} = x$ for $x \in \mathbb{F}_{2^n}$, can be represented as a polynomial of degree $\leq 2^n - 1$. In other words, we may define the (discrete) Fourier transform for functions in $\mathcal{F}_2$ in terms of Lagrange interpolation.

Definition 6.3 For $f \in \mathcal{F}_2$, the (discrete) Fourier Transform of f is defined to be

$$A_k = \sum_{x \in \mathbb{F}_{2^n}^*} f(x)x^{-k}, \quad k = 1, \ldots, 2^n - 1, \quad A_0 = f(0). \tag{6.10}$$

Note that $A_{2^n-1} = \sum_{x \in \mathbb{F}_{2^n}^*} f(x)$ because $x^{2^n-1} = 1$ for all $x \in \mathbb{F}_{2^n}^*$. The inverse formula of Eq. (6.10) is given as follows:

$$f(x) = \sum_{k=0}^{2^n-1} A_k x^k, \quad x \in \mathbb{F}_{2^n}^*. \tag{6.11}$$

Therefore, any function in $\mathcal{F}_2$ can be represented as a polynomial in x, where the coefficients are computed by Eq. (6.10). Thus when we speak of functions from $\mathbb{F}_{2^n}$ to $\mathbb{F}_2$, we assume that their polynomial functions are given. If two such functions are equal, then their polynomial representations are equal. Hence, we have the following lemma.

Lemma 6.4 *Let* $f(x) = \sum_{i=0}^{2^n-1} c_i x^i, c_i \in \mathbb{F}_{2^n}$ *and* $g(x) = \sum_{i=0}^{2^n-1} d_i x^i, d_i \in \mathbb{F}_{2^n}$ *be two polynomial functions from* $\mathbb{F}_{2^n}$ *to* $\mathbb{F}_2$. *Then* $f(x) = g(x)$ *if and only if* $c_i = d_i, 0 \le i \le 2^n - 1$.

Next, we will show that a function in $\mathcal{F}_2$ can be represented as a sum of trace functions.

Theorem 6.5 (TRACE REPRESENTATION OF FUNCTIONS) *Any nonzero function* $f(x)$ *from* $\mathbb{F}_{2^n}$ *to* $\mathbb{F}_2$ *can be represented as*

$$f(x) = \sum_{k \in \Gamma_2(n)} Tr_1^{n_k}(A_k x^k) + A_{2^n-1} x^{2^n-1}, \quad A_k \in \mathbb{F}_{2^{n_k}}, A_{2^n-1} \in \mathbb{F}_2, \tag{6.12}$$

where $\Gamma_2(n)$ *is the set consisting of all coset leaders modulo* $2^n - 1$ *and* n_k *is the size of the coset* C_k.

Proof. Let T consist of all functions represented by Eq. (6.12). For $f(x)$, we first consider a single trace term $g(x) = Tr_1^m(A_k x^k)$ of $f(x)$ for each $k \in \Gamma_2(n)$. Here we write $m = n_k$ for simplicity. Therefore, $|C_k| = m$, a divisor of n, x^k maps x from $\mathbb{F}_{2^n}$ to $\mathbb{F}_{2^m}$, and $Tr_1^n(A_k x)$ maps the images of x^k from $\mathbb{F}_{2^m}$ to $\mathbb{F}_2$. In other words, we have the following composition:

$$\mathbb{F}_{2^n} \xrightarrow{x^k} \mathbb{F}_{2^m} \xrightarrow{Tr_1^m(A_k x)} \mathbb{F}_2,$$

denoted by $g(x) = Tr_1^m(A_k x) \circ x^k, x \in \mathbb{F}_{2^n}$ which is a composition of $Tr_1^m(A_k x)$ and x^k. Thus $Tr_1^m(A_k x^k)$ maps $x \in \mathbb{F}_{2^n}$ to an element in $\mathbb{F}_2$. Consequently, $Tr_1^m(A_k x^k)$ is a function from $\mathbb{F}_{2^n}$ to $\mathbb{F}_2$. For the term $A_{2^n-1} x^{2^n-1}$, we have

$$A_{2^n-1} x^{2^n-1} = \begin{cases} 0 & \text{if } x = 0 \\ A_{2^n-1} & \text{if } x \in \mathbb{F}_{2^n}^*. \end{cases} \tag{6.13}$$

Thus $A_{2^n-1} x^{2^n-1}$ is also a function in $\mathcal{F}_2$. Therefore, $f(x)$ is a sum of functions from $\mathbb{F}_{2^n}$ to $\mathbb{F}_2$, so that $f(x)$ is a function from $\mathbb{F}_{2^n}$ to $\mathbb{F}_2$. Hence we have established that

$$T \subset \mathcal{F}_2. \tag{6.14}$$

In the following, we will show that $|T| = |\mathcal{F}_2|$. Note that there are 2^{2^n} functions in $\mathcal{F}_2$, so we only need to count the cardinal number of T. Let $\Delta_2(m)$ be the set consisting of all coset leaders modulo $2^n - 1$ with size m, where $m \mid n$. Therefore $\Gamma_2(n) = \cup_{m|n} \Delta_2(m)$. For each divisor m of n, $|\Delta_2(m)|$ is equal to $I_2(m)$, the number of irreducible polynomials over $\mathbb{F}_2$ of degree m for $m > 1$ (see Section 3.6 in Chapter 3). For $m = 1$, $\Delta_2(1) = 1$ and $I_2(1) = 2$. We put $2^n - 1$, the index of A_{2^n-1}, into $\Delta_2(1)$, still denoted by $\Delta_2(1)$. After this modification, $|\Delta_2(1)| = I_2(1) = 2$. Consequently, we can decompose $f(x)$ as follows:

$$f(x) = \sum_{m|n} \sum_{k \in \Delta_2(m)} Tr_1^m(A_k x^k), \ A_k \in \mathbb{F}_{2^m}, x \in \mathbb{F}_{2^n}. \tag{6.15}$$

According to Lemma 6.4, different sequences $\{A_k\}_{k \in \Gamma_2(n)}$ produce different functions. Thus

$$|T| = \prod_{m|n} |\{\{A_k\}_{k \in \Delta_2(m)} \text{ such that } A_k \in \mathbb{F}_{2^m}\}|.$$

Because there are 2^m ways to choose A_k for every $k \in \Delta_2(m)$, there are $(2^m)^{\Delta_2(m)} = 2^{m\Delta_2(m)}$ ways to choose the sequence $\{A_k\}_{k \in \Delta_2(m)}$. Hence

$$|T| = \prod_{m|n} 2^{m\Delta_2(m)} = \prod_{m|n} 2^{mI_2(m)} = 2^{\sum_{m|n} mI_2(m)}.$$

Again from the formula for counting the number of the irreducible polynomials over $\mathbb{F}_2$ of degrees dividing n in Section 3.6 of Chapter 3, we have $\sum_{m|n} mI_2(m) = 2^n$. Therefore, we have $|T| = 2^{2^n}$, which establishes the assertion. $\qquad \square$

6.4.2 Method from functions to sequences

Next we will show that each function in $\mathcal{F}_2$ produces a binary sequence with period $N \mid (2^n - 1)$.

Lemma 6.5 *Let $f(x) \in \mathcal{F}_2$, as defined by Eq. (6.15), let α be a primitive element of $\mathbb{F}_{2^n}$, and let $\mathbf{a} = \{a_i\}$ whose elements are given by*

$$a_i = f(\alpha^i), i = 0, 1, \ldots. \tag{6.16}$$

Then the period of $\mathbf{a}$ is a factor of $2^n - 1$.

(Note. $\mathbf{a}$, defined by (6.16), is called an evaluation of the function $f(x)$ at α (in the terminology of algebraic geometry).)

Proof. With the notion used in the proof of Theorem 6.5, notice that for each $k \in \Delta_2(m)$, the evaluation of a single trace term $Tr_1^m(A_k x^k)$ of $f(x)$ at α is a binary sequence with period dividing $2^m - 1$, where m is the size of the coset containing k, so $m \mid n$. Therefore, the period of this evaluation divides $2^n - 1$. For the term $A_{2^n-1} x^{2^n-1}$, the evaluation produces the constant sequence $(A_{2^n-1}, A_{2^n-1}, \ldots)$. This will be combined with the constant sequence $(A_0, A_0, \ldots)$. The sum of these two constant sequences has period 1. Thus, the period of **a** is equal to the least common multiple of these numbers (see Corollary 4.5 in Section 4.4 of Chapter 4. Because $2^n - 1$ is a multiple of each of these periods, the period of **a** divides $2^n - 1$. $\qquad\square$

Thus, for any function $f \in \mathcal{F}_2$, by evaluating $f(\alpha^t), t = 0, 1, \ldots$, we get a sequence over $\mathbb{F}_2$ with period dividing $2^n - 1$, so the evaluation is a sequence in $\mathcal{S}_2$. From the proof of Lemma 6.5, let $g(x)$ be the function that contains all trace terms for which the indices k of A_k are nonzero elements in $\Gamma_2(n)$. Then

$$f(x) = A_0 + g(x) + A_{2^n-1} x^{2^n-1}, \quad A_0, A_{2^n-1} \in \mathbb{F}_2. \qquad (6.17)$$

Note that the sequence generated from Lemma 6.5 is an evaluation of $f(x)$ on the multiplicative group of $\mathbb{F}_{2^n}$; that is, $\mathbb{F}_{2^n}^* = \{\alpha^i \mid i = 0, 1, \ldots, 2^n - 2\}$, which is a cyclic group of order $2^n - 1$. In particular, $A_{2^n-1} x^{2^n-1} = A_{2^n-1}$ for all $x \in F_{2^n}^*$. From Eq. (6.17), we have $f(x) = (A_0 + A_{2^n-1}) + g(x) = c + g(x), x \in \mathbb{F}_{2^n}^*$ where $c = A_0 + A_{2^n-1} \in \mathbb{F}_2$. Because there are 2^{2^n-2} ways to choose $g(x)$ and 2 ways to choose c, the functions in $\mathcal{F}_2$ can only produce 2^{2^n-1} different sequences. Let $\mathcal{F}_2^-$ consist of all functions in $\mathcal{F}_2$ where $A_0 = 0$ in the representation (6.17) or equivalently (6.12). We will establish the one-to-one correspondence between $\mathcal{S}_2$ and $\mathcal{F}_2^-$. (Sometimes we also say this is the one-to-one correspondence between $\mathcal{S}_2$ and $\mathcal{F}_2$ if there is no risk of confusion.)

6.4.3 From sequences to functions: Method 1

Lemma 6.6 (LIFT OF COSETS) *Let n be the order of 2 modulo N where $N < 2^n - 1$, $v = \frac{2^n-1}{N}$, $C_s(N)$ is the coset containing s modulo N, and $vC_s(N) = \{vj \mid j \in C_s(N)\}$ where the multiplication is performed in $\mathbb{Z}$. Let $C_{vs}(2^n - 1)$ be the coset containing vs modulo $2^n - 1$. Then $vC_s(N) = C_{vs}(2^n - 1)$.*

Proof. Note that $vj < 2^n - 1$ for each $j \in C_s(N)$. Let $m = |C_s(N)|$ and $k = |C_{vs}(2^n - 1)|$. Then

$$C_s(N) = \{s, 2s, \ldots, 2^{m-1}s\},$$

where the elements in C_s are reduced modulo N and

$$C_{vs}(2^n - 1) = \{vs, 2vs, \ldots, 2^{k-1}vs\},$$

where the elements in $C_{vs}(2^n - 1)$ are reduced modulo $2^n - 1$. From the definition of m and k, we have $s - s2^m = Nt$ and $vs - vs2^k = (2^n - 1)r$. Because $Nv = 2^n - 1$, it is immediate that $m = k$. Together with the definition of $vC_s(N)$, the assertion follows. □

Let $g(x)$ be the trace representation of **a** with $a_t = g(\beta^t), t = 0, 1, \ldots,$ where $\beta \in \mathbb{F}_{2^n}$ with order $N \mid (2^n - 1)$. We set $g(0) = 0$. Because $\beta \in \mathbb{F}_{2^n}$ and α is a primitive element in $\mathbb{F}_{2^n}$, there exists a positive integer v such that $\beta = \alpha^v$ where $v = \frac{2^n-1}{N}$. Thus, we have

$$a_t = g(\beta^t) = g(\alpha^{vt}) = g((\alpha^t)^v), t = 0, 1, \ldots.$$

If $N = 2^n - 1$, then $g(x)$, for $x \in < \beta >= \mathbb{F}_{2^n}^*$, maps all nonzero elements in $\mathbb{F}_{2^n}$ to $\mathbb{F}_2$ with $g(0) = 0$. Thus, $g(x)$ is a functions from $\mathbb{F}_{2^n}$ to $\mathbb{F}_2$. If $N < 2^n - 1$, then $v > 1$. Let $f(x) = g(x^v)$. Note that x^v maps the elements in $\mathbb{F}_{2^n}^* = <\alpha>$ to $<\beta> = \{\beta^i \mid i = 0, 1, \ldots, N - 1\}$ and $g(x)$ maps $<\beta>$ to $\mathbb{F}_2$. In other words, we have the following composition of these two maps:

$$\mathbb{F}_{2^n}^* \xrightarrow{x^v} < \beta > \xrightarrow{g(x)} \mathbb{F}_2$$

with $g(0) = f(0) = 0$. Thus $f(x)$ is a map from $\mathbb{F}_{2^n}$ to $\mathbb{F}_2$; that is, $f(x) \in \mathcal{F}_2^-$ for which the coset leaders associated with the trace representation of $f(x)$ are obtained from the coset leaders associated with the trace representation of $g(x)$ by multiplying by v. Thus we have established the following result.

Theorem 6.6 *With the above notation, the following map*

$$\eta : \mathcal{S}_2 \longrightarrow \mathcal{F}_2^- \tag{6.18}$$

$$\mathbf{a} \mapsto g(x^v) = f(x) \tag{6.19}$$

*is a one-to-one correspondence between $\mathcal{S}_2$ and $\mathcal{F}_2^-$ where $g(x)$ is the trace representation of **a**. Conversely, for a given function $f(x)$ in $\mathcal{F}_2^-$, an evaluation of $f(x)$ at α is a sequence in $\mathcal{S}_2$; that is,*

$$\eta^{-1}(f(x)) = \mathbf{a},$$

where $a_i = f(\alpha^i), i = 0, 1, \ldots.$

We display this one-to-one correspondence in Figure 6.1. From now on, we use the notation $\mathbf{a} \leftrightarrow f(x)$ to represent the fact that $f(x) = g(x^v)$ where $g(x)$ is the trace representation of **a**. Note that evaluations of both $g(x)$ and $f(x) = g(x^v)$ produce the same sequence **a**, because $a_i = g(\beta^i) = g(\alpha^{vi}) = f(\alpha^i), \forall i$. The set consisting of the exponents that appear in the trace terms of $f(x)$ is said to be a null spectrum set of $f(x)$ or **a**. Note that from Lemma 6.6, the exponents

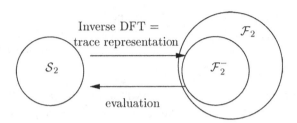

Figure 6.1. One-to-one correspondence.

that appear in the trace terms of $f(x)$ are equal to the exponents that appear in the trace terms of $g(x)$ multiplied by v.

Example 6.8 Let $\mathbf{a} = (01111)$. Then we have $N = 5 \,|\, (2^4 - 1)$ and $v = 3$. Let $\beta = \alpha^3$ be an element in $\mathbb{F}_{2^4}$ of order 5 where α is a primitive element of $\mathbb{F}_{2^4}$ that satisfies $\alpha^4 + \alpha + 1 = 0$. Then the trace representation of $\mathbf{a}$ is given by $g(x) = Tr(x)$; that is, $a_i = Tr(\beta^i)$. Thus, $f(x) = Tr(x^3)$ is a function from $\mathbb{F}_{2^4}$ to $\mathbb{F}_2$.

6.4.4 From sequences to functions: Method 2

We can directly obtain the function $f(x)$ of the correspondence from $\mathbb{F}_{2^n}$ to $\mathbb{F}_2$ by computing the trace representation of a given sequence, $\mathbf{a}$, in S_2, when $N < 2^n - 1$, as follows. Let α be a primitive element of $\mathbb{F}_{2^n}$. We associate $\mathbf{a}$ with a function $f(x)$ in the following fashion:

$$f(\alpha^i) = a_i, i = 0, 1, \dots, 2^n - 2, f(0) = 0. \qquad (6.20)$$

We then apply the DFT, Eq. (6.10), to $f(x)$, which gives

$$A_k = \sum_{t=0}^{2^n-2} a_t \alpha^{-tk}, k = 1, \dots, 2^n - 2, 2^n - 1. \qquad (6.21)$$

Furthermore, let $v = \frac{2^n-1}{N}$. We have

$$A_k = 0, \forall k \not\equiv 0 \ (\mathrm{mod}\ v)$$

and

$$f(x) = \sum_{j=0}^{N-1} A_{jv} x^{jv}.$$

In other words, the Fourier spectrum A_k is equal to zero for all the indexes k that are not multiples of v. (See Exercise 10 for the proof of this result.) The polynomial form or the trace representation of $f(x)$, obtained in this way, is a function from $\mathbb{F}_{2^n}$ to $\mathbb{F}_2$. For this method, we do not need the lift process in Method 1. From now on, when we say the trace representation of a sequence in $\mathcal{S}_2$ is a function in $\mathcal{F}_2$, we always mean that the trace representation of the sequence is computed in terms of a primitive element in $\mathbb{F}_{2^n}$ as shown in Eq. (6.21). Note that this is the case, because we usually use a primitive polynomial to generate $\mathbb{F}_{2^n}$.

For example, if we compute the DFT of $\mathbf{a} = (01111)$, in Example 6.8, using the primitive element α, then we directly obtain that the trace representation of $\mathbf{a}$ is $Tr(x^3)$.

Remark 6.3 If $N = 2^n - 1$, then $A_k, k = 1, \ldots, 2^n - 2$, computed using Eq. (6.21) are equal to those computed using the second version of the DFT of the sequence $\mathbf{a}$ (see Table 6.2) when the primitive elements involved are the same. For $k = 2^n - 1$ in Eq. (6.21) and A_0 in the second version of thee DFT of the sequence, respectively, we have $A_{2^n-1} = \sum_{t=0}^{2^n-2} a_t = A_0$. So, two methods give the same trace representation of $\mathbf{a}$. (Note that A_0 in the second version of the DFT of the sequence is different from A_0 in the DFT of the function where $A_0 = f(0)$, although we use the same notation.)

6.4.5 Examples

Example 6.9 (a) For $n = 4$, let $f(x)$ be a function from $\mathbb{F}_{2^4}$ to $\mathbb{F}_2$. According to Theorem 6.5, we have

$$f(x) = Tr_1^4(A_1 x) + Tr_1^4(A_3 x^3) + Tr_1^4(A_7 x^7) + Tr_1^2(A_5 x^5) + A_0 + A_{15} x^{15},$$

$$A_1, A_3, A_7 \in \mathbb{F}_{2^4}, A_5 \in \mathbb{F}_{2^2}, \text{ and } A_0, A_{15} \in \mathbb{F}_2. \quad (6.22)$$

From this representation, there are $2^4 \cdot 2^4 \cdot 2^4 \cdot 2^2 \cdot 2 \cdot 2 = 2^{16}$ different functions. Let α be a primitive element in $\mathbb{F}_{2^4}$; $\mathbf{a}$ the evaluation of $f(x)$ at α, that is, $a_i = f(\alpha^i), i = 0, 1, \ldots, 14$; and $\mathbf{a}_j$ the evaluation of $Tr(A_j x^j)$ at α for $j \in \Gamma_2(4) = \{0, 1, 3, 5, 7\}$. Then

$$\mathbf{a} = \mathbf{a}_1 + \mathbf{a}_3 + \mathbf{a}_7 + \mathbf{a}_5 + \mathbf{a}_0, \quad (6.23)$$

where $\mathbf{a}_0 = (c, c, \ldots), c \in \mathbb{F}_2$, a constant sequence. Note that $\mathbf{a}_j$ is a sequence in $G(f_j)$ where $f_j(x)$ is the minimal polynomial of $\alpha^j, 0 \neq j \in \Gamma_2(4)$, and $|G(f_i)| = 2^4$ for $i \in \{1, 3, 7\}$ and $|G(f_5)| = 2^2$. In the following, we list representatives of shift-distinct classes for these sets. The other elements in these sets can be obtained by the shift operator.

$$G(f_0) = \{(0), (1)\}$$
$$G(f_1) = \{(0), (000100110101111)\}$$
$$G(f_3) = \{(0), (01111), (00110), (10100)\}$$

$$G(f_7) = \{(0), (011110101100100)\}$$
$$G(f_5) = \{(0), (011)\}.$$

Thus, there are $2^4 \cdot 2^4 \cdot 2^4 \cdot 2^2 \cdot 2 = 2^{15}$ sequences constructed by evaluation of $f(x)$ at α when the A_i's run through $\mathbb{F}_{2^4}$. Or equivalently, any sequence evaluated using a function from $\mathbb{F}_{2^4}$ to $\mathbb{F}_2$ is a sum of sequences where each summand is taken from a different $G(f_j)$.

(b) Let $n = 5$, α be a primitive element in $\mathbb{F}_{2^5}$ with $\alpha^5 + \alpha^3 + 1 = 0$, and $f(x)$ be a function from $\mathbb{F}_{2^5}$ to $\mathbb{F}_2$. According to Theorem 6.5, we have

$$\begin{aligned} f(x) = Tr(A_1 x) &+ Tr(A_3 x^3) + Tr(A_5 x^5) + Tr(A_7 x^7) + Tr(A_{11} x^{11}) \\ &+ Tr(A_{15} x^{15}) + A_0 + A_{31} x^{31}, \end{aligned} \tag{6.24}$$
$$A_i \in \mathbb{F}_{2^5}, i \notin \{0, 31\}, \text{ and } A_0, A_{31} \in \mathbb{F}_2$$

where $Tr(x)$ is the trace function from $\mathbb{F}_{2^5}$ to $\mathbb{F}_2$. For $\mathbf{a} \leftrightarrow f(x)$, we have

$$\mathbf{a} = \mathbf{a}_1 + \mathbf{a}_3 + \mathbf{a}_5 + \mathbf{a}_7 + \mathbf{a}_{11} + \mathbf{a}_{15} + \mathbf{a}_0, \tag{6.25}$$

where $\mathbf{a}_i \in G(f_i)$, where f_i is the minimal polynomial of α^i, $i \in \Gamma_2(5) = \{0, 1, 3, 5, 7, 11, 15\}$ (see Table 3.3 in Section 3.3 in Chapter 3 for the f_i's). The shift-distinct classes of $G(f_i)$ are listed as follows:

$$G(f_0) = \{(0), (1)\}$$
$$G(f_1) = \{(0), (1000010101110110001111100110100)\}$$
$$G(f_3) = \{(0), (1001001100001011010100011101111)\}$$
$$G(f_5) = \{(0), (1110110011100001101010010001011)\}$$
$$G(f_7) = \{(0), (1111101110001010110100001100100)\}$$
$$G(f_{11}) = \{(0), (1110100010010101100001110011011)\}$$
$$G(f_{15}) = \{(0), (1001011001111100011011101010000)\}.$$

Thus there are $(2^5)^6 \cdot 2 = 2^{31}$ sequences constructed by evaluation of $f(x)$ at α when the A_i's run through $\mathbb{F}_{2^5}$. (Note. The sequences in $G(f_i)$, $0 \neq i \in \Gamma_2(5)$, are all the shift-distinct m-sequences of period 31.)

Let q be a prime or a power of a prime, $\mathcal{S}_q$ be the set consisting of all sequences over $\mathbb{F}_q$ of period $N \mid (q^n - 1)$, $\mathcal{F}_q$ be the set consisting of all functions from $\mathbb{F}_{q^n}$ to $\mathbb{F}_q$, and α be a primitive element of $\mathbb{F}_{q^n}$. Then we have the following results whose proofs are similar to those for $q = 2$, so we omit them here. For $f \in \mathcal{F}_q$, the discrete Fourier transform of f is defined as

$$A_k = -\sum_{x \in \mathbb{F}_{q^n}} f(x) x^{-k}, \ k = 1, \ldots, q^n - 1, \ A_0 = f(0). \tag{6.26}$$

The inverse formula of (6.26) is given as follows:

$$f(x) = \sum_{k=0}^{q^n-1} A_k x^k, x \in \mathbb{F}_{q^n}. \tag{6.27}$$

Theorem 6.7 *Any function in $\mathcal{F}_q$ can be represented as*

$$f(x) = \sum_{k \in \Gamma_q(n)} Tr_1^{n_k}(A_k x^k) + A_{q^n-1} x^{q^n-1}, \tag{6.28}$$

$$A_k \in \mathbb{F}_{q^{n_k}}, A_{q^n-1} \in \mathbb{F}_q, x \in \mathbb{F}_{q^n},$$

where $\Gamma_q(n)$ is the set consisting of all coset leaders modulo $q^n - 1$ and n_k is the size of the coset C_k. Furthermore, $\mathbf{a} = \{a_i\}$, an evaluation of $f(x)$ at α, that is, $a_i = f(\alpha^i), \forall i$, belongs to S_q. Conversely, given $\mathbf{a} \in S_q$, let $f(x)$ be the trace representation of $\mathbf{a}$, computed using a primitive element of $\mathbb{F}_{q^n}$. Then $f(x)$ belongs to $\mathcal{F}_q$.

6.5 Hadamard transform and convolution transform

In this section, we will restrict ourselves to the case that $\mathbf{a}$ is a sequence over $\mathbb{F}_p$ of period $N \mid (p^n - 1)$ where p is a prime, that is, $\mathbf{a} \in S_p$, and $f(x)$ is its trace representation computed using a primitive element of $\mathbb{F}_{p^n}$. We will denote $q = p^n$ (note that here the meaning of q is different from that in the previous sections). In this section, we define the Hadamard transform and the convolution transform of $\mathbf{a}$ in terms of its trace representation $f(x)$, a function in $\mathcal{F}_p$.

Recall the following notation that was introduced in Chapter 2 (also in Chapter 5): $\mathbb{C}$ is the complex field, ω is a primitive pth root of unity, and the additive character of $\mathbb{F}_p$ is defined as $\chi(x) = \omega^x$, $x \in \mathbb{F}_p$. For $x \in \mathbb{C}$, x^* represents the complex conjugate of x.

6.5.1 Hadamard transform

Definition 6.4 *With the notation above,*

1. For $\mathbf{a} \leftrightarrow f(x)$, the Hadamard transform of $\mathbf{a}$ or $f(x)$ is defined by

$$\widehat{f}(\lambda) = \sum_{x \in \mathbb{F}_q} \chi(Tr(\lambda x)) \chi^*(f(x)) \tag{6.29}$$

$$= \sum_{x \in \mathbb{F}_q} \omega^{Tr(\lambda x) - f(x)} \quad (q = p^n). \tag{6.30}$$

In particular, if $p = 2$, the Hadamard transform of $\mathbf{a}$ or $f(x)$ is defined by

$$\widehat{f}(\lambda) = \sum_{x \in \mathbb{F}_q} (-1)^{Tr(\lambda x) + f(x)}. \tag{6.31}$$

2. *For a function $F(x)$ from $\mathbb{F}_q$ to $\mathbb{C}$, the Hadamard transform of $F(x)$ is defined as*

$$\widehat{F}(\lambda) = \sum_{x \in \mathbb{F}_q} \chi(Tr(\lambda x)) F^*(x). \tag{6.32}$$

Note that $\widehat{f}(\lambda)$ may not be an integer if $p > 2$. From the definition and the properties of trace functions, the following lemma on the inverse Hadamard transform is immediate.

Lemma 6.7 *The inverse Hadamard transform of $f(x)$ is given as follows:*

$$\chi(f(x)) = \frac{1}{q} \sum_{\lambda \in F_q} \chi(Tr(\lambda x)) \widehat{f}^*(\lambda). \tag{6.33}$$

In particular, if $p = 2$,

$$(-1)^{f(x)} = \frac{1}{q} \sum_{\lambda \in F_q} (-1)^{Tr(\lambda x))} \widehat{f}(\lambda). \tag{6.34}$$

We have the following relationship between the Hadamard transform of $f(x)$ and the crosscorrelation between the sequence $\mathbf{a}$, an evaluation of f, and the m-sequence with trace representation $Tr(x)$.

Proposition 6.3 *Let α be a primitive element in $\mathbb{F}_{2^n}$ and f be a function from $\mathbb{F}_{p^n}$ to $\mathbb{F}_p$ with $f(0) = 0$. Let $\mathbf{a} = \{a_i\}$ and $\mathbf{c} = \{c_i\}$ whose elements are given by $a_i = f(\alpha^i)$ and $c_i = Tr(\alpha^i)$, $i = 0, 1, \ldots$, respectively. Then*

$$C_{\mathbf{c},\mathbf{a}}(\tau) + 1 = \widehat{f}(\lambda), \text{ where } \lambda = \alpha^\tau, \tau = 0, 1, \ldots, q - 2. \tag{6.35}$$

Proof. From the definition of the crosscorrelation function of the sequences $\mathbf{c}$ and $\mathbf{a}$, we have

$$C_{\mathbf{c},\mathbf{a}}(\tau) = \sum_{i=0}^{q-2} \chi(c_{i+\tau} - a_i)$$

$$= \sum_{i=0}^{q-2} \chi(Tr(\alpha^{i+\tau}) - f(\alpha^i))$$

$$= \sum_{x \in \mathbb{F}_q} \chi(Tr(\lambda x) - f(x)) - 1, \ \lambda = \alpha^\tau$$

$$= \widehat{f}(\lambda) - 1.$$

The identity (6.35) shows that the crosscorrelation between the m-sequence having the trace representation $Tr(\lambda x)$ and an arbitrary sequence in $\mathcal{S}_p$ is one

Table 6.3. *The Hadamard transform and its inverse (first version)*

Hadamard Transform (HT):
$\widehat{f}(\lambda) = \sum_{x \in \mathbb{F}_q} \chi(Tr(\lambda x)) \chi^*(f(x)) = \sum_{x \in \mathbb{F}_q} \omega^{Tr(\lambda x) - f(x)}$
Inverse of HT: $\omega^{f(x)} = \frac{1}{q} \sum_{\lambda \in F_q} \omega^{Tr(\lambda x)} \widehat{f}^*(\lambda)$
HT for $p = 2$: $\widehat{f}(\lambda) = \sum_{x \in \mathbb{F}_q} (-1)^{Tr(\lambda x) + f(x)}$
Inverse of HT: $(-1)^{f(x)} = \frac{1}{q} \sum_{\lambda \in F_q} (-1)^{Tr(\lambda x)} \widehat{f}(\lambda)$

less than the Hadamard transform of the trace representation of the sequence **a**. Note that the period of **a** is N, which is a factor of $q - 1$, and the m-sequence **c** has period $q - 1$. Thus $\mathrm{lcm}(N, q - 1) = q - 1$.

We feature the Hadamard transform and the inverse Hadamard transform in Table 6.3. Another version of the Hadamard transform is defined in Table 6.4.

However, the first version has a close connection to the concept of non-linearity of boolean functions, which has important applications in the design of stream ciphers and block ciphers. It is also directly connected to cross-correlation between an m-sequence with a linear function as its trace representation and an arbitrary sequence with period dividing $q - 1$. Thus, in this book, we will use the first version of the definition of the Hadamard transform.

Example 6.10 Let $\mathbb{F}_{2^3}$ be defined by $\alpha^3 + \alpha + 1 = 0$. Let $\mathbf{a} = (1110010)$, an m-sequence over $\mathbb{F}_2$ of period 7. Thus $\mathbf{a} \leftrightarrow f(x) = Tr(x^3)$. The Hadamard transform of $f(x)$ is given by

$$\widehat{f}(\lambda) = \sum_{x \in \mathbb{F}_{2^3}} (-1)^{Tr(\lambda x) + Tr(x^3)}.$$

We compute the values of the Hadamard transform of $f(x)$ as follows.

Table 6.4. *The second version of the Hadamard transform*

Hadamard Transform (HT): $\widehat{f}(\lambda) = \frac{1}{\sqrt{q}} \sum_{x \in \mathbb{F}_q} \omega^{Tr(\lambda x) - f(x)}$
Inverse of HT: $\omega^{f(x)} = \frac{1}{\sqrt{q}} \sum_{\lambda \in \mathbb{F}_q} \omega^{Tr(\lambda x)} \widehat{f}^*(\lambda)$

$\widehat{f}(0) = \sum_{x \in F_{2^3}} (-1)^{Tr(x^3)} = 0$	$\widehat{f}(1) = \sum_{x \in F_{2^3}} (-1)^{Tr(x+x^3)} = -4$
$\widehat{f}(\alpha) = \sum_{x \in F_{2^3}} (-1)^{Tr(\alpha x+x^3)} = 0$	$\widehat{f}(\alpha^3) = \sum_{x \in F_{2^3}} (-1)^{Tr(\alpha^3 x+x^3)} = 4$
$\widehat{f}(\alpha^2) = \sum_{x \in F_{2^3}} (-1)^{Tr(\alpha^2 x+x^3)} = 0$	$\widehat{f}(\alpha^6) = \sum_{x \in F_{2^3}} (-1)^{Tr(\alpha^6 x+x^3)} = 4$
$\widehat{f}(\alpha^4) = \sum_{x \in F_{2^3}} (-1)^{Tr(\alpha^4 x+x^3)} = 0$	$\widehat{f}(\alpha^5) = \sum_{x \in F_{2^3}} (-1)^{Tr(\alpha^5 x+x^3)} = 4$

Thus

$$(\widehat{f}(0), \widehat{f}(1), \widehat{f}(\alpha), \ldots, \widehat{f}(\alpha^6)) = (0, -4, 0, 0, 4, 0, 4, 4).$$

Let $\mathbf{c} = \{Tr(\alpha^i)\} = (1001011)$, an m-sequence of period 7. From Eq. (6.35), the Hadamard transform of $f(x)$ can be computed in terms of the crosscorrelation between $\mathbf{c}$ and $\mathbf{a}$:

$$\widehat{f}(\lambda) = 1 + C_{\mathbf{c},\mathbf{a}}(\tau) = 1 + \sum_{i=0}^{6} (-1)^{c_{i+\tau}+a_i},$$

where $\lambda = \alpha^\tau$. The latter method does not involve any computation in an extension field. Thus it is frequently used in the practical computation of the Hadamard transform of functions or sequences.

6.5.2 Convolution transform

We introduced the convolution between two functions or two sequences in Section 1.8. In this section, we interpret it as two functions from $\mathbb{F}_q$ to $\mathbb{F}_p$ $(q = p^n)$ in terms of the additive character of $\mathbb{F}_q$.

Definition 6.5 (a) *Let* $\mathbf{a} \leftrightarrow f(x)$ *and* $\mathbf{b} \leftrightarrow g(x)$. *Then the convolution of* $\mathbf{a}$ *and* $\mathbf{b}$ *or the convolution of* $f(x)$ *and* $g(x)$ *is defined by*

$$f * g(w) = \sum_{x \in \mathbb{F}_q} \chi(f(x+w))\chi^*(g(x))$$

$$= \sum_{x \in \mathbb{F}_q} \omega^{f(x+w)-g(x)}.$$

In particular, if $p = 2$, *the above transform becomes*

$$f * g(w) = \sum_{x \in \mathbb{F}_q} (-1)^{f(x+w)+g(x)}.$$

(b) *For two functions* $F(x)$ *and* $G(x)$ *from* $\mathbb{F}_q$ *to* $\mathbb{C}$, *the convolution of* F *and* G *is defined as*

$$F * G(w) = \sum_{x \in \mathbb{F}_q} F(x+w)G^*(x).$$

For the convolution of $f(x)$ and $g(x)$, if we take $g(x) = f(x)$, then

$$f * f(w) = \sum_{x \in F_q} \omega^{f(x+w) - f(x)}.$$

This can be interpreted as the autocorrelation of $f(x)$ through the additive group of F_q, called the additive autocorrelation of $f(x)$. Similarly, when $f \neq g$, we interpret $f * g(w)$ as the crosscorrelation between $f(x)$ and $g(x)$ through the additive group of F_q, called the additive crosscorrelation between f and g. However, the sequence given by $f(x + w)$ is not a shift of the sequence given by $f(x)$ under the additive group of F_q for most cases of $w \in F_q$. The following example shows a modification of the definition of $f(x + w)$.

Example 6.11 With **a** in Example 6.10, $f(x) = Tr(x^3)$. Let $g(x) = Tr(x)$. We compute f and g as follows.

F_{2^3} Labels of elements	F_{2^3} $(\alpha^2, \alpha, 1)$	$f(x) = Tr(x^3)$	$g(x) = Tr(x)$
t_0	$0 = 000$	0	0
t_1	$1 = 001$	1	1
t_2	$\alpha = 010$	1	0
t_3	$\alpha^3 = 011$	0	1
t_4	$\alpha^2 = 100$	1	0
t_5	$\alpha^6 = 101$	0	1
t_6	$\alpha^4 = 110$	1	0
t_7	$\alpha^5 = 111$	0	1

Define

$$A_{f,g}(w) = \sum_{i=0}^{7} (-1)^{f(t_i + t_\tau) + g(t_i)}, \ w = t_\tau, \tau = 0, 1, \ldots, 7,$$

where $t_i + t_\tau$ is reduced modulo 2^3. By computation,

$$\{A_{f,g}(t_\tau)\}_{\tau=0}^{7} = (-4, 4, -4, 4, -4, 4, -4, 4)$$

and

$$\{A_{f,f}(t_\tau)\}_{\tau=0}^{7} = (8, -4, 0, 0, 0, 0, 0, -4).$$

Let $\mathbf{u} = \{u_i\}$ and $\mathbf{v} = \{v_i\}$ be two binary sequences whose elements are given by $u_i = f(t_i)$ and $v_i = g(t_i)$, $i = 0, 1, \ldots, 7$; that is, $\mathbf{u} = (01101010)$ and $\mathbf{v} = (01010101)$. Then we have $\{f(t_i + w)\}_{i=0}^{7} = L^\tau(\mathbf{u})$ and

$$A_{f,g}(t_\tau) = \sum_{i=0}^{7} (-1)^{u_{i+\tau} + v_i}, \ w = t_\tau, \tau = 0, 1, \ldots, 7 \qquad (6.36)$$

$$A_{f,f}(t_\tau) = \sum_{i=0}^{7} (-1)^{u_{i+\tau}+u_i}, \ w = t_\tau, \tau = 0, 1, \ldots, 7, \qquad (6.37)$$

where $i + \tau$ is reduced modulo 2^3.

So, for a given function from $\mathbb{F}_q$ to $\mathbb{F}_p$, we can construct two sequences, one given by the multiplicative group of $\mathbb{F}_q$ and the other given by the additive group of $\mathbb{F}_q$. We formalize this method in the next section.

6.6 Correlation of functions

In this section, we will formally define two types of auto- or crosscorrelation of functions. Let α be a primitive element of $\mathbb{F}_q$. Then $\{1, \alpha, \ldots, \alpha^{n-1}\}$ is a basis for $\mathbb{F}_q$ over $\mathbb{F}_p$ (recall $q = p^n$). For any element x in $\mathbb{F}_q$, we have

$$x = x_0 + x_1\alpha + \cdots + x_{n-1}\alpha^{n-1}, \quad x_i \in \mathbb{F}_p. \qquad (6.38)$$

We define a map ρ from $\mathbb{F}_q$ to $\mathbb{Z}_q$ as follows:

$$\rho(x) = x_0 + x_1 p + \cdots + x_{n-1}p^{n-1}, \quad x \in \mathbb{F}_q, \qquad (6.39)$$

where x is represented by Eq. (6.38). Then, ρ is a one-to-one correspondence between $\mathbb{F}_q$ and $\mathbb{Z}_q$. We arrange an ordering of the elements of $\mathbb{F}_q$ as follows:

$$\mathbb{F}_q = \{t_i \mid i = \rho(t_i) i = 0, 1, \ldots, q-1\}.$$

Let $f \in \mathcal{F}_p$, a function from $\mathbb{F}_q$ to $\mathbb{F}_p$. We associate $f(x)$ with two sequences $\mathbf{a}$ and $\mathbf{u}$, whose elements are given by

$$a_i = f(\alpha^i), i = 0, 1, \ldots,$$
$$u_i = f(t_i), i = 0, 1, \ldots.$$

By this definition, $\mathbf{a}$ is a sequence over $\mathbb{F}_p$ with a period dividing $q - 1$, as we discussed in Section 6.4. The sequence $\mathbf{u}$, defined by the linear space structure of $\mathbb{F}_q$, has a period dividing p^n. The sequence $\mathbf{a}$ was called an evaluation of $f(x)$ at α in Section 6.4. Here, we will call the sequence $\mathbf{u}$ an evaluation of $f(x)$ in the additive group of $\mathbb{F}_q$ with the natural number order ρ and denote this by $\mathbf{a} \leftrightarrow f(x)$ and $\mathbf{u} \overset{\rho}{\leftrightarrow} f(x)$.

Definition 6.6 *Let $f(x)$ and $g(x)$ be two functions from $\mathbb{F}_q$ to $\mathbb{F}_p$. We define*

$$C_{f,g}(\lambda) = \sum_{x \in \mathbb{F}_q} \omega^{f(\lambda x) - g(x)}, \lambda \in \mathbb{F}_q; \qquad (6.40)$$

$$A_{f,g}(w) = \sum_{x \in \mathbb{F}_q} \omega^{f(\sigma(x+w)) - g(x)}, \quad w \in \mathbb{F}_q. \qquad (6.41)$$

where $\sigma(x + w) = \rho^{-1}(\rho(x) + \rho(w))$. $C_{f,g}(\lambda)$ *is called the* (multiplicative) crosscorrelation function *between* f *and* g, *and the* (multiplicative) autocorrelation function *of* f *if* $g = f$, *with* $C_{f,f}(\lambda)$ *denoted by* $C_f(\lambda)$. $A_{f,g}(w)$ *is called the* crosscorrelation function *between* f *and* g *under the additive group, and the autocorrelation function of* f *under the additive group if* $g = f$, *denoted by* $A_f(w)$.

Let $\mathbf{a} \leftrightarrow f(x)$ and $\mathbf{u} \overset{\rho}{\leftrightarrow} f(x)$; let $\mathbf{b} \leftrightarrow g(x)$ and $\mathbf{v} \overset{\rho}{\leftrightarrow} g(x)$. From the definition, we have the following relationships between the correlation defined in Chapter 5 and the additive auto- or crosscorrelation, defined above.

Proposition 6.4 *With the above notation,*

(a)

$$C_f(\lambda) - 1 = \sum_{i=0}^{p^n-2} \omega^{a_{i+\tau}-a_i}, \ \lambda = \alpha^\tau, \forall \tau, \tag{6.42}$$

$$C_f(0) = \sum_{x \in \mathbb{F}_{p^n}} \omega^{f(x)}$$

$$C_{f,g}(\lambda) - 1 = \sum_{i=0}^{p^n-2} \omega^{a_{i+\tau}-b_i}, \ \lambda = \alpha^\tau, \ \forall \tau, \tag{6.43}$$

$$C_{f,g}(0) = \sum_{x \in \mathbb{F}_{p^n}} \omega^{g(x)}$$

$$\widehat{f}(\lambda) = C_{Tr,f}(\lambda), \lambda \in \mathbb{F}_q. \tag{6.44}$$

The expression (6.42) is equal to $C_{\mathbf{a}}(\tau)$, *the autocorrelation of* $\mathbf{a}$, *if the period of* $\mathbf{a}$ *is* $p^n - 1$. *Otherwise, it is equal to a multiple of* $C_{\mathbf{a}}(\tau)$. *Similarly, if the least common multiple of the periods of* $\mathbf{a}$ *and* $\mathbf{b}$ *is equal to* $p^n - 1$, *then Eq. (6.43) is equal to* $C_{\mathbf{a},\mathbf{b}}(\tau)$, *the crosscorrelation of* $\mathbf{a}$ *and* $\mathbf{b}$.

(b)

$$A_f(\lambda) = \sum_{i=0}^{p^n-1} \omega^{u_{i+\tau}-u_i}, \lambda = t_\tau, \tau = 0, 1, \ldots, \tag{6.45}$$

$$A_{f,g}(w) = \sum_{i=0}^{p^n-1} \omega^{u_{i+\tau}-v_i}, \ w = t_\tau, \tau = 0, 1, \ldots. \tag{6.46}$$

where $i + \tau$ *is reduced modulo* p^n. *Equation (6.45) is equal to* $C_{\mathbf{u}}(\tau)$, *the autocorrelation of* $\mathbf{u}$, *if the period of* $\mathbf{u}$ *is* p^n. *Otherwise, it is equal to a multiple of* $C_{\mathbf{u}}(\tau)$. *Similarly, if the least common multiple of the periods of* $\mathbf{u}$ *and* $\mathbf{v}$ *is equal to* p^n, *then Eq. (6.46) is equal to* $C_{\mathbf{u},\mathbf{v}}(\tau)$, *the crosscorrelation between* $\mathbf{u}$ *and* $\mathbf{v}$.

If $p = 2$, the auto- or crosscorrelation functions under the additive group have the following simple forms.

$$A_f(w) = \sum_{i=0}^{2^n-1}(-1)^{u_{i+\tau}-u_i}, \ w = t_\tau, \tau = 0, 1, \ldots,$$

$$A_{f,g}(w) = \sum_{i=0}^{2^n-1}(-1)^{u_{i+\tau}-v_i}, \ w = t_\tau, \tau = 0, 1, \ldots.$$

Example 6.12 (a) Let $f(x) = Tr(x^3)$ and $g(x) = Tr(x)$, the functions from $\mathbb{F}_8$ to $\mathbb{F}_2$ in Example 6.11. We illustrate the multiplicative and additive auto- or crosscorrelations of these two functions in the following table.

$\mathbf{a} \leftrightarrow f(x)$	$\mathbf{u} \overset{\rho}{\leftrightarrow} f(x)$
$\mathbf{a} = (1110100)$, per$(\mathbf{a}) = 7$	$\mathbf{u} = (01101010)$, per$(\mathbf{u}) = 8$
$\mathbf{b} \leftrightarrow g(x)$	$\mathbf{v} \overset{\rho}{\leftrightarrow} g(x)$
$\mathbf{b} = (1001011)$, per$(\mathbf{b}) = 7$	$\mathbf{v} = (01)$, per$(\mathbf{v}) = 2$, a divisor of 8
$\lambda = \alpha^\tau$, $\widehat{f}(\alpha^\tau) = C_{b,a}(\tau) + 1$:	$w = t_\tau$, $A_{f,g}(w) = C_{u,v}(\tau)$:
$(-4, 0, 0, 4, 0, 4, 4)$	$(-4, 4, -4, 4, -4, 4, -4, 4)$
$C_f(\lambda) = C_a(\tau) + 1$:	$A_f(w) = C_u(\tau)$:
$(8, 0, 0, 0, 0, 0, 0, 0)$	$(8, -4, 0, 0, 0, 0, 0, -4)$

(b) Let $\mathbb{F}_{2^4}$ be defined by $\alpha^4 + \alpha + 1 = 0$ (see Example 3.6 in Chapter 3 for the table of this field). Thus, α is a primitive element of $\mathbb{F}_{2^4}$. Let $f(x) = Tr(\alpha x^3)$ where $Tr(x)$ is the trace function from $\mathbb{F}_{2^4}$ to $\mathbb{F}_2$. Let

$$a_i = f(\alpha^i), i = 0, 1, \ldots,$$
$$u_i = f(t_i), i = 0, 1, \ldots.$$

Then

$$\mathbf{a} = \{a_i\} = (00101)$$
$$\mathbf{u} = \{u_i\} = (0001100001110001),$$

where $\mathbf{a}$, the evaluation of $f(x)$ on the multiplicative group of $\mathbb{F}_{2^4}$, has period 5, and $\mathbf{u}$, the evaluation of $f(x)$ on the additive group of $\mathbb{F}_{2^4}$, has period 16, which is the maximum. Furthermore, their respective autocorrelations are given by

τ	0	1	2	3	4
$C_a(\tau)$	5	-3	1	1	-3

and

τ	0	1	2	3	4	5	6	7	8	9	10	11	12	13	14	15
$C_{\mathbf{u}}(\tau)$	16	4	-4	-8	0	4	4	0	0	0	4	4	0	-8	-4	4

From these examples, it is seen that auto- or crosscorrelation of f and g and the auto- or crosscorrelation of f and g under the additive group are completely different. For example, in case (a), f has 2-level autocorrelation, but has 3-valued autocorrelation under the additive group. In case (b), f has multivalued autocorrelation for both the multiplicative and additive groups. It is worth pointing out that the Hadamard transform measures the correlation in terms of the multiplicative group of $\mathbb{F}_q$. On the other hand, the convolution transform measures the correlation in terms of the additive group of $\mathbb{F}_q$. However, these two types of correlation are related, as will be shown in the next section.

Remark 6.4 It would seem that the construction of 2-level autocorrelation sequences over $\mathbb{F}_2$ with period $2^n - 1$, or equivalently, of functions from $\mathbb{F}_{2^n}$ to $\mathbb{F}_2$ whose evaluations are 2-level autocorrelation sequences, constitutes a very challenging problem. However, it is relatively easy to construct functions from $\mathbb{F}_{2^n}$ to $\mathbb{F}_2$ with ideal 2-level additive autocorrelation, because this occurs for a large family of functions, the so-called bent functions, which will be introduced in Chapter 10. In other words, for any bent function, all out-of-phase additive autocorrelation values are equal to zero.

6.7 Laws of the Hadamard transform and convolution transform

In this section, we will discuss some properties of the Hadamard transform and the convolution transform. We denote by $\widehat{g}^*$ the complex conjugate of $\widehat{g}$; that is, $\widehat{g}^* = (\widehat{g})^*$.

Proposition 6.5 *Let $f(x)$ be a function from $\mathbb{F}_q$ to $\mathbb{F}_p$ $(q = p^n)$.*

1. *Involution law:* $\frac{1}{q}\widehat{\widehat{f}} = f$.
2. *Additive law:* $\widehat{f + g} = \widehat{f} + \widehat{g}$.
3. *Convolution law:* $\widehat{f * g} = \widehat{f} \cdot \widehat{g}^*$. *In particular, if $p = 2$, $\widehat{f * g} = \widehat{f} \cdot \widehat{g}$. In other words, if $p = 2$, the Hadamard transform of the convolution of f and g is equal to the product of their Hadamard transforms.*
4. *Parseval energy formula:* $\sum_{x \in \mathbb{F}_q} \widehat{f}^2(x) = q^2$ *where* $\widehat{f}^2(x) = \widehat{f}(x)\widehat{f}^*(x)$.

Proof. Assertions 1 and 2 are immediate from the definition of the Hadamard transform. In the following, we will prove assertions 3 and 4. Notice that $(f * g)(x)$ is a function from $\mathbb{F}_q$ to the complex field, $\mathbb{C}$. We have

$$\widehat{f * g}(\lambda) = \sum_x \chi(Tr(\lambda x))((f * g)^*(x))$$

$$= \sum_x \chi(Tr(\lambda x)) \left(\sum_{w \in \mathbb{F}_q} \chi(f(x + w))\chi^*(g(w)) \right)^*$$

$$= \sum_x \chi(Tr(\lambda x)) \sum_{w \in \mathbb{F}_q} \chi^*(f(x + w))\chi(g(w))$$

$$= \sum_{w \in \mathbb{F}_q} \chi(g(w)) \sum_{x \in \mathbb{F}_q} \chi(Tr(\lambda x))\chi^*(f(x + w))$$

$$= \sum_{w \in \mathbb{F}_q} \chi(g(w))\chi(-Tr(\lambda w)) \sum_{y \in \mathbb{F}_q} \chi(Tr(\lambda y))\chi^*(f(y)), \quad (x + w = y)$$

$$= \sum_{w \in \mathbb{F}_q} \chi(g(w))\chi^*(Tr(\lambda w)) \cdot \widehat{f}(\lambda)$$

$$= \widehat{f}(\lambda) \left(\sum_{w \in \mathbb{F}_q} \chi(Tr(\lambda w))\chi^*(g(w)) \right)^*$$

$$= \widehat{f}(\lambda)\widehat{g}^*(\lambda).$$

Thus, assertion 3 is established. For assertion 4, from the definition of $\widehat{f}$, by manipulation, we have

$$\sum_{\lambda \in \mathbb{F}_q} \widehat{f}^2(\lambda) = \sum_{\lambda \in \mathbb{F}_q} \sum_{x \in \mathbb{F}_q} \chi(Tr(\lambda x) - f(x)) \left(\sum_{y \in \mathbb{F}_q} \chi(Tr(\lambda y) - f(y)) \right)^*$$

$$= \sum_{\lambda \in \mathbb{F}_q} \sum_{x, y \in \mathbb{F}_q} \chi(Tr(\lambda x) - f(x) - Tr(\lambda y) + f(y))$$

$$= \sum_{x, y \in \mathbb{F}_q} \chi(f(y) - f(x)) \sum_{\lambda \in \mathbb{F}_q} \chi(Tr(\lambda(x - y)))$$

$$= q \sum_{x \in \mathbb{F}_q} \chi(f(x) - f(x)) = q^2,$$

where the next to last identity comes from $\sum_{\lambda \in \mathbb{F}_q} \chi(Tr(\lambda \beta)) = q$ if $\beta = 0$, while otherwise it is equal to zero.   □

Proposition 6.6 (PARSEVAL'S EQUATION)

$$\sum_{x \in \mathbb{F}_q} \chi(g(x))\chi^*(f(x)) = \frac{1}{q} \sum_{x \in \mathbb{F}_q} \widehat{g}^*(x)\widehat{f}(x). \tag{6.47}$$

In particular, if $g(x) = f(\lambda x)$,

$$\sum_{x \in F_q} \chi(f(\lambda x)) \chi^*(f(x)) = \frac{1}{q} \sum_{x \in F_q} \widehat{f}^*(\lambda x) \widehat{f}(x). \qquad (6.48)$$

When $p = 2$, $\widehat{f}(\lambda)$ *is a real number for any* $\lambda \in \mathbb{F}_{2^n}$, *and Eq. (6.48) becomes*

$$\sum_{x \in F_q} (-1)^{f(x) + f(\lambda x)} = \frac{1}{q} \sum_{x \in F_q} \widehat{f}(x) \widehat{f}(\lambda x). \qquad (6.49)$$

In other words, when $p = 2$, *the autocorrelation of the sequence* **a** *or the function* $f(x)$ *is equal to the autocorrelation of its Hadamard transform up to a factor of* $\frac{1}{q}$.

Proof.

$$\frac{1}{q} \sum_{y \in F_q} \widehat{g}^*(y) \widehat{f}(y)$$

$$= \frac{1}{q} \sum_{y \in F_q} \left(\sum_x \chi(Tr(yx)) \chi^*(g(x)) \right)^* \sum_w \chi(Tr(yw)) \chi^*(f(w))$$

$$= \frac{1}{q} \sum_{x,w} \chi(g(x)) \chi^*(f(w)) \sum_y \chi(Tr(y(w - x)))$$

$$= \frac{1}{q} q \sum_x \chi(g(x)) \chi^*(f(x)),$$

which establishes the assertion. □

This is a remarkable formula which was used to establish the 2-level autocorrelation property for all the newly discovered sequences which will be presented in Chapter 9. In the sequence version, the Parseval formula states that if the respective crosscorrelation functions of two sequences of period $q - 1$ with a common m-sequence of period $q - 1$ having the linear trace representation are equal, so are their respective autocorrelation functions. However, it is not easy to construct two functions that have equal Hadamard transforms, which will be seen through some examples in Chapter 9.

Proposition 6.7 *Recall that the additive autocorrelation of* f *is given by*

$$V_f(w) = \sum_{x \in F_q} \omega^{f(x+w) - f(x)}, \ w \in F_q. \qquad (6.50)$$

Then

$$\widehat{f}^2(\lambda) = \widehat{V}_f(\lambda). \qquad (6.51)$$

Table 6.5. *Relationships of multiplicative and additive correlation
of functions*

$f : \mathbb{F}_q \to \mathbb{F}_p \ (q = p^n)$	
Cyclic order	ρ order
$\mathbb{F}_q^* = \{1, \alpha, \ldots, \alpha^{q-2}\}$	$\mathbb{F}_q = \{t_0, t_1, \ldots, t_{q-1}\}$
$\mathbf{a} = \{a_i\} : a_i = f(\alpha^i)$	$\mathbf{u} = \{u_i\} : u_i = f(t_i)$
Period of $\mathbf{a}$: a factor of $q - 1$	Period of $\mathbf{u}$: a factor of q
Multiplicative autocorrelation of f $$C_f(\lambda) = \sum_{x \in \mathbb{F}_q} \omega^{f(\lambda x) - f(x)}$$ $$= \sum_{i=0}^{q-2} \omega^{a_{i+\tau} - a_i} + 1$$ $$= C_{\mathbf{a}}(\tau) + 1, \ \text{per}(\mathbf{a}) = q - 1$$ $$\lambda = \alpha^\tau, \tau = 0, 1, \ldots$$	Additive autocorrelation (AC) of f = convolution of itself $$V_f(w) = f * f(w)$$ $$= \sum_{x \in \mathbb{F}_q} \omega^{f(x+w) - f(x)}$$ $$A_f(w) = \sum_{i=0}^{q-1} \omega^{u_{i+\tau} - u_i}$$ $$= C_{\mathbf{u}}(\tau), \ \text{per}(\mathbf{u}) = q$$ $$w = t_\tau, \tau = 0, 1, \ldots$$
Ideal 2-level autocorrelation of $\mathbf{a}$: $$C_{\mathbf{a}}(\tau) = \begin{cases} p^n - 1 & \text{if } \tau \equiv 0 \ (\text{mod } p^n - 1) \\ -1 & \text{if } \tau \not\equiv 0 \ (\text{mod } p^n - 1) \end{cases}$$ when $\text{per}(\mathbf{a}) = p^n - 1$	Ideal 2-level autocorrelation of $\mathbf{u}$: $$C_{\mathbf{u}}(\tau) = \begin{cases} p^n & \text{if } \tau \equiv 0 \ (\text{mod } p^n) \\ 0 & \text{if } \tau \not\equiv 0 \ (\text{mod } p^n) \end{cases}$$ when $\text{per}(\mathbf{u}) = p^n$
Hadamard transform and crosscorrelation $$\widehat{f}(\lambda) = C_{Tr, f}(\lambda) = \sum_{x \in \mathbb{F}_q} \omega^{Tr(\lambda x) - f(x)}$$ $$= 1 + C_{\mathbf{b}, \mathbf{a}}(\tau),$$ where $\mathbf{b} = \{Tr(\alpha^i)\}$	Hadamard transform and additive AC $$\widehat{f}^2(\lambda) = \widehat{V}_f(\lambda)$$

*In other words, the square of the Hadamard transform of f or $\mathbf{a}$ is equal to
the Hadamard transform of the convolution of f with itself, or equivalently,
the square of the Hadamard transform of f or $\mathbf{a}$ is equal to the Hadamard
transform of the additive autocorrelation function of f. Conversely,*

$$V_f(w) = \frac{1}{q} \sum_{\lambda \in \mathbb{F}_q} \omega^{Tr(w\lambda)} \widehat{f}^2(\lambda). \tag{6.52}$$

Proof. By setting $g(x) = f(x)$ in the convolution law in Proposition 6.5, the
result follows immediately.   □

In Table 6.5, we summarize the relationships among three transforms and
two types of correlation functions related to functions from $\mathbb{F}_{p^n}$ to $\mathbb{F}_p$.

6.8 The matrix representation of the DFT and the Hadamard transform

Let $\mathbf{a} = \{a_t\}$ be a sequence over $\mathbb{F}_q$ of period $N \mid (q^n - 1)$ where q is a prime or a power of a prime, and let α be an element in $\mathbb{F}_{q^n}$ of order N. Then we can write the DFT of $\mathbf{a}$ (with respect to α) in Eq. (6.1) in the following matrix form:

$$(A_0, A_1, A_2, \ldots, A_{N-1}) = (a_0, a_1, a_2, \ldots, a_{N-1})M, \tag{6.53}$$

where

$$M = \begin{bmatrix} 1 & 1 & 1 & \cdots & 1 \\ 1 & \alpha & \alpha^2 & \cdots & \alpha^{N-1} \\ 1 & \alpha^2 & \alpha^{2\cdot 2} & \cdots & \alpha^{2(N-1)} \\ \vdots & & & & \\ 1 & \alpha^{N-1} & \alpha^{2(N-1)} & \cdots & \alpha^{(N-1)(N-1)} \end{bmatrix}. \tag{6.54}$$

The inverse DFT (see Table 6.1) is given by

$$(a_0, a_1, a_2, \ldots, a_{N-1}) = -(A_0, A_1, A_2, \ldots, A_{N-1})M^{-1}, \tag{6.55}$$

where M^{-1} is the inverse of M whose entries are given by

$$M^{-1} = \begin{bmatrix} 1 & 1 & 1 & \cdots & 1 \\ 1 & \alpha^{-1} & \alpha^{-2} & \cdots & \alpha^{-(N-1)} \\ 1 & \alpha^{-2} & \alpha^{-2\cdot 2} & \cdots & \alpha^{-2(N-1)} \\ \vdots & & & & \\ 1 & \alpha^{-(N-1)} & \alpha^{-2(N-1)} & \cdots & \alpha^{-(N-1)(N-1)} \end{bmatrix}. \tag{6.56}$$

Using the matrix representation, the DFT of the sequence $\mathbf{a}$ in Example 6.1 is given by

$$(A_0, A_1, \ldots, A_6) = (0, 1, 0, 0, 0, 1, 0)M,$$

where

$$M = \begin{bmatrix} 1 & 1 & 1 & 1 & 1 & 1 & 1 \\ 1 & \alpha & \alpha^2 & \alpha^3 & \alpha^4 & \alpha^5 & \alpha^6 \\ 1 & \alpha^2 & \alpha^4 & \alpha^6 & \alpha & \alpha^3 & \alpha^5 \\ 1 & \alpha^3 & \alpha^6 & \alpha^2 & \alpha^5 & \alpha & \alpha^4 \\ 1 & \alpha^4 & \alpha & \alpha^5 & \alpha^2 & \alpha^6 & \alpha^3 \\ 1 & \alpha^5 & \alpha^3 & \alpha & \alpha^6 & \alpha^4 & \alpha^2 \\ 1 & \alpha^6 & \alpha^5 & \alpha^4 & \alpha^3 & \alpha^2 & \alpha \end{bmatrix}$$

$$\implies$$

$$(A_0, A_1, \ldots, A_6) = (0, \alpha^6, \alpha^5, 1, \alpha^3, 1, 1).$$

Its inverse DFT is given by

$$(a_0, a_1, \ldots, a_6) = (0, \alpha^6, \alpha^5, 1, \alpha^3, 1, 1)M^{-1},$$

where

$$
M^{-1} = \begin{bmatrix}
1 & 1 & 1 & 1 & 1 & 1 & 1 \\
1 & \alpha^6 & \alpha^5 & \alpha^4 & \alpha^3 & \alpha^2 & \alpha \\
1 & \alpha^5 & \alpha^3 & \alpha & \alpha^6 & \alpha^4 & \alpha^2 \\
1 & \alpha^4 & \alpha & \alpha^5 & \alpha^2 & \alpha^6 & \alpha^3 \\
1 & \alpha^3 & \alpha^6 & \alpha^2 & \alpha^5 & \alpha & \alpha^4 \\
1 & \alpha^2 & \alpha^4 & \alpha^6 & \alpha & \alpha^3 & \alpha^5 \\
1 & \alpha & \alpha^2 & \alpha^3 & \alpha^4 & \alpha^5 & \alpha^6
\end{bmatrix}.
$$

Next we look at the matrix representation of the Hadamard transform of sequences. Let $\mathbf{a} \leftrightarrow f(x)$ where $f(x)$ is a function from $\mathbb{F}_q$ to $\mathbb{F}_p$ where $q = p^n$; that is, $a_i = f(\alpha^i)$, $i = 0, 1, \ldots$, where α is a primitive root of $\mathbb{F}_q$. Let $\mathbf{c} = \{c_i\}$ whose elements are given by $c_i = \omega^{Tr(\alpha^i)}$ where $Tr(x)$ is the trace function from $\mathbb{F}_q$ to $\mathbb{F}_p$. Then the matrix representation of the Hadamard transform of $\mathbf{a}$ or $f(x)$ (see Table 6.3) is given by

$$(\widehat{f}(0), \widehat{f}(1), \widehat{f}(\alpha), \ldots, \widehat{f}(\alpha^{N-1})) = (1, \omega^{-a_0}, \omega^{-a_1}, \ldots, \omega^{-a_{N-1}})H, \qquad (6.57)$$

where $N = q - 1$ and H is a $q \times q$ matrix whose entries are given by

$$
H = \begin{bmatrix}
1 & 1 & 1 & \cdots & 1 \\
1 & c_0 & c_1 & \cdots & c_{N-1} \\
1 & c_1 & c_2 & \cdots & c_0 \\
\vdots & & & & \\
1 & c_{N-1} & c_0 & \cdots & c_{N-2}
\end{bmatrix}. \qquad (6.58)
$$

Note that $\frac{1}{\sqrt{q}}H$ is a unitary matrix; that is, $H(H^T)^* = qI$ where I is the $q \times q$ identity matrix. We take the conjugates of both sides of Eq. (6.57) and then multiply the equation by H. It follows that

$$(1, \omega^{a_0}, \omega^{a_1}, \ldots, \omega^{a_{N-1}}) = \frac{1}{q}(\widehat{f}^*(0), \widehat{f}^*(1), \widehat{f}^*(\alpha), \ldots \widehat{f}^*(\alpha^{N-1}))H, \qquad (6.59)$$

which is the matrix representation of the inverse of the Hadamard transform. In particular, for $p = 2$, we have the Hadamard transform and its inverse as follows:

$$(\widehat{f}(0), \widehat{f}(1), \widehat{f}(\alpha), \ldots, \widehat{f}(\alpha^{N-1})) = (1, (-1)^{a_0}, \ldots, (-1)^{a_{N-1}})H$$

$$(1, (-1)^{a_0}, \ldots, (-1)^{a_{N-1}}) = \frac{1}{q}(\widehat{f}(0), \widehat{f}(1), \widehat{f}(\alpha), \ldots, \widehat{f}(\alpha^{N-1}))H.$$

In this case, H is the Hadamard matrix introduced in Chapter 2. Hence, this transformation is named the Hadamard transform.

For example, for the sequence $\mathbf{a} = (1110010) \leftrightarrow f(x) = Tr(x^3)$ in Example 6.10, we have $\mathbf{c} = (1, -1, 1, -1, 1, 1, -1, -1)$. Thus the matrix representations of its Hadamard transform and the inverse Hadamard transform are given by

$$(\widehat{f}(0), \widehat{f}(1), \widehat{f}(\alpha), \ldots, \widehat{f}(\alpha^6)) = (1, -1, -1, -1, 1, 1, -1, 1)H$$
$$(1, (-1)^{a_0}, \ldots, (-1)^{a_6}) = \frac{1}{8}(0, -4, 0, 0, 4, 0, 4, 4)H,$$

where the Hadamard matrix H is given by

$$H = \begin{bmatrix} 1 & 1 & 1 & 1 & 1 & 1 & 1 & 1 \\ 1 & -1 & 1 & -1 & 1 & 1 & -1 & -1 \\ 1 & 1 & -1 & 1 & 1 & -1 & -1 & -1 \\ 1 & -1 & 1 & 1 & -1 & -1 & -1 & 1 \\ 1 & 1 & 1 & -1 & -1 & -1 & 1 & -1 \\ 1 & 1 & -1 & -1 & -1 & 1 & -1 & 1 \\ 1 & -1 & -1 & -1 & 1 & -1 & 1 & 1 \\ 1 & -1 & -1 & 1 & -1 & 1 & 1 & -1 \end{bmatrix}.$$

Exercises for Chapter 6

1. Let $\mathbf{a} = (0110011)$ be a binary sequence with period 7. Compute the spectral sequence of $\mathbf{a}$.

2. Let $\mathbb{F}_{2^3}$ be defined by $\alpha^3 + \alpha + 1 = 0$. Let $\mathbf{a}$ be a binary sequence with period 7 whose spectral sequence is given by $\mathbf{A} = (0, 1, 1, \alpha^3, 1, \alpha^6, \alpha^5)$. Determine the elements in the first period of $\mathbf{a}$.

3. Let $q = 2$, $n = 8$, and $N = 17$. Determine all cyclotomic cosets modulo 17 with respect to 2.

4. Let $q = 2$, $n = 12$, and $N = 45$. Determine all cyclotomic cosets modulo 45 with respect to 2.

5. Let $q = 2$, $n = 6$, and $N = 21$. Let $\mathbb{F}_{2^6}$ be defined by $\alpha^6 + \alpha + 1 = 0$. Let $\mathbf{a}$ be a binary sequence of period 21 whose elements are given by

$$\mathbf{a} = (010000001010000010000).$$

Verify the following three spectra of the sequence:

$$A_5 = \alpha^{37}, A_7 = 1, A_9 = \alpha^9.$$

(Write out your computational steps.)

6. Let $q = 2$, $n = 4$, $N = 15$, and a binary sequence $\mathbf{a}$ of period 15 be given by:

$$\mathbf{a} = (100101111000110).$$

(This is a modified de Bruijn sequence.) Compute the spectral sequence $\{A_k\}$ and the trace representation of this sequence.

7. Let α be a primitive element of $\mathbb{F}_{2^5}$ and $\mathbf{a}$ be a binary sequence of period 31 that has the following trace representation:

$$a_t = Tr(a\alpha^t) + Tr(b\alpha^{5t}), a, b \in \mathbb{F}_{2^5}, t = 0, 1, \ldots.$$

Assume that $\mathbb{F}_{2^5}$ is defined by $\alpha^5 + \alpha^3 + 1 = 0$. Find a choice for a and b such that the sequence $\mathbf{a}$ is balanced; that is, there are 16 1's and 15 0's in one period.

8. Design a binary sequence of period 15 satisfying the following requirements:
   (a) There are at least 8 nonzero terms in its spectral sequence.
   (b) It is balanced; that is, there are eight 1's and seven 0's in one period.
   Compute the autocorrelation function of the sequence that you designed, and determine the linear span and the minimal polynomial of the sequence.

9. Let α be a primitive element in $\mathbb{F}_{2^4}$ with $\alpha^4 + \alpha + 1 = 0$.
   (a) Let $\mathbf{a} = (011110101100100)$ with the trace representation $f(x) = Tr(x^7)$. Determine the Hadamard transform of $\mathbf{a}$ or $f(x)$.
   (b) Let $\mathbf{b} = (000100110101111)$ with the trace representation $g(x) = Tr(x)$. Compute $f * ptg(w), w \in \mathbb{F}_{2^4}$.
   (c) With the above $\mathbf{a}$ and $\mathbf{b}$, determine the crosscorrelation between $\mathbf{a}$ and $\mathbf{b}$ by using the connection between the crosscorrelation and the Hadamard transform.
   (d) Let $\mathbf{a} = (01100)$ be a binary sequence with period 5. Determine the trace representation of $\mathbf{a}$ and the Hadamard transform of $\mathbf{a}$.

10. Let $\{a_t\}$ be a binary sequence of period $N \mid (2^n - 1)$, and let the DFT of $\{a_t\}$ be given by

$$A_k = \sum_{t=0}^{N-1} a_t \beta^{-tk}, 0 \le k < N,$$

where β is an element in $\mathbb{F}_{2^n}$ of order N. Let α be a primitive element of $\mathbb{F}_{2^n}$. We write $\beta = \alpha^v$ where $v = \frac{2^n-1}{N}$. Let

$$B_k = \sum_{t=0}^{2^n-2} a_t \alpha^{-tk}, 0 \le k < 2^n - 1.$$

Show that
   (a) $\{B_k\}$ has period N.
   (b) $B_k = 0$ if $k \not\equiv 0 \pmod{v}$, and $A_j = B_{jv}, 0 \le j < N$.

11. A function in $\mathcal{F}_2$ is called a bent function if and only if $\widehat{f}(\lambda) = \sqrt{q}$ for all $\lambda \in \mathbb{F}_q$ where $q = 2^n$. (Note. Bent functions exist only for n even.) Let $\delta(w) = f * f(w)$. Show that $\delta(w) = 0$ for all $w \ne 0 \iff \widehat{f}(\lambda) = \pm\sqrt{2^n}$ for all $\lambda \in \mathbb{F}_q$.

12. Let α be a primitive element in $\mathbb{F}_{2^5}$ with $\alpha^5 + \alpha^2 + 1 = 0$. Let $\mathbf{a} \leftrightarrow f(x) = Tr(x^5)$. Determine the Hadamard transform of $\mathbf{a}$ or $f(x)$.

13. Let α be a primitive element in $\mathbb{F}_{2^6}$ with $\alpha^6 + \alpha + 1 = 0$. Let $\mathbf{a} = (011010100)$ be a binary sequence with period 9. Compute the Hadamard transform of $\mathbf{a}$. (Hint: Use the connection between crosscorrelation and the Hadamard transform.)

14. With the same $\mathbb{F}_{2^5}$ as in problem 12, let the elements of $\mathbf{a}$ be defined by $a_i = Tr(\alpha^{3i} + \alpha^{15i}), i = 0, 1, \ldots, 30$; that is, $\mathbf{a} \leftrightarrow f(x) = Tr(x^3 + x^{15})$. Compute $\widehat{f}(\lambda)$,

the autocorrelation of **a** and the autocorrelation of $\widehat{f}(\lambda)$. (The autocorrelation of $\widehat{f}(\lambda)$ is defined as $C(w) = \sum_{x \in \mathbb{F}_{2^5}} \widehat{f}(x)\widehat{f}(wx)$.) What do you observe?

15. Let $f(x)$ be a function from $\mathbb{F}_q$ to $\mathbb{F}_2$ where $q = 2^n$. We define two functions as follows:

$$\widehat{f}^+(\lambda) = \sum_{x \in \mathbb{F}_q}(-1)^{Tr(\lambda x)} f(x)$$

and

$$\delta^+(w) = \sum_{x \in \mathbb{F}_q}(f(x) + f(x + w)).$$

Show that

$$\widehat{f}(\lambda) = -2\widehat{f}^+(x)$$

and

$$\delta(w) = 2^n - 2\delta^+(w).$$

7

Cyclic Difference Sets and Binary Sequences with Two-Level Autocorrelation

In this chapter, we introduce cyclic difference sets and their relationship to binary sequences with 2-level autocorrelation. In terms of this relation, we exhibit the balance, constant-on-cosets, and 2-tuple balance property of binary sequences with 2-level autocorrelation and derive some constraints on their Fourier spectra and an achievable upper bound for the linear span of binary 2-level autocorrelation sequences.

7.1 Cyclic difference sets and their relationship to binary sequences with two-level autocorrelation

7.1.1 Cyclic difference sets

Definition 7.1 *A cyclic difference set modulo v, also called a cyclic (v, k, λ) difference set, is a set $D = \{d_1, d_2, \ldots, d_k\}$ of k integers distinct modulo v, such that the congruence $d_i - d_j \equiv t \pmod{v}$ has exactly λ solution pairs (d_i, d_j) of elements of D, for each integer t, $1 \le t \le v - 1$.*

Note. Since there are $k(k - 1)$ choices of $d_i \ne d_j$ from D, giving each of $v - 1$ values of t exactly λ times, we have $k(k - 1) = \lambda(v - 1)$ as a necessary condition for a cyclic (v, k, λ) difference set to exist.

Example 7.1 (a) $D = \{0, 1, 2, 4\}$ is a cyclic $(7, 4, 2)$ difference set, as seen in the following difference table

mod 7	0	1	2	4
0	0	1	2	4
1	6	0	1	3
2	5	6	0	2
4	3	4	5	0

Each of $1, 2, 3, 4, 5, 6$ occurs twice as a difference modulo 7.

(b) $D = \{2, 3, 5, 11\}$ is a cyclic $(13, 4, 1)$ difference set, as seen in the following difference table

mod 13	2	3	5	11
2	0	1	3	9
3	12	0	2	8
5	10	11	0	6
11	4	5	7	0

Each of $1, 2, 3, 4, 5, 6, 7, 8, 9, 10, 11, 12$ occurs once as a difference modulo 13.

If a constant c is added to each element of a cyclic (v, k, λ) difference set $D = \{d_1, d_2, \ldots, d_k\}$, the new set $D' = D + c = \{d_1 + c, d_2 + c, \ldots, d_k + c\}$ is again a (v, k, λ) difference set, with the same difference table, because $(d_i + c) - (d_j + c) \equiv d_i - d_j \pmod{v}$. This translate D' of D is often considered the same difference set as D.

If b is any constant relatively prime to v (i.e., $\gcd(b, v) = 1$), and $D = \{d_1, d_2, \ldots, d_k\}$ is a cyclic (v, k, λ) difference set, then $\overline{D} = bD = \{bd_1, bd_2, \ldots, bd_k\}$ is again a (v, k, λ) difference set, because $bd_i - bd_j \equiv b(d_i - d_j) \equiv bt \pmod{v}$, and because each t, $1 \le t \le v - 1$, occurs λ times, and bt is a permutation of the nonzero values of t modulo v for any b with $\gcd(b, v) = 1$, the difference table for $\overline{D}$ will again contain each $t \not\equiv 0 \pmod{v}$ exactly λ times.

Combining these two operations, from a given cyclic (v, k, λ) difference set D, we get as many as $v\phi(v)$ difference sets with the same parameters in the form $bD + c$, where there are v choices of c modulo v and $\phi(v)$ choices of b modulo v, because by definition the Euler phi-function $\phi(v)$ is the number of integers distinct modulo v and relatively prime to v.

Definition 7.2 *If D is a cyclic (v, k, λ) difference set, and m is a positive integer such that $mD \equiv D + c \pmod{v}$, then m is called a multiplier of the difference set D.*

Note. Since 1 is a multiplier, and if both m_1 and m_2 are multipliers, then $m_1 m_2$ is a multiplier, the multipliers of a given (v, k, λ) cyclic difference set D form a subgroup M of the multiplicative group $\mathbb{Z}_v^*$ modulo v, called the multiplier group of D.

Theorem 7.1 *If m is a multiplier of the cyclic (v, k, λ) difference set D, so that $mD \equiv D + c$, then there is a translate $D' = D + a$ of D such that $mD' = D'$, provided that $\gcd(m - 1, v) = 1$.*

Proof. Since $\gcd(m - 1, v) = 1, m - 1$ has a multiplicative inverse, $(m - 1)^{-1}$, modulo v, take $a \equiv -c(m - 1)^{-1} \pmod{v}$. Then $mD' \equiv m(D + a) \equiv mD + ma \equiv D + c + ma \equiv D + a \equiv D' \pmod{v}$, since $c + ma \equiv a \pmod{v}$ iff $c + (m - 1)a \equiv 0 \pmod{v}$ iff $a \equiv -c(m - 1)^{-1} \pmod{v}$. $\qquad\square$

Note. The theorem is still true without the condition that $\gcd(m - 1, v) = 1$, but the proof is considerably more difficult. See McFarland and Rice (1978). This result is quoted by Beth, Jungnickel, and Lenz (1999, p. 305) as follows:

> 2.6. Theorem: Let D be an abelian (v, k, λ)-difference set. Then there exists a translate of D which is fixed by every numerical multiplier of D. In particular, every cyclic difference set has a translate fixed by all its multipliers.

Theorem 7.2 (THE MULTIPLIER THEOREM) *If D is a cyclic (v, k, λ) difference set, and p is a prime divisor of $k - \lambda$ with $p > \lambda$, then p is a multiplier of D.*

This theorem, first proved by Marshall Hall, Jr., can be found in many books on combinatorial designs. H. Ryser conjectured that the condition $p > \lambda$ can be dropped, but this has never been proved. However, there is a stronger form of the multiplier theorem which weakens the condition $p > \lambda$, as follows:

Fact 7.1 *If D is a cyclic (v, k, λ) difference set and d is a divisor of $k - \lambda$ with $\gcd(d, v) = 1$ and $d > \lambda$, then the integer t is a multiplier of D if for each prime divisor p of d there is a positive integer j such that $p^j \equiv t \pmod{v}$.*

For details of all the results mentioned above, see Ryser, 1963.

Note. The multiplier theorem is used in two ways: As an aid to constructing a cyclic (v, k, λ) difference set when one exists and to prove that no such difference set exists when there is none.

7.1.2 The relationship to binary sequences with two-level autocorrelation

Construction

Given a cyclic (v, k, λ) difference set $D = \{d_1, d_2, \ldots, d_k\}$, we form the periodic binary sequence of period v, $S = \{s_1, s_2, \ldots, s_v\}$, where $s_j = a$ if $j \in D$ and $s_j = b$ if $j \notin D$. Let $S_\tau = \{s_{1+\tau}, s_{2+\tau}, \ldots, s_{v+\tau}\}$ where subscripts are reduced modulo v. Comparing S with $S_0 = S$ (that is, comparing S with itself), we find a opposite a k times, and b opposite b $v - k$ times, and there are no other cases. However, comparing S with S_τ, $\tau \not\equiv 0 \pmod{v}$, we find a opposite a λ times (since $d_i - d_j \equiv \tau \pmod{v}$ exactly λ times); and the remaining $k - \lambda$ times that a occurs in S, a must be opposite b in S_τ. Similarly, in the $k - \lambda$ times that a in S_τ is not opposite a in S, it must be opposite b in S. Thus, comparing S with S_τ term-by-term, the pair (a, a) occurs λ times, the pair (a, b) occurs $k - \lambda$ times, the pair (b, a) occurs $k - \lambda$ times, and the pair (b, b) must occur all the remaining times, namely $v - \lambda - 2(k - \lambda) = v - 2k + \lambda$ times.

Therefore, the unnormalized autocorrelation of the sequence S is given by

$$C(\tau) = \sum_{j=1}^{v} s_j s_{j+\tau} = \begin{cases} ka^2 + (v - k)b^2, & \tau \equiv 0 \ (\text{mod } v) \\ \lambda a^2 + (v - 2k + \lambda)b^2 + 2(k - \lambda)ab, & \tau \not\equiv 0 \ (\text{mod } v). \end{cases}$$

Thus, the sequence S, corresponding to the (v, k, λ) cyclic difference set D, has a two-valued autocorrelation function $C(\tau)$, with one value when $\tau \equiv 0$ (mod v) and the other value for all $\tau \not\equiv 0$ (mod v). The two most frequent choices of values for a and b are (1) $a = 1, b = 0$, and (2) $a = +1, b = -1$. With $a = 1, b = 0$, we have

$$C(\tau) = \begin{cases} k, & \tau \equiv 0 \ (\text{mod } v) \\ \lambda, & \tau \not\equiv 0 \ (\text{mod } v), \end{cases}$$

corresponding to the fact that when the difference set D is translated by any amount $\tau \not\equiv 0$ (mod v) relative to itself, there will be λ hits, corresponding to the λ times that two elements of D differ by τ (mod v); while if it is unshifted ($\tau \equiv 0$ (mod v)), each of the k elements of D will line up with itself.

With $a = +1, b = -1$, we get

$$C(\tau) = \begin{cases} k + (v - k) = v, & \tau \equiv 0 \ (\text{mod } v) \\ \lambda + (v - 2k + \lambda) - 2(k - \lambda) = v - 4(k - \lambda), & \tau \not\equiv 0 \ (\text{mod } v). \end{cases}$$

In this case, the normalized correlation values are:

$$\frac{1}{v} C(\tau) = \begin{cases} 1, & \tau \equiv 0 \ (\text{mod } v) \\ 1 - \frac{4n}{v}, & \tau \not\equiv 0 \ (\text{mod } v), \end{cases}$$

where $n = k - \lambda$ is a basic parameter of the difference set D.

Definition 7.3 *If D is a cyclic (v, k, λ) difference set, $D = \{d_1, d_2, \ldots, d_k\}$, we define the* complementary difference set D^* *to consist of the $v - k$ values modulo v not in D.*

It is easily seen that D^* is also a cyclic difference set modulo v, but with parameters $(v', k', \lambda') = (v, v - k, v - 2k + \lambda)$. Note that for $D^*, n' = k' - \lambda' = (v - k) - (v - 2k + \lambda) = k - \lambda = n$; so the two parameters v and $n(= k - \lambda)$ are the same for both D and D^*. Thus, the binary sequence of $+1$'s and -1's has the same autocorrelation for D and D^*, which also follows immediately from the fact that in this case the sequence S^* for D^* is the negative of the sequence S for D, and in the calculation of $C(\tau)$, we see that $(-s_j)(-s_{j+\tau}) = s_j s_{j+\tau}$.

Another way to calculate the correlation between two binary sequences, each taking the values a and b, is to count the number of agreements, A, and the number of disagreements, D. Then the unnormalized correlation is given by $C(\tau) = A - D$, whereas the normalized correlation is given by

$$\frac{1}{v}C(\tau) = \frac{A - D}{A + D},$$

where the length, or period, of the sequences being compared is v, which must be the total number of agreements plus disagreements. This is also the value we get for $C(\tau) = \sum_{j=1}^{v} s_j s_{j+\tau}$ when the two values is the sequence $S = \{s_j\}$ are $a = +1$ and $b = -1$.

In this discussion, we have seen how any cyclic (v, k, λ) difference set D leads to a binary sequence S of period v having a 2-level autocorrelation function. The converse is also true. Given any binary sequence S (say, taking the values a and b) with period v that has a 2-level autocorrelation function, we can form the corresponding cyclic difference set D modulo v by putting j in D iff $s_j = a$. If this occurs for k terms in the sequence S, then D has parameters (v, k, λ), where $\lambda = k(k - 1)/(v - 1)$.

Definition 7.4 *A cyclic (v, k, λ) difference set D is said to be of* Hadamard type *if, for some positive integer t, $v = 4t - 1$, $k = 2t - 1$, $\lambda = t - 1$. (Here, $n = k - \lambda = t$. This is the largest value that n can have for given v, for any cyclic (v, k, λ) difference set, $n = (v + 1)/4$.)*

Note. The binary sequence of $+1$'s and -1's corresponding to a cyclic difference set D of Hadamard type can be used to construct a cyclic Hadamard matrix, of order $v + 1 = 4t$, as derived in Chapter 2.

Definition 7.5 *A cyclic (v, k, λ) difference set with parameters $v = n^2 + n + 1$, $k = n + 1$, $\lambda = 1$, is referred to as* the finite projective plane of order n. *(Here, $n = k - 1 < \sqrt{v}$, which is the smallest value that n can have for given v.)*

Notes

1. Cyclic difference sets of Hadamard type are known to exist when $v = 4t - 1$ is a prime, when $v = 4t - 1$ is a product of twin primes (e.g., $v = 4t - 1 = 35 = 5 \cdot 7$), and when $4t$ is a power of 2. It is conjectured that there are no other cases. (The smallest other value of v which has not yet been ruled out is $v = 3439$. See Song and Golomb (1994) and Kim and Song (1999).

2. The only known constructions for the finite projective plane of order n occur when n is a prime or a power of a prime. It is known that there are no planes of order 6, order 10, order 14, and infinitely many others, but there are also infinitely many values of n that have not been ruled out, starting at $n = 12$.

7.2 More results about *C*

In this section, we return to the binary sequences whose elements are taken from the binary field $GF(2)$ and their autocorrelations are defined through the character function $\chi(x) = (-1)^x$. We will use the following notation in the rest of this chapter.

- U, the set of all binary sequences of odd period N that contain $(N + 1)/2$ ones and $(N - 1)/2$ zeros in each period;
- M, the subset of U consisting of those sequences that are constant-on-cosets; and
- C, the set consisting of all binary sequences of period N with the 2-level autocorrelation property.

In Section 5.7 of Chapter 5, we discussed the case $N = 2^n - 1$ in which we assumed that $C \subset M$. Here, we discuss some more results about C.

First, we discuss some operators under which the autocorrelations of sequences are invariant. (Some of them were already explained in Section 7.1 in the term of cyclic difference sets, and some of them were mentioned in previous chapters. Here we collect them together.) Then, using the multiplier theorem, we show that any binary 2-level autocorrelation sequence of period N is constant-on-cosets if $N = 2^n - 1$ or $N \equiv 7 \pmod{8}$. Finally, we show that any binary 2-level autocorrelation sequence satisfies the 2-tuple balance property introduced in Section 5.6 of Chapter 5.

7.2.1 Invariance of autocorrelation

Proposition 7.1 *Let* **a** *be a binary sequence with period N. Then the range of the autocorrelation function of* **a** *under each of the following operations is invariant. Precisely*

(a) *Complement: If we complement the elements in* **a**, *the resulting sequence is denoted as* $\mathbf{b} = \{b_i\}$; *that is*, $b_i = a_i + 1, i = 0, 1, \ldots$ *Then*

$$C_{\mathbf{b}}(\tau) = C_{\mathbf{a}}(\tau), \tau = 0, 1, \ldots.$$

(b) *Shift operator: If* $\mathbf{b} = L^k(\mathbf{a})$, *that is*, $b_i = a_{i+k}, i = 0, 1, \ldots$, *which is a sequence obtained by cyclically shifting k terms of* **a**, *then*

$$C_{\mathbf{b}}(\tau) = C_{\mathbf{a}}(\tau + k), \tau = 0, 1, \ldots.$$

(c) *Decimation: If* $\mathbf{b} = \mathbf{a}^{(t)}$ *with* $\gcd(t, N) = 1$, *that is*, $b_i = a_{it}, i = 0, 1, \ldots$, *then*

$$C_{\mathbf{b}}(\tau) = C_{\mathbf{a}}(t\tau), \tau = 0, 1, \ldots.$$

(d) Mixed operations: If $\mathbf{b} = (L^k(a))^{(t)}$ with $\gcd(t, N) = 1$ where $b_i = a_{k+ti}$, $i = 0, 1, \ldots$, then

$$C_{\mathbf{b}}(\tau) = C_{\mathbf{a}^{(t)}}(\tau + kt^{-1}) = C_{\mathbf{a}}(t\tau + k), \tau = 0, 1, \ldots.$$

In other words, if we shift first and then perform decimation, then the range of the autocorrelation of the resulting sequence equals the range of the autocorrelation of the original sequence.

Proof. Assertions (a) and (b) are clear. For assertion (c), notice that

$$\{(a_{it+t\tau}, a_{it}) \mid 0 \le i < N\} = \{(a_{i+t\tau}, a_i) \mid 0 \le i < N\}$$

because t is coprime to N. Thus we have

$$C_{\mathbf{b}}(\tau) = \sum_{i=0}^{N-1}(-1)^{b_{i+\tau}+b_i} = \sum_{i=0}^{N-1}(-1)^{a_{ti+t\tau}+a_{it}}$$

$$= \sum_{j=0}^{N-1}(-1)^{a_{j+t\tau}+a_j} = C_{\mathbf{a}}(t\tau),$$

which establishes assertion (c). For assertion (d), if t is coprime to N, then the order of applying the shift operator and the decimation operator can be changed, as shown in the following formula:

$$\mathbf{b} = (L^k(\mathbf{a}))^{(t)} = L^{kt^{-1}}(\mathbf{a}^{(t)}),\tag{7.1}$$

where t^{-1} is the inverse of t modulo N. Thus the first identity in assertion (d) is obtained from applying assertion (b) to $\mathbf{a}^{(t)}$ and the second identity is obtained from applying assertion (c) to $L^k(\mathbf{a})$. □

Note. This proposition and Eq. (7.1) are true for any periodic sequence over $\mathbb{F}_q$.

Definition 7.6 *Let* $\mathbf{a}$ *and* $\mathbf{b}$ *be two binary sequences with period* N. *If* $\mathbf{b}$ *is a* t*-decimation of* $\mathbf{a}$ *where* $\gcd(t, N) = 1$, *then we say that* $\mathbf{a}$ *and* $\mathbf{b}$ *are* decimation equivalent. *Otherwise, they are said to be* decimation distinct.

According to Proposition 7.1(c), if two sequences are decimation equivalent, then they have the same range for their autocorrelations. In particular, if a sequence has 2-level autocorrelation, so does its t-decimation as long as t is coprime to the period of the sequence. Furthermore, decimation is an equivalence relation. Thus, sequences in C can be partitioned into different decimation-equivalence classes. Interesting problems about C are how to construct sequences in C, and how many different decimation-equivalence classes there are.

7.2.2 When is C a subset of M?

Recall that in Section 5.7 of Chapter 5, we assumed that C is a subset of M for the case $N = 2^n - 1$. Here, we will prove that C is indeed a subset of M; that is, any 2-level autocorrelation binary sequence possesses the constant-on-cosets property. From Section 7.1, such sequences are in natural correspondence with the cyclic Hadamard difference sets with $v = N$, $k = (N - 1)/2$, and $\lambda = (N - 3)/4$. By Hall's result on multipliers (see Section 7.1), 2 is a multiplier for all Hadamard difference sets with $v = 2^n - 1$. In this case, we have $C \subset M$. However, if 2 is not a multiplier of the corresponding Hadamard difference set, then C may not be a subset of M. In other words, 2 is a multiplier of a Hadamard difference set D with $(N, (N - 1)/2, (N - 3)/4)$ if and only if the corresponding sequence satisfies the constant-on-cosets property. According to the Multiplier Theorem, 2 is a multiplier of D if $k - \lambda = (N + 1)/4$ is a multiple of 2, which implies that $N \equiv 7 \pmod 8$. Note that $2^n - 1 \equiv 7 \pmod 8$ for any $n \geq 3$. We summarize these discussions in the following proposition.

Proposition 7.2 *Assume that $N \mid (2^n - 1)$. Then $C \subset M$ provided that $N \equiv 7 \pmod 8$.*

7.2.3 The balance and two-tuple balance property of C

In Chapter 5, we showed that any m-sequence over $\mathbb{F}_q$, for both $q = 2$ and $q > 2$, satisfies the balance property, constant-on-cosets property, and 2-tuple balance property. As discussed above, any 2-level autocorrelation sequence of period $N = 2^n - 1$ or $N \equiv 7 \pmod 8$ satisfies the constant-on-cosets property. In the following, we show that any binary sequence with 2-level autocorrelation, that is, any sequence in C, satisfies the balance property and 2-tuple balance property. This result can easily be obtained from the fact that the binary 2-level autocorrelation sequences correspond to cyclic Hadamard difference sets (see Section 7.1). Here, we merely list these results in the language of sequences, without proof.

Lemma 7.1 *Let* **a** *be a binary sequence of period N, and let* $w(\mathbf{a})$ *denote the Hamming weight of* **a** *(i.e.,* $w(\mathbf{a})$ *is the number of nonzero elements in one period of* **a***). For any* $\tau \not\equiv 0 \pmod{N}$, *let*

$$T(\tau) = \{(a_k, a_{k+\tau}) \mid 0 \le k < N\}$$

and

$$N_{ij} = |\{k \mid 0 \le k < N, (a_k, a_{k+\tau}) = (i, j)\}|, \quad i, j \in \{0, 1\}.$$

Then

$$N_{01} = N_{10} \tag{7.2}$$
$$N_{01} + N_{11} = w(\mathbf{a}) \tag{7.3}$$
$$N_{00} - N_{11} = N - 2w(\mathbf{a}) \tag{7.4}$$
$$w(\mathbf{a} + L^\tau(\mathbf{a})) = \frac{N - C_\mathbf{a}(\tau)}{2}. \tag{7.5}$$

Note. The formula (7.5) is frequently used in coding and sequence analysis.

Theorem 7.3 *With the same notation as in Lemma 7.1, if* **a** *has 2-level autocorrelation, then*

$$w(\mathbf{a}) = (N + 1)/2, \tag{7.6}$$
$$N_{01} = N_{10} = N_{11} = (N + 1)/4 \text{ and } N_{00} = (N - 3)/4. \tag{7.7}$$

In particular, if $N = 2^n - 1$, *then*

$$w(\mathbf{a}) = 2^{n-1}, \tag{7.8}$$
$$N_{01} = N_{10} = N_{11} = 2^{n-2} \text{ and } N_{00} = 2^{n-2} - 1. \tag{7.9}$$

In other words, if **a** *is a 2-level autocorrelation sequence of period* $N = 2^n - 1$, *then Eq. (7.8) shows that* **a** *is balanced with* 2^{n-1} *1's in each period. The identities (7.9) imply that* **a** *satisfies the 2-tuple balance property, that is, for any* $\tau \not\equiv 0 \pmod{2^n - 1}$, *each nonzero 2-tuple* $(i, j) \in \mathbb{F}_2^2$ *occurs* 2^{n-2} *times in* $T(\tau)$ *and the* $(0, 0)$ *2-tuple occurs* $2^{n-2} - 1$ *times in* $T(\tau)$.

Example 7.2 Let $n = 5$ and $\mathbf{a} \in C$ be given by

$$\mathbf{a} = (1001001000011101010001111011011),$$

which is a quadratic residue sequence of period 31. Note that because **a** is constant-on-cosets, $T(\tau 2^i) = T(\tau), i = 0, 1, \ldots, 4$. Thus, we only need to compute the elements of $T(\tau)$ for $\tau \in \Gamma_2(5)^*$ where $\Gamma_2(n)^*$ refers to excluding 0 from $\Gamma_2(n)$. These values are shown in Table 7.1. From the table, we can

Table 7.1. $T(\tau)$ *for the quadratic residue sequence of period 31*

k	$T(1)$ (a_k, a_{k+1})	$T(3)$ (a_k, a_{k+3})	$T(5)$ (a_k, a_{k+5})	$T(7)$ (a_k, a_{k+7})	$T(11)$ (a_k, a_{k+11})	$T(15)$ (a_k, a_{k+15})
0	(1, 0)	(1, 1)	(1, 0)	(1, 0)	(1, 1)	(1, 1)
1	(0, 0)	(0, 0)	(0, 1)	(0, 0)	(0, 1)	(0, 0)
2	(0, 1)	(0, 0)	(0, 0)	(0, 0)	(0, 1)	(0, 1)
3	(1, 0)	(1, 1)	(1, 0)	(1, 0)	(1, 0)	(1, 0)
4	(0, 0)	(0, 0)	(0, 0)	(0, 1)	(0, 1)	(0, 0)
5	(0, 1)	(0, 0)	(0, 0)	(0, 1)	(0, 0)	(0, 0)
6	(1, 0)	(1, 0)	(1, 1)	(1, 1)	(1, 1)	(1, 1)
7	(0, 0)	(0, 0)	(0, 1)	(0, 0)	(0, 0)	(0, 1)
8	(0, 0)	(0, 1)	(0, 1)	(0, 1)	(0, 0)	(0, 1)
9	(0, 0)	(0, 1)	(0, 0)	(0, 0)	(0, 0)	(0, 1)
10	(0, 1)	(0, 1)	(0, 1)	(0, 1)	(0, 1)	(0, 0)
11	(1, 1)	(1, 0)	(1, 0)	(1, 0)	(1, 1)	(1, 1)
12	(1, 1)	(1, 1)	(1, 1)	(1, 0)	(1, 1)	(1, 1)
13	(1, 0)	(1, 0)	(1, 0)	(1, 0)	(1, 1)	(1, 0)
14	(0, 1)	(0, 1)	(0, 0)	(0, 1)	(0, 0)	(0, 1)
15	(1, 0)	(1, 0)	(1, 0)	(1, 1)	(1, 1)	(1, 1)
16	(0, 1)	(0, 0)	(0, 1)	(0, 1)	(0, 1)	(0, 1)
17	(1, 0)	(1, 0)	(1, 1)	(1, 1)	(1, 0)	(1, 0)
18	(0, 0)	(0, 1)	(0, 1)	(0, 0)	(0, 1)	(0, 0)
19	(0, 0)	(0, 1)	(0, 1)	(0, 1)	(0, 1)	(0, 1)
20	(0, 1)	(0, 1)	(0, 0)	(0, 1)	(0, 1)	(0, 0)
21	(1, 1)	(1, 1)	(1, 1)	(1, 0)	(1, 0)	(1, 0)
22	(1, 1)	(1, 0)	(1, 1)	(1, 1)	(1, 0)	(1, 1)
23	(1, 1)	(1, 1)	(1, 0)	(1, 1)	(1, 1)	(1, 0)
24	(1, 0)	(1, 1)	(1, 1)	(1, 1)	(1, 0)	(1, 0)
25	(0, 1)	(0, 0)	(0, 1)	(0, 0)	(0, 0)	(0, 0)
26	(1, 1)	(1, 1)	(1, 1)	(1, 0)	(1, 1)	(1, 0)
27	(1, 0)	(1, 1)	(1, 0)	(1, 1)	(1, 0)	(1, 1)
28	(0, 1)	(0, 1)	(0, 0)	(0, 0)	(0, 0)	(0, 1)
29	(1, 1)	(1, 0)	(1, 1)	(1, 0)	(1, 0)	(1, 1)
30	(1, 1)	(1, 0)	(1, 0)	(1, 1)	(1, 0)	(1, 0)

verify that **a** satisfies the 2-tuple balance property; that is, each of the 2-tuples (0, 1), (1, 0), and (1,1) occurs 8 times in $T(\tau)$, $\tau = 1, 3, 5, 7, 11, 15$ and the (0, 0) 2-tuple occurs 7 times in those sets.

7.3 Fourier spectral constraints

In this section, we derive the Fourier spectral constraints on binary 2-level autocorrelation sequences with the constant-on-cosets property and an achievable upper bound for their linear spans. We use the notation introduced in the previous section.

7.3.1 Fourier spectral constraints on C

Let $q = 2^n$, $\mathbf{a} = \{a_t\}$ be a binary sequence with period $N \mid (2^n - 1)$, and $\{A_k\}$ its spectral sequence. From Chapter 6, the DFT and the inverse DFT (the first version) are given as follows:

$$A_k = \sum_{t=0}^{N-1} a_t \alpha^{tk}, \quad k = 0, 1, \ldots, N - 1.$$

$$a_t = \sum_{k=0}^{N-1} A_k \alpha^{-kt} = \sum_{k \in \Gamma(N)} Tr(A_k \alpha^{-kt}), \quad t = 0, 1, \ldots, N - 1,$$

where α is a primitive element of $\mathbb{F}_q$, $Tr(x)$ is a trace function from a subfield of $\mathbb{F}_q$ to $\mathbb{F}_2$, which depends on the size n_k of coset C_k, and $\Gamma(N)$ is the set consisting of coset leaders modulo N with respect to the multiplier 2.

Note that $A_0 = \sum_{t=0}^{N-1} a_t \equiv w(\mathbf{a}) \,(\mathrm{mod}\, 2)$. If $w(\mathbf{a}) \equiv 0 \,(\mathrm{mod}\, 2)$, then $A_0 = 0$.

Theorem 7.4 (CHARACTERISTICS OF M) *A binary sequence of period N is constant-on-cosets if and only if its spectral sequence is binary. In other words,* $\mathbf{a} \in M \iff A_k \in \mathbb{F}_2, \forall k$.

Proof. Suppose that $\mathbf{a}$ is constant-on-cosets. It suffices to show that $A_k^2 = A_k$. Note that $a_t^2 = a_t$. Using the inverse of the DFT and the trace property, we have

$$a_{2t} = \sum_{k \in \Gamma(N)} Tr(A_k \alpha^{-2tk}) = \left(\sum_{k \in \Gamma(N)} Tr(A_k^{2^{n-1}} \alpha^{-kt}) \right)^2$$

$$= \sum_{k \in \Gamma(N)} Tr(A_k^{2^{n-1}} \alpha^{-kt}), \quad t = 0, 1, \ldots, N - 1.$$

This last identity comes from the fact that the element inside the squaring operator belongs to $\mathbb{F}_2$. Thus the spectral sequence of $\mathbf{a}^{(2)} = \{a_{2t}\}$ is given by $\{A_k^{2^{n-1}}\}_{k \geq 0}$. Because $\mathbf{a}$ is constant-on-cosets, that is, $a_t = a_{2t}, t = 0, 1, \ldots,$ $N - 1$, then $\mathbf{a} = \mathbf{a}^{(2)}$. Therefore, the spectral sequences of $\mathbf{a}$ and $\mathbf{a}^{(2)}$ should be equal, that is, $A_k = A_k^{2^{n-1}}, k = 0, 1, \ldots, N - 1$ (Proposition 6.1 in Chapter 6). Squaring both sides of this identity and noticing that $A_k^{2^n} = A_k$ because $A_k \in \mathbb{F}_{2^n}$, we get $A_k^2 = A_k \implies A_k \in \mathbb{F}_2$. Conversely, if $A_k \in \mathbb{F}_2$, using Lemma 6.3, we have $A_{2k} = A_k^2 = A_k, \forall k \implies \{A_k\}$ is constant-on-cosets. By exchanging the role of $\mathbf{a}$ with $\{A_k\}$ in the process of deriving that A_k is binary, in other words, considering $\{a_t\}$ as a spectral sequence of $\{A_k\}$, we get $a_t \in \mathbb{F}_2$. Again using Lemma 6.3, $a_t = a_{2t} \implies \mathbf{a} \in M$. $\qquad\square$

From Theorem 7.4, we have the following corollary.

Corollary 7.1 *If* $\mathbf{a} \in C$ *and* $C \subset M$, *then* $A_k \in \{0, 1\}$ *and* $\{A_k\}$ *is constant-on-cosets.*

Theorem 7.5 (CONSTRAINT FOR RECIPROCAL PAIRS) *With the above notation, for* $\mathbf{a} \in C$ *with* $N = 2^n - 1$ *or* $N \equiv 7 \pmod 8$, *so that* $C \subset M$, *we have*

$$A_k A_{-k} = 0, \ k = 1, 2, \ldots, N - 1. \tag{7.10}$$

In other words, for any pair consisting of a monomial x^k *and its reciprocal term* x^{-k}, *at most one of them occurs in the trace representation of* $\mathbf{a}$. *Or equivalently, for any pair consisting of an irreducible polynomial over* $\mathbb{F}_2$ *and its reciprocal, at most one of them occurs as a divisor of the minimal polynomial of* $\mathbf{a}$.

To show this result, we need some preparation. We will use the Hall polynomial to prove Theorem 7.5. If $\mathbf{a} = \{a_t\} \in C$, then the set consisting of the indices t such that $a_t = 0$ for $0 \leq t < N$, denoted by D, is a Hadamard difference set with the parameters (v, k, λ) where $v = N$, $k = (N - 1)/2$, and $\lambda = (N - 3)/4$, which was shown in Section 7.1. The Hall polynomial of D is defined as

$$D(x) = \sum_{t=0}^{N-1} (1 + a_t)x^t. \tag{7.11}$$

Using the Hall polynomial, D is a difference set if and only if its Hall polynomial satisfies

$$D(x)D(x^{-1}) = (k - \lambda) + \lambda \sum_{r=0}^{N-1} x^r. \tag{7.12}$$

We consider the polynomials on both sides of the above identity as polynomials over $\mathbb{F}_{2^n}$ and assume that $N \equiv 7 \pmod 8$. Then $k - \lambda = (N + 1)/4 \equiv 0 \pmod 2$ and $\lambda = (N - 3)/4 \equiv 1 \pmod 2$. Thus, Eq. (7.12) becomes

$$D(x)D(x^{-1}) = \sum_{r=0}^{N-1} x^r. \tag{7.13}$$

Lemma 7.2 *If* $\mathbf{a} \in C$ *where* $N \equiv 7 \pmod 8$, *so that* $C \subset M$, *then*

$$\sum_{k=1}^{N-1} A_k A_{-k} \alpha^{-kr} = 0, r = 0, 1, \ldots, N - 1. \tag{7.14}$$

Proof. Since $\mathbf{a} \in C$, we have $A_0 = 0$. Using the inverse DFT, and by resetting $A_0 = 1$, we have

$$
D(x)D(x^{-1}) = \sum_{t=0}^{N-1}\left(\sum_{k=0}^{N-1} A_k \alpha^{kt} + 1\right)x^t \sum_{t'=0}^{N-1}\left(\sum_{k'=0}^{N-1} A'_k \alpha^{k't'} + 1\right)x^{-t'}
$$

$$
= \sum_{t,t'}\left(\sum_{k,k'} A_k A_{k'} \alpha^{-(kt+k't')}\right)x^{t-t'}
$$

$$
= \sum_{r}\sum_{k,k'} A_k A_{k'} \alpha^{-kr}\left(\sum_{t} \alpha^{-(k+k')t}\right)x^r.
$$

Notice that

$$
\sum_{t=0}^{N-1} \alpha^{-(k+k')t} = \begin{cases} 0 & \text{if } k+k' \neq 0 \\ 1 & \text{if } k+k' = 0 \end{cases}
$$

Consequently,

$$
D(x)D(x^{-1}) = \sum_{r=0}^{N-1}\left(\sum_{k=0}^{N-1} A_k A_{-k} \alpha^{-kr}\right)x^r. \tag{7.15}
$$

According to Eq. (7.13), it follows that

$$
\sum_{k=0}^{N-1} A_k A_{-k} \alpha^{-kr} = 1. \tag{7.16}
$$

Since $A_0 = 1$, Eq. (7.14) is true.   □

Proof of Theorem 7.5. Let $x_k = A_k A_{-k}$. Then $x_k \in \{0, 1\}$ by Theorem 7.4. The equations shown in (7.14) are a system of linear equations in $N - 1$ variables whose coefficient matrix is given by

$$
B = \begin{bmatrix}
1 & 1 & \cdots & 1 \\
\alpha^{-1} & \alpha^{-2} & \cdots & \alpha^{-(N-1)} \\
\alpha^{-2} & \alpha^{-4} & \cdots & \alpha^{-2(N-1)} \\
\vdots & & & \\
\alpha^{-(N-1)} & \alpha^{-(N-1)2} & \cdots & \alpha^{-(N-1)^2}
\end{bmatrix}.
$$

The submatrix consisting of the first $N - 1$ rows is a Vandermonde matrix, so the rank of B is $N - 1$. It follows that $x_k = 0, k = 1, 2, \ldots, N - 1$.   □

Notice that if $-k \in C_k$ (the coset containing k), then $A_{-k} = A_k \Longrightarrow A_k A_{-k} = A_k^2 = 0$ from Theorem 7.5. In this case, the minimal polynomial of α^k is self-reciprocal. In other words, a self-reciprocal polynomial over $\mathbb{F}_2$ of degree $m|n$ can be characterized as a minimal polynomial of an element

Table 7.2. *Spectral constraints*

$\mathbf{a} \in C$ where $N = 2^n - 1$ or $N \equiv 7 \pmod 8$, so $C \subset M$
1. Constant-on-cosets: $A_k \in \{0, 1\}$ and $A_{2k} = A_k, k = 0, 1, \ldots, N - 1$.
2. Zero self-reciprocal: $A_k = 0$, if $-k \in C_k$.
3. Zero product: $A_k A_{-k} = 0$, if $-k \notin C_k$.

α^k in $\mathbb{F}_{2^n}$ with $-k \in C_k$. As an analogue to polynomials, we also call C_{-k} a reciprocal coset of C_k, and A_{-k} a reciprocal spectrum of A_k. If $-k \in C_k$, then C_k is called a self-reciprocal coset and A_k a self-reciprocal spectrum because $A_{-k} = A_k$ in this case. Summarizing the results on the constraints on spectral coefficients of 2-level autocorrelation sequences, we have the three necessary conditions, shown in Table 7.2, for a binary 2-level autocorrelation sequence to exist.

A binary sequence of period N with 2-level autocorrelation can be implemented as a sum of several LFSRs. In the LFSR language, the above result shows that no LFSR with a self-reciprocal minimal polynomial occurs in the set of the LFSRs used for implementation of the sequence.

Example 7.3 We apply the constraints listed in Table 7.2 to the cases $n = 4$ and $n = 5$, respectively, where $N = 2^n - 1$.

(a) We consider Example 6.9 in Chapter 6 where $n = 4$, $N = 2^4 - 1 = 15$, and α is a primitive element of $\mathbb{F}_{2^4}$ with $\alpha^4 + \alpha + 1 = 0$. Let $\mathbf{a} \leftrightarrow f(x)$ (that is, $f(x)$ is the trace representation of $\mathbf{a}$) and $\mathbf{a} \in C$. Then from the first constraint, that is, $\{A_k\}$ is constant-on-cosets, we have

$$f(x) = Tr_1^4(A_1 x) + Tr_1^4(A_3 x^3) + Tr_1^4(A_7 x^7) + Tr_1^2(A_5 x^5), \quad (7.17)$$
$$A_i \in \mathbb{F}_2, i \in \Gamma_2(4)^* = \{1, 3, 5, 7\}.$$

According to the second constraint, the zero self-reciprocal spectral condition, we have $A_3 = A_5 = 0$ because, from $-3 \equiv 12 \pmod{15}$, $-3 \in C_3 = \{3, 6, 12, 9\}$ and $-5 \equiv 10 \pmod{15}$, $-5 \in C_5 = \{5, 10\}$; or equivalently, the minimal polynomials $x^4 + x^3 + x^2 + x + 1$ and $x^2 + x + 1$ of α^3 and α^5, respectively, are self-reciprocal. Because $-1 \equiv 14 \pmod{15}$, $-1 \in C_7$, using the third constraint, only one of A_1 and A_7 is not zero in the trace representation $f(x)$. Therefore, if $\mathbf{a}$ has 2-level autocorrelation, then either $\mathbf{a}$ is generated by $x^4 + x + 1$, which is the minimal polynomial of α, or $\mathbf{a}$ is generated by $x^4 + x^3 + 1$, the minimal polynomial of α^7. However, they generate m-sequences in both cases. Thus, for $n = 4$, there are only two shift-distinct binary sequences of period 15 with 2-level autocorrelation, and both are m-sequences. In other words, there is only one decimation-equivalent class of 2-level autocorrelation sequences of period 15, which is the m-sequences.

(b) Let $n = 5$, $N = 2^5 - 1 = 31$, α be a primitive element in $\mathbb{F}_{2^5}$ with $\alpha^5 + \alpha^3 + 1 = 0$, and $\mathbf{a} \leftrightarrow f(x)$ where $\mathbf{a} \in C$. Applying the constant-on-cosets condition, we have

$$f(x) = Tr(A_1 x) + Tr(A_3 x^3) + Tr(A_5 x^5) + Tr(A_7 x^7) + Tr(A_{11} x^{11})$$
$$+ Tr(A_{15} x^{15}), \quad A_i \in \mathbb{F}_2, i \in \Gamma_2(5)^* = \{1, 3, 5, 7, 11, 15\}. \tag{7.18}$$

Because $-1 \in C_{15}$, $-3 \in C_7$, $-5 \in C_{11}$ (for the pairs of these reciprocal polynomials of degree 5, see Example 3.11(b) in Chapter 3), there are no self-reciprocal polynomials. Thus

$$f(x) = c_1 Tr(x^{r_1}) + c_2 Tr(x^{r_2}) + c_3 Tr(x^{r_3}),$$
$$\text{with all } c_i \in \mathbb{F}_2, r_1 \in \{1, 15\}, r_2 \in \{3, 7\}, r_3 \in \{5, 11\}.$$

By verification, we have only two decimation-distinct classes of sequences with 2-level autocorrelation:

1. $f(x) = Tr(x^r)$, $r \in G_2(5)^*$, which gives m-sequences;
2. $f(x) = Tr(x + x^5 + x^7)$ or $f(x^3)$, whose evaluations are quadratic residue sequences.

For $f(x) = Tr(x) \leftrightarrow \mathbf{a}$ where $a_t = Tr(\alpha^t), t = 0, 1, \ldots,$ we have

$$\mathbf{a} = (1000010101110110001111100110100)$$

$$\updownarrow \text{ DFT}$$

$$\mathbf{A} = (0110100010000000100000000000000).$$

For $f(x) = Tr(x + x^5 + x^7) \leftrightarrow \mathbf{a}$ where $a_t = Tr(\alpha^t + \alpha^{5t} + \alpha^{7t}), t = 0, 1, \ldots,$ we have

$$\mathbf{a} = (1001001000011101010001111011011)$$

$$\updownarrow \text{ DFT (the second version)}$$

$$\mathbf{A} = (0110110111000101011000100100).$$

Note that the spectral sequence of the quadratic residue sequence $\mathbf{a}$ is equal to the complement of $\mathbf{a}$; that is, $\mathbf{A} = 1 + \mathbf{a}$ where $\mathbf{A} = \{A_k\}$.

Notes

1. For any two shift-distinct m-sequences, one can be obtained from the other by a decimation. Thus, there is only one class of m-sequences under decimation equivalence.

2. For C, which contains all binary sequences of period N with 2-level autocorrelation, if $N = 2^n - 1$, from the above example, we know that for $N = 15$ there is only one decimation-equivalence class in C, which is the m-sequences. For $N = 31$, there are two decimation-equivalence classes in C. One class contains six shift-distinct m-sequences, and the other contains two shift-distinct quadratic residue sequences (up to shift equivalence).

7.3.2 An upper bound on linear spans

The linear span of a sequence $\mathbf{a} = \{a_t\}$ is equal to the number of nonzero elements in its spectral sequence $\{A_k\}$. For the case $\mathbf{a} \in C \subset M$, according to the spectral constraints listed in Table 7.2, an upper bound on $\mathbf{a}$ can be obtained by exclusion of the self-reciprocal cosets, or equivalently, the self-reciprocal polynomials over $\mathbb{F}_2$. We state the number of such indices in $\{A_k\}$ as a fact.

Fact 7.2 *Let n be the smallest integer such that $N \mid (2^n - 1)$ and $\mathcal{K} = \{k \in \mathbb{Z}_N$ such that $-k \in C_k\}$. Then*

$$|\mathcal{K}| = 1 + \sum_{i=0}^{r-1} 2^{2^i m}, \tag{7.19}$$

where r is determined by $n = 2^r m$ where m is odd.

Note that for the case when N is prime, there is no k, $0 < k < N$, such that $-k \in C_k$. Together with the constraints in Table 7.2 and Fact 7.2, the following results are immediate.

Theorem 7.6 *Let $n = 2^r m$ where m is odd. If $\mathbf{a} \in C$ with $N = 2^n - 1$ or $N \equiv 7 \pmod 8$, so that $C \subset M$, then $LS(\mathbf{a})$, the linear span of $\mathbf{a}$, is upper-bounded by*

$$LS(\mathbf{a}) \le \frac{1}{2}\left(N - 1 - \sum_{i=0}^{r-1} 2^{2^i m}\right).$$

In particular, if N is a prime, or if n is prime in the case $N = 2^n - 1$, then

$$LS(\mathbf{a}) \le \frac{1}{2}(N - 1).$$

Remark 7.1 Let $\mathbf{a}$ be a quadratic residue sequence of period N with $N \equiv 7 \pmod 8$. Then $\mathbf{a}$ has the constant-on-cosets property. On the other hand, the linear span of **a** is given by

$$LS(\mathbf{a}) = \frac{1}{2}(N - 1). \tag{7.20}$$

(We will leave this as an exercise.) This shows that a quadratic residue sequence of period $N \equiv 7 \pmod 8$ achieves the upper bound. Thus the bound given in Theorem 7.6 is tight. For example, in Example 7.3, when **a** is a quadratic residue sequence of period 31, it has linear span 15, which achieves the upper bound given by Theorem 7.6.

Note

The spectral constraints for binary 2-level autocorrelation sequences and the achievable upper bound for their linear spans were studied by Gong and Golomb (1999). The results presented here are from that paper. The proof for the case $N \equiv 7 \pmod 8$ given here is similar to that for the case $2^n - 1$ in Gong and Golomb (1999). For linear spans of quadratic residue sequences, see No et al. (1996) for the case of period $N = 2^n - 1$ and Ding, Helleseth, and Shan (1998) for general period N a prime.

Exercises for Chapter 7

1. In Remark 7.1, Eq. (7.20), is left as an exercise. Prove that it is true.
2. Determine whether a cyclic difference set D exists for each of the given parameter sets (v, k, λ). In each case, look at the factors of $n = k - \lambda$ to find a multiplier m of D (if D exists), using the original Multiplier Theorem or its generalization. Then list all cyclotomic cosets modulo v relative to m. Look for k-element unions of these cyclotomic cosets. If there are such unions, using the difference table method, determine whether any one of them is a (v, k, λ) cyclic difference set. (If there is no k-element union, or if none of the k-element unions yields the desired difference set, then D fails to exist.)
   (a) $(v, k, \lambda) = (21, 5, 1)$.
   (b) $(v, k, \lambda) = (43, 7, 1)$.
   (c) $(v, k, \lambda) = (73, 9, 1)$.
   (d) $(v, k, \lambda) = (91, 10, 1)$.
   (e) $(v, k, \lambda) = (11, 5, 2)$.
   (f) $(v, k, \lambda) = (29, 8, 2)$.
   (g) $(v, k, \lambda) = (37, 9, 2)$.
   (h) $(v, k, \lambda) = (15, 7, 3)$.
   (i) $(v, k, \lambda) = (25, 9, 3)$.
   (j) $(v, k, \lambda) = (31, 10, 3)$.

8

Cyclic Hadamard Sequences, Part 1

Binary sequences of period N with 2-level autocorrelation have many important applications in communications and cryptology. From Section 7.1, 2-level autocorrelation sequences are in natural correspondence with cyclic Hadamard difference sets with $v = N$, $k = (N - 1)/2$, and $\lambda = (N - 3)/4$. For this reason, they are named cyclic Hadamard sequences. In this chapter, 2-level autocorrelation always means ideal 2-level autocorrelation. There are three classic constructions for binary 2-level autocorrelation sequences that were known before 1997 (including some generalizations along these lines after 1997). One is m-sequences, described in Chapter 5, with period $N = 2^n - 1$. The second construction is based on a number theory approach, including three types of sequences in Chapter 2, which are the quadratic residue sequences, Hall sextic residue sequences, and twin prime sequences. The period of such a sequence is either a prime or a product of twin primes. The third construction is associated with intermediate subfields. The resulting sequences have subfield decompositions and period $N = 2^n - 1$. They include GMW sequences, cascaded GMW sequences, and generalized GMW sequences. Although the resulting sequences are binary, this construction relies heavily on intermediate fields and compositions of functions. As a consequence, it involves sequences over intermediate fields that are not binary sequences. The content of this chapter is organized as follows. In Section 8.1, we investigate general constructions for 2-level autocorrelation sequences over $\mathbb{F}_q$ with subfield decompositions where q is a power of a prime and introduce the subfield reducible concept for sequences. In Section 8.2, we carry out the general constructions in Section 1, yielding four types of GMW sequences, including the original GMW sequences, cascaded GMW sequences, and generalized GMW sequences. Sections 8.3–8.6 are devoted to discussions of the apparent randomness, linear span, shift-distinct property, and implementation of these sequences, respectively.

8.1 Constructions with subfield decomposition

In this section, we discuss two general constructions of 2-level autocorrelation sequences associated with intermediate fields and introduce the concept of subfield decomposition of sequences and their reducibility. We consider sequences over $\mathbb{F}_q$ with period $q^n - 1$ where q is a prime or a power of a prime. In Chapter 5, we introduced the one-to-one correspondence between sequences over $\mathbb{F}_q$ and the functions from $\mathbb{F}_{q^n}$ to $\mathbb{F}_q$ where q is a power of a prime. In this section, we investigate the constructions in terms of their trace representations. Let $\mathbf{u}$ be a sequence over $\mathbb{F}_q$ with period $q^n - 1$ and $f(x)$ be its trace representation. Then $f(x)$ is a function from $\mathbb{F}_{q^n}$ to $\mathbb{F}_q$. When we use the one-to-one correspondence between sequences and functions, we always assume that $f(0) = 0$. Otherwise, we may use $g(x) = f(x) - f(0)$, which satisfies $g(0) = 0$ (note that this transformation does not change the values of the autocorrelation of $\mathbf{u}$). We use the following notation in this section:

- $q = p^r$ where p is a prime and r a positive integer;
- $C_q(n)$, the set consisting of all 2-level autocorrelation sequences over $\mathbb{F}_q$ of period $q^n - 1$;
- α, a primitive element of $\mathbb{F}_{q^n}$; and
- $u_i = f(\alpha^i)$, $i = 0, 1, \ldots$, with period $N \mid (q^n - 1)$; that is, $\mathbf{u} \leftrightarrow f(x)$ where $f(x)$ is the trace representation of $\mathbf{u}$.

8.1.1 Some old and some new

In the following, we interpret the balance, 2-level autocorrelation, and 2-tuple balance property of sequences over $\mathbb{F}_q$ as properties of functions from $\mathbb{F}_{q^n}$ to $\mathbb{F}_q$.

Definition 8.1 *For any $c \in F_q$, if*

$$f(x) = c$$

has q^{n-1} solutions in $\mathbb{F}_{q^n}$, then $f(x)$ is said to be balanced. *This is equivalent to saying that $\mathbf{u}$ is balanced.*

In Section 5.1 of Chapter 5, we defined

$$\chi(x) = \omega^{Tr_{\mathbb{F}_q/\mathbb{F}_p}(x)}, \ x \in \mathbb{F}_q, \tag{8.1}$$

where ω is a primitive pth root of unity and the autocorrelation of $\mathbf{u}$ is given by

$$C_{\mathbf{u}}(\tau) = \sum_{i=0}^{q^n-2} \chi(u_{i+\tau})\chi^*(u_i). \tag{8.2}$$

Lemma 8.1 *If $f(x)$ is a balanced function from $\mathbb{F}_{q^n}$ to $\mathbb{F}_q$, and $\beta \in \mathbb{F}_{q^n}$, then*

$$\sum_{x \in \mathbb{F}_{q^n}} \chi(f(\beta x)) = \begin{cases} 0 & \text{if } \beta \neq 0 \\ q^n & \text{if } \beta = 0. \end{cases}$$

Proof. If $\beta = 0$, the lemma is true. Thus, it suffices to deal with the case $\beta \neq 0$. Since $f(x)$ is a balanced function from $\mathbb{F}_{q^n}$ to $\mathbb{F}_q$, and for each $1 \neq \beta \in \mathbb{F}_{q^n}$ βx permutes the elements of $\mathbb{F}_{q^n}$, we have that $f(\beta x) = c$ has q^{n-1} solutions in $\mathbb{F}_{q^n}$ for each $c \in \mathbb{F}_q$. Applying Lemma 5.1 of Chapter 5,

$$\sum_{x \in \mathbb{F}_{q^n}} \chi(f(\beta x)) = q^{n-1} \sum_{c \in \mathbb{F}_q} \chi(c) = 0. \qquad \square$$

Note. The is an extension of Lemma 5.1 in Chapter 5.

For $q = p$, we introduced $C_f(\lambda)$, the autocorrelation of $f(x)$, in Definition 6.6 of Section 6.5 of Chapter 6. Here, we generalize that definition to general q as follows.

$$C_f(\lambda) = \sum_{x \in \mathbb{F}_{q^n}} \chi(f(\lambda x))\chi^*(f(x))$$
$$= \sum_{x \in \mathbb{F}_{q^n}} \chi(f(\lambda x) - f(x)), \quad \lambda = \alpha^\tau \in \mathbb{F}_{q^n}^* \qquad (8.3)$$

and

$$C_f(0) = \sum_{x \in \mathbb{F}_{q^n}} \chi(f(x)).$$

Thus $C_f(\lambda) = C_{\mathbf{u}}(\tau) + 1$. If $\mathbf{u}$ has 2-level autocorrelation, then $C_f(\lambda) = 0$ for $\lambda \neq 0$ and $\lambda \neq 1$. We will extend this property to the function $f(x)$ by including $C_f(0) = 0$.

Definition 8.2 *If*

$$C_f(\lambda) = \begin{cases} q^n & \text{if } \lambda = 1 \\ 0 & \text{if } \lambda \neq 1, \end{cases}$$

then $f(x)$ is said to be orthogonal.

From the definition, if $f(x)$ is orthogonal, then $\mathbf{u}$ has 2-level autocorrelation. Note that for the case $q = p$, if $\mathbf{u}$ has 2-level autocorrelation, then $\mathbf{u}$ is balanced (see Exercise 4). Thus $C_f(0) = 0$ from Lemma 8.1. Thus we have shown the following result.

Proposition 8.1 *For $q = p$, $f(x)$ is orthogonal if and only if $\mathbf{u}$ has 2-level autocorrelation.*

In Section 5.6 of Chapter 5, we showed that any sequence over $\mathbb{F}_q$ of period $q^n - 1$ with the 2-tuple balance property has 2-level autocorrelation. In the following, we interpret this property for polynomial functions. Note that the trace representation of a τ-shift of $\mathbf{u}$ is given by $f(\lambda x)$ where $\lambda = \alpha^\tau$; that is, $L^\tau(\mathbf{u}) \leftrightarrow f(\lambda x)$. Let

$$T_f(\lambda) = \{(f(x), f(\lambda x)) \mid x \in \mathbb{F}_{q^n}^*\}, 1 \neq \lambda \in \mathbb{F}_{q^n}, \text{ and} \tag{8.4}$$

$$T_{\mathbf{u}}(\tau) = \{(u_i, u_{i+\tau}) \mid 0 \leq i < q^n - 1\}. \tag{8.5}$$

Note that the 2-tuple balance property of the sequence $\mathbf{u}$ is defined on the set $T_{\mathbf{u}}(\tau)$. Since $\mathbf{u}$ is obtained by evaluating $f(x)$ on the multiplicative group $<\alpha>$ of $\mathbb{F}_{q^n}$ (here $<\alpha> = F_{q^n}^* = \{1, \alpha, \ldots, \alpha^{q^n-2}\}$), from the above notations, we have

$$T_f(\lambda) = T_{\mathbf{u}}(\tau) \cup \{(0, 0)\}, \lambda = \alpha^\tau. \tag{8.6}$$

Definition 8.3 *Let $f(x)$ be a function from $\mathbb{F}_{q^n}$ to $\mathbb{F}_q$ with $f(0) = 0$ and let $T_f(\lambda)$ be defined by Eq. (8.4). We say that $f(x)$ satisfies the 2-tuple balance property if $f(x)$ satisfies the following two conditions:*

1. *For $1 \neq \lambda \notin \mathbb{F}_q$, any pair $(\eta, \mu) \in \mathbb{F}_q^2$ occurs q^{n-2} times in $T_f(\lambda)$.*
2. *For $1 \neq \lambda \in \mathbb{F}_q^*$, there exists some $1 \neq \mu \in \mathbb{F}_q$ such that $(\eta, \mu\eta)$ occurs q^{n-1} times in $T_f(\lambda)$ for every $\eta \in \mathbb{F}_q$.*

From the above definition, we have the following proposition.

Proposition 8.2 *Let $f(x)$ be a function from $\mathbb{F}_{q^n}$ to $\mathbb{F}_q$ that satisfies the 2-tuple balance property. Then there exists a function ρ on $\mathbb{F}_q$ such that for any $y \in \mathbb{F}_q$,*

$$f(yx) = \rho(y)f(x), \text{ for all } x \text{ in } \mathbb{F}_{q^n}, \text{ where } \rho(y) \neq 1 \text{ when } y \neq 1.$$

Proof. From Definition 8.3, for each $1 \neq \lambda \in \mathbb{F}_q^*$, there exists some $\mu \in \mathbb{F}_q$ such that

$$f(\lambda x) = \mu f(x), \forall x \in \mathbb{F}_q.$$

Let

$$\rho : \lambda \mapsto \rho(\lambda) = \mu, \lambda \in \mathbb{F}_q^*,$$

and $\rho(i) = i, i \in \{0, 1\}$. Then ρ is a function on $\mathbb{F}_q$ with $\rho(\lambda) \neq 1$ when $\lambda \neq 1$, which establishes the assertion. $\qquad\square$

Table 8.1. *Relationships involving 2-level autocorrelation (AC)*

$q = 2$	$f(x)$ or **u** has the 2-tuple balance property $\Longleftrightarrow f(x)$ is orthogonal $\Longleftrightarrow$ **u** has 2-level AC
$q = p > 2$	$f(x)$ or **u** has the 2-tuple balance property $\Longleftrightarrow f(x)$ is orthogonal $\Longleftrightarrow$ **u** has 2-level AC
$q = p^r, r > 1$	$f(x)$ or **u** has the 2-tuple balance property and $C_f(0) = 0$ $\Longrightarrow f(x)$ is orthogonal $\Longrightarrow$ **u** has 2-level AC

In Table 8.1, we summarize the relationships among the 2-level autocorrelation, orthogonality, and 2-tuple balance property for different q. Note that **u**, or equivalently $f(x)$, being balanced implies $C_f(0) = 0$.

Let $v = q - 1$ and $d = (q^n - 1)/(q - 1)$. Suppose that **u** $\leftrightarrow f(x)$ is a (v, d) interleaved sequence. Hence, **u** can be arranged into a (v, d) array, $A(\mathbf{a}, \mathbf{e})$, whose columns are either shifts of the sequence **a** of period v or zero sequences where the shifts are given by $\mathbf{e} = (e_0, e_1, \ldots, e_{d-1})$ where $e_j = e$ if the jth column is an e-shift of **a**; otherwise $e_j = \infty$ by convention. (Note. If the period of **u**, N, is less than $q^n - 1$, we use $(q^n - 1)/N$ periods of **u** to complete the $v \times d$ array.) In this case, we say that $f(x)$ or **u** admits a (v, d) interleaved structure.

Proposition 8.3 $f(x)$ *or* **u** *admits a* (v, d) *interleaved structure, associated with the base sequence* **a**, *if and only if there exists a function* $g(x)$ *on* $\mathbb{F}_q$ *such that*

$$f(yx) = g(y)f(x), x \in \mathbb{F}_{q^n}, y \in \mathbb{F}_q, \tag{8.7}$$

where the base sequence **a** *is given by the evaluation of* $g(x)$.

Proof. We only need to show sufficiency since the necessity is clear. For $u_i = f(\alpha^i)$ where α is a primitive element of $\mathbb{F}_{q^n}$, let $\beta = \alpha^d \in \mathbb{F}_q$. If $f(x)$ satisfies Eq. (8.7), we have

$$u_{id+j} = f(\alpha^{id+j}) = g(\alpha^{id})f(\alpha^j) = g(\beta^i)f(\alpha^j), 0 \le i < v, 0 \le j < d.$$

Let $\mathbf{a} = \{a_i\}$ whose elements are given by

$$a_i = g(\beta^i), i = 0, 1, \ldots, v - 1$$

and $\mathbf{e} = (e_0, e_1, \ldots, e_{d-1})$ whose elements are given by

$$
e_j = \begin{cases} e & \text{if } f(\alpha^j) \neq 0, f(\alpha^j) = \beta^e \\ \infty & \text{if } f(\alpha^j) = 0. \end{cases}
$$

Then $\mathbf{u}$ is a (v, d) interleaved sequence associated with the base sequence $\mathbf{a}$ and the shift sequence $\mathbf{e}$. Thus, $f(x)$ admits an interleaved structure.　　$\square$

If $f(x)$ satisfies the relation

$$
f(yx) = y^r f(x), \text{ for } y \in \mathbb{F}_{2^m}, x \in \mathbb{F}_{2^n}, m \mid n,
$$

where $\gcd(r, 2^m - 1) = 1$, then we say that $f(x)$ is an r-form function from $\mathbb{F}_{q^n}$ to $\mathbb{F}_q$. From Proposition 8.3, the following result is immediate.

Corollary 8.1 *Any r-form function from $\mathbb{F}_{q^n}$ to $\mathbb{F}_q$ admits a (v, d) interleaved structure.*

8.1.2 Construction of $\mathcal{C}_q(n)$ with subfield decomposition: An algebraic approach

We will consider the case that n is a composite number. We simply use the notation $\mathbb{F}_r$ for $\mathbb{F}_{q^r}$ when we draw a figure to explain a relationship among subfields. Recall that $Tr_m^n(x)$ denotes a trace function from $\mathbb{F}_{q^n}$ to $\mathbb{F}_{q^m}$ where $m \mid n$.

Definition 8.4 (SUBFIELD FACTORIZATION OF FUNCTIONS OR SEQUENCES) *Let $\mathbf{u} = \{u_i\}$ be a sequence over $\mathbb{F}_q$ of period $N \mid (q^n - 1)$ and let $f(x)$ be its trace representation. If there is $m > 1$, a proper factor of n, such that $f(x)$ can be decomposed into a composition of $h(x)$ and $g(x)$ where $h(x)$ is a function from $\mathbb{F}_{q^n}$ to $\mathbb{F}_{q^m}$ and $g(x)$ a function from $\mathbb{F}_{q^m}$ to $\mathbb{F}_q$, i.e.,*

$$
f(x) = g(x) \circ h(x) \tag{8.8}
$$

or in diagram form

$$
\begin{array}{c}
\mathbb{F}_n \\
\downarrow \; h(x) \\
\mathbb{F}_m \\
\downarrow \; g(x) \\
\mathbb{F}_1
\end{array}
$$

then we say that $f(x)$ or $\mathbf{u}$ is subfield reducible, *Eq. (8.8) is called a subfield factorization of $f(x)$ or $\mathbf{u}$. Otherwise, $f(x)$ or $\mathbf{u}$ is said to be* subfield irreducible.

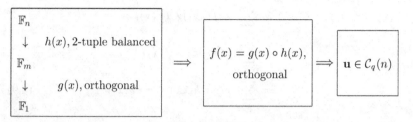

Figure 8.1. Two-tuple balance approach.

From this definition, we know that m-sequences of period $q^n - 1$ are subfield reducible if n is not a prime. Note that the subfield reducibility or irreducibility of functions or sequences is meaningful only for n composite.

Example 8.1 Let $p = 2, n = 6, m = 3$, and $f(x) = Tr_1^6(x^3 + x^5)$. Then $f(x)$ has the following subfield factorization:

$$
\begin{aligned}
&\mathbb{F}_6 \\
&\downarrow \quad h(x) = Tr_3^6(x) \\
&\mathbb{F}_3 \\
&\downarrow \quad g(x) = Tr_1^3(x^3) \\
&\mathbb{F}_1
\end{aligned}
$$

This can be verified as follows:

$$
\begin{aligned}
g(x) \circ h(x) &= Tr_1^3(x^3) \circ Tr_3^6(x) = Tr_1^3\big(Tr_3^6(x)^3\big) \\
&= Tr_1^3\big(\big(x + x^{2^3}\big)^3\big) \\
&= Tr_1^3(x^3 + x^{17} + x^{10} + x^{24}) \\
&= Tr_1^3\big(Tr_3^6(x^3 + x^5)\big) \\
&= Tr_1^6(x^3 + x^5) = f(x).
\end{aligned}
$$

Theorem 8.1 *If $f(x)$ has a subfield decomposition, that is, $f(x) = g(x) \circ h(x)$ where $h(x) : \mathbb{F}_{q^n} \to \mathbb{F}_{q^m}$ satisfies the balance property and the 2-tuple balance property, and $g(x) : \mathbb{F}_{q^m} \to \mathbb{F}_q$ is orthogonal, then $f(x)$ is orthogonal. Consequently, $\mathbf{u}$ has 2-level autocorrelation.*

This result is illustrated in Figure 8.1.

Proof. The proof given below is similar to that of Proposition 5.4 in Chapter 5. For $\lambda \neq 1$,

$$C_f(\lambda) = \sum_{x \in \mathbb{F}_{q^n}} \chi(f(\lambda x))\chi^*(f(x))$$

$$= \sum_{x \in \mathbb{F}_{q^n}} \chi(f(\lambda x) - f(x))$$

$$= \sum_{x \in \mathbb{F}_{q^n}} \chi(g \circ h(\lambda x) - g \circ h(x)) \qquad (8.9)$$

1. $\lambda \notin \mathbb{F}_{q^m}^*$. In this case, substituting condition 1 of Definition 8.3 into Eq. (8.9), we have

$$C_f(\lambda) = q^{n-2} \sum_{\eta, \mu \in \mathbb{F}_{q^m}} \chi(g(\eta) - g(\mu))$$

$$= q^{n-2} \sum_{\eta \in \mathbb{F}_{q^m}} \chi(g(\eta)) \sum_{\mu \in \mathbb{F}_{q^m}} \chi(-g(\mu)) = 0.$$

The last identity follows from applying Lemma 8.1 because $g(x)$ is balanced.

2. $0 \neq \lambda \in \mathbb{F}_{q^m}$. Substituting condition 2 of Definition 8.3 into Eq. (8.9), we have

$$C_f(\lambda) = q^{n-1} \sum_{\eta \in \mathbb{F}_{q^m}} \chi(g(\mu\eta) - g(\eta)), \ 1 \neq \mu \in \mathbb{F}_{q^m}$$

$$= q^{n-1} C_g(\mu) = 0 \qquad \text{(by } C_g(\mu) = 0 \text{ if } \mu \neq 1\text{).}$$

3. $\lambda = 0$. Because h is balanced and g satisfies $C_g(0) = 0$, $C_f(0) = 0$ is clear.

Thus $f(x)$ is orthogonal. Therefore **u** has 2-level autocorrelation. $\qquad\square$

Example 8.2 The function $f(x)$, given in Example 8.1, is orthogonal, or equivalently, **u** whose elements are given by $u_i = f(\alpha^i)$, $i = 0, 1, \ldots$ has 2-level autocorrelation. This is because $h(x) = Tr_3^6(x)$ is the trace representation of an m-sequence over $\mathbb{F}_{2^3}$ of degree 2. Thus it is balanced and satisfies the 2-tuple balance property (Theorem 5.7, Section 5.6, Chapter 5). Furthermore, $g(x) = Tr_1^3(x^3)$ is a trace representation of an m-sequence over $\mathbb{F}_2$ of degree 3, so it is orthogonal. Thus $f(x)$ is orthogonal from the above theorem. Therefore, **u** has 2-level autocorrelation. (Note that **u** is a GMW sequence, which will be introduced later in this chapter.)

From the proof of Theorem 8.1, if $f(x)$ has a subfield factorization as $h(x) \circ g(x)$ where $h(x)$ satisfies the 2-tuple balance property, then $C_f(\lambda) = 0$ for all $\lambda \notin \mathbb{F}_q$, as long as $g(x)$ is balanced. Thus whether **u** has 2-level autocorrelation only depends on whether $g(x)$ has 2-level autocorrelation. This property is illustrated in Figure 8.2 for the case $q = 2$ where $v = 2^m - 1$ and $d = (2^n - 1)/v$. The inner cycle contains the exponents of the nonzero elements of $\mathbb{F}_2$, and the middle cycle and outer cycles contain exponents of the nonzero elements of $\mathbb{F}_{2^m}^*$ and $\mathbb{F}_{2^n}^*$,

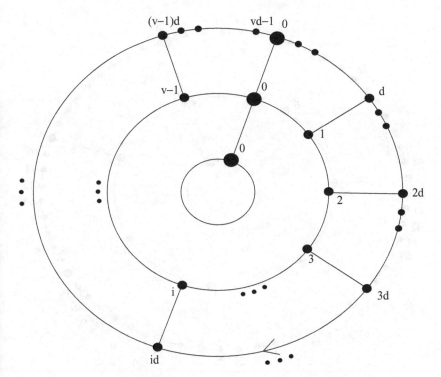

Figure 8.2. Relationship among cyclic groups of extension fields.

respectively, represented as cyclic groups. Precisely, let α be a primitive element of $\mathbb{F}_{2^n}$, $d = (2^n - 1)/(2^m - 1)$ and $\beta = \alpha^d$. Then β is a primitive element of $\mathbb{F}_{2^m}$, and $\alpha^{2^n-1} = 1$ belongs to $\mathbb{F}_2$. The point on the inner cycle represents zero, which is the exponent of 1. The points on the middle cycle and the outer cycle are the exponents of β and α, respectively. The proof of Theorem 8.1 shows that the autocorrelation function of **u** at τ is equal to -1 for all $\tau \neq id$, $i = 0, 1, \ldots, 2^m - 2$; that is, $C_{\mathbf{u}}(\tau) = -1, \tau \neq id, i = 0, 1, \ldots, 2^m - 2$, which only requires that $g(x)$ is balanced. In other words, if $g(x)$ is balanced, then there are $(2^n - 2) - (2^m - 2) = 2^n - 2^m$ points in $\mathbb{F}_{2^n}$ such that $C_{\mathbf{u}}(\tau) = -1$. The condition of $g(x)$ being orthogonal guarantees that $C_{\mathbf{u}}(\tau) = -1$ on the rest of the $q^m - 2$ points that belong to $\mathbb{F}_{2^m}^*$, that is, the points corresponding to the ones on the middle cycle. So, for cryptographic applications, it is desired that functions have no subfield factorizations, as this does not possess much randomness.

We summarize the above discussion into the following corollary.

Corollary 8.2 *Let $f(x) = g(x) \circ h(x)$ where $h(x) : \mathbb{F}_{q^n} \to \mathbb{F}_{q^m}$ is 2-tuple balanced and $g(x) : \mathbb{F}_{q^m} \to \mathbb{F}_q$ is balanced, and let $\mathbf{u} \leftrightarrow f(x)$ and $\mathbf{a} \leftrightarrow g(x)$.*

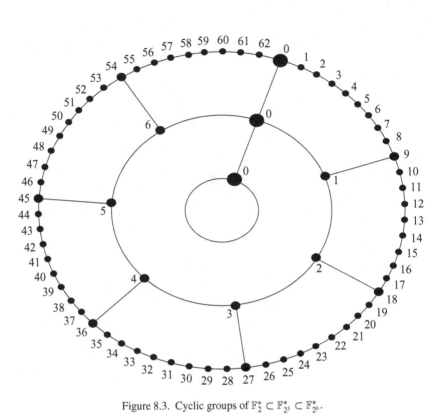

Figure 8.3. Cyclic groups of $\mathbb{F}_2^* \subset \mathbb{F}_{2^3}^* \subset \mathbb{F}_{2^6}^*$.

Then

$$C_{\mathbf{u}}(\tau) = -1, \text{ for all } \tau \not\equiv 0 \ (\text{mod } d),$$

where $d = (q^n - 1)/(q^m - 1)$ *and*

$$C_{\mathbf{u}}(id) = C_{\mathbf{a}}(i), i \not\equiv 0 \ (\text{mod } q^m - 1).$$

Example 8.3 Considering Example 8.2, we replace $Tr_1^3(x^3)$ by $g(x)$, an arbitrary balanced function from $\mathbb{F}_{2^3}$ to $\mathbb{F}_2$; that is, $\mathbf{u} \leftrightarrow g(x) \circ Tr_3^6(x)$. From the proof of Theorem 8.1, for the autocorrelation function of $\mathbf{u}$ we have $C(\tau) = -1$ for all 56 values of $\tau \notin S = \{9, 18, 27, 36, 45, 54\}$, and $g(x)$ being orthogonal only guarantees that $C(\tau) = -1$ for these 6 values; that is, $\tau \in S$. The points are illustrated in Figure 8.3

8.1.3 Construction of $C_q(n)$ with subfield decomposition: A combinatoric approach

In the following, we present another construction for 2-level autocorrelation sequences with subfield decomposition in terms of the interleaved structure. We first investigate a formula for computing correlation functions of interleaved sequences via their array forms. Then we derive a method to arrange a $v \times d$ array by using a 2-level autocorrelation sequence of period v and the zero sequence such that the resulting interleaved sequence has 2-level autocorrelation.

At this point, we may assume that both v and d are arbitrary positive numbers. For simplicity, we will also use the notation of the (Hermitian) dot product of two vectors. Let $\mathbf{a} = (a_0, a_1, \ldots, a_{v-1})$ and $\mathbf{b} = (b_0, b_1, \ldots, b_{v-1})$ be two vectors of $\mathbb{F}_q^v$. We denote by $<\mathbf{a}, \mathbf{b}>$ the (Hermitian) dot product of two vectors $\chi(\mathbf{a}) = (\chi(a_0), \chi(a_1), \ldots, \chi(a_{v-1}))$ and $\chi(\mathbf{b}) = (\chi(b_0), \chi(b_1), \ldots, \chi(b_{v-1}))$ (see Chapter 1); that is,

$$<\mathbf{a}, \mathbf{b}> = (\chi(\mathbf{a}) \cdot \chi(\mathbf{b})) = \sum_{i=0}^{v-1} \chi(a_i - b_i). \tag{8.10}$$

Consider a (v, d) interleaved sequence $\mathbf{u}$ associated with $(\mathbf{a}, \mathbf{e})$ where the shift sequence $\mathbf{e} = (e_0, e_1, \ldots, e_{d-1}) \in \mathbb{Z}_v^d$. When needed, the shift sequence is extended to $(e_0, e_1, \ldots, e_{q^n-2})$ in the same way as in Section 5.4 of Chapter 5; that is, for $k = id + j$ with $0 \le j < d$,

$$e_{id+j} = \begin{cases} e_j + i \in \mathbb{Z}_v & \text{if } e_j \ne \infty, 0 \le j < d \\ \infty & \text{if } e_j = \infty, 0 \le j < d. \end{cases} \tag{8.11}$$

We continue to use the symbol $\mathbf{e}$ to denote the resulting sequence. Note that the shift sequence of $L^{id}(\mathbf{u})$, regarded as a (v, d) interleaved sequence, is equal to $(e_0 + i, e_1 + i, \ldots, e_{d-1} + i)$, denoted by $\mathbf{e} + i$.

Proposition 8.4 *Let $\mathbf{u}$ be a (v, d) interleaved sequence with its array form $A(\mathbf{a}, \mathbf{e})$. For $\tau \ge 0$, we write $\tau = rd + s, 0 \le r, s < d$. Then the array form of $L^\tau(\mathbf{u})$ is given by*

$$A(\mathbf{a}, \mathbf{e}'),$$

where the elements of its shift sequence $\mathbf{e}' = (e_0', e_1', \ldots, e_{d-1}')$ are given by

$$e_j' = \begin{cases} e_{j+s} + r \pmod{v} & \text{if } e_{j+s} \ne \infty, 0 \le j < d \\ \infty & \text{if } e_{j+s} = \infty, 0 \le j < d. \end{cases}$$

Proof. Let $A = (A_0, A_1, \ldots, A_{d-1}) = (a_{i,j})$ and $T = (T_0, T_1, \ldots, T_{d-1})$ be the array forms of $\mathbf{u}$ and $L^\tau(\mathbf{u})$, respectively. From the definition, $A_j = L^{e_j}(\mathbf{a})$,

$0 \leq j < d$ if $e_j \neq \infty$, otherwise $A_j = 0$. Note that the first element in the sequence $L^\tau(\mathbf{u})$ is the entry $a_{r,s}$. From the definition of interleaved sequences, we have $a_{r,d+j} = a_{r+1,j}$ for each j with $d - s \leq j < d$. So T has the following array form:

$$
\begin{bmatrix}
a_{r,s} & \cdots & a_{r,d-1} & a_{r+1,0} & \cdots & a_{r+1,s-1} \\
a_{r+1,s} & \cdots & a_{r+1,d-1} & a_{r+2,0} & \cdots & a_{r+2,s-1} \\
\vdots & & & & & \\
a_{v-1,s} & \cdots & a_{v-1,d-1} & a_{1,0} & \cdots & a_{1,s-1} \\
\vdots & & & & & \\
a_{r-1,s} & \cdots & a_{r-1,d-1} & a_{r,0} & \cdots & a_{r,s-1}
\end{bmatrix}.
$$

Therefore, for $0 \leq j < d - s$ and $A_{s+j} \neq 0$,

$$T_j = L^r(A_{s+j}) = L^r(L^{e_{s+j}}(\mathbf{a})) = L^{r+e_{s+j}}(\mathbf{a}). \tag{8.12}$$

For $d - s \leq j < d$ and $A_{j-(d-s)} \neq 0$,

$$
\begin{aligned}
T_j &= L^{r+1}(A_{j-(d-s)}) = L^{r+1}(L^{e_{j-(d-s)}}(\mathbf{a})) \\
&= L^{r+1+e_{j-(d-s)}}(\mathbf{a}).
\end{aligned} \tag{8.13}
$$

Applying Eq. (8.11), $1 + e_{j-(d-s)} = e_{j-(d-s)+d} = e_{s+j}$ for $d - s \leq j < d$. Substituting this into Eq. (8.13), we get $T_j = L^{r+e_{j+s}}(\mathbf{a})$ with $d - s \leq j < d$. Together with Eq. (8.12), the result follows. $\qquad\square$

With the notation above, the autocorrelation of $\mathbf{u}$ can be computed by

$$C_{\mathbf{u}}(\tau) = \sum_{j=0}^{d-1} <T_j, A_j>. \tag{8.14}$$

In other words, the autocorrelation of $\mathbf{u}$ can be computed in terms of the autocorrelation of $\mathbf{a}$, which is a sum of the dot products of corresponding column vectors of the array forms of the sequence, $\mathbf{u}$, and τ-shift of $\mathbf{u}$, $L^\tau(\mathbf{u})$. This structure is useful for parallel computation of the autocorrelation of $\mathbf{u}$.

When the base sequence of $\mathbf{u}$, $\mathbf{a}$, has 2-level autocorrelation, then for a fixed j, $0 \leq j < d$, if $A_j \neq T_j$, we have $<A_j, T_j> = -1$, which includes the case that one of A_j and T_j is equal to a zero sequence. This contributes -1 to Eq. (8.14). If $A_j = T_j$, this results in a zero column in $T - A$, so it contributes v to Eq. (8.14). This happens in two cases. One case is that both A_j and T_j are nonzero. In this case, $r + e_{j+s}$ and e_j are equal, but not equal to ∞. The other is that both A_j and T_j are zero; that is, both e_{j+s} and e_j are equal to ∞. In the following, we distinguish these two cases. For $s \not\equiv 0 \pmod{d}$, we define the following two sets:

$$T_0(s) = \{e_{j+s} - e_j \text{ such that } e_{j+s} \neq \infty \text{ and } e_j \neq \infty, 0 \leq j < d\} \quad (8.15)$$
$$T_\infty(s) = \{j \text{ such that } e_{j+s} = \infty \text{ and } e_j = \infty, 0 \leq j < d\}. \quad (8.16)$$

Note that these are the same sets as those we introduced for m-sequences in Chapter 5. Let $N_s(r)$ be the number of $r \in \mathbb{Z}_v$ that occur in $T_0(s)$.

Lemma 8.2 *With the above notation, we write* $\tau = rd + s, 0 \leq s < d$. *If* **a** *has 2-level autocorrelation, then*

$$C_{\mathbf{u}}(\tau) = -d + (v + 1)(N_s(r) + |T_\infty(s)|).$$

Proof. Let

$$N = |\{j \text{ such that } A_j = T_j, 0 \leq j < d\}|.$$

Then $N = N_s(r) + |T_\infty(s)|$ and

$$C_{\mathbf{u}}(\tau) = -(d - N) + vN = -d + (v + 1)N,$$

which establishes the result. □

Assume that $v = q^m - 1$ and $d = (q^n - 1)/(q^m - 1), m|n$. From Lemma 8.2, **u** has 2-level autocorrelation if and only if

$$C_{\mathbf{u}}(\tau) = -1 \iff -d + (v + 1)(N_s(r) + |T_\infty(s)|) = -1,$$
$$\tau = rd + s, 0 \leq s < d.$$

This shows that **u** has 2-level autocorrelation if and only if

$$N_s(r) + |T_\infty(s)| = (q^{n-m} - 1)/(q^m - 1) \quad (8.17)$$

for all $(r, s) \neq (0, 0)$ with $0 \leq r < v$ and $0 \leq s < d$. For any $s \not\equiv 0 \pmod{d}$ and $r \in \mathbb{Z}_v$, if the sequence **e** satisfies Eq. (8.17), then we say that **e** has the mixed difference property.

Note that here the mixed difference property is only defined for the case of $v = q^m - 1$ and $d = (q^n - 1)/(q^m - 1)$ where $m|n$. We summarize the above result in the following proposition.

Proposition 8.5 *Let* **u** *be a* (v, d) *interleaved sequence with array form* $A(\mathbf{a}, \mathbf{e})$. *Assume that* **a** *has 2-level autocorrelation (or equivalently* $g(x)$ *is orthogonal). Then* **u** *has 2-level autocorrelation if and only if* **e** *has the mixed difference property.*

In the following proposition, we will establish that the shift sequence of any m-sequence satisfies the mixed difference property.

Proposition 8.6 *If* $\mathbf{e}$ *is a shift sequence of an m-sequence over* $\mathbb{F}_{q^m}$ *of degree* $l = n/m$ *regarded as a* (v, d) *interleaved sequence, then* $\mathbf{e}$ *has the mixed difference property.*

Proof. Note that any m-sequence $\mathbf{u}$ of period $q^n - 1$ has a subfield decomposition $\mathbf{u} = Tr_1^m(\mathbf{b})$ for $m|n$ where $\mathbf{b}$ is an m-sequence over $\mathbb{F}_{q^m}$ of degree $l = n/m$. Thus $\mathbf{u}$, regarded as a (v, d) interleaved sequence, has the same shift sequence as $\mathbf{b}$ (see Chapter 5). According to the property of shift sequences of m-sequences (Fact 5.1, Chapter 5), each element in $\mathbb{Z}_v$ occurs q^{n-2m} times in $T_0(s)$, and $|T_\infty(s)| = (q^{n-2m} - 1)/(q^m - 1)$. Thus we have

$$N_s(r) + |T_\infty(s)| = q^{n-2m} + (q^{n-2m} - 1)/(q^m - 1) = (q^{n-m} - 1)/(q^m - 1).$$

Thus $\mathbf{e}$ has the mixed difference property. $\square$

In Chapter 5, we presented the shift sequences of binary m-sequences of degree 6 and degree 9, regarded as a 7×9 array and a 7×73 array, respectively, which illustrated Proposition 8.6.

When the base sequence has 2-level autocorrelation, a (v, d) interleaved sequence has 2-level autocorrelation iff its shift sequence satisfies the mixed difference property. Therefore, for a given 2-level autocorrelation sequence over $\mathbb{F}_q$ with period $q^m - 1$, constructing a 2-level autocorrelation sequence over $\mathbb{F}_q$ of period $q^n - 1$ is equivalent to finding a way to arrange a $v \times d$ array using the known shorter 2-level autocorrelation sequence and the zero sequence. This is an extraordinary phenomenon. It transfers the construction of 2-level autocorrelation sequences of period $q^n - 1$ with subfield $\mathbb{F}_{q^m}$ decomposition into the construction of two shorter sequences. One is a 2-level autocorrelation sequence of period $v = q^m - 1$ and the other is a sequence with the mixed difference property whose symbols are taken from $\mathbb{Z}_v(\infty) = \mathbb{Z}_v \cup \{\infty\}$. Furthermore, we can apply this result to finding new 2-level autocorrelation sequences. This will be stated clearly in the following two theorems.

Theorem 8.2 *Let* $\mathbf{u} = A(\mathbf{a}, \mathbf{e})$ *and* $\mathbf{v} = A(\mathbf{d}, \mathbf{k})$ *be two* (v, d) *interleaved sequences where both* $\mathbf{a}$ *and* $\mathbf{d}$ *have 2-level autocorrelation (they need not be equal). If there exist some integers r and s such that $e_j \equiv r + k_{j+s}$ (mod v) for which none of e_j and k_{j+s} is equal to* ∞ *(otherwise $e_j = k_{j+s} = \infty$, $0 \le j < d$), then*

$$C_{\mathbf{u}}(\tau) = C_{\mathbf{v}}(\tau + t), t = rd + s, \tau = 0, 1, \ldots.$$

In other words, for two interleaved sequences, if their base sequences have 2-level autocorrelation, then their autocorrelation functions are equal if and

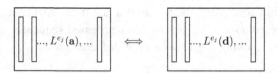

Figure 8.4. Interleaved approach.

only if their shift sequences are equal after applying the shift operation and the affine operation.

The following theorem is immediate from this result.

Theorem 8.3 *Let* **a** *and* **d** *be two sequences over* $\mathbb{F}_q$ *of period* $q^m - 1$ *with 2-level autocorrelation, and let* $A(\mathbf{a}, \mathbf{e})$ *and* $A(\mathbf{d}, \mathbf{e})$ *be array forms of two* (v, d) *interleaved sequences associated with* $(\mathbf{a}, \mathbf{e})$ *and* $(\mathbf{d}, \mathbf{e})$ *respectively. If one of them has 2-level autocorrelation, so does the other.*

This result is illustrated in Figure 8.4 for arrays (where the vertical bars represent the column vectors). In detail,

$$
\begin{array}{cccc}
e_0 & e_1 & \cdots & e_{d-1}
\end{array}
$$

$$[L^{e_0}(\mathbf{a}), L^{e_1}(\mathbf{a}), \ldots, L^{e_{d-1}}(\mathbf{a})] \in \mathcal{C}_q(n)$$

$$\mathbf{a} \in \mathcal{C}_q(m)$$

$$\updownarrow$$

$$
\begin{array}{cccc}
e_0 & e_1 & \cdots & e_{d-1}
\end{array}
$$

$$[L^{e_0}(\mathbf{d}), L^{e_1}(\mathbf{d}), \ldots, L^{e_{d-1}}(\mathbf{d})] \in \mathcal{C}_q(n)$$

$$\mathbf{d} \in \mathcal{C}_q(m)$$

In other words, if one of the interleaved sequences from the above diagram has 2-level autocorrelation, then one can construct new 2-level autocorrelation sequences in the following way. We retain the shift sequence and replace the base sequence with another sequence over $\mathbb{F}_q$ of period $q^m - 1$ with 2-level autocorrelation. Then the resulting interleaved sequence has 2-level autocorrelation. The diagram illustrating Theorem 8.3 not only provides a visual way to view this method, but also provides another way to generate new sequences with 2-level autocorrelation.

Remark 8.1 If $A(\mathbf{a}, \mathbf{e})$ yields a 2-level autocorrelation sequence of period $q^n - 1$ and **a** is a 2-level autocorrelation sequence of period $q^m - 1$, then $A(\mathbf{a}, \mathbf{e})$ and

$A(\mathbf{d}, \mathbf{e})$ are shift equivalent if and only if $\mathbf{a}$ and $\mathbf{d}$ are shift equivalent. Thus, as long as the column sequences of two interleaved sequences are not shift equivalent, the resulting sequences are not shift equivalent.

The constructions presented in Theorems 8.1 and 8.3 are very general. In the next section, we will provide constructions for the function $h(x)$ with the 2-tuple balance property of Theorem 8.1 and the shift sequences with the mixed shift property of Theorem 8.3, respectively.

8.2 GMW constructions

Starting with this section, we restrict ourselves to binary 2-level autocorrelation sequences that correspond to cyclic Hadamard difference sets. In this section, we give two constructions for 2-level autocorrelation sequences, of period $2^n - 1$ with subfield decomposition. One way is to construct $h(x)$ in Theorem 8.1, and the other is to construct shift sequences in Theorem 8.3. These will produce GMW sequences, cascaded GMW sequences, and generalized GMW sequences. All of these sequences correspond to cyclic Hadamard difference sets discovered by Goldon, Mills, and Welch in 1962.

8.2.1 Construction I

For an m-sequence $\mathbf{u}$ of degree n, we assume that $\mathbf{u} \leftrightarrow f(x) = Tr_1^n(x^k)$, where $\gcd(k, 2^n - 1) = 1$. From Chapter 5, if n is a composite number, then we have the following subfield decomposition of $f(x)$:

$$
\begin{array}{l}
\mathbb{F}_n \\
\downarrow \ Tr_m^n(x^k) \\
\mathbb{F}_m \\
\downarrow \ Tr_1^m(x) \\
\mathbb{F}_1
\end{array}
$$

where $Tr_m^n(x^k)$ is a trace representation of an m-sequence over $\mathbb{F}_{2^m}$ of degree $l = n/m$. Thus it satisfies the 2-tuple balance property. On the other hand, $Tr_1^m(x)$ is a trace representation of an m sequence over $\mathbb{F}_2$ of degree m, which has 2-level autocorrelation. We extend this composition below. Note that for the binary case, if $g(x)$ is a function from $\mathbb{F}_{2^m}$ to $\mathbb{F}_2$, then the 2-tuple balance property and orthogonality for $g(x)$ are equivalent. Furthermore, for $\mathbf{a} \leftrightarrow g(x)$, $\mathbf{a}$ has 2-level autocorrelation if and only if $g(x)$ is orthogonal.

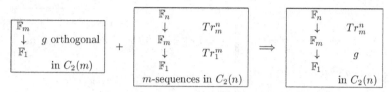

Figure 8.5. Operations of Construction I.

Construction I. Let α be a primitive element of $\mathbb{F}_{2^n}$, $m|n$, and $g(x)$ an orthogonal function from $\mathbb{F}_{2^m}$ to $\mathbb{F}_2$. We construct a sequence $\mathbf{u} = \{u_i\}$ whose elements are given by

$$u_i = f(\alpha^i), i = 0, 1, \ldots,$$

where $f(x)$ is a composition of $Tr_m^n(x^k)$ and $g(x)$,

$$f(x) = g(x) \circ Tr_m^n(x^k), \gcd(k, 2^n - 1) = 1.$$

Theorem 8.4 $\mathbf{u}$ *is a sequence of period* $2^n - 1$ *with 2-level autocorrelation.*

Note that for two distinct values of k in the above construction, the resulting sequences are decimation equivalent. Thus, except when considering the shift-equivalent property of this construction, we set $k = 1$. The above result is illustrated in Figure 8.5

Proof of Theorem 8.4.

Method 1. *Algebraic Approach (using Theorem 8.1).* Because the evaluation of $Tr_m^n(x^k)$ is an m-sequence over $\mathbb{F}_{2^m}$, and any m-sequence is balanced with the 2-tuple balance property, $Tr_m^n(x^k)$ is a balanced function from $\mathbb{F}_{2^n}$ to $\mathbb{F}_{2^m}$ with the 2-tuple balance property. Together with $g(x)$ having 2-level autocorrelation, applying Theorem 8.1, $f(x)$, the composition of $h(x)$ and $g(x)$, is orthogonal. Thus $\mathbf{u}$ has 2-level autocorrelation.

Method 2. *Combinatoric Approach (using Theorem 8.3).* Let $v = 2^m - 1$ and $d = (2^n - 1)/(2^m - 1)$. Note that an evaluation of $Tr_1^n(x^k)$ is an m sequence over $\mathbb{F}_2$ of period $2^n - 1$ which admits a (v, d) interleaved structure. Let $A(\mathbf{a}, \mathbf{e})$ be its array form, and let $\mathbf{d}$ be the evaluation of $g(x)$. Then $\mathbf{d} \in C_2(m)$. Consequently, $\mathbf{u}$ is a (v, d) interleaved sequence with array form $A(\mathbf{d}, \mathbf{e})$. According to Theorem 8.3, $\mathbf{u}$ has 2-level autocorrelation. $\square$

In the following, we present an interleaved structure of $\mathbf{u}$ in detail. Let $\beta = \alpha^d$. Then β is a primitive element of $\mathbb{F}_{2^m}$. Let $\mathbf{b} \leftrightarrow Tr_m^n(x)$ (we set $k = 1$)

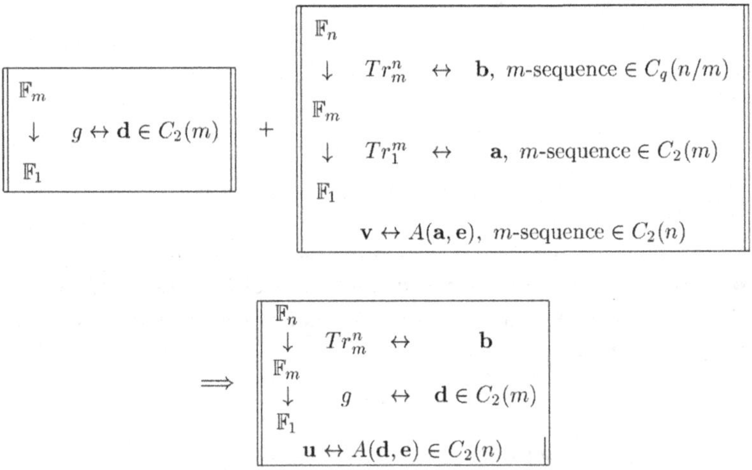

$$
\boxed{
\begin{array}{l}
\mathbb{F}_m \\
\downarrow \quad g \leftrightarrow \mathbf{d} \in C_2(m) \\
\mathbb{F}_1
\end{array}
}
\quad + \quad
\boxed{
\begin{array}{l}
\mathbb{F}_n \\
\downarrow \quad Tr_m^n \quad \leftrightarrow \quad \mathbf{b}, \ m\text{-sequence} \in C_q(n/m) \\
\mathbb{F}_m \\
\downarrow \quad Tr_1^m \quad \leftrightarrow \quad \mathbf{a}, \ m\text{-sequence} \in C_2(m) \\
\mathbb{F}_1 \\
\qquad \mathbf{v} \leftrightarrow A(\mathbf{a}, \mathbf{e}), \ m\text{-sequence} \in C_2(n)
\end{array}
}
$$

$$
\implies
\boxed{
\begin{array}{l}
\mathbb{F}_n \\
\downarrow \quad Tr_m^n \quad \leftrightarrow \quad \mathbf{b} \\
\mathbb{F}_m \\
\downarrow \quad g \quad \leftrightarrow \quad \mathbf{d} \in C_2(m) \\
\mathbb{F}_1 \\
\quad \mathbf{u} \leftrightarrow A(\mathbf{d}, \mathbf{e}) \in C_2(n)
\end{array}
}
$$

Figure 8.6. Interleaved approach for Construction I.

whose elements are given by

$$
b_i = Tr_m^n(\alpha^i), \ i = 0, 1, \ldots
$$

and

$$
\mathbf{a} \leftrightarrow Tr_1^m(x), \qquad a_i = Tr_1^m(\beta^i), \qquad \mathbf{v} \leftrightarrow Tr_1^m(x) \circ Tr_m^n(x)
$$

$$
\mathbf{d} \leftrightarrow g(x), \qquad d_i = g(\beta^i), \qquad \mathbf{u} \leftrightarrow g(x) \circ Tr_m^n(x).
$$

Then the three sequences $\mathbf{b}$, $\mathbf{v}$, and $\mathbf{u}$ are (v, d) interleaved sequences associated with $(\{\beta^i\}, \mathbf{e})$, $(\mathbf{a}, \mathbf{e})$, and $(\mathbf{d}, \mathbf{e})$, respectively. Furthermore, all of them have a common shift sequence $\mathbf{e}$ whose elements are given by

$$
e_j = \begin{cases} e & \text{if } Tr_m^n(\alpha^j) \neq 0, \ Tr_m^n(\alpha^j) = \beta^e \\ \infty & \text{if } Tr_m^n(\alpha^j) = 0. \end{cases} \tag{8.18}
$$

We illustrate the interleaved approach of the decompositions of $\mathbf{u}$ and $\mathbf{v}$ in Figure 8.6 where $q = 2^m$.

8.2.2 Constructions of $g(x)$

There are three types of particular functions for $g(x)$ such that $g(x)$ is orthogonal.

A. *GMW Sequences*: A field chain is given by

$$
\mathbb{F}_2 \subset \mathbb{F}_{2^m} \subset \mathbb{F}_{2^n}
$$

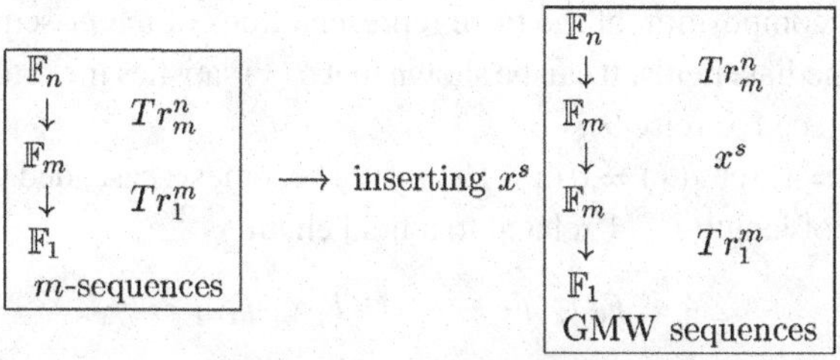

Figure 8.7. Diagram of a GMW construction extended from m-sequences.

and $g(x) = Tr_1^m(x^s)$ with $\gcd(s, 2^m - 1) = 1$. So,

$$f(x) = Tr_1^m(x^s) \circ Tr_m^n(x^k), \tag{8.19}$$

$$u_i = f(\alpha^i) = Tr_1^m(Tr_m^n(\alpha^{ik})^s), i = 0, 1, \ldots. \tag{8.20}$$

Figure 8.7 illustrates how the GMW construction is extended from the m-sequence construction (which is the 2-tuple balance property approach).

From the interleaved approach, a GMW sequence can be constructed from an m-sequence in a (v, d) interleaved structure by replacing the base sequence in the array by an s-decimation of the base sequence where $s \not\equiv 2^i \pmod{2^m - 1}$. The resulting sequence is a GMW sequence. This is illustrated in Figure 8.8.

B. If m is also a composite number, we assume that $r|m$. A *cascaded GMW function* of length t from $\mathbb{F}_{2^m}$ to $\mathbb{F}_{2^r}$, say $\sigma(x)$, is defined as

$$\sigma(x) = x^{s_0} \circ Tr_{F_1/F_0}(x^{s_1}) \circ Tr_{F_2/F_1}(x^{s_2}) \circ \cdots \circ Tr_{F_t/F_{t-1}}(x^{s_t}), x \in \mathbb{F}_t,$$

where the field chain is as follows:

$$\mathbb{F}_{2^r} = F_0 \subset F_1 \subset \cdots \subset F_t = \mathbb{F}_{2^m},$$

where $\circ$ is the function composition operator and the s_i's satisfy $\gcd(s_i, |F_i| - 1) = 1$. In other words, a cascaded GMW function is a

$$\mathbf{a}^{(s)}, \; A(\mathbf{a}, \mathbf{e}) = [L^{e_0}(\mathbf{a}), \cdots, L^{e_j}(\mathbf{a}) \cdots, L^{e_{d-1}}(\mathbf{a})]$$

m-sequence

a replaced by $\mathbf{a}^{(s)}$ $\Longrightarrow$ $A(\mathbf{a}^{(s)}, \mathbf{e}) = [L^{e_0}(\mathbf{a}^{(s)}), \cdots, L^{e_j}(\mathbf{a}^{(s)}), \cdots, L^{e_{d-1}}(\mathbf{a}^{(s)})]$

GMW sequence

Figure 8.8. GMW construction: interleaved approach from m-sequences.

cascaded composition of the trace representations of the m-sequences over the intermediate fields. It can be shown that $\sigma(x)$ satisfies the 2-tuple balance property (see Exercise 9).

B.1 If $r = 1$, set $g(x) = \sigma(x)$. Then $f(x)$ becomes a cascaded GMW function of length $t + 1$ related to a field chain

$$\mathbb{F}_2 = F_0 \subset F_1 \subset \cdots \subset F_t \subset F_{t+1} = \mathbb{F}_{2^n}.$$

Applying Theorem 8.4 inductively, $f(x)$ has 2-level autocorrelation.

B.2 If $r > 1$, let $w(x)$ be an orthogonal function from $\mathbb{F}_{2^r}$ to $\mathbb{F}_2$. Let $g(x) = w(x) \circ \sigma(x)$, a composition of $\sigma(x)$ and $w(x)$ where $\sigma(x)$ is a balanced function with the 2-tuple balance property and $w(x)$ is orthogonal. According to Theorem 8.1, $g(x)$ is orthogonal.

C. $g(x)$ can be selected from any 2-level autocorrelation sequences based on a number theory approach, such as quadratic residue sequences or the Hall sextic residue sequences when their period is equal to $2^m - 1$ (see Chapter 2), as well as those which will be presented in the next chapter.

8.2.3 Examples

From the proof of Construction I, we have two methods to implement these sequences. One method is to use a finite field configuration. The other is to use the interleaved structure. We will use some examples to illustrate these methods. A formal discussion of implementation issues will be given in Section 8.6.

Example 8.4 The smallest composite number n is 4. In this case, we only have one m-sequence of degree 2 that is associated with the subfield $\mathbb{F}_{2^2}$. Thus, Construction I cannot generate any new 2-level sequences of period 15. Next, we look at the case $n = 6$. For this case, there are two shift-distinct m-sequences of degree 3. Thus, it is possible to construct GMW sequences of period 63.

Method 1. Finite field computation. A GMW function related to the field chain $\mathbb{F}_2 \subset \mathbb{F}_{2^3} \subset \mathbb{F}_{2^6}$ is given by

$$f(x) = Tr_1^3(x^3) \circ Tr_3^6(x)$$

(which is the function given in Example 8.2). In this case, we have

$$u_i = Tr_1^3(b_i^3), i = 0, 1, \ldots,$$

where $\mathbf{b} = \{b_i\}$ is an m-sequence over $\mathbb{F}_{2^3}$ of degree 2. In the following, we will give a procedure to compute the elements of $\mathbf{u}$. A general procedure for generating GMW sequences will be presented in Section 8.6. Here we choose $t(x) = x^3 + x^2 + 1$, a primitive polynomial over $\mathbb{F}_2$; β, a root of $t(x)$

in the extension field $\mathbb{F}_{2^3}$; and $w(x) = x^2 + \beta^3 x + \beta$, a primitive polynomial over $\mathbb{F}_{2^3}$.

Procedure:

1. Generate $\mathbb{F}_{2^3}$ using $t(x)$. (In the following, we list the exponents of β for the nonzero elements in $\mathbb{F}_{2^3}$, and the exponent of zero in $\mathbb{F}_{2^3}$ is listed as ∞ as usual.)
2. Compute the m-sequence $\mathbf{b}$ over $\mathbb{F}_{2^3}$ with the minimal polynomial $w(x)$ by selecting an initial state $(b_0, b_1) = (0, \beta^3)$:

$$b_{i+2} = \beta^3 b_{1+i} + \beta b_i, i = 0, 1, \ldots.$$

Thus, we obtain $\mathbf{b}$ as follows, represented as a $(7, 9)$ interleaved sequence:

$$\mathbf{b} \leftrightarrow A(\{\beta^i\}, \mathbf{e}) = \begin{bmatrix} \infty & 3 & 6 & 5 & 5 & 2 & 3 & 5 & 3 \\ \infty & 4 & 0 & 6 & 6 & 3 & 4 & 6 & 4 \\ \infty & 5 & 1 & 0 & 0 & 4 & 5 & 0 & 5 \\ \infty & 6 & 2 & 1 & 1 & 5 & 6 & 1 & 6 \\ \infty & 0 & 3 & 2 & 2 & 6 & 0 & 2 & 0 \\ \infty & 1 & 4 & 3 & 3 & 0 & 1 & 3 & 1 \\ \infty & 2 & 5 & 4 & 4 & 1 & 2 & 4 & 2 \end{bmatrix} \qquad (8.21)$$

3. Map: $x \to x^3$ in $\mathbb{F}_{2^3}$. Then

$$\mathbf{b} \to \mathbf{b}^3 \leftrightarrow \begin{bmatrix} \infty & 2 & 4 & 1 & 1 & 6 & 2 & 1 & 2 \\ \infty & 5 & 0 & 4 & 4 & 2 & 5 & 4 & 5 \\ \infty & 1 & 3 & 0 & 0 & 5 & 1 & 0 & 1 \\ \infty & 4 & 6 & 3 & 3 & 1 & 4 & 3 & 4 \\ \infty & 0 & 2 & 6 & 6 & 4 & 0 & 6 & 0 \\ \infty & 3 & 5 & 2 & 2 & 0 & 3 & 2 & 3 \\ \infty & 6 & 1 & 5 & 5 & 3 & 6 & 5 & 6 \end{bmatrix}$$

4. Map: $x \to Tr_1^3(x)$. Then $\mathbf{u}$ is given by $u_i = Tr_1^3(b_i^3)$. Note that $Tr_1^3(\beta^i) = 0$ if $i \in \{3, 5, 6\}$ and $Tr_1^3(\beta^i) = 1$ if $i \in \{0, 1, 2, 4\}$. Thus the elements of $\mathbf{u}$ are given by

$$\mathbf{u} \leftrightarrow \begin{bmatrix} 0 & 1 & 1 & 1 & 1 & 0 & 1 & 1 & 1 \\ 0 & 0 & 1 & 1 & 1 & 1 & 0 & 1 & 0 \\ 0 & 1 & 0 & 1 & 1 & 0 & 1 & 1 & 1 \\ 0 & 1 & 0 & 0 & 0 & 1 & 1 & 0 & 1 \\ 0 & 1 & 1 & 0 & 0 & 1 & 1 & 0 & 1 \\ 0 & 0 & 0 & 1 & 1 & 1 & 0 & 1 & 0 \\ 0 & 0 & 1 & 0 & 0 & 0 & 0 & 0 & 0 \end{bmatrix} \qquad (8.22)$$

Read out the above array starting at the left-most top entry, from left to right, row by row. We have

$$\mathbf{u} = 011110111001111010010110111010001101$$
$$011001101000111010001000000.$$

Method 2. Interleaved approach. In this case, $v = 7$ and $d = 9$. Let $c(x) = x^6 + x + 1$ be a primitive polynomial over $\mathbb{F}_2$ of degree 6 and α a root of $c(x)$ in $\mathbb{F}_{2^6}$. Let $\beta = \alpha^9$. Then β is a primitive element of $\mathbb{F}_{2^3}$ whose minimal polynomial is $t(x) = x^3 + x^2 + 1$ (see Table 3.7 in Appendix B of Chapter 3).

Procedure:

1. Using $(0, 0, 0, 0, 0, 1)$ as an initial state, generate an m-sequence $\mathbf{v}$ by $c(x)$, and represent it in a $(7, 9)$ interleaved structure:

$$\mathbf{v} \leftrightarrow A(\mathbf{a}, \mathbf{e}) = \begin{bmatrix} 0 & 0 & 0 & 0 & 0 & 1 & 0 & 0 & 0 \\ 0 & 1 & 1 & 0 & 0 & 0 & 1 & 0 & 1 \\ 0 & 0 & 1 & 1 & 1 & 1 & 0 & 1 & 0 \\ 0 & 0 & 1 & 1 & 1 & 0 & 0 & 1 & 0 \\ 0 & 1 & 0 & 1 & 1 & 0 & 1 & 1 & 1 \\ 0 & 1 & 1 & 0 & 0 & 1 & 1 & 0 & 1 \\ 0 & 1 & 0 & 1 & 1 & 1 & 1 & 1 & 1 \end{bmatrix}$$

(This is the m-sequence of degree 6 in Example 5.8 of Chapter 5.)

2. Generate an m-sequence of degree 3, using $t(x)$, which is constant-on-cosets, denoted by $\mathbf{a}$. Then $\mathbf{a} = (1110100)$.

3. Compute the shift sequence of $\mathbf{v}$ with respect to $\mathbf{a}$. We obtain

$$\mathbf{e} = (\infty, 3, 6, 5, 5, 2, 3, 5, 3).$$

4. Compute $\mathbf{d}$, a 3-decimation of $\mathbf{a}$; that is,

$$\mathbf{d} = \mathbf{a}^{(3)} = (1001011).$$

5. Fill the array $A(\mathbf{d}, \mathbf{e})$ using $\mathbf{d}$, the zero sequence, and the shift sequence $\mathbf{e}$. In other words, in $A(\mathbf{a}, \mathbf{e})$, we replace $\mathbf{a}$ by $\mathbf{d} = \mathbf{a}^{(3)}$ while the shift sequence is retained. The resulting sequence is the GMW sequence $\mathbf{u}$, given by (8.22).

This process is illustrated below.

---

m-sequence **v**

(∞	3	6	5	5	2	3	5	3)
(0,	$L^3(\mathbf{a})$,	$L^6(\mathbf{a})$,	$L^5(\mathbf{a})$,	$L^5(\mathbf{a})$,	$L^2(\mathbf{a})$,	$L^3(\mathbf{a})$,	$L^5(\mathbf{a})$,	$L^3(\mathbf{a}))$

replacing $L^{e_j}(\mathbf{a})$ by $L^{e_j}(\mathbf{d}) \Longrightarrow$

GMW-sequence **u**

| (0, | $L^3(\mathbf{d})$, | $L^6(\mathbf{d})$, | $L^5(\mathbf{d})$, | $L^5(\mathbf{d})$, | $L^2(\mathbf{d})$, | $L^3(\mathbf{d})$, | $L^5(\mathbf{d})$, | $L^3(\mathbf{d}))$ |

$(\mathbf{d} = \mathbf{a}^{(3)})$

For verification, the randomness profile of the GMW sequence of period 63 is as follows.

- Period 63;
- Balance, ideal 2-tuple distribution;
- 2-level autocorrelation;
- Linear span: for $s = 3$, $LS(\mathbf{u}) = 12$; and
- The number of shift-distinct binary GMW sequences of period 63 is 6, as there are 6 shift-distinct *m*-sequences of period 63.

Example 8.5 Let $n = 8$. Then $m = 4$ is the only choice for m such that the resulting sequence is not an *m*-sequence of period 255. Thus $v = 15$ and $d = 17$. There is only one coset leader $s = 7$ modulo 15 satisfying $1 < s < 15$ and $\gcd(s, 15) = 1$. Thus, a GMW sequence of period 255 is given by

$$f(x) = Tr_1^4(x^7) \circ Tr_4^8(x).$$

Let **v** be an *m*-sequence of degree 8 generated by $c(x) = x^8 + x^4 + x^3 + x^2 + 1$, a primitive polynomial over $\mathbb{F}_2$. Then **v** admits a (15, 17) interleaved structure $A(\mathbf{a}, \mathbf{e})$ where the base sequence is

$$\mathbf{a} = (000100110101111),$$

which is constant-on-cosets and generated by $x^4 + x + 1$, and the shift sequence is given by

$$\mathbf{e} = (\infty, 2, 4, 2, 8, 12, 4, 0, 19, 9, 14, 8, 5, 0, 3, 2).$$

The process of generating the GMW sequence **u** is to replace **a** in the array $A(\mathbf{a}, \mathbf{e})$ by $\mathbf{a}^{(7)}$, a 7-decimation of **a**, where the shift sequence **e** is retained.

$\mathbf{a}^{(7)}$	$A(\mathbf{a}, \mathbf{e})$ m-sequence		$A(\mathbf{a}^{(7)}, \mathbf{e})$ GMW sequence
0	000001000111000010		011111101111010011
1	010111000000011001		010110011000111111
1	001001101111001000		011100111001001011
1	001010110110101110		000000011111011110
1	010110000111111011		011111111000111101
0	011110101110100011		001001101001000101
1	000011011100011110	$\implies$	010101010000000111
0	011100110001011101		010101001111101001
1	001000101001010101		001010111110101001
1	011101110110011111		000011010111101101
0	011111010100110011		010110000111100011
0	010100011000001111		000011001000110001
1	010101011111100101		001001110110011001
0	000010011111111100		001010100000111010
0	001011110001101001		011100100110010111

Example 8.6 Considering the m-sequence of degree 9 given in Example 5.11 in Section 5.3 of Chapter 5, we choose $s = 3$. Then a GMW sequence of period 511 can be constructed from the array $A(\mathbf{a}, \mathbf{e})$ by replacing **a** by $\mathbf{a}^{(3)}$.

$$A(\mathbf{a}^{(3)}, \mathbf{e})$$
GMW sequence of period 511 for $s = 3$

111110111001111110000111101010111000010100101010100010011100111010000010
110001001001101100111001000000110000110111110001101111011001010001000001100
101101000110010110010011110111100001101110010011000000000100111001010001
010011111111101000010100011101011001111010111001100010011000000110100001
100010110110000100101101011101101001001101011010111100101010100010101101
001111110000010010111110101010001000100011001001111100101110011000000110
011100001111111010101010110111010001011001110000011110110110011010110101111

Example 8.7 Let $n = 10$ and $m = 5$. Then $v = 2^m - 1 = 31$ and $d = 2^5 + 1 = 33$. Let α be a primitive element in $\mathbb{F}_{2^{10}}$ with the minimal polynomial $t(x) = x^{10} + x^3 + 1$, and let $\beta = \alpha^d$. Then β is a primitive element in $\mathbb{F}_{2^5}$ with the minimal polynomial $t(x) = x^5 + x^4 + x^3 + x^2 + 1$. Let **v** be an m-sequence generated by $t(x)$. Then **v** admits a $(v, d) = (31, 33)$ interleaved structure $A(\mathbf{a}, \mathbf{e})$,

Table 8.2. *M-sequence of period 1023 of degree 10*

$\mathbf{v} \leftrightarrow A(\mathbf{a}, \mathbf{e})$
0 0 0 0 0 0 0 1 0 0 0 0 0 0 1 0 0 1 0 0 0 1 0 0 0 0 0 1 1 0 0 1 0
0 1 1 0 1 0 0 0 0 1 0 0 1 0 1 0 1 0 0 0 0 1 1 1 1 0 1 0 1 1 1 0 1
0 1 1 0 1 1 0 1 1 0 0 0 0 0 0 0 0 1 1 0 0 0 0 0 1 1 0 1 1 0 0 1 1
0 0 0 0 1 0 1 0 1 1 0 1 0 1 1 1 0 0 0 1 1 0 1 1 1 1 1 1 0 0 0 1 0
0 0 1 1 1 1 0 0 1 1 1 1 0 1 1 0 1 1 0 1 0 0 0 0 0 0 0 1 0 1 0 0 0
0 1 0 1 1 0 1 0 1 0 1 0 0 0 1 1 1 1 1 0 1 1 1 1 0 0 1 0 0 1 0 1 1
0 0 0 0 0 1 0 0 1 1 0 0 1 0 0 0 1 0 1 0 0 0 1 1 0 1 1 0 1 1 1 0 0
0 0 0 0 1 1 1 1 0 0 0 1 1 1 0 1 1 1 1 1 1 0 0 1 0 0 0 0 1 1 0 0
0 1 0 1 1 0 1 1 1 0 1 0 0 0 0 1 1 0 1 0 1 0 1 1 0 0 1 1 1 1 0 0 1
0 1 1 0 1 1 0 0 1 0 0 0 0 0 1 0 0 0 1 0 0 1 0 0 1 1 0 0 0 0 0 0 1
0 1 1 0 0 0 1 0 1 0 0 1 1 1 0 1 1 0 0 1 1 1 0 0 0 1 0 1 1 1 1 1 1
0 1 0 1 0 0 0 1 0 1·1 1 0 1 1 0 1 0 1 1 0 0 0 0 1 1 0 0 1 1 0 1 1
0 1 0 1 0 0 0 0 0 1 1 1 0 1 0 0 1 1 1 1 0 1 0 0 1 1 0 1 0 1 0 0 1
0 0 1 1 1 0 0 0 0 0 1 1 1 1 1 0 0 1 1 1 0 0 1 1 0 1 1 1 1 0 1 0 0
0 1 0 1 0 1 0 1 1 0 1 1 1 1 1 0 0 0 0 1 0 0 1 1 1 0 1 0 0 0 1 1 1
0 1 0 1 1 1 1 1 0 1 1 0 1 0 0 1 0 0 0 0 1 0 0 0 0 1 0 1 0 0 1 0 1
0 1 1 0 0 0 1 1 1 0 0 1 1 1 1 1 1 0 1 1 0 0 0 0 1 0 0 0 1 1 0 1
0 0 1 1 1 0 0 1 0 0 1 1 1 1 0 0 0 0 1 1 0 1 1 1 0 1 1 0 0 0 1 1 0
0 0 1 1 1 1 0 1 1 1 1 1 0 1 0 0 1 0 0 1 0 1 0 0 0 0 0 0 1 1 0 1 0
0 0 1 1 0 0 1 0 1 1 1 0 1 0 0 1 0 1 1 0 1 0 0 0 1 0 0 0 1 0 1 1 0
0 1 1 0 1 0 0 1 0 1 0 0 1 0 0 0 1 1 0 0 0 0 1 1 1 0 1 1 0 1 1 1 1
0 0 0 0 0 1 0 1 1 1 0 0 1 0 1 0 1 1 1 0 0 1 1 1 0 1 1 1 1 0 1 1 1 0
0 1 1 0 0 1 1 1 0 1 0 1 0 1 1 1 0 1 1 1 1 0 1 1 0 0 1 0 1 0 0 0 1
0 0 1 1 0 1 1 0 0 0 1 0 0 0 0 1 1 1 0 0 1 0 1 1 1 1 1 0 0 1 0 1 0
0 1 1 0 0 1 1 0 0 1 0 1 0 1 0 1 0 0 1 1 1 1 1 1 0 0 1 1 0 0 0 1 1
0 1 0 1 1 1 1 0 0 1 1 0 1 0 1 1 0 1 0 0 1 1 0 0 0 1 0 0 1 0 1 1 1
0 0 0 0 1 0 1 1 1 1 0 1 0 1 0 1 0 1 0 1 1 1 1 1 1 1 1 0 1 0 0 0 0
0 1 0 1 0 1 0 0 1 0 1 1 1 1 0 0 0 1 0 1 0 1 1 1 0 1 1 1 0 1 0 1
0 0 1 1 0 1 1 1 0 0 1 0 0 0 1 1 1 0 0 0 1 1 1 1 1 1 1 1 1 0 0 0
0 0 0 0 1 1 1 0 0 0 0 1 1 1 1 1 1 0 1 1 1 0 0 0 1 0 0 1 1 1 1 1 0
0 0 1 1 0 0 1 1 1 1 1 0 1 0 1 1 0 0 1 0 1 1 0 0 1 0 0 1 0 0 1 0 0

as shown in Table 8.2. The base sequence **a** is given by

$$\mathbf{a} = (1111101110001010110100001100100),$$

which is generated from $t(x)$ in the constant-on-coset phase, and the shift
sequence is given as follows:

Shift sequence of m-sequence $\mathbf{v}$									
∞	23	15	20	30	10	9	17	29	13
20	21	18	21	3	9	27	12	26	21
9	7	11	11	5	22	11	4	6	27
18	14	23							

Table 8.3. *GMW sequence of period 1023 with* $s = 15$

$A(\mathbf{a}^{(15)}, \mathbf{e})$
0 1 1 0 1 0 0 1 1 0 0 0 0 0 1 0 1 1 0 0 0 1 0 0 0 0 0 0 1 1 0 1 1
0 1 0 0 1 0 0 0 1 1 0 0 1 0 0 0 1 0 1 0 0 0 1 1 1 1 1 0 1 1 1 1 1
0 1 1 0 0 1 0 1 1 1 0 1 0 1 0 0 1 1 1 1 0 0 0 0 1 1 0 1 0 1 0 0 1
0 0 0 1 0 0 1 0 0 0 1 1 0 1 1 1 1 1 1 1 0 1 1 0 1 1 1 0 1 0 1 0
0 1 1 1 1 1 0 0 0 1 1 1 0 1 1 0 1 0 1 1 0 0 1 1 0 0 1 0 0 1 0 0 1
0 1 0 1 0 1 1 0 1 0 1 0 1 0 0 1 0 1 1 0 1 1 0 0 0 1 0 0 0 0 1 1 1
0 1 0 0 0 0 1 1 0 1 0 1 1 1 0 1 0 0 0 1 1 0 1 1 0 1 1 0 1 0 1 0 1
0 1 0 1 1 1 0 1 0 0 1 1 1 1 0 0 1 1 0 1 0 1 0 0 1 1 0 0 0 1 1 0 1
0 1 1 1 1 0 1 1 1 0 1 1 0 1 0 1 0 0 1 1 1 1 1 1 0 1 1 1 1 0 0 0 1
0 0 1 1 0 1 0 0 1 0 1 1 1 1 0 0 0 0 1 0 0 0 0 1 1 0 0 1 0 1 1 0
0 0 1 1 0 0 1 1 0 1 1 1 1 1 0 1 1 0 0 1 1 1 0 0 1 0 0 1 0 1 1 1 0
0 1 0 1 0 0 0 1 0 1 1 0 1 0 1 0 1 1 1 0 0 0 0 0 0 0 0 1 1 1 1 1 1
0 0 1 0 0 0 0 1 0 1 0 0 1 0 1 0 0 1 1 0 0 1 1 1 1 1 1 0 0 0 1 0 0
0 0 1 0 1 1 0 1 0 0 0 1 1 1 0 0 0 1 0 1 0 0 1 1 0 0 1 1 1 0 1 1 0
0 1 1 1 0 1 1 1 1 1 0 0 0 1 1 0 0 0 0 1 0 1 1 1 0 1 0 0 0 0 1 1
0 1 1 0 1 1 1 0 0 1 0 0 0 0 0 1 0 1 0 0 1 0 0 0 0 1 0 1 0 0 0 1 1
0 0 1 0 1 0 1 0 1 1 0 1 1 1 1 1 1 1 0 1 1 1 1 1 0 1 1 0 0 1 1 1 0
0 0 0 1 0 1 0 1 1 1 1 1 0 1 0 0 0 1 1 1 0 1 1 1 0 0 1 0 1 0 0 1 0
0 0 0 1 1 1 1 0 0 1 1 0 0 0 0 1 1 1 0 0 1 1 1 1 1 0 1 0 1 1 0 0 0
0 0 1 0 0 1 1 0 1 0 0 0 1 0 0 1 1 1 1 0 1 0 1 1 1 0 1 1 1 1 1 0 0
0 1 0 0 1 1 1 1 0 0 0 0 1 0 1 1 0 0 1 0 1 1 1 1 1 0 1 1 0 0 1 1 1
0 0 0 0 0 1 1 1 1 1 0 0 0 0 1 1 1 0 0 0 1 1 0 0 0 1 0 1 1 1 0 0 0
0 1 1 0 0 0 1 0 0 0 0 1 0 1 1 1 0 1 1 1 1 1 0 0 1 0 0 0 1 0 0 0 1
0 1 1 1 0 0 0 0 0 0 1 0 0 0 0 0 1 0 0 0 0 1 1 1 1 1 1 1 1 1 0 1 1
0 0 0 0 1 1 0 0 0 1 0 1 0 1 1 0 0 0 1 1 0 1 0 0 1 1 0 1 1 0 0 1 0
0 1 0 1 1 0 1 0 1 1 1 1 1 1 1 0 1 0 1 1 0 0 0 1 0 0 1 1 0 1 0 1
0 0 0 1 1 0 0 1 1 0 1 0 0 0 1 0 0 1 0 0 0 0 1 1 1 1 1 1 0 0 0 0 0
0 1 0 0 0 1 0 0 1 0 0 1 1 1 1 0 1 0 0 1 0 1 1 1 0 0 1 1 0 1 1 0 1
0 0 1 1 1 1 1 1 0 0 1 0 1 0 1 1 1 0 1 0 1 0 0 0 0 1 0 0 1 1 1 0 0
0 0 0 0 1 0 1 1 1 0 0 1 0 1 0 1 1 0 1 1 1 0 0 0 1 0 0 0 0 1 0 1 0
0 0 1 1 1 0 0 0 1 1 1 0 1 0 0 0 0 0 1 0 0 1 0 0 0 0 0 1 0 0 1 0 0

The GMW sequence with $s = 15$ can be obtained by replacing $\mathbf{a}$ with $\mathbf{a}^{(15)}$ while the shift sequence is retained. Thus the array form of this GMW sequence is given by $A(\mathbf{a}^{15}, \mathbf{e})$, as shown in Table 8.3. We also can replace $\mathbf{a}$ by a quadratic residue sequence of period 31, which is given by

$$\mathbf{d} = (1001001000011101010001111011011).$$

The array form of this new sequence is given by $A(\mathbf{d}, \mathbf{e})$, as shown in Table 8.4.

Table 8.4. *Generalized GMW sequence of period 1023 from*
QR sequence **d**

$A(\mathbf{d}, \mathbf{e})$
0 1 1 0 1 0 0 1 1 1 0 1 0 1 1 0 1 1 1 1 0 0 1 1 0 1 1 0 1 1 0 0 1
0 1 0 1 1 1 0 0 1 0 1 1 0 1 0 0 0 1 1 1 0 0 1 1 1 1 1 0 0 0 0 1 1
0 0 1 1 0 1 1 0 1 1 1 1 0 1 0 1 1 0 0 1 1 0 1 1 0 1 1 1 0 1 0 0 0
0 1 0 1 0 1 1 0 0 0 1 1 1 1 1 1 1 1 1 1 0 0 0 0 0 0 0 0 1 1 1 1
0 1 0 1 1 0 1 1 0 1 1 0 1 0 0 1 1 0 1 0 1 1 1 0 1 1 0 0 1 1 0 1
0 0 0 0 0 1 0 1 1 0 0 1 1 1 0 0 0 1 1 1 0 1 0 0 0 1 0 0 1 0 1 0 0
0 1 1 1 0 0 1 1 0 0 1 1 1 1 0 1 0 0 0 1 1 1 1 1 0 1 0 1 0 1 0 1
0 1 1 1 1 1 0 1 0 0 1 0 0 0 0 0 1 0 0 0 0 0 0 1 1 0 1 1 1 0 1 1
0 1 1 0 0 0 1 0 1 1 0 1 1 1 1 1 0 0 1 1 1 1 0 0 1 1 0 1 0 0 1 1 1
0 0 1 1 0 0 0 1 0 1 1 1 1 1 1 0 0 1 0 1 0 0 0 0 0 1 0 1 1 0 1 1 0
0 0 0 1 0 0 0 1 0 1 1 1 0 1 1 0 1 1 0 1 0 1 1 1 1 0 1 0 0 1 0 1 0
0 1 1 1 0 1 0 0 0 1 1 0 1 0 0 0 0 1 1 0 0 0 1 1 0 0 1 1 1 0 1 0 1
0 0 1 0 1 1 1 1 0 0 0 0 1 0 1 1 0 1 0 0 1 0 1 1 1 1 1 0 0 0 1 1 0
0 0 0 0 1 1 1 1 1 1 0 1 1 1 0 1 0 0 0 1 1 0 1 1 0 0 1 1 0 0 1 1 0
0 1 1 1 1 1 1 1 1 1 1 0 0 0 1 1 0 1 0 0 1 1 0 0 0 0 0 0 0 0 0 0 1
0 0 1 0 0 0 1 0 1 0 0 0 0 0 0 1 1 1 0 0 1 1 1 1 0 1 1 0 1 1 0 1 0
0 0 1 0 1 1 0 0 0 1 0 1 1 1 0 0 1 0 1 1 0 1 1 1 1 0 1 0 1 1 1 1 0
0 0 0 1 0 1 1 1 1 1 1 0 0 0 0 1 1 1 1 0 1 1 0 0 1 0 0 1 1 1 0 1 0
0 0 0 0 1 0 1 0 0 1 0 0 0 0 1 1 0 1 1 0 1 0 1 1 1 0 1 1 1 0 0 0 0
0 1 1 0 0 1 0 0 1 0 0 0 1 0 1 0 1 1 0 0 0 1 1 1 1 0 1 1 0 1 1 0 1
0 1 0 0 0 1 1 1 0 0 0 0 0 0 1 1 0 0 1 0 1 1 1 1 0 1 1 1 1 0 0 1 1
0 1 0 0 0 1 1 0 0 1 0 1 0 1 1 1 1 0 0 1 1 0 0 0 1 1 0 0 1 1 0 0 1
0 0 1 1 1 0 1 0 0 0 1 1 0 1 0 1 0 1 1 1 1 1 0 0 1 1 0 1 0 0 0 0 0
0 1 0 1 1 0 0 0 1 0 1 1 0 1 1 0 0 0 0 1 0 1 1 1 0 0 1 1 1 0 0 1 1
0 0 0 1 1 1 0 0 1 1 1 0 1 0 1 0 0 0 0 0 0 1 0 0 1 1 0 0 1 0 1 0 0
0 1 0 0 1 0 1 1 1 0 0 1 1 1 0 1 1 1 0 1 1 0 0 0 1 0 0 1 1 1 1 0 1
0 0 0 1 0 0 0 1 1 0 1 0 1 0 1 0 1 0 1 0 0 0 1 1 1 1 1 1 0 1 1 0 0
0 0 1 0 1 1 0 1 0 0 0 1 0 1 1 0 1 0 1 1 0 1 0 0 0 0 0 1 0 1 0 0 0
0 0 1 1 1 0 1 0 1 0 1 0 1 0 1 1 1 0 1 0 1 0 0 0 0 0 0 0 1 1 1 1 0
0 1 1 0 0 0 0 1 1 1 0 0 0 0 0 0 0 0 1 0 0 0 0 0 1 0 0 0 0 0 0 1 1
0 1 0 0 1 0 0 0 0 1 0 0 1 0 0 0 1 1 0 0 0 1 0 0 0 1 0 1 0 1 1 1 1

From Construction I, we have four types of 2-level autocorrelation sequences of period $2^n - 1$ with subfield factorization. We summarize these
classes in Table 8.5 where $g(x)$ in type 3 and $q(x)$ in type 4 would be chosen as quadratic residue sequences, or the Hall sextic residue sequences, or
any of the 2-level autocorrelation sequences that will be introduced in the next
chapter.

Table 8.5. *Prototypes of 2-level AC from Construction I (GMW Construction)*

Type 1: GMW sequences
$\mathbb{F}_2 \subset \mathbb{F}_{2^m} \subset \mathbb{F}_{2^n}$ $Tr_1^m(x^s) \circ Tr_m^n(x^k)$
Type 2: Cascaded GMW
$\mathbb{F}_2 = F_0 \subset F_1 \subset \cdots \subset F_t \subset F_t = \mathbb{F}_{2^n}$ $Tr_{F_1/F_0}(x^{s_1}) \circ Tr_{F_2/F_1}(x^{s_2}) \circ \cdots \circ Tr_{F_t/F_{t-1}}(x^{s_t}), x \in F_t$
Generalized GMW
Type 3: $\mathbb{F}_2 \subset \mathbb{F}_{2^m} \subset \mathbb{F}_{2^n}$ $g(x) \circ Tr_m^n(x^k)$ Type 4: $\mathbb{F}_2 = F_0 \subset F_1 \subset \cdots \subset F_t \subset F_t = \mathbb{F}_{2^n}$ $q(x) \circ Tr_{F_2/F_1}(x^{s_2}) \circ \cdots \circ Tr_{F_t/F_{t-1}}(x^{s_t}), x \in F_t$ where both $g(x)$ and $q(x)$ are from any 2-level autocorrelation sequences not listed in Type 1 and Type 2

8.2.4 Construction II

We consider the following composition:

$$\mathbb{F}_n$$
$$\downarrow \quad h(x)$$
$$\mathbb{F}_m$$
$$\downarrow \quad Tr_1^m(x)$$
$$\mathbb{F}_1$$

Theorem 8.5 *The sequence* **u**, *in Construction II, is a binary sequence of period* $2^n - 1$ *with 2-level autocorrelation.*

Construction II. Let α be a primitive element of $\mathbb{F}_{2^n}$, and let $h(x)$ be an r-form function from $\mathbb{F}_{2^n}$ to $\mathbb{F}_{2^m}$; that is,

$$h(yx) = y^r h(x), \text{ for } y \in \mathbb{F}_{2^m}, x \in \mathbb{F}_{2^n}$$

where $\gcd(r, 2^m - 1) = 1$. Suppose that $T_1^m(x) \circ h(x)$ is orthogonal and $g(x)$ is an orthogonal function from $\mathbb{F}_{2^m}$ to $\mathbb{F}_2$. We construct a sequence $\mathbf{u} = \{u_i\}$ whose elements are given by

$$a_i = f(\alpha^i), i = 0, 1, \ldots,$$

where

$$f(x) = g(x) \circ h(x).$$

This result can be illustrated by the following composition diagram:

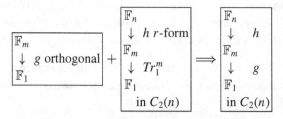

Proof of Theorem 8.5. Here we cannot apply Theorem 8.1, because $h(x)$ may not satisfy the 2-tuple balance property. However, the result is easy to establish using Theorem 8.3, the interleaved approach. Let $v = 2^m - 1$ and $d = (2^n - 1)/(2^m - 1)$. From Corollary 8.1 in Section 8.3, $h(x)$ admits a (v, d) interleaved structure. Let $\mathbf{v} \leftrightarrow Tr_1^m(x) \circ h(x)$ and $\mathbf{u} \leftrightarrow g(x) \circ h(x)$. Then both $\mathbf{v}$ and $\mathbf{u}$ are (v, d) interleaved sequences whose array forms are given by $A(\mathbf{a}, \mathbf{e})$ and $A(\mathbf{d}, \mathbf{e})$, where

$$\mathbf{a} \leftrightarrow Tr_1^m(x)$$
$$\mathbf{d} \leftrightarrow g(x^r)$$

and the common shift sequence $\mathbf{e}$ is given by

$$e_j = \begin{cases} e & \text{if } h(\alpha^j) \neq 0, h(\alpha^j) = \beta^e \\ \infty & \text{if } h(\alpha^j) = 0, \end{cases} \tag{8.23}$$

where α is a primitive element of $\mathbb{F}_{2^n}$ and $\beta = \alpha^d$, a primitive element of $\mathbb{F}_{2^m}$. According to Theorem 8.3, $\mathbf{u}$ has 2-level autocorrelation. $\square$

Note that $Tr_m^n(x^k)$ is an r-form function where $r = k \pmod{2^m - 1}$. Thus Construction I can be considered a special case of Construction II.

Remark 8.2 There are only two known constructions for $h(x)$ in Construction II.

(a) $h(x)$ is a single trace term; that is, $h(x) = Tr_m^n(x^k)$. In this case, Construction II degenerates into Construction I.
(b) $h(x)$ is a cascaded GMW function of length t. In this case, $f(x)$ produces the GMW sequences of Type 4 listed in Table 8.5. Thus, Construction II also degenerates into Construction I.

Up to now, no r-form function has been found that does not fall into one of the above two cases. However, in the nonbinary case, there exist such functions for ternary sequences that do not belong to classes corresponding to the above two constructions.

Together with Theorem 8.2, the above results show that there are only two known constructions for sequences over $\mathbb{Z}_v(\infty)(= \mathbb{Z}_v \cup \{\infty\})$ of length d with the mixed difference property. One is constructed from m-sequences and the other from GMW sequences or cascaded GMW sequences.

Remark 8.3 These two constructions are also valid for sequences over $\mathbb{F}_q$ for q a prime or a power of a prime, because they follow from Theorem 8.1 and Theorem 8.3.

8.2.5 Subfield factorization

According to the definition of subfield factorization of sequences or functions, any type of GMW sequence is reducible. For a field chain $\mathbb{F}_2 \subset E \subset K \subset \mathbb{F}_{2^n}$, we will call $Tr_{K/E}(x^s)$ a GMW factor of $f(x)$ or $\mathbf{u}$ where $\gcd(s, |K|) = 1$ and $s > 1$.

From known results, for any binary 2-level autocorrelation sequence with period $2^n - 1$ where n is a composite number, its possible subfield factorization has a form of either Type 2 or Type 4 in Table 8.5. Another interesting phenomenon is that any known function from $\mathbb{F}_{2^n}$ to $\mathbb{F}_{2^m}$ ($m|n$) with the 2-tuple balance property is equal to either the trace function or a GMW function of length $t > 0$. In other words, for binary sequences, the known functions with the 2-tuple balance property are either the trace function or one that can be factorized into GMW factors in subfields.

We propose some questions related to these concerns as research problems in the exercises for this chapter. A proof or disproof of these problems will lead to either the complete classification of binary 2-level autocorrelation sequences of period $2^n - 1$ with subfield decomposition or to the discovery of new binary 2-level autocorrelation sequences with subfield decomposition.

In the rest of this chapter, we will discuss the statistical properties, linear span, shift-distinct property, and implementation of the four types of GMW sequences constructed using Construction I.

8.3 Statistical properties of GMW sequences of all types

We continue to use the notation introduced in previous sections. We first show that all four types of GMW sequences over $\mathbb{F}_q$ satisfy the balance property (i.e., R-1) and the constant-on-cosets property and that cascaded GMW sequences also satisfy the 2-tuple balance property. We then discuss their ideal k-tuple distribution.

Proposition 8.7 *Let q be a prime or a power of a prime, and let* **u** *be a GMW sequence over $\mathbb{F}_q$ of period $q^n - 1$ belonging to one of the four types of GMW sequences. Then*

(a) **u** *is balanced.*
(b) **u** *satisfies the constant-on-cosets property.*
(c) For $q = 2$, **u** *has the 2-tuple balance property. For $q > 2$, if* **u** *is a Type 1 or Type 2 GMW sequence, then it has the 2-tuple balance property.*

Proof. The validity of the first two properties is easy to see. For the third property, in the binary case, the 2-tuple balance property is equivalent to 2-level autocorrelation. For $q > 2$, we have presented this in Remark 8.2 of the last section. □

In the following, we assume that $\mathbf{u} \leftrightarrow f(x)$ is a binary 2-level autocorrelation sequence from Construction I where $f(x) = g(x) \circ Tr_m^n(x^k)$ where $g(x)$ is an orthogonal function from $\mathbb{F}_{2^m}$ to $\mathbb{F}_2$. (Note. The following result is also true for nonbinary sequences.)

Lemma 8.3 *The sequence* **u** *satisfies the ideal l-tuple distribution ($l = n/m$). In other words, each nonzero l-tuple $(c_0, \ldots, c_{l-1}) \in \mathbb{F}_2^l$ occurs 2^{n-l} times in* $\mathbf{u} = (u_0, u_1, \ldots, u_{2^n-2})$ *(cyclically), and the zero l-tuple occurs $2^{n-l} - 1$ times.*

Proof. Note that

$$u_i = g(b_i^s), i = 0, 1, \ldots,$$

where $\mathbf{b} = (b_0, b_1, \ldots, b_{q^l-2})$ is an m-sequence over $\mathbb{F}_q (q = 2^m)$ of degree l whose elements are given by

$$b_i = Tr_m^n(\alpha^i), i = 0, 1, \ldots,$$

where α is a primitive element in $\mathbb{F}_{2^n}$. From Chapter 5, the m-sequence **b** satisfies the ideal l-tuple distribution. In other words, each nonzero l-tuple $(\lambda_0, \ldots, \lambda_{l-1}) \in \mathbb{F}_q^l$ occurs once in **b**. Because $g(x)$ has 2-level autocorrelation, it is balanced; that is, $g(x) = c, c \in \mathbb{F}_2$ has 2^{m-1} solutions in $\mathbb{F}_{2^m}$. Therefore, for each l-tuple $(c_0, \ldots, c_{l-1}) \in \mathbb{F}_2^l$, there are $2^{m-1} \cdot 2^{m-1} \cdots 2^{m-1}$ solutions for

$$((g(\lambda_0), \ldots, g(\lambda_{l-1})) = (c_0, \ldots, c_{l-1}).$$

Thus each nonzero l-tuple $(c_0, \ldots, c_{l-1}) \in \mathbb{F}_2^l$ occurs $2^{(m-1)l} = 2^{n-l}$ times in **u**. It is immediate that the zero l-tuple occurs $2^{(m-1)l} - 1 = 2^{n-l} - 1$ times in **u** since the zero l-tuple does not appear in **b**. □

Since the ideal l-tuple distribution implies the ideal k-tuple distribution for every k with $1 \le k \le l$, we have established the following result.

Theorem 8.6 *If $\mathbf{u} \leftrightarrow f(x)$, a GMW sequence, has a subfield decomposition of $f(x) = g(x) \circ Tr_m^n(x^k)$ where $g(x)$ is an orthogonal function from $\mathbb{F}_{2^m}$ to $\mathbb{F}_2$, then $\mathbf{u}$ has the ideal k-tuple distribution for each k with $1 \le k \le l = n/m$.*

Example 8.8 Consider the GMW sequences in Examples 8.4 and 8.6.

1. The GMW sequence in Example 8.4 has $l = 2$. Thus it is balanced and has ideal 2-tuple distribution. Precisely, there are 32 1's and 31 0's in one period of a GMW sequence of period 63. Each pair (0, 1), (1, 0), (1, 1) occurs $2^{6-2} = 16$ times, and (0, 0) occurs 15 times in the sequence.
2. For the GMW sequence in Example 8.6, we have $l = 3$. Thus, it is balanced and has ideal k-tuple distribution, $k = 2, 3$. In detail, there are 256 1's and 255 0's. Each nonzero pair $(i, j), i, j \in \mathbb{F}_2$ occurs 128 times, and (0, 0) occurs 127 times. Each nonzero triple $(i, j, k), i, j, k \in \mathbb{F}_2$ occurs 64 times, and (0, 0, 0) occurs 63 times.

8.4 Linear spans of GMW sequences of all types

For binary m-sequences of degree n, the linear span is n. In this section, we will show the linear span of binary 2-level autocorrelation sequences constructed from Construction I, that is, the four types of GMW sequences listed in Table 8.5.

8.4.1 Linear spans of GMW sequences (type 1)

Theorem 8.7 *Let $\mathbf{u}$ be a binary GMW sequence whose elements are given by*

$$u_i = f(\alpha^i), i = 0, 1, \ldots,$$

$$f(x) = Tr_1^m(x^s) \circ Tr_m^n(x^k), \gcd(k, 2^n - 1) = 1, \gcd(s, 2^m - 1) = 1.$$

Then $LS(\mathbf{u})$, the linear span of GMW sequence $\mathbf{u}$, is given by

$$LS(\mathbf{u}) = ml^{w(s)},$$

where $w(s)$ is the Hamming weight of s.

Proof. The linear span of $\mathbf{u}$ is equal to the number of nonzero terms in the (Fourier) spectral sequence of the sequence. Since $f(x)$ is the trace representation of $\mathbf{u}$, the coefficients of $f(x)$ form the spectral sequence of $\mathbf{u}$. Thus the linear span of **u** is equal to $w(f)$, the number of nonzero coefficients of $f(x)$; that is, $LS(\mathbf{u}) = w(f)$ (see Theorem 6.3 of Chapter 6). Since $\gcd(s, 2^m - 1) = 1$, we have

$$f(x) = t(x) + t(x)^2 + \cdots + t(x)^{2^{m-1}}, \quad t(x) = Tr_m^n(x^k)^s,$$

where $t(x) \not\equiv t(x)^{2^i} \pmod{x^{2^n} - x}$. Notice that the values of k do not change the linear span of **u**. We may set $k = 1$ for computation. Thus, $t(x)$ is a power of a trace function. According to Theorem 3.15 in Section 3.6 of Chapter 3, the number of nonzero coefficients in $t(x)$ is given by $l^{w(s)}$; that is, $w(t(x)) = l^{w(s)}$. For any $0 \le j < m$, $w(t(x)^{2^j}) = l^{w(s)}$. Thus we get $w(f) = ml^{w(s)}$, which establishes the assertion. $\qquad\qquad\square$

8.4.2 Linear spans of cascaded GMW sequences (type 2)

Proposition 8.8 *Let* **u** *be a cascaded GMW sequence of length t whose elements are given by*

$$u_i = f(\alpha^i), i = 0, 1, \ldots,$$
$$f(x) = Tr_{F_1/F_0}(x^{s_1}) \circ Tr_{F_2/F_1}(x^{s_2}) \circ \cdots \circ Tr_{F_t/F_{t-1}}(x^{s_t}), x \in \mathbb{F}_t,$$

where a field chain is given by

$$\mathbb{F}_2 = F_0 \subset F_1 \subset \cdots \subset F_t = \mathbb{F}_{2^n},$$

and the s_i's satisfy $\gcd(s_i, q_i - 1) = 1$ where $q_i = |F_i|, i = 0, 1, \ldots, t$. Let $n_i = [F_i : F_{i-1}] > 1$, the dimension of F_i regarded as a linear space over $\mathbb{F}_{i-1}, i = 1, \ldots, t$. If

$$s_i = \sum_{j=0}^{n_i-1} s_{ij} q_{i-1}^j, s_{ij} \in \mathbb{F}_2,$$

then the linear span of **u** *is given by*

$$LS(\mathbf{u}) = n_1 n_2^{w(s_1)} n_2^{w(s_1)} \cdots n_t^{w(s_{t-1})},$$

where $w(s_i)$ is the Hamming weight of $s_i, i = 1, \ldots, t - 1$.

This result can be established by repeatedly applying Theorem 3.15 in Chapter 3 to $\sigma_i(x) \circ Tr_{F_{i+1}/F_i}(x)$ where $\sigma_i(x)$ is a cascaded GMW function from F_i to $F_0 = \mathbb{F}_2$. For details, see Gong (1996).

Remark 8.4 Unlike the GMW case, that is, where the field chain has length 2, for general values of the s_i's, no formula has been found for which the linear span of the sequence is directly related to the s_i's. In the last subsection of this chapter, we will give an algorithm to compute the linear span of cascaded

GMW sequences that also yields the trace representation (a polynomial form) of cascaded GMW functions.

8.4.3 Linear spans of generalized GMW sequences (types 3 and 4)

To represent the function $g(x)$ in Construction I in polynomial form, applying Theorem 3.17 in Section 3.6 of Chapter 3, the following result follows.

Theorem 8.8 *Let* **u** *be a generalized binary GMW sequence whose elements are given by*

$$u_i = f(\alpha^i), i = 0, 1, \ldots,$$

$$f(x) = g(x) \circ Tr_m^n(x^k), \gcd(k, 2^n - 1),$$

where $n = lm$ *and* $g(x)$ *is a function from* $\mathbb{F}_{2^m}$ *to* $\mathbb{F}_2$ *with 2-level autocorrelation:*

$$g(x) = \sum_{i=0}^{2^m-1} c_i x^i, c_i \in \mathbb{F}_{2^m}.$$

Then the linear span of **u** *is given by*

$$LS(\mathbf{u}) = \sum_{c_i \neq 0} l^{w(i)}.$$

Remark 8.5 Representing $g(x)$ in polynomial form can be done using the Fourier transform, because we can regard $g(x)$ as a trace representation of a binary sequence of period $2^m - 1$ (see Chapter 6 for details). Note that $g(x)$ is balanced. Thus the term x^{2^m-1} does not appear (see Exercise 8). According to Remark 3.2 in Section 3.6 of Chapter 3, the maximum linear span for a generalized GMW sequence of period $2^n - 1$ is given by

$$(l + 1)^m - l^m \text{ (where } n = lm).$$

Thus we have the following bound for the linear spans of the GMW sequences from Construction I.

Proposition 8.9 *With the notation in Theorem 8.8, the linear span of a cascaded GMW sequence or a generalized GMW sequence of Type 3 or Type 4 is upper-bounded by*

$$LS(\mathbf{u}) < (l + 1)^m - l^m.$$

Note that for a 2-level autocorrelation sequence of period $2^n - 1$, the maximum linear span would be $(2^n - 2)/2$ (see Chapter 7). Compared to this bound, the upper bound in Proposition 8.9 is quite small.

Example 8.9 Consider Example 8.7 where $n = 10$, $m = 5$, and $l = 2$.

(a) $\mathbf{u} \leftrightarrow f(x) = Tr_1^5(x^{15}) \circ Tr_5^{10}(x)$; that is, $g(x) = Tr_1^5(x^{15})$. This is a GMW sequence. Since $w(15) = 4$, according to Theorem 8.7, the linear span of $\mathbf{u}$ is given by

$$LS(\mathbf{u}) = 5 \cdot 2^{w(15)} = 5 \cdot 2^4 = 80$$

and the trace representation of $\mathbf{u}$ is given by

$$f(x) = Tr_1^{10}(x^{15} + x^{23} + x^{27} + x^{29} + x^{77} + x^{85} + x^{89} + x^{147}),$$

where the exponents are computed by function $\tau_{15}(\mathbf{t})$, $\mathbf{t} \in \mathbb{Z}_2^4$, defined in Eq. (3.8) in Theorem 3.15 of Chapter 3.

(b) $\mathbf{u} \leftrightarrow f(x) = g(x) \circ Tr_5^{10}(x)$ where $g(x) = Tr_1^5(x + x^5 + x^7)$, which is the trace representation of the quadratic residue sequence

$$\mathbf{a} = (1001001000011101010001111011011).$$

Applying Theorem 8.8, the linear span of $\mathbf{u}$ is given by

$$\begin{aligned}
LS(\mathbf{u}) &= 5 \cdot 2^{w(1)} + 5 \cdot 2^{w(5)} + 5 \cdot 2^{w(7)} \\
&= 5 \cdot 2 + 5 \cdot 2^2 + 5 \cdot 2^3 \\
&= 10 + 20 + 40 \\
&= 70
\end{aligned}$$

and we have

$$g(Tr_5^{10}(x)) = Tr_1^{10}(x + x^5 + x^7 + x^9 + x^{19} + x^{25} + x^{69}).$$

If we choose $g(x) = Tr_1^5(x^3 + x^{11} + x^{15})$, which is the trace representation of the 3-decimation of the quadratic residue sequence $\mathbf{a}$, and applying Theorem 8.8, the linear span of $\mathbf{u}$ is increased to 140. This shows that decimation of the evaluation of $g(x)$ does not change the autocorrelation of the resulting sequence, but it changes the linear span dramatically, as the linear span only depends on the Hamming weight of the exponents of the monomial trace term when the field chain is fixed.

8.4.4 Algorithm for computing linear spans of cascaded GMW sequences (type 2)

In general, we do not know how to represent the linear span of a cascaded GMW sequence in a way that it directly related to the s_i's, similar to the result in Proposition 8.8. However, for any choice of s_i's, we have the following algorithm to compute the linear span of the resulting sequence. This algorithm also simultaneously outputs the trace representation of the sequence, which is much more efficient than using the Fourier transform for computing the trace representation of the sequence.

If S is a subset of $\mathbb{Z}$ or $\mathbb{Z}_k$, depending on the context, we denote $rS = \{rs | s \in S\}$ where multiplication is performed in $\mathbb{Z}$ or $\mathbb{Z}_k$, respectively. Let $\mathbf{u}$ be a cascaded GMW sequence and $\mathbf{u} \leftrightarrow f(x)$.

Algorithm 8.1 AN ALGORITHM FOR COMPUTING LINEAR SPANS OF CASCADED GMW SEQUENCES

Input:

    1. n, a composite number;

    2. Parameters for a field chain:

      - $n = n_1 n_2 \cdots n_t, n_i > 1$

      - $\mathbb{F}_2 = F_0 \subset F_1 \subset \cdots \subset F_t = \mathbb{F}_{2^n}$

      - $|F_i| = 2^{c_i}, [F_i : F_{i-1}] = n_i, i = 1, 2, \ldots, t$

      - $c_{i+1} = n_{i+1} c_i, i = 1, 2, \ldots, t - 1$ *(Note the divisibility:* $c_1 | c_2 | \cdots | c_r.$*)*

    3. Selection of s_i: $\gcd(s_i, |F_i| - 1) = 1, i = 1, \ldots, t - 1.$

Output: $LS(\mathbf{u})$, *the linear span of* $\mathbf{u}$, *and the exponents of the trace terms of* $f(x)$.

Procedure

1. Initialization:

   (a) Set $G = \{1, 2, \ldots, 2^{c_1 - 1}\}$

   (b) Set
$$S = \{1, s_1, \ldots, s_{t-1}\}$$
$$N = \{1, n_1, \ldots, n_t\}$$
$$C = \{1, c_1, \ldots, c_t\}$$

   (c) $j = 1$

2. While $(j < t)$

   (a) Set $s = s_j, m = c_j, l = n_{j+1}$

   (b) Compute $sG = \{sg \pmod{2^m - 1} \mid g \in G\}$

   (c) For each $r \in sG, r = 2^{i_1} + \cdots + 2^{i_k}, k = w(r)$

      Compute $\tau_r(\mathbf{t}) = 2^{i_1 + m t_1} + \cdots + 2^{i_k + m t_k}$, *for all* $\mathbf{t} = (t_1, \cdots, t_k) \in \mathbb{Z}_l^k$

      Set $M(r) = \{\tau_r(\mathbf{t}) \mid \mathbf{t} \in \mathbb{Z}_l^k\}$

      Set $T = \bigcup_{r \in sG} M(r)$

      Set $G = T$

   (d) If $(j < t)$, *set* $j = j + 1$, *otherwise set* $j = t$

3. Return $|G|$ *and* G.

From the output of the algorithm, we have

$$f(x) = \sum_{i \in G} x^i$$

so that $LS(\mathbf{u}) = |G|$.

8.5 Shift-distinct sequences from GMW constructions

Sequences from the GMW construction are related to a field chain of length t. When $t > 2$, it is not easy to prove the criterion for cyclically shift-distinct GMW sequences of all types. This was established by Gong, Dai, and Golomb in (2000). Here, we present them as facts. These results also facilitate counting the number of cyclic Hadamard difference sets that correspond to the four types of GMW sequences.

Note that all the four types of GMW sequences, listed in Table 8.5, can be merged into two different types. One is of Type 2, cascaded GMW sequences associated with a field chain of length t, because Type 1 is a particular case of Type 2 for $t = 2$. The other is of Type 4, because Type 3 is a particular case of Type 4. Thus, in the following, results are only presented for the GMW sequences of Types 2 and 4, respectively.

Let a field chain be given by

$$\mathbb{F}_{2^r} = F_0 \subset F_1 \subset \cdots F_{t-1} \subset F_t = \mathbb{F}_{2^n}, t \geq 2, \qquad (8.24)$$

where $n = n_1 n_2 \cdots n_t$, $|F_i| = 2^{c_i}$, and $[F_i : F_{i-1}] = n_i, i = 1, 2, \ldots, t$.

Fact 8.1 *Let $\mathbf{u}$ and $\mathbf{v}$ be two cascaded GMW sequences of period $2^n - 1$ (Types 1 or 2).*

(a) *If $\mathbf{u}$ and $\mathbf{v}$ are associated with two different field chains between $\mathbb{F}_2$ and $\mathbb{F}_{2^n}$, then they are shift distinct.*

(b) *If both $\mathbf{u}$ and $\mathbf{v}$ are associated with the same field chain Eq. (8.24) where $r = 1$, let*

$$f(x) = Tr_{F_1/F_0}(x^{s_1}) \circ Tr_{F_2/F_1}(x^{s_2}) \circ \cdots \circ Tr_{F_t/F_{t-1}}(x^{s_t}), x \in \mathbb{F}_t$$
$$g(x) = Tr_{F_1/F_0}(x^{k_1}) \circ Tr_{F_2/F_1}(x^{k_2}) \circ \cdots \circ Tr_{F_t/F_{t-1}}(x^{k_t}), x \in \mathbb{F}_t,$$

where s_i and k_i are coset leaders modulo $2^{c_i} - 1$ coprime to $2^{c_i} - 1$. Then $\mathbf{u}$ and $\mathbf{v}$ are shift equivalent if and only if $s_i = k_i, i = 1, \ldots, t$. Furthermore, the total number of shift-distinct cascaded GMW sequences associated with the field chain (8.24), up to decimation equivalence, is given by

$$N_{GMW}(n, t) = \prod_{1 \leq i \leq t-1} \left(\frac{\phi(2^{c_i} - 1)}{c_i} - 1 \right),$$

where $\phi(x)$ is the Euler ϕ-function, which represents how many numbers in the range from 1 and x are coprime to x.

Note that $\frac{\phi(2^{c_i}-1)}{c_i} - 1$ is the number of all m-sequences of degree c_i obtained by decimation from a given m-sequence, not including the given sequence.

Remark 8.6 There is only one m-sequence of degree n of period $2^n - 1$, up to decimation equivalence.

Fact 8.2 *Let $\mathbf{u}$ and $\mathbf{v}$ be two GMW sequences of Type 4 (including Type 3).*

(a) If $\mathbf{u}$ and $\mathbf{v}$ are associated with two different field chains between $\mathbb{F}_2$ and $\mathbb{F}_{2^n}$, then they are shift distinct.

(b) If both $\mathbf{u}$ and $\mathbf{v}$ are associated with the same field chain as shown in Eq. (8.24), for $x \in \mathbb{F}_t$, let

$$f(x) = h_1(x) \circ Tr_{F_1/F_0}(x^{s_1}) \circ Tr_{F_2/F_1}(x^{s_2}) \circ \cdots \circ Tr_{F_t/F_{t-1}}(x^{s_t})$$
$$g(x) = h_2(x) \circ Tr_{F_1/F_0}(x^{k_1}) \circ Tr_{F_2/F_1}(x^{k_2}) \circ \cdots \circ Tr_{F_t/F_{t-1}}(x^{k_t}),$$

where s_i and k_i are coset leaders modulo $2^{c_i} - 1$ coprime to $2^{c_i} - 1$, and $h_i(x), i = 1, 2$, are two orthogonal functions from $\mathbb{F}_{2^r} = F_0$ to $\mathbb{F}_2$. Then $\mathbf{u}$ and $\mathbf{v}$ are shift equivalent if and only if $s_i = k_i, i = 1, \ldots, t$, and $h_1(x) = h_2(x)$.

Example 8.10 We apply these results to the examples in Section 8.2.

(a) $n = 6$. Because there is only one m-sequence of degree 2, namely (110), the smallest field chain that generates GMW sequences of period 63 is given by

$$\mathbb{F}_2 \subset \mathbb{F}_{2^3} \subset \mathbb{F}_{2^6}.$$

There are two shift-distinct m-sequences of degree 3. From Fact 8.1, there is only one GMW sequence of period 63 up to decimation equivalence.

(b) $n = 8$. A possible field chain that can generate GMW sequences of period 255 is given by

$$\mathbb{F}_2 \subset \mathbb{F}_{2^4} \subset \mathbb{F}_{2^8}.$$

There are two shift-distinct m-sequences of degree 4. From Fact 8.1, there is only one GMW sequence of period 255 up to decimation equivalence.

(c) $n = 9$. By a similar argument to the case of $n = 6$, there is only one GMW sequence of period 511 up to decimation equivalence.

(d) $n = 10$. A possible field chain that can generate GMW sequences of period 1023 is given by

$$\mathbb{F}_2 \subset \mathbb{F}_{2^5} \subset \mathbb{F}_{2^{10}}.$$

There are six shift-distinct m-sequences of degree 5 and two shift-distinct quadratic residue sequences of period 31. According to Facts 8.1 and 8.2, there are five shift-distinct GMW sequences of Type 1, and two shift-distinct GMW sequences of Type 3, up to decimation equivalence. Thus the GMW construction can generate seven shift-distinct GMW sequences of period 1023 in total, up to decimation equivalence.

(e) If $n = 12 = 2^2 \times 3$, then $M = \frac{\varphi(2^{12}-1)}{12} = 144$; that is, we have 144 shift-distinct m-sequences of degree 12. Note that there is only one m-sequence of degree 12, up to decimation equivalence. For GMW sequences, there are three different field chains of length 2 and one field chain of length 3. We list these data in the following table together with the number of shift-distinct GMW sequences up to decimation equivalence.

GMW sequences of Types 1 and 2 of period 4095

Field chain of length 2	$N_{\text{GMW}}(12, 2) =$ $\frac{\phi(2^m-1)}{m} - 1$
$\mathbb{F}_2 \subset \mathbb{F}_{2^3} \subset \mathbb{F}_{2^{12}}$	1
$\mathbb{F}_2 \subset \mathbb{F}_{2^4} \subset \mathbb{F}_{2^{12}}$	1
$\mathbb{F}_2 \subset \mathbb{F}_{2^6} \subset \mathbb{F}_{2^{12}}$	5
Field chain of length 3	$N_{\text{GMW}}(12, 3) =$ $\left(\frac{\phi(2^6-1)}{6} - 1\right)\left(\frac{\varphi(2^3-1)}{3} - 1\right)$
$\mathbb{F}_2 \subset \mathbb{F}_{2^3} \subset \mathbb{F}_{2^6} \subset \mathbb{F}_{2^{12}}$	5

Thus, for $n = 12$, all sequences from GMW constructions are of Types 1 and 2, that is, GMW sequences of length 2 and cascaded GMW sequences of length 3. Therefore, the total number of 2-level autocorrelation sequences generated by the GMW construction is a sum of the above cases. In other words, there are a total of 12 shift-distinct GMW sequences of period 2047 up to decimation equivalence.

8.6 Implementation aspects of GMW constructions

From the proof of Construction I, there are two ways to generate the four types of GMW sequences. One is to apply the finite field configuration and the other is to use the interleaved structure with precomputation. We have shown these

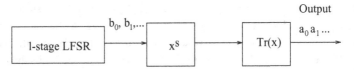

Figure 8.9. Finite field configuration of a GMW generator.

two methods in Example 8.4 for generating GMW sequences of period 63. In the following, we will present these two methods for arbitrary n.

8.6.1 Finite field configuration

We first represent a sequence from Construction I as a function of an m-sequence over $\mathbb{F}_{2^m}$ as we did in Example 8.4. Let α be a primitive element in $\mathbb{F}_{2^n}$. Let $\mathbf{u} = \{u_i\}$ be a sequence of period $2^n - 1$ from Construction I and $\mathbf{b} = \{b_i\}$ be an m-sequence over $\mathbb{F}_{2^m}$ of degree $l = n/m$. Then the elements of $\mathbf{u}$ are given by

$$u_i = g(b_i), i = 0, 1, \ldots. \tag{8.25}$$

In particular, if $g(x) = Tr_1^m(x^s)$, that is, if $\mathbf{u}$ is a GMW sequence, then u_i can be obtained by first raising the elements of m-sequence $\{b_i\}$ to a power s, and then applying a trace function from $\mathbb{F}_{2^m}$ to $\mathbb{F}_2$. In other words, we have

$$u_i = Tr_1^m(b_i^s), i = 0, 1, \ldots. \tag{8.26}$$

The finite field configuration for the implementation of GMW sequences is shown in Figure 8.9 where $Tr(x)$ represents the trace function from $\mathbb{F}_{2^m}$ to $\mathbb{F}_2$.

In Figures 8.10–8.13, we present functional blocks of finite field configurations for GMW sequences of all types where $q = 2^m$. In these figures, LFSR represents the l-stage LFSR over $\mathbb{F}_q$ where $q = 2^m$, and GMW denotes a GMW factor $Tr(x^s)$ where the extension fields for the trace function $Tr(x)$ are determined from the context.

To implement m-sequences over $\mathbb{F}_{2^m}$ at the hardware level, we need to find primitive polynomials over the extension field $\mathbb{F}_{2^m}$. A method of finding such polynomials was already discussed in Section 3.5 of Chapter 3. In the following, we give an algorithm to perform computations of the above implementations of GMW sequences of Type 1. Other types of GMW sequences can be computed using a similar approach.

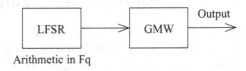

Figure 8.10. Finite field configuration of GMW generator of type 1.

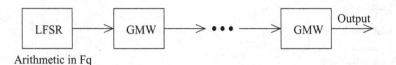

Arithmetic in Fq

Figure 8.11. Finite field configuration of a cascaded GMW generator of type 2.

Algorithm 8.2 AN ALGORITHM FOR COMPUTING GMW SEQUENCES USING FINITE FIELDS

Input:

- *n, a positive integer, and* $1 < m|n$.
- $t(x) = x^m + t_{m-1}x^{m-1} + \cdots + t_1x + t_0, t_i \in \mathbb{F}_2$, *a primitive polynomial over* $\mathbb{F}_2$ *for generating* $\mathbb{F}_{2^m}$ ($t(x)$ *can be chosen as an irreducible polynomial*).
- $w(x) = x^l + w_{l-1}x^{l-1} + \cdots + w_1x + w_0, w_i \in \mathbb{F}_{2^m}$ is   a *primitive polynomial over* $\mathbb{F}_{2^m}$ *of degree* $l = n/m$.
- $B = (b_0, b_1, \ldots, b_{l-1}), b_i \in \mathbb{F}_{2^m}$, *a nonzero vector.*
- *s:* $1 < s < 2^m - 1$, *coprime to* $2^m - 1$.

Output: $\mathbf{u} = u_0, u_1, \ldots$, *a binary GMW sequence of period* $2^n - 1$.

Procedure $(n, m, t(x), w(x), B, s, \mathbf{u})$

1. *Generate a finite field* $\mathbb{F}_{2^m}$ *using* $t(x)$.
2. *Use* $(b_0, b_1, \ldots, b_{l-1})$ *as the initial state of an LFSR with characteristic polynomial* $w(x)$ *to generate a sequence* $\mathbf{b} = \{b_i\}$:
   for $i = 0, 1, \ldots, l - 1$, *compute:* $d_i = b_i^s, u_i = Tr_1^m(d_i)$;
   for $i = l, l + 1, \ldots, 2^n - 2$, *compute:*

$$b_i = \sum_{j=0}^{l-1} w_j b_{j+i}$$

$$d_i = b_i^s$$

$$u_i = Tr_1^m(d_i), i = 0, 1, \ldots;$$

   all computations are performed in $\mathbb{F}_{2^m}$.
3. *Return* $\mathbf{u} = u_0, u_1, \ldots, u_{2^n-2}$.

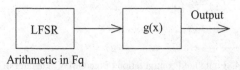

Arithmetic in Fq

Figure 8.12. Finite field configuration of a GMW generator of type 3.

Arithmetic in Fq

Figure 8.13. Finite field configuration of a GMW generator of type 4.

Example 8.11 Considering the GMW sequence in Example 8.4, note that for any $x \in \mathbb{F}_{2^3}$, $Tr_1^3(x) = x_0 + x_1 Tr_1^3(\beta) + x_2 Tr_1^3(\beta^2)$, $x_i \in \mathbb{F}_2$ where $\beta^3 + \beta^2 + 1 = 0$. Since $Tr(\beta^{2^i}) = 1$, $i = 0, 1$, and 2, $Tr_1^3(x) = x_0 + x_1 + x_2$. Therefore, this GMW generator can be implemented by a finite field configuration, shown in Figure 8.14, where all arithmetic is performed in $\mathbb{F}_{2^3}$ except for the last block.

8.6.2 Interleaved approach with precomputation

In Examples 8.4–8.7 in Section 8.2, we have shown the process for generating GMW sequences of periods 63, 255, 511, and 1023 using their interleaved structures. We will now present it in the following algorithm. As before, we only present the case for GMW sequences (Type 1). For other types of GMW sequences, the process is similar to that for Type 1.

Algorithm 8.3 AN ALGORITHM FOR GENERATING GMW SEQUENCES BY INTERLEAVED STRUCTURE

Input:

- *n, positive integer and* $1 < m|n$.
- $c(x) = x^n + c_{n-1}x^{n-1} + \cdots + c_1 x + c_0$, $c_i \in \mathbb{F}_2$, *a primitive polynomial over* $\mathbb{F}_2$ *of degree n.*
- $V = (v_0, v_1, \ldots, v_{n-1})$, *a nonzero binary vector.*
- *s:* $1 < s < 2^m - 1$, *coprime to* $2^m - 1$.

Output: $\mathbf{u} = u_0, u_1, \ldots$, *a binary GMW sequence of period* $2^n - 1$.

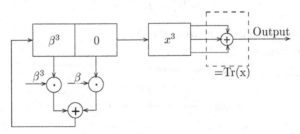

Figure 8.14. Finite field configuration of the GMW generator of period 63.

Precomputation: Finding the base sequence and shift sequence of the m-sequence generated by $c(x)$.

1. Set $v = 2^m - 1$ and $d = (2^n - 1)/v$.
2. Generate an m-sequence $\mathbf{v}$ with characteristic polynomial $c(x)$ and initial state V, and arrange it into a (v, d) array $A = (A_0, A_1, \ldots, A_{d-1})$ where the A_j's are column vectors of A.
3. Find the smallest j_0 such that the first bit in A_{j_0} is not zero (i.e., the first bit in A_i is equal to zero for $i = 0, 1, \ldots, j_0 - 1$, but the first bit in A_{j_0} is not zero), and set $\mathbf{t} = A_{j_0}$.
4. Find the constant-on-coset phase of $\mathbf{t}$ using Theorem 5.6 in Section 5.5 of Chapter 5; that is, compute

$$a_i = t_i + t_{2i} + \cdots + t_{2^{m-1}i}, i = 0, 1, \ldots, m - 1,$$

and find k using the shifting operator such that

$$(a_0, a_1, \ldots, a_{m-1}) = (t_k, t_{k+1}, \ldots, t_{k+m-1}).$$

Set $a_i = t_{k_i}, i = m, \ldots, v - 1$.
//$\mathbf{a}$ is the base sequence which is constant-on-cosets. (Note. This step is not necessary. One may choose any nonzero A_j as the base sequence.)
5. Compute the shift for each column in A with respect to $\mathbf{a}$; that is,
   for $j = j_0 + 1, \ldots, d - 2$ **do**
      if $A_j = \mathbf{0}$, set $e_j = \infty$;
      else find e by shifting such that $A_j = L^e(\mathbf{a})$, and set $e_j = e$.
6. Return $\mathbf{a} = (a_0, a_1, \ldots, a_{v-1})$ and $\mathbf{e} = (e_0, e_1, \ldots, e_{d-1})$

Procedure $(v, d, \mathbf{a}, \mathbf{e}, s, \mathbf{u})$

1. Set $\mathbf{d} = \{d_i\}$ with $d_i = a_{is}, i = 0, 1, \ldots, v - 1$.
2. Fill the array $T = (T_0, T_1, \ldots, T_{d-1})$ by:
      if $e_j = \infty$, set $T_j = \mathbf{0}$;
      else set $T_j = L^{e_j}(\mathbf{d})$.
3. Set $\mathbf{u} = R_0 || R_1 || \cdots || R_{v-1}$ where R_i is the ith row vector of T and the symbol "$||$" denotes concatenation.
4. Return $\mathbf{u}$

8.6.3 Some analysis of two algorithms

1. **Constant-on-Coset Phase versus Arbitrary Initial States:** A particular GMW sequence defined by Eq. (8.19) satisfies the constant-on-cosets

property, i.e., $u_{2i} = u_i, i = 0, 1, \ldots$. However, the GMW sequences generated by both algorithms may be shifts of the sequence defined by Eq. (8.19) because we take arbitrary initial states for **b** in Algorithm 8.2 and for **v** in Algorithm 8.3, respectively. If we need to generate the one with the constant-on-cosets property, we should modify selections of the initial states for LFSR sequences involved in these two algorithms as follows.

(a) In Algorithm 8.2, choose $b_i = Tr_m^n(\alpha^i), i = 0, 1, \ldots, l - 1$ as an initial state of **b** where α is a root of $w(x)$ in the extension $\mathbb{F}_{2^n}$.

(b) In Algorithm 8.3, choose $v_i = Tr_1^n(\alpha^i), i = 0, 1, \ldots, n - 1$ as an initial state of **v** where α is a root of $c(x)$ in the extension $\mathbb{F}_{2^n}$.

2. **Finding primitive polynomials over $\mathbb{F}_{2^m}$:** Algorithm 8.2 involves finding primitive polynomials over the intermediate subfield $\mathbb{F}_{2^m}$ of degree l with $n = lm$ for generating the m-sequence **b**. Essentially, the first step for implementation of GMW sequence generators for large n is to find primitive polynomials over $\mathbb{F}_{2^m}$. In Section 3.5 of Chapter 3, we provided a method to compute primitive polynomials over $\mathbb{F}_{2^m}$ in terms of primitive polynomials over $\mathbb{F}_2$. The primitive polynomials over $\mathbb{F}_{2^m}$ of degree l when $n = lm \leq 32$ were also listed in Appendix B of that chapter.

3. **Relationships among three LFSRs involved in two algorithms:**

(a) LFSR 1 with characteristic polynomial $c(x)$ in Algorithm 8.3, which is a primitive polynomial over $\mathbb{F}_2$ of degree n, generates shift-equivalent binary m-sequences of period $2^n - 1$.

(b) $t(x)$ in Algorithm 8.2, which is a primitive polynomial over $\mathbb{Z}_2$ of degree m, generates the finite field $\mathbb{F}_{2^m}$.

(c) LFSR 2 with characteristic polynomial $w(x)$ in Algorithm 8.2, which is a primitive polynomial over $\mathbb{F}_{2^m}$ of degree $l = n/m$, generates shift-equivalent m-sequences over $\mathbb{F}_{2^m}$ of degree l.

Let α be a root of $c(x)$ in the extension $\mathbb{F}_{2^n}$. Then $w(x)$ is the minimal polynomial of α over $\mathbb{F}_{2^m}$ (Note. This is the method provided in Chapter 3 for computing primitive polynomials over $\mathbb{F}_{2^m}$ in terms of primitive polynomials over $\mathbb{F}_2$.) Let $\beta = \alpha^d$ where $d = \frac{2^n-1}{2^m-1}$. Then we can choose $t(x)$ as the minimal polynomial of β over $\mathbb{F}_2$ (although this is not necessary). In this way, $t(x)$ is the minimal polynomial of **a** in Algorithm 8.3.

4. **Tradeoff of m and l:** From Theorems 8.6 and 8.7, large l (and therefore small m) provides better statistical properties, but small linear spans. Conversely, small l (large m) will have poor statistical properties, but large linear spans.

5. **Tradeoff of Two Algorithms:** For moderate n, the interleaved method is much more efficient than the first method since it only involves binary sequences and all computation is performed modulo 2 and using the shift operator. However, for large n, because we need huge storage for the interleaved method, it is pratical to use the finite field configuration.

8.6.4 Profile of randomness of GMW sequences

The randomness properties of GMW sequences in comparison with m-sequences are listed as follows.

Profile of randomness of GMW sequences compared to m-sequences

	GMW sequences	m-sequences
Period	$2^n - 1$	$2^n - 1$
R-1	Yes	Yes
k-tuple distribution	$1 \leq k \leq l = n/m$	$1 \leq k \leq n$
R-2	length $1 \leq k \leq l = n/m$	$1 \leq k \leq n$
R-3	Yes	Yes
Linear span LS	$ml^2 = nl \leq LS \leq ml^{m-1}$ $= nl^{m-2}$	n
The number of shift-distinct sequences, $N(x, n)$	$N(\text{GMW}, n) =$ $\frac{\phi(2^n-1)}{n} \sum_{m\|n} \left(\frac{\phi(2^m-1)}{m} - 1 \right)$	$N(m\text{-}seq, n) =$ $\frac{\phi(2^n-1)}{n}$
Implementation	For small n, software implementation with pre-computation by the interleaved structure. For large n, using finite field arithmetics and LFSR over non-binary field.	LFSR (binary)
Parameters l and m	Large l and small m, better k-tuple distribution and smaller linear span; small l and large m, poorer k-tuple distribution and large linear span.	Not applicable

Note

For further discussion of orthogonal functions, see MacDonald (1999). Using intermediate fields to construct 2-level autocorrelation sequences goes back to the original work of Gordon, Mills, and Welch in 1962. In that paper, they established Construction I, that is, Theorem 8.4 using the treatment of cyclic Hadamard difference sets. After that, the research along this line has been to find more constructions for orthogonal functions $g(x)$. The proof of Construction I can be derived from the work of Gordon, Mills, and Welch (1962). It also explicitly follows from repeatedly using the result of Klapper, Chan, and Goresky in 1993, which is similar to the proof of Theorem 8.1. The historical processes were as follows. The construction of GMW sequences published by Scholtz and Welch in 1984 showed that one may choose $g(x) = Tr_1^m(x^s)$, the trace representation of m-sequences of degree m, and called these GMW sequences. The randomness properties, shift-distinct property, linear span, and implementation using finite field configuration were also discussed in that paper. In 1993, Klapper, Chan, and Goresky showed that $g(x)$ can be taken as a cascaded GMW function, with the resulting sequence called a cascaded GMW sequence (Type 2) in that paper. This extension was also explored by No in his Ph.D. thesis, later published in 1996. Some partial results on GMW sequences of Type 3 were investigated by No et al. in 1997. The interleaved approach was discussed by Gong (1995), and Theorem 8.2 for the case of base sequences being m-sequences was also presented in that work. Presenting Construction I in this clear way appeared in Gong (2002). Construction II was discussed by Klapper (1995). The linear span of cascaded GMW sequences was discussed in Klapper, Chan, and Goresky (1993) and Gong (1996). The result on counting shift-distinct cascaded GMW sequences was obtained by Gong, Dai, and Golomb in 2000. For shift-distinct GMW sequences of types 3 and 4, see Dal, Gong, and Ye (2001).

Exercises for Chapter 8

1. For $n = 6$, investigate spectral sequences, or the trace representation of possible 2-level autocorrelation sequences of period 63.
2. Compute the spectral sequence of a quadratic residue sequence of period 23. What is the linear span of the sequence?
3. Generate a quadratic residue sequence of period 127. Can you show theoretically that the linear span of the quadratic sequence is equal to 63? If so, generalize this result to any quadratic sequences of period $2^n - 1$.
4. Let $\mathbf{a}$ be a sequence over $\mathbb{F}_p$ of period $p^n - 1$ where p is a prime. Show that if $\mathbf{a}$ has 2-level autocorrelation, then $\mathbf{a}$ is balanced.
5. For $\mathbf{a} \leftrightarrow f(x)$, a sequence over $\mathbb{F}_q$ of period $q^n - 1$, construct a counterexample in which $\mathbf{a}$ has 2-level autocorrelation, but $C_f(0) \neq 0$.
6. Compute the elements in $T_f(s)$ and $T_\infty(s)$ for m-sequences of degrees 8 and 10. Verify Lemma 8.6 using these sequences.
7. Find all sequences defined on $\mathbb{Z}_7 \cup \{\infty\}$ with the mixed shift property. Classify them according to the rule in Theorem 8.2. How many classes do you get?
8. Let $\mathbf{a} \leftrightarrow f(x)$ where $\mathbf{a}$ is a balanced binary sequence of period $2^n - 1$ and $f(0) = 0$. We may write $f(x) = \sum_{i=0}^{2^n-1} c_i x^i$. Show that $c_{2^n-1} = 0$.

9. Let m be a composite number, and assume that $r|m$. A *cascaded GMW function* of length t from $\mathbb{F}_{2^m}$ to $\mathbb{F}_{2^r}$, say $\sigma(x)$, is given by

$$\sigma(x) = Tr_{F_1/F_0}(x^{s_1})^{s_0} \circ Tr_{F_2/F_1}(x^{s_2}) \circ \cdots \circ Tr_{F_t/F_{t-1}}(x^{s_t}), \ x \in \mathbb{F}_t,$$

where the field chain is

$$\mathbb{F}_{2^r} = F_0 \subset F_1 \subset \cdots \subset F_t = \mathbb{F}_{2^m},$$

where $\circ$ is the function composition operator and the s_i's satisfy $\gcd(s_i, |F_i| - 1) = 1$. Show that $\sigma(x)$ satisfies the 2-tuple balance property.

10. Generate a GMW sequence of period 63 that is associated with a primitive element α in $\mathbb{F}_{2^6}$ such that $\alpha^6 + \alpha^5 + \alpha^2 + \alpha + 1 = 0$.

11. Compute the run distribution for span n distribution for GMW sequences of period 63, 255, 511, and 1023 in Examples 8.5–8.8 and the GMW sequence of Type 3 of period 1023 in Example 8.8.

12. Design two shift-distinct binary sequences of period 63 with 2-level autocorrelation, where one of them has linear span at least 10. Which constructions do you prefer to use? What is the crosscorrelation of these two sequences?

13. Let m be a proper factor of n. Let $\mathbf{e}$ be a shift sequence of a GMW sequence regarded as a (v, d) interleaved sequence where $v = 2^m - 1$ and $d = (2^n - 1)/v$. We construct a sequence $\mathbf{k} = (k_0, k_1, \ldots, k_{d-1})$ by the following rule:

$$k_j = \begin{cases} re_j \ (\text{mod } v) & \text{if } e_j \not\equiv \infty \\ \infty & \text{if } e_j = \infty. \end{cases}$$

Prove that $\mathbf{k}$ satisfies the mixed shift property. Furthermore, show that $\mathbf{k}$ is a shift sequence of a binary m-sequence of period $2^n - 1$ regarded as a (v, d) interleaved sequence. Use a GMW sequence of period 255 to explain your results.

14. Design a binary sequence of period 1023 with 2-level autocorrelation and linear span at least 50. How do you implement this generator?

15. Research Problem: Prove that a GMW sequence of period $2^n - 1$ will not satisfy the span n property. (The span n property means that each nonzero n-tuple occurs exactly once in one period of the sequence.)

16. Let $f(x)$ be a trace representation of a GMW sequence $\mathbf{u}$ of period $2^{14} - 1$ where $f(x)$ is given by

$$f(x) = g(x) \circ Tr_7^{14}(x), \ x \in \mathbb{F}_{14}.$$

(a) Compute the linear span of $\mathbf{u}$ for $g(x) = Tr_1^7(x^s)$ where $r = 3, 7, 15,$ and 43.

(b) Determine the maximum linear span for GMW sequences of period $2^{14} - 1$.

(c) How many GMW sequences of period $2^{14} - 1$ can you generate, up to decimation equivalence?

17. Design a cascaded GMW sequence of period 4095 associated with the following field chain:

$$\mathbb{F}_2 \subset \mathbb{F}_{2^3} \subset \mathbb{F}_{2^6} \subset \mathbb{F}_{2^{12}}.$$

What is the maximum linear span for the cascaded GMW sequences associated with the above field chain? Compare this bound with the linear span of the sequence that you designed.

18. Find the implementations of both the finite field configuration and the interleaved structure for the GMW sequences of period 63 that you designed in Exercise 12. Compare it with an m-sequence of period 63.

19. Research Problem: Prove or disprove that $h(x)$ in Construction II is equal to either the trace function or a cascaded GMW function of length $t > 1$.

20. Research Problem: Prove or disprove the following three statements:

    (a) For any binary 2-level autocorrelation sequences with period $2^n - 1$ where n is a composite number, a possible subfield factorization takes the form of either Type 2 or Type 4 in Table 8.5.

    (b) Any function from $\mathbb{F}_{2^n}$ to $\mathbb{F}_{2^m}$ $(m|n)$ with the 2-tuple balance property is either a trace function or a GMW function of length $t > 0$, that is, a function whose subfield factorization consists entirely of GMW factors.

    (c) For $v = 2^m - 1$ and $d = (2^n - 1)/v$, a d dimensional vector whose elements are taken from $\mathbb{Z}_v(\infty)$, satisfying the mixed difference property is equal to either a shift sequence of an m-sequence over $\mathbb{F}_2$ of degree n (or equivalently, an m-sequence over $\mathbb{F}_{2^m}$ of degree $l = n/m$) regarded as a (v, d) interleaved sequence or a shift sequence of a cascaded GMW sequence of length t for $t > 1$. (Note. All known results indicate that the above three statements are equivalent.)

9

Cyclic Hadamard Sequences, Part 2

Before 1997, only two essentially different constructions that were not based on a number theory approach were known for cyclic Hadamard difference sets with parameter $(2^n - 1, 2^{n-1} - 1, 2^{n-2} - 1)$ or, equivalently, for binary 2-level autocorrelation sequences of period $2^n - 1$ for arbitrary n. One is the Singer construction, which gives m-sequences, and the other is the GMW construction, which produces four types of GMW sequences. Exhaustive searches had been done for $n = 7, 8$, and 9 in 1971, 1983, and 1992, respectively. However, there was no explanation for several of the sequences found for these lengths that did not follow from then-known constructions. In this chapter, we will describe the remarkable progress in finding new constructions for 2-level autocorrelation sequences of period $2^n - 1$ since 1997. (An exhaustive search was also done for $n = 10$ in 1998.) The order of presentation of these remarkable constructions will follow the history of the developments of this research. Section 9.1 presents constructions of 2-level autocorrelation sequences having multiple trace terms. In Section 9.2, the hyper-oval constructions are introduced. Section 9.3 shows the Kasami power construction. In the last section, we introduce the iterative decimation-Hadamard transform, a method of searching for new sequences with 2-level autocorrelation.

9.1 Multiple trace term sequences

In this section, we present 3-term sequences, 5-term sequences, and the Welch-Gong transformation sequences. These constructions were initially found by computer search, and their 2-level autocorrelation property was verified for $n \leq 23$, before general proofs of their validity were discovered.

9.1.1 Three-term sequences

Construction A

For odd $n \geq 5$, and $n = 2m + 1$, with period $p = 2^n - 1$, the binary sequence

$$a_i = Tr(\alpha^i) + Tr(\alpha^{q_1 i}) + Tr(\alpha^{q_2 i}), \ i = 0, 1, \ldots$$

has two-level autocorrelation, where α is a primitive element of $\mathbb{F}_{2^n}$ and

$$q_1 = 2^m + 1, q_2 = 2^m + 2^{m-1} + 1.$$

(Alternatively, we may take $q_2 \equiv q_1^2 \pmod{p}$ to obtain the same sequence $\{a_i\}$.)

In the following, we present the construction from an implementation perspective.

Algorithm 9.1 THREE-TERM SEQUENCE GENERATOR

Precomputation:

1. *Select $n = 2m + 1$, an odd number.*
2. *Select $f_0(x) = x^n + \sum_{i=0}^{n-1} c_i x^i$, primitive over $\mathbb{F}_2$; pick α a root of $f(x)$; and generate $\mathbb{F}_{2^n}$ from $f_0(x)$.*
3. *Compute the decimation numbers:*
   $q_1 = 1 + 2^m$
   $q_2 = 1 + 2^{m-1} + 2^m$
4. *Compute minimal polynomials $f_1(x)$ and $f_2(x)$ for α^{q_1} and α^{q_2}:*
   $f_1(x) = x^n + \sum_{i=0}^{n-1} d_i x^i$
   $f_2(x) = x^n + \sum_{i=0}^{n-1} e_i x^i$
5. *Compute three initial states:*
   $s_i = Tr(\alpha^i), 0 \leq i \leq n - 1$; *set* $S_0 = (s_0, s_1, \ldots, s_{n-1})$
   $t_i = Tr(\alpha^{q_1 i}), 0 \leq i \leq n - 1$; *set* $T_0 = (t_0, t_1, \ldots, t_{n-1})$
   $v_i = Tr(\alpha^{q_2 i}), 0 \leq i \leq n - 1$; *set* $U_0 = (u_0, u_1, \ldots, u_{n-1})$

Input: n, f_i, $i = 0, 1, 2$; S_0, T_0, V_0

Output: $\mathbf{a} = a_0, a_1, \ldots$, a binary 3-term sequence of period $2^n - 1$

procedure$(n, f_0, f_1, f_2, S_0, T_0, U_0, \mathbf{a})$:

Compute:

$s_{n+i} = \sum_{j=0}^{n-1} c_j s_{j+i}, \ t_{n+i} = \sum_{j=0}^{n-1} d_j t_{j+i}, \ and \ u_{n+i} = \sum_{j=0}^{n-1} e_j u_{j+i},$
$i \geq 0$

$a_i = s_i + t_i + u_i, i \geq 0$

Return $\mathbf{a}$

Table 9.1. *Profile of 3-term sequences*

Period	$2^n - 1$
Balance	2^{n-1} 1's and $2^{n-1} - 1$ 0's
Autocorrelation	(ideal) 2-level
$N(T3, n)$, the number of shift distinct 3-term sequences	$\phi(2^n - 1)/n$
Linear span	$3n$

A profile of the randomness of 3-term sequences is presented in Table 9.1. According to the construction, 3-term sequences can be implemented as sums of three LFSRs whose characteristic polynomials are equal to the minimal polynomials of α^{q_i}, $i = 0, 1$, and 2 (set $q_0 = 1$), as shown in Figure 9.1.

Example 9.1 (a) $n = 5 = 2 \times 2 + 1 \implies m = 2$. Let $f_0(x) = x^5 + x^3 + 1$ and α be a root of $f_0(x)$ in $\mathbb{F}_{2^5}$. In this case, $q_1 = 1 + 2^2 = 5$, $q_2 = 1 + 2 + 2^2 = 7$, and the minimal polynomials of α^5 and α^7 are $f_1(x) = x^5 + x^4 + x^3 + x + 1$ and $f_2(x) = x^5 + x^4 + x^3 + x^2 + 1$, respectively. Therefore, for $i = 0, 1, \ldots,$ we have

$$\mathbf{s} = \{s_i\}, s_i = Tr(\alpha^i), S_0 = (1, 0, 0, 0, 0), s_{5+i} = s_{3+i} + s_i;$$

$$\mathbf{s}^{(5)} = \mathbf{t} = \{t_i\}, t_i = Tr(\alpha^{5i}), T_0 = (1, 1, 1, 0, 1),$$

$$t_{5+i} = t_{4+i} + t_{3+i} + t_{1+i} + t_i;$$

$$\mathbf{s}^{(7)} = \mathbf{u} = \{u_i\}, u_i = Tr(\alpha^{7i}), U_0 = (1, 1, 1, 1, 1),$$

$$u_{5+i} = u_{4+i} + u_{3+i} + u_{2+i} + u_i.$$

The LFSR implementation of **a** is shown in Figure 9.2.

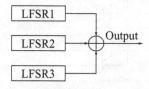

Figure 9.1. LFSR implementation of 3-term sequences.

s =	10000101011101100011111100110100
t =	11101100111000011010100010001011
u =	11111011100010101101000011001001 00
a = **s** + **t** + **u** =	10010010000111010100011110110110 11

This 3-term sequence is balanced and has 2-level autocorrelation and linear span 15. However, this is not a new sequence, because it is also a quadratic residue sequence of period 31. In other words, for add $n = 5$, the 3-term sequences degenerate to the quadratic residue sequences. For odd $n \geq 7$, new 2-level autocorrelation sequences are generated.

(b) $n = 7 = 2 \times 3 + 1 \implies m = 3$. Let $f_0(x) = x^7 + x + 1$ and α be a root of $f_0(x)$ in $\mathbb{F}_{2^7}$. In this case, $q_1 = 1 + 2^3 = 9$, $q_2 = 1 + 2^2 + 2^3 = 13$, and the minimal polynomials of α^9 and α^{13} are $f_1(x) = x^7 + x^5 + x^4 + x^3 + 1$ and $f_2(x) = x^7 + x^6 + x^5 + x^2 + 1$, respectively. Thus, for $i = 0, 1, \ldots,$ we have

$$\mathbf{s} = \{s_i\}, \, s_i = Tr(\alpha^i), \, S_0 = (1, 0, 0, 0, 0, 0, 0),$$

$$s_{7+i} = s_{1+i} + s_i;$$

$$\mathbf{s}^{(9)} = \mathbf{t} = \{t_i\}, \, t_i = Tr(\alpha^{9i}), \, T_0 = (1, 0, 0, 1, 0, 1, 1),$$

$$t_{7+i} = t_{5+i} + t_{4+i} + t_{3+i} + s_i;$$

$$\mathbf{s}^{(13)} = \mathbf{u} = \{u_i\}, \, u_i = Tr(\alpha^{13i}), \, U_0 = (1, 1, 1, 0, 1, 0, 0),$$

$$u_{7+i} = u_{6+i} + u_{5+i} + u_{2+i} + u_i.$$

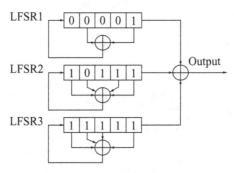

Figure 9.2. LFSR implementation of 3-term sequences of period 31.

Table 9.2. *Calculation of 3-term sequence* **a** *of period 127*

s =	1 0 0 0 0 0 0 1 0 0 0 0 0 1 1 0 0 0 0 1 0 1 0 0 0 1 1 1 1 0 0 1 0 0 0 1 0 1 1 0 0 1 1 1 0 1 0 1 0 0 1 1 1 1 1 0 1 0 0 0 0 1 1 1 0 0 0 1 0 0 1 0 0 1 1 0 1 1 0 1 0 1 1 0 1 1 1 1 0 1 1 0 0 0 1 1 0 1 0 0 1 0 1 1 1 0 1 1 1 0 0 1 1 0 0 1 0 1 0 1 0 1 1 1 1 1 1
t =	1 0 0 1 0 1 1 1 0 1 1 0 1 1 1 0 0 1 1 0 1 1 0 0 1 0 1 0 1 1 0 1 0 1 1 1 1 1 0 1 1 1 1 0 0 0 0 1 1 0 0 0 1 0 0 0 1 0 1 0 0 1 1 0 0 1 1 1 1 0 1 0 1 0 1 0 0 0 1 1 1 1 1 1 1 0 0 1 0 0 0 0 0 0 1 0 1 1 0 0 0 0 0 1 1 1 0 1 0 0 0 0 1 0 0 1 1 1 0 0 0 1 1 0 1 0 0
u =	1 1 1 0 1 0 0 0 1 1 0 0 0 1 0 0 1 1 1 0 0 1 0 1 0 0 1 1 0 1 0 0 1 0 1 1 1 1 0 1 0 1 1 1 0 1 1 1 0 0 0 1 1 1 1 0 0 1 1 0 0 1 0 0 1 0 0 0 1 0 1 0 1 0 1 1 0 1 1 0 0 1 1 1 1 1 1 1 0 1 1 0 1 0 1 0 0 0 0 0 0 0 1 1 0 1 1 1 1 1 0 0 0 0 0 1 0 1 1 0 0 0 0 1 0 0 0 0
a = s + t + v =	1 1 1 1 1 1 1 0 1 0 1 0 1 1 0 0 1 0 0 1 1 1 0 1 1 1 1 0 0 0 0 0 1 1 0 1 0 1 1 0 1 1 1 0 0 0 1 1 1 0 1 0 1 0 0 0 0 1 0 0 0 1 0 1 1 1 1 0 0 0 1 0 0 1 1 1 1 0 0 0 1 1 1 0 1 0 0 1 0 0 0 0 1 0 1 1 1 0 0 0 1 1 0 0 1 0 0 1 0 0 0 1 0 0 1 0 0 1 0 1 0 0 1 1 0 1 1

These three sequences together with the 3-term sequence **a** are listed in Table 9.2, and the LFSR implementation of **a** is shown in Figure 9.3 (all minimal polynomials of elements in $\mathbb{F}_{2^n}$ for $5 \leq n \leq 10$ are given in Appendix C in Chapter 3.) The 3-term sequence **a** of period 127 is balanced with 64 1's and 63 0's and has 2-level autocorrelation and linear span 21. There are 18 3-term sequences that are shift distinct. (For odd $n \geq 7$, the number of new shift-distinct 3-term sequences is equal to the number of shift-distinct m-sequences.)

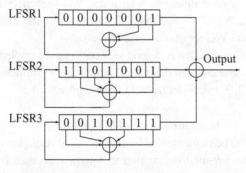

Figure 9.3. LFSR implementation of 3-term sequences of period 127.

9.1.2 Five-term and Welch-Gong transformation sequences

Construction B

Let $n \not\equiv 0 \bmod 3$, α a primitive element of $\mathbb{F}_{2^n}$, and $t(x) = x + x^{q_1} + x^{q_2} + x^{q_3} + x^{q_4}$, $x \in \mathbb{F}_{2^n}$, where the q_i's are given by

$n = 3k - 1$	$q_1 = 2^k + 1$
	$q_2 = 2^{2k-1} + 2^{k-1} + 1$
	$q_3 = 2^{2k-1} - 2^{k-1} + 1$
	$q_4 = 2^{2k-1} + 2^k - 1$
$n = 3k - 2$	$q_1 = 2^{k-1} + 1$
	$q_2 = 2^{2k-2} + 2^{k-1} + 1$
	$q_3 = 2^{2k-2} - 2^{k-1} + 1$
	$q_4 = 2^{2k-1} - 2^{k-1} + 1$

The function defined by

$$f(x) = Tr(t(x + 1) + 1), x \in \mathbb{F}_{2^n} \qquad (9.1)$$

is called the Welch-Gong transformation of $Tr(t(x))$, or the WG transformation for short. Note that $f(x)$ is a function from $\mathbb{F}_{2^n}$ to $\mathbb{F}_2$. Let $\mathbf{a} = \{a_i\}$ and $\mathbf{b} = \{b_i\}$ whose elements are given by

$$a_i = Tr(t(\alpha^i)), i = 0, 1, \ldots, \qquad (9.2)$$
$$b_i = f(\alpha^i) = Tr(t(\alpha^i + 1) + 1), i = 0, 1, \ldots. \qquad (9.3)$$

Then both $\mathbf{a}$ and $\mathbf{b}$ have 2-level autocorrelation, and $\mathbf{b}$ is called a Welch-Gong transformation sequence of $\mathbf{a}$, or a *WG sequence* for short.

A profile of the randomness of 5-term and WG sequences is presented in Table 9.3, where $N(T5, n)$ and $N(WG, n)$ represent the numbers of shift-distinct 5-term and WG sequences, respectively.

According to Construction B, 5-term sequences can be implemented as sums of five LFSRs whose characteristic polynomials are the minimal polynomials of α^{q_i}, $i = 0, 1, 2, 3$, and 4 (set $q_0 = 1$). (See Figure 9.4.)

Implementation of WG sequences

WG sequences can be implemented by a Galois field configuration, in general, or by applying an irregular decimation to 5-term sequences for small n. An irregular decimation is given in the description of Method 1 below.

Table 9.3. *Profile of 5-term sequences and WG sequences*

Period	$2^n - 1$
Balance	2^{n-1} 1's and $2^{n-1} - 1$ 0's
Autocorrelation	(ideal) 2-level
$N(T5, n)$ or $N(WG, n)$	$\phi(2^n - 1)/n$

Linear span	$LS(\mathbf{a}) = 5n;\ LS(\mathbf{b}) = n(2^{\lceil n/3 \rceil} - 3)$
Trace representation	$\mathbf{a} \leftrightarrow g(x) = Tr(t(x))$
	$\mathbf{b} \leftrightarrow f(x) = \sum_{i \in I} Tr(x^i)$ where
	$I = I_1 \cup I_2$ for $n = 3k - 1$ where
	$I_1 = \{2^{2k-1} + 2^{k-1} + 2 + i \mid 0 \le i \le 2^{k-1} - 3\}$
	$I_2 = \{2^{2k} + 3 + 2i \mid 0 \le i \le 2^{k-1} - 2\}$
	and where $I = \{1\} \cup I_3 \cup I_4$
	for $n = 3k - 2$ where
	$I_3 = \{2^{k-1} + 2 + i \mid 0 \le i \le 2^{k-1} - 3\}$
	$I_4 = \{2^{2k-1} + 2^{k-1} + 2 + i \mid 0 \le i \le 2^{k-1} - 3\}$

Method 1: Trinomial Decimation. For small n, the WG sequence $\mathbf{b}$ can be implemented using the following result. Let α be a primitive element of $\mathbb{F}_{2^n}$ and $\mathbf{b}$ the WG sequence from $\mathbf{a}$. Then the elements of $\mathbf{b}$ can be obtained by performing an irregular decimation on $\mathbf{a}$ as follows

$$b_0 = a_0, \text{ and } b_i = \begin{cases} a_{\tau(i)} & \Longleftrightarrow & n \text{ even} \\ a_{\tau(i)} + 1 & \Longleftrightarrow & n \text{ odd}, \end{cases}$$

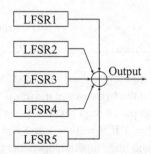

Figure 9.4. LFSR implementation of 5-term sequences.

or equivalently, $b_i \equiv a_{\tau(i)} + n \pmod{2}$, for $i > 0$, where $\tau(i)$ is determined by

$$\alpha^{\tau(i)} = \alpha^i + 1. \tag{9.4}$$

Algorithm 9.2 WG SEQUENCE GENERATOR FOR SMALL n

    Input:

              – $n \not\equiv 0 \bmod 3$
              – $h(x)$, primitive polynomial over $\mathbb{F}_2$ of degree n
              – α, a root of $h(x)$ in $\mathbb{F}_{2^n}$
              – 5-term sequence $\mathbf{a}$ with $a_i = Tr(t(\alpha^i))$.

    Output: $\mathbf{b} = \{b_i\}$, a WG sequence of period $2^n - 1$

Procedure $(n, \mathbf{a}, \mathbf{b})$

1. *Generate the trinomial table of $\mathbb{F}_{2^n}$ with defining polynomial $h(x)$: listing $\tau(i)$ such that $\alpha^{\tau(i)} = \alpha^i + 1$.*
2. *Compute $b_0 = a_0$, $b_i = a_{\tau(i)} + n \pmod{2}$, $i = 1, \ldots, 2^n - 2$.*
3. *Return $\{b_i\}$.*

Method 2: Finite Field Configuration. In general, WG sequences can be implemented by a finite field configuration (see Figure 9.5). From Construction B, only the exponents q_3 and q_4 require more multiplications in $\mathbb{F}_{q^n}$. More precisely, we may rewrite x^{q_3} and x^{q_4} and then count how many multiplications are needed:

$n = 3k - 1$	$x^{q_3} = P^{2^{k-1}} \cdot x$, $P = x^{2^k - 1}$ $\implies$ $k + 1$ multiplications
	$x^{q_4} = P \cdot x^{2^{2k-1}}$ $\implies$ 1 multiplication using the result P from x^{q_3}
$n = 3k - 2$	$x^{q_3} = Q^{2^{k-1}} \cdot x$, $Q = x^{2^{k-1}-1}$ $\implies$ $(k-1) + 1 = k$ multiplications
	$x^{q_4} = Q \cdot x^{2^{2k-1}}$ $\implies$ 1 multiplication using the result Q from x^{q_3}

On the other hand, for computing x^{q_1} and x^{q_2}, 1 and 2 multiplications, respectively, are required in $\mathbb{F}_{2^n}$ for the two cases of n. Hence, for the implementation shown in Figure 9.5, each output bit requires $(k + 1 + \delta) + 3 = k + 4 + \delta$ multiplications in $\mathbb{F}_{2^n}$ where $\delta = 1$ if $n = 3k - 1$ and $\delta = 0$ if $n = 3k - 1$. Here we consider that squaring and trace function operators have a low cost that can be ignored.

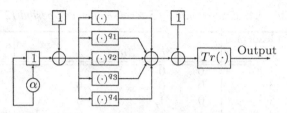

Figure 9.5. Galois field configuration of WG generators.

Example 9.2 A 5-term sequence of period 127 For $n = 5$, a 5-term sequence **a** degenerates to a quadratic sequence and a WG sequence **b** to an m-sequence. For $n \geq 7$, Construction B produces new sequences. Let $n = 7 = 3 \times 3 - 2 \implies k = 3$, let $f(x) = x^7 + x + 1$, and let α be a root of $f(x)$ in $\mathbb{F}_{2^7}$. Then

$$q_1 = 2^2 + 1 = 5, \quad q_2 = 2^4 + 2^2 + 1 = 21,$$

$$q_3 = 2^4 - 2^2 + 1 = 13, \text{ and } q^4 = 2^5 - 2^2 + 1 = 29.$$

We compute the minimal polynomials $f_{\alpha^i}(x)$ of α^i, $i \in \{5, 21, 13, 29\}$ (or by look-up table in Appendix C of Chapter 3) as follows:

$$f_{\alpha^5}(x) = x^7 + x^3 + x^2 + x + 1$$

$$f_{\alpha^{21}}(x) = x^7 + x^6 + x^3 + x + 1$$

$$f_{\alpha^{13}}(x) = x^7 + x^6 + x^5 + x^2 + 1$$

$$f_{\alpha^{29}}(x) = x^7 + x^4 + 1.$$

Sequences generated by $f_{\alpha^i}(x)$ with the constant-on-cosets property (recall that, from Section 5.6 of Chapter 5, a binary sequence $\mathbf{a} = \{a_i\}$ is constant-on-cosets if $a_{2i} = a_i$, for all $i = 0, 1, \dots$) are given by

$$\mathbf{s} = \{s_i\}, \, s_i = Tr(\alpha^i), \, S_0 = (1, 0, 0, 0, 0, 0, 0),$$

$$s_{7+i} = s_{1+i} + s_i;$$

$$\mathbf{s}^{(5)} = \{t_i\}, \, t_i = Tr(\alpha^{5i}), \, T_0 = (1, 0, 0, 0, 0, 1, 0),$$

$$t_{7+i} = t_{3+i} + t_{2+i} + t_{1+i} + t_i;$$

$$\mathbf{s}^{(21)} = \{u_i\}, \, u_i = Tr(\alpha^{21i}), \, U_0 = (1, 1, 1, 1, 1, 0, 1),$$

$$u_{7+i} = u_{6+i} + u_{3+i} + u_{1+i} + u_i;$$

Table 9.4. *Computation of 5-term sequence of period 127*

Coset leaders i	s_i	t_i	u_i	v_i	w_i	a_i
1	0	0	1	1	0	0
3	0	0	1	0	1	0
5	0	1	0	0	0	1
7	1	1	0	0	1	1
9	0	1	1	1	1	0
11	0	0	1	0	0	1
13	1	0	1	1	1	0
15	0	0	1	0	1	0
19	1	1	0	0	1	1
21	1	0	0	1	0	0
23	0	1	1	1	0	1
27	1	0	0	1	1	1
29	0	0	0	1	1	0
31	1	1	0	0	0	0
43	1	0	1	1	1	0
47	1	1	0	1	0	1
55	0	1	0	0	0	1
63	1	1	1	0	0	1

$$s^{(13)} = \{v_i\}, \; v_i = Tr(\alpha^{13i}), \; V_0 = (1, 1, 1, 0, 1, 0, 0),$$

$$v_{7+i} = v_{6+i} + v_{5+i} + v_{2+i} + v_i;$$

$$s^{(29)} = \{w_i\}, \; w_i = Tr(\alpha^{29i}), \; W_0 = (1, 0, 0, 1, 0, 0, 1),$$

$$w_{7+i} = w_{4+i} + w_i$$

where $i = 0, 1, \ldots$. Thus, $\mathbf{a} = \mathbf{s} + \mathbf{t} + \mathbf{u} + \mathbf{v} + \mathbf{w}$, whose elements are given in Table 9.4. (We only list the values of the coset leaders.) By using $a_{2^i k} = a_k, i = 0, 1, \ldots, 6$ where

$$k \in \Gamma_2(7) = \{0, 1, 3, 5, 7, 9, 11, 13, 15, 19, 21, 23, 27, 29, 31, 43, 47, 55, 63\},$$

which is the set of all coset leaders modulo 127, we obtain the 5-term sequence:

$\mathbf{a} = 1\,0\,0\,0\,0\,1\,0\,1\,0\,0\,1\,1\,0\,0\,1\,0\,0\,0\,0\,1\,1\,0\,1\,1\,0\,1\,0\,1\,1\,0\,0\,0\,0\,1\,0\,0\,0\,0\,1$
$0\,1\,0\,0\,0\,1\,0\,1\,1\,0\,1\,1\,1\,0\,0\,1\,1\,1\,1\,0\,1\,0\,1\,0\,1\,0\,0\,1\,1\,0\,1\,0\,0\,0\,1\,0\,1\,1\,1$
$0\,0\,1\,0\,0\,0\,0\,0\,0\,1\,1\,1\,0\,1\,1\,1\,1\,1\,0\,1\,1\,0\,1\,1\,1\,0\,0\,0\,0\,1\,1\,1\,1\,1\,1\,0\,1\,0\,0$
$1\,1\,1\,0\,0\,1\,1\,0\,1\,1$

The LFSR implementation of the 5-term sequence $\mathbf{a}$ is shown in Figure 9.6.

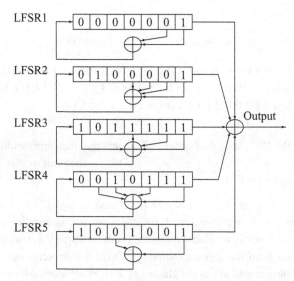

Figure 9.6. LFSR implementation of 5-term sequence of period 127.

A WG sequence of period 127

For the WG sequence **b**, using Algorithm 9.2, we compute the trinomial table for $\mathbb{F}_{2^7}$ defined by $\alpha^7 + \alpha + 1 = 0$; that is, the table contains $\{\tau(i)\}$ where $\alpha^{\tau(i)} = \alpha^i + 1, i = 1, 2, \ldots, 126$.

Trinomial table of $\mathbb{F}_{2^7}$ defined by $\alpha^7 + \alpha + 1 = 0$

–	7	14	63	28	54	126	1	56	90
108	87	125	55	2	31	112	43	53	29
89	57	47	82	123	105	110	66	4	19
62	15	97	77	86	109	106	46	58	100
51	75	114	17	94	68	37	22	119	122
83	40	93	18	5	13	8	21	38	104
124	88	30	3	67	95	27	64	45	107
91	79	85	78	92	41	116	33	73	71
102	118	23	50	101	72	34	11	61	20
9	70	74	52	44	65	111	32	117	103
39	84	80	99	59	25	36	69	10	35
26	96	16	115	42	113	76	98	81	48
121	120	49	24	60	12	6			

From the trinomial table, we obtain

$$b_0 = a_0, b_i = a_{\tau(i)} + 1, i = 1, \ldots, 126,$$

b = 1 0 0 0 0 0 0 1 0 1 0 0 0 0 1 1 0 1 1 1 0 0 0 1 0 1 0 0 1 0 1 1 0 0 1 0 1 0 1
0 0 0 0 1 0 1 1 0 0 0 1 0 0 1 0 1 1 1 0 1 1 0 1 1 0 0 0 1 1 0 0 1 1 1 0 1 1 0
0 1 0 0 0 0 0 1 1 0 0 0 1 1 1 1 0 1 0 1 0 1 1 1 0 1 0 0 1 0 0 1 1 1 1 1 1 1 0
0 1 1 1 1 0 1 1 1 1

By applying the DFT, with α of order 127, we get the trace representation $f(x)$ of **b**, $f(x) = Tr(x + x^3 + x^7 + x^{19} + x^{29})$, which is the same as that computed by the formula in Table 9.3.

For $n = 7$, since $p = 2^7 - 1 = 127 \equiv 3 \pmod 4$ and $127 = 4 \times 5^2 + 27$ where 127 is prime, we therefore have m-sequences, quadratic residue sequences, and Hall sextic residue sequences. From Example 9.1, we have 3-term sequences, and from the above example, we have 5-term sequences and WG sequences. These are all six of the classes of 2-level autocorrelation sequences of period 127 that were obtained by exhaustive search in 1971 by Baumert. In Table 9.5, we list all six of these sequences up to decimation equivalence, together with their trace representations. We only list the values of the coset leaders, because the rest of the elements can be computed by the constant-on-cosets property.

Note. In Table 9.5, the quadratic residue sequence B and the Hall sextic residue sequence C are computed by taking 3 as a primitive element modulo 127. In the last row, the numbers are the exponents in the trace terms. The finite field $\mathbb{F}_{2^7}$ is defined by the primitive polynomial $x^7 + x + 1$ for computing the trace representations of the quadratic residue sequence and the Hall sextic residue sequence.

Example 9.3 A 5-term sequence of period 255 Let $n = 8 = 3 \times 3 - 1 \implies k = 3$. Let $f(x) = x^8 + x^4 + x^3 + x^2 + 1$ and α be a root of $f(x)$ in $\mathbb{F}_{2^8}$. Then

$$q_1 = 2^3 + 1 = 9, \quad q_2 = 2^5 + 2^2 + 1 = 37,$$

$$q_3 = 2^5 - 2^2 + 1 = 29, \text{ and } q^4 = 2^5 + 2^3 - 1 = 39$$

and from Appendix C in Chapter 3 the minimal polynomials $f_{\alpha^i}(x)$ of α^i, $i \in \{9, 37, 29, 39\}$ are:

$$f_{\alpha^9}(x) = x^8 + x^7 + x^5 + x^4 + x^3 + x^2 + 1$$
$$f_{\alpha^{37}}(x) = x^8 + x^6 + x^4 + x^3 + x^2 + x + 1$$
$$f_{\alpha^{29}}(x) = x^8 + x^7 + x^3 + x^2 + 1$$
$$f_{\alpha^{39}}(x) = x^8 + x^4 + x^3 + x + 1.$$

Table 9.5. *A complete list of 2-level autocorrelation sequences of period 127*

i Coset leaders	$A = \{a_i\}$ m-seq	$B = \{b_i\}$ QR	$C = \{c_i\}$ Hall	$D = \{d_i\}$ 3-term	$E = \{e_i\}$ 5-term	$F = \{f_i\}$ WG
0	1	1	1	1	1	1
1	0	0	0	1	0	0
3	0	1	0	1	0	0
5	0	1	0	1	1	0
7	1	1	0	0	1	1
9	0	0	1	0	0	1
11	0	0	1	0	1	0
13	1	0	1	1	0	0
15	0	0	1	0	0	1
19	1	0	0	1	1	1
21	1	0	1	1	0	0
23	0	1	0	1	1	1
27	1	1	0	0	1	0
29	0	1	1	0	0	0
31	1	0	1	0	0	1
43	1	1	1	0	0	1
47	1	0	0	1	1	0
55	0	1	1	0	1	1
63	1	1	0	1	1	1
Trace spectra	1	3, 5, 7, 23, 27, 29, 43, 55, 63	5, 27, 63	1, 9, 13	1, 5, 21, 13, 29	1, 3, 7, 19, 29
Linear span	7	63	21	21	35	35

The sequences generated by $f_{\alpha^i}(x)$ with the constant-on-cosets property are given by

$$\mathbf{s} = \{s_i\}, \ s_i = Tr(\alpha^i), \ S_0 = (0, 0, 0, 0, 0, 1, 0, 0),$$

$$s_{8+i} = s_{4+i} + s_{3+i} + s_{2+i} + s_i;$$

$$\mathbf{t} = \{t_i\}, \ t_i = Tr(\alpha^{9i}), \ T_0 = (0, 1, 1, 0, 1, 0, 0, 1),$$

$$t_{8+i} = t_{7+i} + t_{5+i} + t_{4+i} + t_{3+i} + t_{2+i} + t_i;$$

$$\mathbf{u} = \{u_i\}, \ u_i = Tr(\alpha^{37i}), \ U_0 = (0, 0, 0, 0, 0, 1, 0, 0),$$

$$u_{8+i} = u_{6+i} + u_{4+i} + u_{3+i} + u_{2+i} + u_{1+i} + u_i;$$

$$\mathbf{v} = \{v_i\}, \ v_i = Tr(\alpha^{29i}), \ V_0 = (0, 1, 1, 1, 1, 0, 1, 1),$$

$$v_{8+i} = v_{7+i} + v_{3+i} + v_{2+i} + v_i;$$

$$\mathbf{w} = \{w_i\}, \ w_i = Tr(\alpha^{39i}), \ W_0 = (0, 1, 1, 1, 1, 1, 1, 1),$$

$$w_{8+i} = w_{4+i} + w_{3+i} + w_{1+i} + w_i.$$

Table 9.6. *Calculation of the 5-term sequence* **a** *of period 255*

Coset i	s_i	t_i	u_i	v_i	w_i	a_i	Coset i	s_i	t_i	u_i	v_i	w_i	a_i
0	0	0	0	0	0	0	39	1	0	1	0	0	0
1	0	1	0	1	1	1	43	1	1	1	0	1	0
3	0	0	0	1	1	0	45	0	0	1	1	1	1
5	1	0	1	0	1	1	47	1	1	1	1	0	0
7	0	1	0	1	1	1	51	0	0	0	0	0	0
9	1	1	1	0	0	1	53	1	0	1	0	0	0
11	1	0	0	0	1	0	55	1	1	1	1	0	0
13	0	1	0	1	1	1	59	0	1	0	1	0	0
15	1	1	0	0	0	0	61	1	1	0	0	1	1
17	0	0	0	0	0	0	63	1	1	1	1	1	1
19	0	1	0	0	1	0	85	0	0	0	0	0	0
21	1	0	0	0	0	1	87	1	1	1	1	1	1
23	0	1	1	0	1	1	91	1	0	0	1	1	1
25	0	1	1	1	0	1	95	1	0	0	1	1	1
27	0	1	1	1	1	0	111	0	1	0	1	1	1
29	1	0	1	0	0	0	119	0	0	0	0	0	0
31	0	0	0	0	0	0	127	0	0	1	0	0	1
37	0	1	1	1	1	0							

In Table 9.6, we list the elements of the above five sequences and the 5-term sequence $\mathbf{a} = \mathbf{s} + \mathbf{t} + \mathbf{u} + \mathbf{v} + \mathbf{w}$ where indices are taken from the set of coset leaders modulo 255. By using the constant-on-cosets property, we have the 5-term sequence **a** of period 255 as follows:

$\mathbf{a} = 0\,1\,1\,0\,1\,1\,0\,1\,1\,1\,1\,0\,0\,1\,1\,0\,1\,0\,1\,0\,1\,1\,0\,1\,0\,1\,1\,0\,1\,0\,0\,0\,1\,1\,0\,1\,1\,0\,0$
$0\,1\,0\,1\,0\,0\,1\,1\,0\,0\,0\,1\,0\,1\,0\,0\,0\,1\,0\,0\,0\,0\,1\,0\,1\,1\,1\,1\,1\,0\,1\,1\,0\,1\,0\,0\,1\,0\,0$
$0\,1\,1\,1\,0\,0\,1\,0\,0\,1\,0\,0\,1\,1\,1\,1\,0\,1\,0\,0\,0\,0\,1\,0\,0\,0\,1\,1\,0\,1\,0\,1\,0\,1\,1\,1\,0\,0\,0$
$1\,0\,0\,0\,0\,1\,1\,0\,1\,1\,1\,1\,0\,1\,1\,1\,0\,1\,0\,0\,0\,1\,1\,1\,0\,0\,0\,1\,1\,0\,0\,0\,0\,1\,0\,0\,0\,0\,0$
$0\,0\,1\,1\,1\,1\,1\,0\,0\,1\,0\,1\,1\,0\,0\,1\,0\,1\,1\,1\,0\,0\,0\,0\,1\,1\,1\,1\,1\,0\,1\,0\,0\,1\,1\,1\,0\,1\,0$
$0\,0\,1\,0\,0\,1\,0\,0\,0\,0\,0\,0\,1\,1\,0\,1\,1\,0\,1\,1\,1\,0\,0\,1\,1\,0\,0\,1\,1\,1\,0\,1\,0\,0\,0\,0\,1\,0\,1$
$1\,1\,0\,1\,0\,1\,0\,0\,0\,1\,1\,1\,1\,1\,0\,1\,1\,1\,1\,1\,1$

The LFSR implementation of the 5-term sequence **a** is shown in Figure 9.7.

A WG sequence of period 255

For the WG sequence **b**, using Algorithm 9.2, we compute the trinomial table for $\mathbb{F}_{2^8}$ defined by $\alpha^8 + \alpha^4 + \alpha^3 + \alpha^2 + 1 = 0$, shown in Table 9.7, which contains the sequence $\{\tau(i)\}$ where $\alpha^{\tau(i)} = \alpha^i + 1, i = 1, 2, \ldots, 254$. Thus,

$$b_0 = a_0, b_i = a_{\tau(i)}, i = 0, 1, \ldots, 254,$$

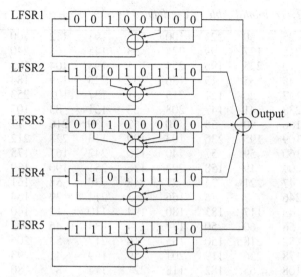

Figure 9.7. LFSR implementation of 5-term sequences of period 255.

and **b** is given by

$$\begin{aligned}
\mathbf{b} = {}& 0\,1\,1\,1\,1\,1\,1\,1\,1\,0\,1\,1\,1\,0\,1\,1\,1\,0\,0\,1\,1\,1\,1\,0\,1\,1\,0\,1\,1\,1\,1\,1\,0\,0\,1\,0\,0\,1 \\
& 0\,1\,0\,1\,0\,1\,0\,0\,0\,1\,1\,1\,0\,0\,0\,1\,1\,1\,0\,1\,0\,1\,1\,1\,0\,1\,1\,0\,0\,0\,1\,1\,1\,0\,0\,0\,0\,1\,0 \\
& 0\,1\,1\,1\,0\,0\,1\,0\,0\,1\,1\,0\,0\,0\,0\,1\,0\,0\,1\,1\,1\,1\,1\,0\,0\,0\,0\,0\,0\,0\,1\,0\,1\,1\,1\,0\,0\,1\,1 \\
& 1\,0\,0\,1\,0\,1\,1\,1\,0\,0\,0\,1\,1\,1\,1\,0\,1\,0\,1\,0\,1\,1\,0\,1\,1\,1\,1\,0\,1\,0\,0\,0\,0\,0\,0\,1\,0\,0\,1 \\
& 0\,0\,1\,0\,1\,0\,1\,1\,0\,0\,0\,1\,1\,0\,0\,1\,0\,0\,1\,0\,1\,1\,0\,0\,0\,0\,0\,1\,0\,1\,1\,0\,0\,1\,0\,0\,1\,1\,1 \\
& 1\,1\,0\,1\,1\,1\,0\,0\,0\,0\,1\,0\,0\,0\,1\,0\,1\,0\,1\,0\,0\,1\,0\,0\,1\,1\,0\,1\,0\,1\,1\,0\,1\,0\,0\,1\,0\,1\,1 \\
& 1\,0\,0\,1\,0\,0\,1\,1\,0\,0\,1\,0\,1\,0\,1\,0\,0\,0\,0\,0\,0.
\end{aligned}$$

By applying the DFT, with respect to the primitive element α, to **b**, we can obtain the trace representation of **b** as follows:

$$\mathbf{b} \leftrightarrow f(x) = Tr(x^{13} + x^{19} + x^{21} + x^{29} + x^{39}),$$

which is the same as that computed from the formula in Table 9.3.

For $n = 8$, we have m-sequences, GMW sequences, 5-term sequences, and WG sequences. These are all the 2-level autocorrelation sequences of period 255 that were obtained by exhaustive search by Cheng in 1983. In Table 9.8, we list all four of these sequences where the m-sequence and the GMW sequence are those in Example 8.6 in Chapter 8, and the 5-term and WG sequences are taken from the above example.

Table 9.7. *Trinomial table of* $\mathbb{F}_{2^8}$ *defined by* $\alpha^8 + \alpha^4 + \alpha^3 + \alpha^2 + 1 = 0$

–	25	50	223	100	138	191	112	200	120
21	245	127	99	224	33	145	68	240	92
42	10	235	196	254	1	198	104	193	181
66	45	35	15	136	32	225	179	184	106
84	157	20	121	215	31	137	101	253	197
2	238	141	147	208	63	131	83	107	82
132	186	90	55	70	162	30	216	17	130
64	109	195	236	103	199	113	228	212	174
168	160	59	57	40	170	242	167	175	203
62	209	19	158	202	176	251	190	139	13
4	47	221	74	27	248	39	58	161	71
126	246	7	76	166	243	214	122	164	153
9	43	117	183	180	194	110	12	140	239
69	56	60	250	177	144	34	46	5	98
128	52	218	150	135	16	217	53	206	188
143	178	226	119	201	159	169	41	93	155
81	108	65	182	118	227	114	87	80	156
85	211	229	232	79	88	95	134	151	37
124	29	163	123	38	249	61	204	149	219
97	6	247	28	125	72	23	49	26	75
8	154	94	89	187	207	148	205	54	91
241	171	78	233	116	44	67	146	142	189
252	102	237	3	14	36	152	165	77	172
231	230	173	213	244	22	73	222	51	129
18	210	86	115	234	11	111	192	105	185
133	96	220	48	24					

9.2 Hyperoval constructions

We first introduce some concepts from projective geometry to give the definition of hyperovals, although we will not need to use this afterward.

Let $q = 2^n$. A projective plane of order q is a 2-$(q^2 + q + 1, q + 1, 1)$ design, denoted $PG(2, q)$, where the points are the 1-dimensional subspaces of $\mathbb{F}_q^3 = \{(c_0, c_1, c_2) \mid c_i \in \mathbb{F}_q\}$ and the blocks (lines) are the 2-dimensional subspaces of $\mathbb{F}_q^3$. (A projective plane of order q is really a (v, k, λ) cyclic difference set with $v = q^2 + q + 1, k = q + 1,$ and $\lambda = 1$.) A set of $q + 1$ points, no three on a line, is called an oval, and a set of $q + 2$ points, no three on a line, is a hyperoval. Segre (1955) showed that

Lemma 9.1 *Every hyperoval in $PG(2, q)$ may be written in the form:*

$$D(f) = \{(1, t, f(t)) \mid t \in \mathbb{F}_q\} \cup \{(0, 1, 0), (0, 0, 1)\},$$

Table 9.8. *A complete list of 2-level autocorrelation sequences of period 255*

i Coset leaders	$A = \{a_i\}$ m-seq	$B = \{b_i\}$ GMW	$C = \{c_i\}$ 5-term	$D = \{d_i\}$ WG
0	0	0	0	0
1	0	1	1	1
3	0	1	0	1
5	1	1	1	1
7	0	0	1	1
9	1	1	1	0
11	1	0	0	1
13	0	0	1	0
15	1	1	0	1
17	0	0	0	0
19	0	0	0	1
21	1	1	1	1
23	0	0	1	0
25	0	1	1	1
27	0	0	0	1
29	1	1	0	1
31	0	1	0	1
37	0	1	0	0
39	1	0	0	0
43	1	0	0	0
45	0	1	1	0
47	1	0	0	0
51	0	0	0	0
53	1	0	0	0
55	1	0	0	1
59	0	1	0	0
61	1	1	1	1
63	1	0	1	0
85	0	0	0	0
87	1	1	1	1
91	1	1	1	0
95	1	0	1	0
111	0	0	1	1
119	0	0	0	0
127	0	1	1	0
Trace spectrum	1	7, 13, 37, 11	1, 9, 37, 29, 39	13, 19, 21, 29, 39
Linear span	8	32	40	40

where f is a permutation of $\mathbb{F}_q$ satisfying $f(0) = 0$, $f(1) = 1$, and for each $\alpha \in \mathbb{F}_q$ such that the polynomial

$$f_\alpha(x) = \frac{f(x + \alpha) + f(\alpha)}{x}, \quad f_\alpha(0) = 0$$

is also a permutation polynomial of $\mathbb{F}_q$.

If $f(x) = x^k$, the hyperoval is called a monomial hyperoval.

Example 9.4 Let $q = 4$, $\mathbb{F}_q = \{0, 1, \alpha, \alpha^2 \mid \alpha^2 + \alpha + 1 = 0\}$, and $f(x) = x^2$. Then the hyperoval defined by $f(x)$ is given by

$$D(f) = \{(1, 0, 0), (1, 1, 1), (1, \alpha, \alpha^2), (1, \alpha^2, \alpha)\} \cup \{(0, 1, 0), (0, 0, 1)\}.$$

Known examples of monomial hyperovals $D(x^k)$ in $PQ(2, q)$ include:

- The regular hyperoval $D(x^2)$.
- The translate hyperovals $D(x^{2^i})$, where $\gcd(n, i) = 1$, $1 < i < \frac{n}{2}$.
- The Segre hyperoval $D(x^6)$, where $n \geq 5$ is odd.
- The Glynn type 1 hyperoval $D(x^k)$, where $n \geq 7$ is odd, and $k = \sigma + \gamma$ where σ and γ are defined by

$$\sigma = 2^{\frac{(n+1)}{2}}, \text{ and} \tag{9.5}$$

$$\gamma = \begin{cases} 2^{\frac{3n+1}{4}} & \text{if } n \equiv 1 \bmod 4 \\ 2^{\frac{n+1}{4}} & \text{if } n \equiv 3 \bmod 4. \end{cases} \tag{9.6}$$

- The Glynn type 2 hyperoval $D(x^{3\sigma+4})$, where $n \geq 11$ is odd and σ is given by Eq. (9.5).

Glynn conjectured that the hyperovals listed above together with the equivalent hyperovals $D(x^k)$, k, $\frac{1}{k}$, $1 - k$, $\frac{1}{1-k}$, $\frac{k-1}{k}$, $\frac{k}{k-1}$ modulo $q - 1$ are all monomial hyperovals. This conjecture still remains open at this time. Using the geometry of hyperovals, Maschietti proved that the hyperoval $D(x^k)$ gives a $(2^n - 1, 2^{n-1} - 1, 2^{n-2} - 1)$ cyclic Hadamard difference set. Thus, from Theorem 7.1 in Chapter 7, these cyclic Hadamard difference sets yield ideal 2-level autocorrelation sequences of period $2^n - 1$.

9.2.1 Construction

In the following, we consider a map $\tau : F_{2^n} \to F_{2^n}$, defined by

$$\tau(x) = x + x^k.$$

Maschietti established 1998 that

$$D(x^k) = \{(1, x, x^k)| \; x \in F_q\} \cup \{(0, 1, 0), (0, 0, 1)\}$$

in $PG(2, q)$ is a hyperoval if and only if $\gcd(k, q - 1) = 1$ and τ is a two-to-one map from $\mathbb{F}_q$ into itself. Let $Im(\tau) = \{x + x^k \; | \; x \in \mathbb{F}_q\}$. Therefore,

$$D_k = Im(\tau) \backslash \{0\}$$

is a $(2^n - 1, 2^{n-1} - 1, 2^{n-2} - 1)$ difference set in $F_{2^n}^*$.

We are now ready to introduce the general hyperoval construction for 2-level autocorrelation sequences.

Hyperoval construction

Let $\mathbf{a} = \{a_i\}$ whose elements are given by

$$a_i = \begin{cases} 0, & i \in Im(\tau) \backslash \{0\} \\ 1, & \text{otherwise} \end{cases}$$

where $\tau(x) = x + x^k$. For $k = 6$, we call $\mathbf{a}$ a Segre hyperoval sequence; for k a Glynn type 1 exponent, we call $\mathbf{a}$ a Glynn type 1 hyperoval sequence, and for k a Glynn type 2 exponent, we call $\mathbf{a}$ a Glynn type 2 hyperoval sequence. Maschietti's result showed that $\mathbf{a}$ has 2-level autocorrelation.

Note. The regular and the translate hyperovals produce m-sequences.

Example 9.5 Construct a Segre hyperoval sequence of period 31. Let $n = 5$. First we compute the Segre 2-to-1 map whose values are listed in Table 9.9, where $\mathbb{F}_{2^5}$ is defined by $h(x) = x^5 + x^3 + 1$, primitive, and α is a root of $h(x)$ in $\mathbb{F}_{2^5}$. In the table, the column labeled "x in exp" lists the exponent i for $x = \alpha^i$. For the column labeled "x in vector rep." the vector $(x_0, x_1, x_2, x_3, x_4)$ represents the coefficients of $x = x_0 + x_1\alpha + x_2\alpha^2 + x_3\alpha^3 + x_4\alpha^4$, and for the column labeled "$x + x^6$" the numbers are exponents of $x + x^6$ in α. For example, for $x = \alpha^6$, we have $x + x^6 = \alpha^{19}$. By using this 2-to-1 map, the Segre hyperoval sequence of period 31 is constructed as:

$Im(\tau)$	–	1	2	–	4	5	–	7	8	9	10	–	–	–	14	–
	↓	↓	↓	↓	↓	↓	↓	↓	↓	↓	↓	↓	↓	↓	↓	↓
a	1	0	0	1	0	0	1	0	0	0	0	1	1	1	0	1

$Im(\tau)$	16	–	18	19	20	–	–	–	–	25	–	–	28	–	–
	↓	↓	↓	↓	↓	↓	↓	↓	↓	↓	↓	↓	↓	↓	↓
a	0	1	0	0	0	1	1	1	1	0	1	1	0	1	1.

Table 9.9. *Segre 2-to-1 map in* $\mathbb{F}_{2^5}$

x in exp	x in vector rep.	$x + x^6$	x in exp	x in vector rep.	$x + x^6$
–	0 0 0 0 0	–	15	0 1 1 0 0	14
0	1 0 0 0 0	–	16	0 0 1 1 0	2
1	0 1 0 0 0	4	17	0 0 0 1 1	28
2	0 0 1 0 0	8	18	1 0 0 1 1	20
3	0 0 0 1 0	25	19	1 1 0 1 1	16
4	0 0 0 0 1	16	20	1 1 1 1 1	5
5	1 0 0 1 0	9	21	1 1 1 0 1	10
6	0 1 0 0 1	19	22	1 1 1 0 0	9
7	1 0 1 1 0	1	23	0 1 1 1 0	7
8	0 1 0 1 1	1	24	0 0 1 1 1	14
9	1 0 1 1 1	10	25	1 0 0 0 1	8
10	1 1 0 0 1	18	26	1 1 0 1 0	5
11	1 1 1 1 0	20	27	0 1 1 0 1	19
12	0 1 1 1 1	7	28	1 0 1 0 0	4
13	1 0 1 0 1	18	29	0 1 0 1 0	25
14	1 1 0 0 0	2	30	0 0 1 0 1	28

This is just the 3-term sequence that appeared in Example 9.1, which is also equal to the quadratic residue sequence. In other words, there are three constructions for 2-level autocorrelation sequences of period 31 with linear span 15, namely, the quadratic residue construction, the 3-term construction, and the Segre hyperoval construction.

For $n = 7$, sequences constructed by the hyperoval construction are also equal to known sequences. For $n = 9$, the hyperoval construction explains two 2-level autocorrelation sequences found by exhaustive search in 1992. Starting from $n = 11$, the hyperoval construction produces new classes of sequences with 2-level autocorrelation.

In the following subsection, we will provide another proof for the 2-level autocorrelation property of hyperoval sequences. This method also leads to a proof for the conjectured sequences with 2-level autocorrelation that will be described in Section 3.

9.2.2 Hadamard transform of hyperoval sequences

Let $q = 2^n$ and let **a** be a binary sequence of period $2^n - 1$. For a function $f(x) : F_{2^n} \to F_2$, recall the concepts of the Hadamard transform and the inverse

Hadamard transform of $f(x)$ from Chapter 6. Here we reproduce them as follows.

$$\widehat{f}(\lambda) = \sum_{x \in F_{2^n}} (-1)^{Tr(\lambda x) + f(x)}$$

$$(-1)^{f(\lambda)} = \frac{1}{2^n} \sum_{x \in F_{2^n}} (-1)^{Tr(\lambda x)} \widehat{f}(x)$$

A fundamental tool for studying Hadamard transforms is to use Parseval's equality, which gives

$$C_f(\lambda) = \sum_{x \in F_q} (-1)^{f(x) + f(\lambda x)} = \frac{1}{q} \sum_{x \in F_q} \widehat{f}(x) \widehat{f}(\lambda x). \tag{9.7}$$

From Proposition 6.3 in Chapter 6, if $f(x)$ is a trace representation of $\mathbf{a}$, that is, $a_i = f(\alpha^i)$, $i \geq 0$, where α is a primitive element of $\mathbb{F}_q$, then

$$C_{\mathbf{a}}(\tau) = \sum_{x \in \mathbb{F}_q} (-1)^{f(x) + f(\lambda x)} - 1 = C_f(\lambda) - 1$$

where $\lambda = \alpha^\tau$. Thus $\mathbf{a}$ has 2-level autocorrelation if and only if $C_f(\lambda) = 0$ for $\tau \not\equiv 0 \pmod{q-1}$. According to Parseval's equality (9.7),

$$C_f(\lambda) = 0 \iff \sum_{x \in F_q} \widehat{f}(x) \widehat{f}(\lambda x) = 0, \forall \lambda \notin \mathbb{F}_2.$$

In other words, a sequence has 2-level autocorelation if and only if its Hadamard transform has 2-level autocorrelation. A monomial (trace) function, $S_v(x) = Tr(x^v)$, where $\gcd(v, q-1) = 1$ and $0 < v < q-1$, is the trace representation of an m-sequence. This function is also referred to as a Singer function, because it gives the Singer Hadamard difference set. From the above discussion, to show that hyperoval sequences have 2-level autocorrelation, it suffices to show that the Hadamard transform of a hyperoval sequence is equal to the Hadamard transform of one of the m-sequences; that is,

$$\widehat{f}(\lambda) = \widehat{S}_v(\lambda^t), \text{ for some } v \text{ and } t. \tag{9.8}$$

If so,

$$\sum_x \widehat{S}_v(x) \widehat{S}_v(\lambda^t x) = \sum_x (-1)^{Tr(x^v) + Tr(\lambda^t x^v)} = 0.$$

The last identity follows from the autocorrelation of an m-sequence defined by $Tr(x^v)$. Together with Eq. (9.8), the 2-level autocorrelation property of hyperoval sequences can be established. We now establish Eq. (9.8).

Lemma 9.2 *Let* **a** *be a hyperoval sequence defined by a Segre, Glynn type 1, or Glynn type 2 hyperoval, and let* $f(x)$ *be its trace representation. Then*

$$\widehat{f}(\lambda) = \widehat{S}_k(\lambda^t), \text{ where } t = \frac{k-1}{k}, \; \forall \lambda \in \mathbb{F}_{2^n}.$$

Proof. By the construction, we have

$$(-1)^{f(x)} = \begin{cases} 1, & \text{if } x \in Im(\tau) \backslash \{0\}, \text{ recalling } Im(\tau) = \{x + x^k \,|x \in \mathbb{F}_q\}. \\ -1, & \text{otherwise.} \end{cases}$$

Thus, we may rewrite the Hadamard transform of $f(x)$ as

$$\widehat{f}(\lambda) = \sum_{x \in Im(\tau)} (-1)^{Tr(\lambda x)} - \sum_{x \notin Im(\tau)} (-1)^{Tr(\lambda x)} = 2 \times \sum_{x \in Im(\tau)} (-1)^{Tr(\lambda x)}.$$

The last identity follows from $\sum_{x \in Im(\tau)}(-1)^{Tr(\lambda x)} + \sum_{x \notin Im(\tau)}(-1)^{Tr(\lambda x)} = 0$.
Hence

$$\widehat{f}(\lambda) = \sum_{x \in F_q}(-1)^{Tr(\lambda(x^k + x))} \quad \text{(by the fact that } \tau \text{ is a 2-to-1 map)}$$

$$= \sum_{x \in F_q}(-1)^{Tr(\lambda x^k + \lambda x)}$$

$$= \sum_{y \in F_q}(-1)^{Tr(y^k + \lambda\lambda^{-k^{-1}}y)}$$

$$\text{(by } y^k = \lambda x^k, \gcd(k, q - 1) = 1 \implies x = \lambda^{-k^{-1}}y)$$

$$= \widehat{S}_k\left(\lambda^{\frac{k-1}{k}}\right). \qquad \square$$

Theorem 9.1 *Let* **a** *be a hyperoval sequence. Then* **a** *has 2-level autocorrelation.*

Proof. Let $\mathbf{a} \leftrightarrow f(x)$. From Lemma 9.2, we have

$$\widehat{f}(\lambda) = \widehat{S}_k(\lambda^t), \text{ where } t = \frac{k-1}{k}, \text{ and } \gcd(t, q - 1) = 1.$$

Consequently, for $\tau \neq 0$, we have

$$C_{\mathbf{a}}(\tau) + 1 = \sum_{x \in F_q}(-1)^{f(x) + f(\lambda x)}$$

$$= \sum_{x \in F_q} \widehat{f}(x)\widehat{f}(\lambda x)$$

$$= \sum_{x} \widehat{S}_k(x^t)\widehat{S}_k((\lambda x)^t)$$

$$= \sum_{y \in \mathbb{F}_q} \widehat{S}_k(y)\widehat{S}_k(\lambda^t y) \quad \text{(set } y = x^t)$$

Table 9.10. *Profile of the hyperoval sequences*

• Period $2^n - 1$, n odd, $n \geq 5$ for Segre type, $n \geq 7$ for Glynn type 1, and $n \geq 11$ for Glynn type 2
• 2-level autocorrelation
• Balance property
• $N(H, n)$, the number of shift distinct hyperoval sequences for each type, $N(H, n) = \frac{\varphi(2^n - 1)}{n}$

$$= \sum_x (-1)^{Tr(x^k) + Tr(\lambda^{tk} x^k)} \quad \text{(by Parseval's equality)}$$
$$= 0.$$

The last identity follows from $\lambda^{tk} \neq 0 \Longrightarrow C_a(\tau) + 1 = 0 \Longrightarrow C_a(\tau) = -1$.

$\square$

9.2.3 Profile of randomness of hyperoval sequences

In Table 9.10, we first present the randomness profile of hyperoval sequences and then formulae for computing the linear spans of these sequences. We omit a proof of the results on linear spans. Let l_n represent the number of trace terms in the trace representation of a hyperoval sequence of period $2^n - 1$. Then $\{l_n\}$ satisfies a linear recursive relation as shown in Table 9.11. Moreover, the linear span of the sequence is equal to nl_n. Prototypes of linear spans of hyperoval sequences for $5 \leq n \leq 41$ are listed in Table 9.12.

Example 9.6 Let $n = 9$. We have

$$\sigma = 2^{(9+1)/2} = 2^5 = 32, \quad \gamma = 2^{(3 \times 9 + 1)/4} = 2^7 = 128 \Longrightarrow$$

$$k = \sigma + \gamma = 32 + 128 = 160 \in C_5 = \{5, 10, 20, 40, 80, 160, 320, 129, 258\}$$

for Glynn type 1, and

$$k = 3\sigma + 4 = 100 \in C_{25} = \{25, 50, 100, 200, 400, 289, 67, 134, 268\}$$

for Glynn type 2, where the C_i's are cosets modulo 511 with coset leaders i. Let $f(x) = x^9 + x^4 + 1$, a primitive polynomial over $\mathbb{F}_2$, be a defining polynomial for $\mathbb{F}_{2^9}$, and let α be a root of $f(x)$ in $\mathbb{F}_{2^9}$; that is, $f(\alpha) = 0$. In Table 9.13, we

Table 9.11. *Formulae for the linear spans of hyperoval sequences*

Type	k	n odd	Linear span	Recursive formula
Segre	6	$n \geq 5$	nl_n	$l_n = l_{n-2} + l_{n-4} + 1, n \geq 7$ Initial state: $l_2 = 0, l_3 = l_4 = 1, l_5 = 3$
Glynn 1	$\sigma + \gamma$	$n \geq 7$	nl_n	$l_n = l_{n-2} + l_{n-4} + l_{n-6}$ $+ l_{n-8} + 1, n \geq 13$ Initial state: $l_5 = 1, l_7 = 3, l_9 = 7, l_{11} = 13$
Glynn 2	$3\sigma + 4$	$n \geq 11$	nl_n	$l_n = l_{n-2} + 3l_{n-4} - l_{n-6}$ $- l_{n-8} + 1, n \geq 11$ Initial state: $l_3 = 1, l_5 = 1, l_7 = 5, l_9 = 7$

Table 9.12. *Prototype of l_n for $5 \leq n \leq 41$*

n	l_n		
	Segre	Glynn 1	Glynn 2
5	3	1	1
7	5	3	5
9	9	7	7
11	15	13	21
13	25	25	37
15	41	49	89
17	67	95	173
19	109	183	383
21	177	353	777
23	287	681	1665
25	465	1313	3441
27	753	2531	7277
29	1219	4879	15159
31	1973	9405	31885
33	3193	18129	66645
35	5167	34945	139865
37	8361	67359	292757
39	13529	129839	613823
41	21891	250273	1285585

Table 9.13. *2-to-1 maps of monomial hyperovals in* $\mathbb{F}_{2^9}$

x	Segre τ_6	coset	Glynn 1 τ_{160}	coset	Glynn 2 τ_{100}	coset
1	508	127	381	223	221	183
3	129	5	140	35	339	117
5	61	61	59	59	25	25
7	238	119	101	83	195	27
9	319	127	304	19	168	21
11	419	61	125	125	18	9
13	269	27	350	175	479	255
15	450	23	32	1	44	11
17	366	183	495	255	376	47
19	442	183	284	57	413	119
21	405	87	392	35	252	63
23	507	255	470	183	411	111
25	12	3	270	29	392	35
27	15	15	253	191	454	55
29	350	175	176	11	235	183
31	309	107	394	43	451	31
35	507	255	4	1	484	79
37	314	117	106	53	266	21
39	440	55	301	91	93	93
41	267	23	328	41	475	223
43	192	3	415	127	280	35
45	98	35	175	175	476	119
47	308	77	163	53	331	93
51	371	119	502	223	157	117

x	Segre τ_6	coset	Glynn 1 τ_{160}	coset	Glynn 2 τ_{100}	coset
53	8	1	306	83	19	19
55	162	37	470	183	245	175
57	36	9	421	93	486	111
59	468	117	147	77	158	79
61	171	171	43	43	272	17
63	343	175	255	255	447	255
73	365	219	365	219	292	73
75	112	7	22	11	318	125
77	477	239	362	171	136	17
79	282	53	415	127	424	53
83	141	53	408	51	135	29
85	228	57	171	171	159	125
87	162	37	103	103	252	63
91	60	15	174	87	257	3
93	210	77	133	21	97	19
95	75	75	434	91	395	47
103	112	7	159	125	417	29
107	401	57	466	93	493	223
109	280	35	436	109	102	51
111	458	87	218	109	109	109
117	356	75	417	29	134	25
119	321	13	253	191	387	15
123	205	107	456	57	390	27
125	341	171	355	59	110	55

list 2-to-1 maps of Segre, and Glynn types 1 and 2, respectively, for indices taken on coset leaders. For the column under x, we list i where $x = \alpha^i$, i a coset leader modulo 511. In the column under $\tau_k(x) = x + x^k$, we list t such that $x + x^k = \alpha^t$ for $x = \alpha^i$, and in the column under coset, the number listed is the corresponding leader of the coset containing t such that $x + x^k = \alpha^t$. For example, for the first row, $x = \alpha$, $\tau_6(x) = x + x^6 = \alpha^{508} \implies t = 508$, and $508 \in C_{127} = \{508, 505, 499, 487, 463, 415, 319, 127, 254\}$ where 127 is a coset leader modulo 511. Note that the map $\tau_k(x) = x + x^k$ is also a 2-to-1 map on the set of coset leaders, and there are a total of 58 nonzero cosets modulo 511.

From Table 9.13, we construct Segre and Glynn types 1 and 2 sequences, which are listed in Table 9.14. (As usual, we only list the values at the coset leaders. The other values can be obtained by the constant-on-cosets property.) For $n = 9$, we have m-sequences; because 3 divides 9 and 9 is odd, we have GMW sequences; and we also have 3-term sequences. We also list these three sequences in Table 9.14 where m-sequence A and GMW sequence B are those from Example 5.11 in Chapter 5 and Example 8.6 in Chapter 8, respectively; and 3-term sequence C is given by $\{Tr(T3(\alpha^i))\}$, where $T3(x) = x + x^{17} + x^{25}$. The trace spectra of these sequences are listed in Table 9.15.

In the case $n = 9$, Glynn type 1 sequences and Glynn type 2 sequences are decimation equivalent. We will show this as follows. Let $G_i(x)$ be the trace representation of Glynn type i sequences in Table 9.15. Let

$$G_1 = \{1, 5, 9, 13, 19, 37, 43\}, \text{ and}$$
$$G_2 = \{17, 23, 37, 43, 45, 75, 87\}.$$

Then

$$G_1(x) = \sum_{i \in G_1} Tr(x^i) = Tr(x + x^5 + x^9 + x^{13} + x^{19} + x^{37} + x^{43})$$
$$G_2(x) = \sum_{i \in G_2} Tr(x^i) = Tr(x^{17} + x^{23} + x^{37} + x^{43} + x^{45} + x^{75} + x^{87}).$$

To show that E and F are decimation equivalent, we only need to show that there exists some integer r such that $G_1 = rG_2$. In the following, we show that $G_1 = 45^{-1}G_2$. Note that

$$45^{-1} \equiv 159 \equiv 125 \times 2^7 \pmod{511}. \tag{9.9}$$

We compute

$$159G_2 = \{159i \pmod{511} \mid i \in G_2\} = \{148, 80, 262, 194, 1, 172, 36\}$$

Table 9.14. *A complete list of 2-level autocorrelation sequences of period 511*

Coset i	A	B	C	D	E	F	Coset i	A	B	C	D	E	F
0	1	1	1	1	1	1							
1	0	1	1	0	0	1	63	0	0	1	1	1	0
3	0	1	0	0	1	0	73	0	1	1	1	1	0
5	1	0	1	0	1	1	75	1	0	0	0	1	1
7	0	1	1	0	1	1	77	1	0	0	0	0	1
9	1	0	1	0	1	0	79	1	0	0	1	1	0
11	0	1	0	1	0	0	83	1	0	0	1	0	1
13	0	1	1	0	1	1	85	0	1	0	1	1	1
15	1	1	0	0	1	0	87	0	1	1	0	0	1
17	0	0	0	1	1	0	91	0	1	0	1	0	1
19	1	0	1	1	0	0	93	0	1	1	1	0	0
21	0	1	0	1	0	0	95	1	0	0	1	1	1
23	1	1	1	0	1	1	103	0	1	0	1	0	1
25	1	0	0	1	1	0	107	0	0	1	0	1	1
27	1	0	1	0	1	0	109	1	1	0	1	0	0
29	1	0	0	1	0	0	111	1	0	0	1	1	0
31	0	1	0	1	1	0	117	1	0	1	0	1	0
35	1	0	1	0	0	0	119	0	1	0	0	1	0
37	0	1	0	0	1	1	123	0	1	0	1	1	1
39	1	1	1	1	1	1	125	0	1	0	1	0	0
41	1	0	1	1	0	1	127	0	1	1	0	0	1
43	1	0	1	1	0	1	171	1	1	0	0	0	1
45	0	0	0	1	1	1	175	1	1	0	0	0	0
47	1	0	1	1	1	0	183	1	0	1	0	0	0
51	1	0	0	1	0	0	187	1	0	1	1	1	1
53	0	0	0	0	0	0	191	0	0	1	0	0	1
55	0	1	1	0	0	0	219	1	0	0	0	0	1
57	0	1	1	0	0	1	223	1	1	1	1	0	0
59	0	0	1	1	0	1	239	1	0	1	0	1	1
61	0	1	1	0	1	1	255	0	1	0	0	0	0

Table 9.15. *Trace representations of the sequences in Table 9.14*

Types	Trace spectra	Linear span
m-sequence A	1	9
GMW B	3, 17, 129	27
3-term C	1, 17, 25	27
Segre D	1, 5, 7, 9, 19, 25, 37, 77, 117	81
Glynn 1 E	1, 5, 9, 13, 19, 37, 43	63
Glynn 2 F	17, 23, 37, 43, 45, 75, 87	63

and

$$37 \times 4 \equiv 148 \ (\text{mod } 511), \quad 5 \times 2^4 \equiv 80 \ (\text{mod } 511),$$
$$13 \times 2^8 \equiv 262 \ (\text{mod } 511), \quad 19 \times 2^6 \equiv 194 \ (\text{mod } 511),$$
$$43 \times 2^2 \equiv 172 \ (\text{mod } 511), \quad 9 \times 2^2 \equiv 36 \ (\text{mod } 511).$$

Therefore, $G_1 = 45^{-1} G_2$. Together with Eq. (9.9), we have

$$G_2(x^{125}) = G_1(x).$$

In other words, from the Glynn type 2 sequence F listed in Table 9.14, by applying the 125-decimation operator to it, we obtain the Glynn type 1 sequence E. These are all the five different classes of 2-level autocorrelation sequences of period 511 up to decimation equivalence. The sequences C, D, and E were obtained by exhaustive search before general constructions for 3-term sequences and hyperoval sequences were known.

9.3 Kasami power function construction

This is a remarkable contribution to this topic. Within a few months of the discovery of WG sequences in 1998, as described in Section 9.2, No, Chung, and Yun found another representation of these sequences using the Kasami power function (KPF) experimentally. Shortly after that, Dobbertin extended this approach to a general construction of 2-level autocorrelation sequences. (Because we have already shown the one-to-one correspondence between 2-level autocorrelation sequences and Hadamard difference sets, we need not distinguish between them. Sometimes, we may use both terms interchangeably.)

In this section, we first introduce the Kasami power function and its almost perfect nonlinear property. We then present the Kasami power function construction for WG sequences and Dobbertin's construction in the following two subsections.

Definition 9.1 *The power function x^d on F_q ($q = 2^n$) is called a Kasami power function and d a Kasami exponent iff*

$$d = 2^{2k} - 2^k + 1 \ \text{where } k < n \text{ and } \gcd(k, n) = 1.$$

In this section, d will always take this value. In 1969, Kasami showed that, in a coding context, for n odd, the Hadamard transform of $f(x) = Tr(x^d)$ has the following 3-valued distribution:

$$\widehat{f}(\lambda) \in \left\{0, \pm 2^{\frac{n+1}{2}}\right\}. \tag{9.10}$$

In 1998, for the special case of $3k \equiv 1 \pmod{n}$, Dillon determined the values of λ such that $\widehat{f}(\lambda) = 0$. Precisely, let n be odd and $3k \equiv 1 \pmod{n}$. Then for any $\lambda \in \mathbb{F}_{2^n}$,

$$\widehat{S}_d(\lambda)^2 = \begin{cases} 0 & \Leftrightarrow Tr(\lambda^{2^k+1}) = 0 \\ 2^{n+1} & \Leftrightarrow Tr(\lambda^{2^k+1}) = 1. \end{cases} \tag{9.11}$$

Equivalently, for all $\lambda \in F_q$,

$$\widehat{S}_d(\lambda)^2 = q(-S_{2^k+1} + 1)(\lambda), \tag{9.12}$$

where the constant function $1(\lambda) = 1 \ \forall \lambda \in \mathbb{F}_q$.

Definition 9.2 *Let $F(x)$ be a function on F_q. $F(x)$ is called an* almost perfect nonlinear *function (APN) if, for all $a, b \in \mathbb{F}_q^*, b \in \mathbb{F}_q$, the equation*

$$F(x) + F(x + a) = b \tag{9.13}$$

has either no solutions or exactly two solutions in $\mathbb{F}_q$.

In other words, $F(x)$ is APN if and only if $F(x) + F(x + a)$ is a 2-to-1 map on $\mathbb{F}_q$ for any $a \in \mathbb{F}_q^*$. This is also equivalent to $F(x) + F(x + 1)$ is a 2-to-1 map on $\mathbb{F}_q$. If $F(x) = x^d$, then Eq. (9.13) becomes

$$(x + a)^d + x^d = b,$$

which has either no solutions or two solutions in F_q. It is known that for n odd, if the Hadamard transform of $f(x) = Tr(x^d)$ satisfies Eq. (9.10), then x^d is an APN function. Together with Kasami's result on the Hadamard transform of $Tr(x^d)$ in Eq. (9.10), we know that x^d is an APN function for n odd. Therefore, $x^d + (x + 1)^d$ is a 2-to-1 map on $\mathbb{F}_q$. However, for the case of n even, it is not easy to show that x^d is APN on $\mathbb{F}_q$. This was established by Dobbertin (1999).

Fact 9.1 *The Kasami power function is APN on $\mathbb{F}_q$, or equivalently, $x^d + (x + 1)^d$ is a 2-to-1 map on $\mathbb{F}_q$.*

9.3.1 The Kasami power function construction
for WG sequences

Let

$$\sigma(x) = (x + 1)^d + x^d, \quad 3k \equiv 1 \pmod{n}.$$

Then $\sigma(x)$ is a 2-to-1 map on $\mathbb{F}_q$ ($q = 2^n$) from Fact 9.1. Let α be a primitive element of F_q, and define

$$N = \begin{cases} Im(\sigma) = \{\sigma(x) \mid x \in \mathbb{F}_q\} & \text{if } n \text{ is even} \\ \mathbb{F}_q \setminus Im(\sigma) & \text{if } n \text{ is odd.} \end{cases} \tag{9.14}$$

Let $\mathbf{a} = \{a_i\}$ whose elements are given by

$$a_i = \begin{cases} 0, & \text{if } \alpha^i \in N \\ 1, & \text{otherwise.} \end{cases} \tag{9.15}$$

Then $\mathbf{a}$ has 2-level autocorrelation. This result was verified by No, Chung, and Yun (1998) for $5 \leq n \leq 23$. However, it happens that $\mathbf{a}$ is just the WG sequence constructed in Section 9.2. Dillon (1999) gave a proof for the case of n odd, and then Dillon and Dobbertin (2004) showed this result for the case of n even. (In Dillon and Dobbertin (2004), the authors were mainly devoted to establishing the 2-level autocorrelation property of the Kasami power function sequences defined in the next subsection.) All of these proofs have a similar flavor, that is, showing that the Hadamard transform of the sequence is equal to a decimation of one of the Hadamard transforms of the Singer function $S_v(x)(= Tr(x^v))$, as was done for the hyperoval sequences. In the following, we will use $N(x)$ to denote the trace representation of the sequence $\mathbf{a}$ in Eq. (9.15).

Lemma 9.3 *Suppose that $n > 5$ is odd and $3k \equiv 1 \pmod{n}$. Then for all $\lambda \in \mathbb{F}_q$,*

$$\widehat{N}(\lambda) = \widehat{S_{2^k+1}}(\lambda^{d^{-1}}), \forall \lambda \in \mathbb{F}_q.$$

Proof. Since n is odd, we have $\gcd(d, 2^n - 1) = 1$. Recall that $\sigma(x) = (x + 1)^d + x^d$ is a 2-to-1 map on $\mathbb{F}_q$. Applying Eq. (9.15), and the involution and convolution laws in Chapter 6, we compute the Hadamard transform of $N(x)$ for $\lambda \in \mathbb{F}_q^*$ as follows:

$$\widehat{N}(\lambda^d) = \sum_{x \in \mathbb{F}_q} \chi(Tr(\lambda^d x))\chi(N(x)).$$

Since n is odd, according to Eq. (9.14) we have

$$\chi(N(x)) = \begin{cases} -1, & y \in Im(\sigma) = \{(x + 1)^d + x^d \mid x \in \mathbb{F}_q\} \\ 1, & \text{otherwise.} \end{cases}$$

Therefore, for $\lambda \neq 0$,

$$\widehat{N}(\lambda^d) = -\sum_{x \in Im(\sigma)} \chi(Tr(\lambda^d x)) + \sum_{x \notin Im(\sigma)} \chi(Tr(\lambda^d x))$$

$$= -2 \sum_{x \in Im(\sigma)} \chi(Tr(\lambda^d x))$$

$$\left(\text{because } \sum_{x \in Im(\sigma)} \chi(Tr(\lambda^d x)) + \sum_{x \notin Im(\sigma)} \chi(Tr(\lambda^d x)) = 0\right)$$

$$= -\sum_{x\in\mathbb{F}_q} \chi(Tr(\lambda^d((x+1)^d + x^d)))$$

(because $(x+1)^d + x^d$ is a 2-to-1 map)

$$= -\sum_{x\in\mathbb{F}_q} \chi(Tr((\lambda x + \lambda)^d + (\lambda x)^d))$$

$$= -\sum_{y\in\mathbb{F}_q} \chi(Tr((y+\lambda)^d + y^d))$$ (variable change by setting $y = \lambda x$)

$$= -\sum_{y\in\mathbb{F}_q} S_d(y+\lambda)S_d(y)$$

$$= -S_d * S_d(\lambda)$$

$$= -\widehat{S_d * S_d}(\lambda)$$ (Involution Law in Chapter 6)

$$= -\widehat{S_d^2}$$ (Proposition 4 in Chapter 6)

$$= -(-\widehat{S_{2^k+1}} + \widehat{1})(\lambda)$$ ((9.12))

$$= \widehat{S_{2^k+1}}(\lambda)$$ (by $\widehat{1}(\lambda) = \sum_{x\in\mathbb{F}_q}(-1)^{Tr(\lambda x)+1} = 0$ if $\lambda \neq 0$).

For $\lambda = 0$, $|N| = 2^{n-1}$ implies $\widehat{N}(0) = 0$. Thus the equation holds for all $\lambda \in \mathbb{F}_q$. $\square$

Theorem 9.2 **a**, *defined by Eq. (9.15), has 2-level autocorrelation (i.e., N is a cyclic Hadamard difference set).*

The odd case follows from Lemma 9.3 by arguments similar to those used for Theorem 9.1. For the even case, the proof follows a similar approach to the proof for the odd case. However, for n even, the Hadamard transform is equal to a sum of Hadamard transforms of $Tr(\gamma x^v)$ for some $\gamma \in \mathbb{F}_q$. We will not give a detailed explanation here. The reader is referred to Dillon and Dobbertin (2004).

Since **a** is a WG sequence, the profile of randomness of **a** was given in Section 9.2. Although this construction does not give new sequences, the construction given by Dobbertin, which will be introduced in the coming subsection, was inspired by this representation.

9.3.2 The Kasami power function construction for B_k

Construction

Let $\gcd(k, n) = 1$, $k < n$ and

$$\Delta_k(x) = (x+1)^d + x^d + 1, x \in \mathbb{F}_q, \tag{9.16}$$

where d is the Kasami exponent, defined at the beginning of this section. Then $\Delta_k(x)$ is a 2-to-1 map on $\mathbb{F}_q$ from Fact 9.1. Let

$$B_k = Im(\Delta_k) = \{\Delta_k(x)|x \in \mathbb{F}_q\}$$
$$C_k = \{x \in \mathbb{F}_q | x^{2^k+1} \in B_k\}.$$

We define two sequences $\mathbf{b} = \{b_i\}$ and $\mathbf{c} = \{c_i\}$ associated with the sets B_k and C_k by

$$b_i = \begin{cases} 0, & \alpha^i \in B_k \\ 1, & \text{otherwise} \end{cases} \quad \text{and} \quad c_i = \begin{cases} 0, & \alpha^i \in C_k \\ 1, & \text{otherwise}. \end{cases} \tag{9.17}$$

Then the two sequences $\mathbf{c}$ and $\mathbf{b}$ are related by $c_i = b_{i(2^k+1)}, 0 < i < 2^n - 1$. In other words, $\mathbf{c}$ is the $(2^k + 1)$-decimation of $\mathbf{b}$. If n is odd, then $\gcd(2^k + 1, 2^n - 1) = 1$, which implies that

$$\boxed{\mathbf{b} \text{ has 2-level auto-correlation } \Leftrightarrow \mathbf{c} \text{ has 2-level auto-correlation.}}$$

If n is even, then $\gcd(2^k + 1, 2^n - 1) > 1$ and we need more features of $\mathbf{b}$ to determine its 2-level auto-correlation property.

　　Dobbertin conjectured that $\mathbf{b}$ has 2-level autocorrelation based on the following evidences:

1. If $k' = 2$ for $kk' \equiv 1 \pmod{n}$, then $\mathbf{b}$ is a 3-term sequence.
2. If $k' = 3$ for $kk' \equiv 1 \pmod{n}$, then $\mathbf{b}$ is a 5-term sequence.
3. The Kasami power function construction for WG sequences.

　　We will not give a complete proof for the result that $\mathbf{b}$ has 2-level autocorrelation. Instead, we will present an outline of the proof given by Dillon and Dobbertin to demonstrate the power of the Parseval equation. The proof has the same flavor as that of Lemma 9.2 for hyperoval sequences and Lemma 9.3 for the KPF construction of WG sequences. We will use $C_k(x)$ to denote the trace representation of $\{c_i\}$.

Lemma 9.4 *Let $n > 5$ be odd. Then*

$$\widehat{C_k}(\lambda) = \widehat{S_3}\left(\lambda^{\frac{2^k+1}{3}}\right) \text{ for all } \lambda \in \mathbb{F}_q.$$

The following functions will be used in the proof.

1. A multiplicity function for a map $\eta : \mathbb{F}_q \rightarrow \mathbb{F}_q$,

$$M_\eta(v) = |\{x \in \mathbb{F}_q | \eta(x) = v\}|, v \in \mathbb{F}_q;$$

that is, M_η is the size of the pre-image of v, and

$$M_\eta : \mathbb{F}_q \to \mathbb{C},$$

where $\mathbb{C}$ is the complex number field.

2. Dickson polynomials, defined by

$$D_{2^k+1}(x) = x^{2^k+1} + D_{2^k-1}(x)$$

$$D_{2^k-1}(x) = \sum_{i=0}^{k-1} x^{2^k+1-2^i}.$$

If $\gcd(k, n) = 1$, then

$$M_{D_3} = \begin{cases} M_{D_{2^k-1}} & \text{for } k \text{ even} \\ M_{D_{2^k+1}} & \text{for } k \text{ odd.} \end{cases} \tag{9.18}$$

(Note that $D_3(x) = x + x^3$.)

3. MCM polynomials: Define

$$f_k(x) = \sum_{i=0}^{k-1} x^{(2^k+1)2^i-2^k}. \tag{9.19}$$

If $\gcd(k, n) = 1$ and k is odd, then f_k is a permutation polynomial on $\mathbb{F}_q$ (Cohen and Matthews 1994). If $\gcd(k, n) = 1$ and k is even, then f_k is a 2-to-1 map on $\mathbb{F}_q$.

An outline of a proof of Lemma 9.4

Since $\gcd(k, n) = 1$ and n is odd, either k or $n - k$ is even. Without loss of generality, we may assume that k is even. From the assertion about MCM polynomials, in this case, f_k is a 2-to-1 map on $\mathbb{F}_q$.

Step 1. It can easily be verified that

$$f_k\big((z^2 + z)^{\frac{1}{2^k+1}}\big) = \Delta_k(z)^{\frac{1}{2^k+1}} \implies Im(f_k) = C_k.$$

Consequently,

$$C_k(v) = M_{f_k(v)} - 1, \quad \text{for all } v \in \mathbb{F}_q$$

$$\iff \widehat{C_k}(\lambda) = (\widehat{M_{f_k} - 1})(\lambda) = \widehat{M_{f_k}}(\lambda) - \widehat{1}(\lambda) = \widehat{M_{f_k}}(\lambda) \text{ for } \lambda \neq 0.$$

In other words,

$$\widehat{C_k}(\lambda) = \widehat{M_{f_k}}(\lambda), \ \lambda \in \mathbb{F}_q^*. \tag{9.20}$$

Step 2. Establish that

$$\widehat{M_{f_k}}(\lambda) = \widehat{M_{D_{2^k-1}}}(\lambda^{2^k+1}), \ \lambda \in \mathbb{F}_q^*; \tag{9.21}$$

that is, the multiplicity of MCM polynomial f_k is equal to the multiplicity of the Dickson polynomial with the variable raised to the power $2^k + 1$.

Step 3. For $\lambda \neq 0$, we may compute $\widehat{C_k}(\lambda)$ as follows.

$$
\begin{aligned}
\widehat{C_k}(\lambda) &= \widehat{M_{f_k}}(\lambda) &&\text{(by (9.20))} \\
&= \widehat{M_{D_{2^k-1}}}(\lambda^{2^k+1}) &&\text{(by (9.21))} \\
&= \widehat{M_{D_3}}(\lambda^{2^k+1}) &&\text{(by (9.18))} \\
&= \sum_{v \in \mathbb{F}_q} \chi(Tr(\lambda^{2^k+1}v))M_{D_3}(v) \\
&= \sum_{x \in \mathbb{F}_q} \chi(Tr(\lambda^{2^k+1}D_3(x))) \\
&= \sum_{x \in \mathbb{F}_q} \chi(Tr(\lambda^{2^k+1}(x^3 + x))) \\
&= \sum_{x \in \mathbb{F}_q} \chi\left(Tr\left((\lambda^{\frac{2^k+1}{3}}x)^3 + \lambda^{2 \cdot \frac{2^k+1}{3}}(\lambda^{\frac{2^k+1}{3}}x)\right)\right) \\
&= \sum_{y \in \mathbb{F}_q} \chi\left(Tr(y^3 + \lambda^{2 \cdot \frac{2^k+1}{3}}y)\right) &&\left(y = \lambda^{\frac{2^k+1}{3}}x\right) \\
&= \widehat{S_3}\left(\lambda^{2 \cdot \frac{2^k+1}{3}}\right) = \widehat{S_3}\left(\lambda^{\frac{2^k+1}{3}}\right). &&\square
\end{aligned}
$$

Theorem 9.3 **b** *has 2-level autocorrelation.*

The odd case follows directly from Lemma 9.4 and an argument similar to the one used in the proof of Theorem 9.1. The even case has a flavor similar to the odd case, but is more complicated. We omit the detailed discussion here.

Profile of the Kasami power construction B_k

1. Period: $2^n - 1$.
2. Balanced.
3. 2-level autocorrelation.
4. $N(KPF, n)$, the number of nonequivalent classes:

$$
N(KPF, n) = \frac{\phi(n)}{2} \cdot \frac{\phi(2^n - 1)}{n}.
$$

Here, when $n/2 < k < n$, $n - k$ gives the same class as k with $1 < k < n/2$.
5. The linear span of **b** is given by

$$
LS = n(2F_{k_1} - 1), \quad k_1 = \min(k', n - k'), n \geq 6, \tag{9.22}
$$

where $kk' \equiv 1 \pmod{n}$ and F_i is the ith Fibonacci number, given by

$$F_0 = F_1 = 1, \ F_{i+2} = F_{i+1} + F_i, \ i \geq 0.$$

6. The trace representation is $Tr(T^{-1}(x^{-1}))$ where $T(x)$ is defined by

$$T(x) = \frac{\sum_{i=1}^{k'} x^{2^{ik}} + \epsilon}{x^{2^k+1}}$$

and

$$\epsilon = \begin{cases} 0 & \text{if } k' \text{ is odd} \\ 1 & \text{if } k' \text{ is even.} \end{cases}$$

$T(x)$ is a permutation polynomial on $\mathbb{F}_q$ and $T^{-1}(x)$ is its inverse. Let $R(x) = T^{-1}(x)$. Then $R(x)$ is given by the following recursive formula:

$$R(x) = \sum_{i=1}^{k'} A_i(x) + V_{k'}(x), \tag{9.23}$$

where A_i and V_i are iteratively defined by

$$A_1(x) = x$$
$$A_2(x) = x^{2^k+1}$$
$$A_{i+2}(x) = x^{2^{(i+1)k}} A_{i+1}(x) + x^{2^{(i+1)k}-2^{ik}} A_i(x), i \geq 1$$

and

$$V_1(x) = 0$$
$$V_2(x) = x^{2^k-1}$$
$$V_{i+2}(x) = x^{2^{(i+1)k}} V_{i+1}(x) + x^{2^{(i+1)k}-2^{ik}} V_i(x), i \geq 1.$$

In Table 9.16, we list the first five terms of A_i and V_i.

Example 9.7 We consider the following special cases.

1. $k' = 2 \Longrightarrow 2k \equiv 1 \pmod{n}$; n must be odd $\Longrightarrow k = (n+1)/2 \Longrightarrow n = 2k - 1$. Note that $B_k = B_{n-k}$. Thus, the trace representation for **b** is given by

$$\mathbf{b} \leftrightarrow f(x) = Tr(R(x)), \ \text{where} \ R(x) = A_1 + A_2 + V_2 = x + x^{2^k+1} + x^{2^k-1},$$

which is a permutation polynomial on $\mathbb{F}_q$. Let $Tr(g(x))$ be the trace representation of the three-term sequence; that is,

$$g(x) = x + x^{q_1} + x^{q_2}.$$

where $q_1 = 1 + 2^m$, and $q_2 = 1 + 2^m + 2^{m-1}$ where $m = k + 1$. Since q_1 and $2^k + 1$ are in the same coset, $Tr(x^{q_1}) = Tr(x^{2^k+1})$. As mentioned in

Table 9.16. $\{A_i\}$ and $\{V_i\}$ for $i \leq 5$

$A_1(x) = x$
$A_2(x) = x^{2^k+1}$
$A_3(x) = x^{2^{2k}} A_2(x) + x^{2^{2k}-2^k} A_1(x)$
$\quad = x^{2^{2k}+2^k+1} + x^{2^{2k}-2^k+1}$
$A_4(x) = x^{2^{3k}} A_3(x) + x^{2^{3k}-2^{2k}} A_2(x)$
$\quad = x^{2^{3k}+2^{2k}+2^k+1} + x^{2^{3k}+2^{2k}-2^k+1} + x^{2^{3k}-2^{2k}+2^k+1}$
$A_5(x) = x^{2^{4k}} A_4(x) + x^{2^{4k}-2^{3k}} A_3(x)$
$\quad = x^{2^{4k}+2^{3k}+2^{2k}+2^k+1} + x^{2^{4k}+2^{3k}+2^{2k}-2^k+1} + x^{2^{4k}+2^{3k}-2^{2k}+2^k+1}$
$\quad\quad + x^{2^{4k}-2^{3k}+2^{2k}+2^k+1} + x^{2^{4k}-2^{3k}+2^{2k}-2^k+1}$
$V_1(x) = 0$
$V_2(x) = x^{2^k-1}$
$V_3(x) = x^{2^{2k}} V_2(x) + x^{2^{2k}-2^k} V_1(x)$
$\quad = x^{2^{2k}+2^k-1}$
$V_4(x) = x^{2^{3k}} V_3(x) + x^{2^{3k}-2^{2k}} V_2(x)$
$\quad = x^{2^{3k}+2^{2k}+2^k-1} + x^{2^{3k}-2^{2k}+2^k-1}$
$V_5(x) = x^{2^{4k}} V_4(x) + x^{2^{4k}-2^{3k}} V_3(x)$
$\quad = x^{2^{4k}+2^{3k}+2^{2k}+2^k-1} + x^{2^{4k}+2^{3k}-2^{2k}+2^k-1} + x^{2^{4k}-2^{3k}+2^{2k}+2^k-1}$

Section 9.2, $q_2 \equiv q_1^2 \pmod{2^n - 1}$. By simple calculation, $(1 + 2^k)^{-1} \equiv 2^k - 1 \pmod{2^n - 1}$. Thus $Tr(R(x)) = Tr(g(x^{2^k-1}))$; that is, this construction also produces 3-term sequences. Thus for $k' = 2$, this construction does not produce new sequences.

2. $k' = 3 \implies 3k \equiv 1 \pmod{n} \implies n \not\equiv 0 \pmod 3$. From Table 9.16, we have

$$R(x) = A_1 + A_2 + A_3 + V_3$$
$$= x + x^{2^k+1} + x^{2^{2k}+2^k+1} + x^{2^{2k}-2^k+1} + x^{2^{2k}+2^k-1}$$

and the trace representation of **b** is as follows:

$$Tr(R(x)) = Tr(x + x^{2^k+1} + x^{2^{2k}+2^k+1} + x^{2^{2k}-2^k+1} + x^{2^{2k}+2^k-1}),$$

which gives the 5-term sequences constructed in Section 9.2. (*Note.* Here the exponents in the monomial trace terms of a 5-term sequence are represented in a unified form for both cases $n = 3k - 1$ and $n = 3k - 2$.) In addition, Dobbertin showed that

3. The trace representation of **a**, defined by Eq. (9.15), is equal to

$$Tr(R(x + 1) + 1),$$

where $Tr(R(x))$ is the trace representation of **b** for $k' = 3$. Because the trace representation of a sequence is unique, from the above assertion 2 we have $N = WG$.

4. $k = 2$. This case yields the same sequence as the one constructed from the Segre hyperoval construction; that is, $B_2 = Segre$.

In the following, we present some examples of new sequences constructed from the Kasami power function construction B_k.

Example 9.8 Let $n = 11$. In this case, from previous known constructions, we have

m-seq	Segre
3-term	Glynn type 1
5-term	Glynn type 2
WG	

From the KPF construction B_k, because $k'k \equiv 1 \pmod{11}$, we have $k' = 2, 3, 4$, and 5, corresponding to the following classes:

$$k' = 2 \leftrightarrow k \equiv 6 (\text{mod } 11), \leftrightarrow n - k = 5 \leftrightarrow \text{3-term}$$
$$k' = 3 \leftrightarrow k \equiv 4 (\text{mod } 11), \qquad \leftrightarrow \text{5-term}$$
$$k' = 4 \leftrightarrow k \equiv 3 (\text{mod } 11), \qquad \leftrightarrow \text{new}$$
$$k' = 5 \leftrightarrow k \equiv 9 (\text{mod } 11), \leftrightarrow n - k = 2 \leftrightarrow \text{Segre}$$

Thus, we obtain one more class of 2-level autocorrelation sequences. We list all eight of these classes of 2-level autocorrelation sequences of period 2047 in Table 9.23 in the Appendix of this chapter.

For $n = 12$, since 12 is even and $12 \equiv 0 \pmod{3}$, we only have m-sequences and GMW sequences (4 types) from previous constructions. There are 13 classes of those sequences. However, from the B_k construction where $k'k \equiv 1 \pmod{12}$, we have one case $k = 5$ that gives one more class of 2-level autocorrelation sequences of period 4095. Thus, there are a total of 14 classes of 2-level autocorrelation sequences of period 4095 from the constructions known so far. These sequences are listed in Table 9.24 in the Appendix.

9.4 The iterative decimation Hadamard transform

From Sections 9.2 and 9.3 of this chapter, all newly discovered 2-level autocorrelation sequences of period $2^n - 1$ have a common remarkable property; that is, their Hadamard transform is equal to the Hadamard transform of some m-sequence when n is odd. This leads to a new construction of sequences with

2-level autocorrelation. In this section, we will discuss this relation in a much broader setting. First, elements of sequences can be taken from $\mathbb{F}_p$ where p is a prime, and we write $q = p^n$. Recall that $\chi(x)$ is the additive character of $\mathbb{F}_p$, defined by $\chi(x) = \omega^x$, $x \in \mathbb{F}_p$ where ω is a primitive pth root of unity.

9.4.1 Definition

Definition 9.3 *Let* $h(x) : \mathbb{F}_q \longrightarrow \mathbb{F}_p$ *be orthogonal over* $\mathbb{F}_p$, **a** *a sequence over* $\mathbb{F}_p$ *with period* $N \mid (q-1)$, *and* $f(x)$ *the trace representation of* **a**. *Thus,* $f(x)$ *is a function from* $\mathbb{F}_q$ *to* $\mathbb{F}_p$. *For an integer* $0 < v < q - 1$, *and* $\lambda \in \mathbb{F}_q$, *we define*

$$\widehat{f_h}(v)(\lambda) = \sum_{x \in \mathbb{F}_q} \chi(h(\lambda x)) \chi^*(f(x^v)) = \sum_{x \in \mathbb{F}_q} \omega^{h(\lambda x) - f(x^v)}. \qquad (9.24)$$

$\widehat{f_h}(v)(\lambda)$ *is called* the first-order decimation-Hadamard transform (DHT) *of* $f(x)$ *(or* **a***) with respect to* $h(x)$, *or* the first-order DHT *for short.*

We may consider the first-order DHT as the Hadamard transform of $f(x^v)$ with respect to $h(x)$. Thus, the inverse Hadamard transform of $f(x^v)$ is given by

$$\chi(f(\lambda^v)) = \frac{1}{q} \sum_{y \in \mathbb{F}_q} \chi(h(\lambda y))(\widehat{f_h}(v)(y))^*$$

$$= \frac{1}{q} \sum_{x, y \in \mathbb{F}_q} \omega^{h(\lambda y) - h(yx) + f(x^v)}, \ \lambda \in \mathbb{F}_q \qquad (9.25)$$

Note that $f(x^v)$ is the trace representation of the v-decimation of sequence **a**. If we take $h(x) = Tr(x)$, then $\widehat{f_h}(v)(\lambda)$, the first-order DHT, is the Hadamard transform of $f(x^v)$, defined in Chapter 6. Hence, the computing process of $\widehat{f_h}(v)(\lambda)$ consists of first applying the decimation operation (corresponding to the sequence) and then the Hadamard transform. Note that here the trace function $Tr(x)$ can be replaced by any orthogonal function.

Definition 9.4 *With the above notation, let* $0 < t < q - 1$. *We define*

$$\widehat{f_h}(v, t)(\lambda) = \sum_{y \in \mathbb{F}_q} \chi(h(\lambda y))(\widehat{f_h}(v)(y^t))^*$$

$$= \sum_{x, y \in \mathbb{F}_q} \omega^{h(\lambda y) - h(y^t x) + f(x^v)}, \ \lambda \in \mathbb{F}_q \qquad (9.26)$$

$\widehat{f_h}(v, t)(\lambda)$ *is called the* second-order decimation-Hadamard transform *of* $f(x)$ *(with respect to* $h(x)$*), or the* second-order DHT *for short.*

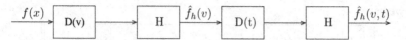

Figure 9.8. Processing of the second-order DHT.

The second-order DHT applies the first-order DHT twice, as illustrated in Figure 9.8.

Remark 9.1 If $t = 1$, then the second-order DHT divided by $1/q$ is just the inverse Hadamard transform of $f(x^v)$ given by Eq. (9.25).

In general, for any integer pair (v, t), for $x \in \mathbb{F}_q$, a value of $\widehat{f_h}(v, t)(x)$, the second-order DHT may be just a complex number. However, if it satisfies the following condition

$$\widehat{f_h}(v, t)(x) \in \{q\omega^i \mid i = 0, \ldots, p - 1\}, \forall x \in \mathbb{F}_q,$$

then we can construct a function, say $g(x)$, from $\mathbb{F}_q$ to $\mathbb{F}_p$, whose elements are given by

$$\chi(g(x)) = \frac{1}{q}\widehat{f_h}(v, t)(x), x \in \mathbb{F}_q. \tag{9.27}$$

In this case, we say that (v, t) is realizable and $g(x)$ is a realization of $f(x)$. In particular, if $p = 2$, then (v, t) is realizable if and only if

$$\widehat{f_h}(v, t)(x) \in \{\pm q\}, \forall x \in \mathbb{F}_q$$

and the realization $g(x)$ is determined by

$$(-1)^{g(x)} = \frac{1}{q}\widehat{f_h}(v, t)(x), x \in \mathbb{F}_q.$$

Before we present the significance of this realization, we give some examples to explain these concepts.

Example 9.9 Let $p = 2$, $n = 4$, $h(x) = f(x) = Tr(x)$, $\mathbb{F}_{2^4}$ be defined by $t(x) = x^4 + x + 1$, and α a root of $t(x)$ in $\mathbb{F}_{2^4}$. Let

$$f(x) \leftrightarrow \mathbf{a} = 000100110101111.$$

Since the first-order DHT of $f(x)$ (or $\mathbf{a}$) is given by

$$\widehat{f_h}(v)(\lambda) = \sum_{x \in \mathbb{F}_{2^4}} (-1)^{Tr(\lambda x) + Tr(x^v)},$$

we have

v	$\{\widehat{f_h}(v)(\alpha^i)\}, i = 0, 1, \ldots, s-1$	$s = \dfrac{15}{\gcd(v,15)}$
3	$8, 0, 0, 0, 0$	5
5	$0, 0, 0$	3
7	$0, 0, 0, 4, 0, 8, 4, -4, 0, 4, 8, -4, 4, -4, -4$	15

The second-order DHTs, $\widehat{f_h}(7, 7)$ and $\widehat{f_h}(7, 5)$, are given by

$$\widehat{f_h}(7, t)(\lambda) = \sum_{x, y \in \mathbb{F}_{2^4}} (-1)^{Tr(\lambda y) + Tr(y^t x) + Tr(x^7)}, \quad t \in \{5, 7\}$$

and

$\{\widehat{f_h}(7, 7)(\alpha^i)\} = -16, -16, -16, 16, -16, 24, 16, 8, -16, 16, 24, 8, 16, 8, 8$
$\{\widehat{f_h}(7, 5)(\alpha^i)\} = 16, -16, -16.$

Thus, $(7, 7)$ is not a realizable pair, but $(7, 5)$ is a realizable pair that realizes the sequence 011 of period 3.

Example 9.10 Let $p = 2$, $n = 5$, and $h(x) = f(x) = Tr(x)$. Let α be a primitive element in $\mathbb{F}_{2^5}$ with minimal polynomial $t(x) = x^5 + x^3 + 1$, and the m-sequence $\mathbf{a}$ of degree 5 is defined as the evaluation of $Tr(x)$ at α; that is,

$$\mathbf{a} = 1000010101110110001111100110100.$$

Case 1. $(v, t) = (3, 3)$. In this case, the first DHT of $f(x)$ is

$$\widehat{f_h}(3)(\lambda) = \sum_{x \in \mathbb{F}_{2^5}} (-1)^{Tr(\lambda x) + Tr(x^3)}$$

and

$\{\widehat{f_h}(3)(\alpha^i)\} = -8, 0, 0, 0, 0 - 8, 0, 8, 0, -8, -8, 8, 0, 8, 8, 0$
$\phantom{\{\widehat{f_h}(3)(\alpha^i)\} =} 0, 0, -8, 8, -8, 8, 8, 0, 0, 8, 8, 0, 8, 0, 0.$

The second-order DHT of $f(x)$ is

$$\widehat{f_h}(3, 3)(\lambda) = \sum_{x, y \in \mathbb{F}_{2^5}} (-1)^{Tr(\lambda y) + Tr(y^3 x) + Tr(x^3)}$$

and

$\{\widehat{f_h}(3, 3)(\alpha^i)\} = -32, 32, 32, -32, 32, 32, -32, 32, 32, 32, 32, -32, -32,$
$\phantom{\{\widehat{f_h}(3, 3)(\alpha^i)\} =} -32, 32, -32, 32, -32, 32, 32, 32, -32, -32, -32, -32,$
$\phantom{\{\widehat{f_h}(3, 3)(\alpha^i)\} =} 32, -32, -32, 32, -32, -32.$

Thus $(3, 3)$ is realizable. The realization is given by

$$(-1)^{g(x)} = \frac{1}{32}\widehat{f_h}(3, 3)(x) \Longrightarrow g(x) = Tr(x + x^5 + x^7)$$

or the sequence

$$1001001000011101010001111011011$$

which is a quadratic residue sequence of period 31.

Case 2. We take $(v, t) = (15, 7)$. In this case, values of the first DHT $\widehat{f_h}(15)$ are:

$$\{\widehat{f_h}(15)(\alpha^i)\} = 12, 8, 8, -8, 8, 4, -8, 4, 8, 4, 4, -4, -8, -4, 4, 0,$$
$$8, -8, 4, 4, 4, -4, -4, 0, -8, 4, -4, 0, 4, 0, 0.$$

The second-order DHT $\widehat{f_h}(15, 7)$ is:

$$\{\widehat{f_h}(15, 7)(\alpha^i)\} = -112, 0, 0, -16, 0, -32, -16, 16, 0, -32, -32, 0, -16, 0,$$
$$16, 48, 0, -16, -32, 16, -32, 0, 0, 48 - 16, 16, 0, 48, 16,$$
$$48, 48.$$

Thus $(15, 7)$ is not realizable.

From the definition of the second-order DHT and the inverse Hadamard transform, if $g(x)$ is a realization of $f(x)$, then we have the following two important observations:

Observation 1. Equality of the Hadamard transform of $f(x)$ and its realization $g(x)$:

$$\widehat{g}(\lambda) = \widehat{f_h}(v)(\lambda^t), \forall \lambda \in \mathbb{F}_q \tag{9.28}$$

Observation 2. Equality of the autocorrelation function of $f(x)$ and its realization: Let (v, t) be a realizable pair of $f(x)$ with $\gcd(v, q - 1) = 1$ and $\gcd(t, q - 1) = 1$, and $g(x)$ a realization. From Observation 1 and the Parseval equality, the autocorrelation of $f(x)$ is equal to the autocorrelation of $g(x)$. Thus, $g(x)$ is orthogonal if and only if $f(x)$ is orthogonal. In the sequence version, we have the following result.

$$\text{If } \omega^{g(x)} = \frac{1}{q}\widehat{f_h}(v, t)(x)$$

and $\mathbf{a} \leftrightarrow f(x)$ and $\mathbf{b} \leftrightarrow g(x)$, then

$\mathbf{b}$ has 2-level autocorrelation if and only if $\mathbf{a}$ does.

Thus, in order to find new sequences with 2-level autocorrelation, one may start with any known 2-level autocorrelation sequence, and then apply the second-order DHT to it for solving for realizable pairs. In general, it is not easy to find realizable pairs. Moreover, when it is found that (v, t) is a realizable pair and $g(x)$ is its realization, there are still two cases that may occur.

Case 1. $g(x) \neq f(x^s)$ or $\mathbf{b} \neq \mathbf{a}^{(s)}$ for any $0 < s < q - 1$ with $\gcd(s, q - 1) = 1$. In this case, $\mathbf{a}$ and $\mathbf{b}$ are decimation distinct. Thus, the second-order DHT provides a method to search for sequences with 2-level autocorrelation.

Case 2. If $g(x) = f(x^s)$, then we say that (v, t) is self-realizable. A self-realizable pair does not produce any other class of 2-level autocorrelation sequences except for the class given by $f(x)$ itself.

Note. As we pointed out before,

$$\widehat{f_h}(v, 1)(x) = \chi(f(x^v)), x \in \mathbb{F}_q.$$

Thus $(v, 1)$ is realizable for all $0 < v < q - 1$ with $\gcd(v, q - 1) = 1$, and the realization is $f(x^v)$, a trivial case.

From the definition of the second-order DHT, we allow that the two functions $h(x)$ and $f(x)$ (or equivalently, two sequences) can be any orthogonal functions (or any 2-level autocorrelation sequences). When $f(x) = h(x)$, as in Examples 9.9 and 9.10, $\widehat{f_h}(v, t)(x)$ is said to be symmetric and is denoted by $\widehat{f}(v, t)(x)$ without the subscript h. This is the most interesting case for binary sequences, because all newly discovered binary sequences with 2-level autocorrelation can be obtained by performing the symmetric DHT for $h(x) = Tr(x)$.

9.4.2 Examples of realizable pairs for small n

We will provide some examples of realizations of 2-level autocorrelation sequences in this subsection.

Example 9.11 For $p = 2$ and $n = 5$, we have two 2-level autocorrelation sequences of period 31 up to decimation. One is an m-sequence, and the other is a quadratic residue sequence, denoted as $\mathbf{c}$ and $\mathbf{a}$, respectively. If α is a primitive element in $\mathbb{F}_{2^5}$ with the minimal polynomial $t(x) = x^5 + x^3 + 1$, then $\mathbf{c}$ is the same as the m-sequence in Example 9.10. We reproduce them as follows:

$$\mathbf{c} = 1000010101110110001111100110100$$
$$\mathbf{a} = 1001001000011101010001111011011$$

There are seven coset leaders modulo 31. We list the first-order DHT of the pair (h, f) in Table 9.17 where $h, f \in \{Tr(x), QR(x) = Tr(x + x^5 + x^7)\}$, the realizable pair and the realization represented by the trace representation. (Here

Table 9.17. *Realizable pairs for n = 5*

(h, f)	Coset i	First-order DHT $\widehat{f_h}(\alpha^i)$							Realizable pair (v, t)	Realization trace spectra
$(Tr(x),$ $Tr(x))$		0	1	3	5	7	11	15		
	3	−8	0	0	−8	8	8	0	$(3, 3)$ $(3, 5)$	1, 5, 7 11
	5	−8	0	0	8	−8	8	0	$(5, 3)$	7
	7	−8	0	8	0	8	0	−8	$(7, 11)$	5
	11	−8	−8	0	8	0	0	8	$(11, 7)$ $(11, 11)$	3 1, 5, 7
$(Tr(x),$ $QR(x))$	1	−8	0	−8	0	8	0	8	$(1, 3)$ $(1, 11)$	11 3

only exponents in the trace representation are listed. This rule will be applied to all examples.)

There are six realizable pairs for the symmetric DHT for $(h(x), f(x)) = (Tr(x), Tr(x))$. For the case of $h(x) = Tr(x)$ and $f(x) = QR(x) = Tr(x + x^5 + x^7)$, notice that $f(x^i) = f(x)$, $i \in \{5, 15\}$, and there are four realizable pairs. From Table 9.17, we have the following interesting exponential sum equalities:

$(11, 11):$ $\quad \frac{1}{32} \sum_{x,y \in \mathbb{F}_{2^5}} (-1)^{Tr(\lambda y + y^{11}x) + Tr(x^{11})} = (-1)^{Tr(\lambda + \lambda^5 + \lambda^7)}$

$(1, 3):$ $\quad \frac{1}{32} \sum_{x,y \in \mathbb{F}_{2^5}} (-1)^{Tr(\lambda y + y^3 x) + Tr(x + x^5 + x^7)} = (-1)^{Tr(\lambda^{11})}$

$(1, 11):$ $\quad \frac{1}{32} \sum_{x,y \in \mathbb{F}_{2^5}} (-1)^{Tr(\lambda y + y^{11}x) + Tr(x + x^5 + x^7)} = (-1)^{Tr(\lambda^3)}.$

Example 9.12 (a) $p = 2$, $n = 6$. Let α be a primitive element of $\mathbb{F}_{2^6}$ with the minimal polynomial $x^6 + x + 1$. We set $h(x) = Tr(x) \leftrightarrow \mathbf{c}$, an m-sequence evaluated at α, and $f(x) = Tr_1^3[Tr_3^6(x)^3] = Tr_1^6(x^3 + x^5) \leftrightarrow \mathbf{a}$, a GMW sequence:

$\mathbf{c} = 00000100001100010100111101000111001001011011011001101010111111$

$\mathbf{a} = 01111011100111101001011011101000110101100110100011101000100000$

(Note that these are the same m-sequence and GMW sequence presented in Example 8.4 in Chapter 8.) By computation, $(13, 23)$ is a realizable pair where the first-order DHT, $\widehat{f_h}(13)$, is:

Coset leaders i	0	1	21	9	27	3	5	7	11	13	15	23	31
$f_h(13)(\alpha^i)$	16	0	0	0	16	0	-8	-8	0	8	0	16	-8

This realization is given by

$$g(x) = Tr(x^{11} + x^{15}) \leftrightarrow \mathbf{b} :$$

$\mathbf{b} = 00000101001001110101110100101110001100111111001001011001110100$

In other words, we have the following exponential sum equality:

$$\frac{1}{64} \sum_{x,y \in \mathbb{F}_{2^6}} (-1)^{Tr(\lambda y + y^{23} x) + Tr(x^{13 \cdot 3} + x^{13 \cdot 5})} = (-1)^{Tr(\lambda^{11} + \lambda^{15})}.$$

Note that $g(x) = f(x^5)$; that is, $\mathbf{b} = \mathbf{a}^{(5)}$. So, this is a self-realizable pair.

(b) $p = 2$ and $n = 8$. Let α be a primitive element in $\mathbb{F}_{2^8}$ with the minimal polynomial $x^8 + x^4 + x^3 + x^2 + 1$, $h(x) = Tr(x) \leftrightarrow \mathbf{c}$, an m-sequence of period 255, and

$$f(x) = Tr_1^4(Tr_4^8(x)^7) = Tr_1^8(x^7 + x^{11} + x^{13} + x^{37}) \leftrightarrow \mathbf{a},$$

a GMW sequence. (Both sequences are listed in Table 9.8.) By computation, $(11, 23)$ is a realizable pair. In the following, the first DHT $\widehat{f_h}(11)$ is listed by its coset leaders whose order is the same as that in Table 9.8 in Section 9.2.

$\{\widehat{f_h}(11)(\alpha^i)\}$: $0, 0, 0, 16, -16, 0, -16, 0, 16, 64, 16, 0, -16, -16, 0, 0, 32,$
$0, 0, 0, -16, -16, 0, -16, 0, 16, -16, 16, 0, 0, -16, 32, 16,$
$0, -16.$

The realization

$$g(x) = Tr(x + x^{47} + x^{53} + x^{61}) \leftrightarrow \mathbf{b} :$$

0 1 1 1 1 1 1 0 1 1 1 1 1 1 0 1 1 0 1 1 1 0 1 1 1 0 1 1 0 0 1 0 1 1 0 0 1 0 1
0 1 0 0 1 1 1 1 0 1 1 0 0 1 0 1 1 0 0 0 1 1 1 0 1 1 1 1 1 0 0 0 0 1 0 0 1 1 0
0 1 1 0 0 0 0 0 1 0 1 1 1 0 1 0 0 0 1 1 1 1 0 1 0 1 1 1 0 0 1 0 1 0 0 1 0 1 0
0 1 0 1 0 1 0 0 0 1 0 1 1 1 0 1 1 1 1 0 1 0 1 0 1 0 0 1 0 0 0 0 1 1 0 1 0 0 1
0 1 1 1 1 1 0 0 0 1 0 1 0 0 0 0 1 0 0 0 1 1 1 1 1 0 0 0 1 1 0 0 0 0 0 0 1 0 1
1 1 1 1 0 0 0 1 0 0 1 1 1 1 0 1 1 0 0 0 0 1 1 0 0 1 0 0 0 0 1 1 0 0 0 1 1 0 1
0 0 1 0 0 0 1 0 0 1 1 0 0 0 0 1 0 0 1 0 0

In other words, we have

$$\frac{1}{256} \sum_{x,y\in\mathbb{F}_{2^8}} (-1)^{Tr(\lambda y+y^{23}x)+Tr(x^{53}+x^{47}+x^{31}+x^{19})} = (-1)^{Tr(\lambda+\lambda^{47}+\lambda^{53}+\lambda^{61})},$$

where $Tr(x^{53} + x^{47} + x^{31} + x^{19}) = f(x^{11})$. We also have $g(x) = f(x^{29})$, which shows that $(11, 23)$ is self-realizable.

9.4.3 Hadamard equivalence

From the two observations in Section 9.4.1, we may introduce the following classification for 2-level autocorrelation sequences.

Let $g(x)$ be a function realized by $f(x)$ with respect to the orthogonal function $h(x)$, and let $f(x) \leftrightarrow \mathbf{a}$ and then $g(x) \leftrightarrow \mathbf{b}$. Then we say that $f(x)$ (or $\mathbf{a}$) and $g(x)$ (or $\mathbf{b}$) are equivalent under the Hadamard transform $h(x)$) (Hadamard equivalent for short), denoted by $f \sim_h g$ or $\mathbf{a} \sim_h \mathbf{b}$. Therefore, f and g are Hadamard equivalent; that is, $f \sim_h g$, if and only if there exists a pair (v, t) with $\gcd(v, q - 1) = 1$ and $\gcd(t, q - 1) = 1$ such that Eq. (9.28) is true. Let C be the set that contains all 2-level sequences of period $2^n - 1$ (the same notation used in Chapter 8). Let $H_{(f,h)}$ be the set consisting of all sequences that are Hadamard equivalent to f under h; that is,

$$H_{(f,h)} = \{g \in C \mid g \sim_h f\}.$$

From Lemmas 9.2, 9.3, and 9.4, we find that the hyperoval sequences $(f(x))$, the WG sequence $(N(x))$, and the Kasami power function sequences $(C_k(x))$ are all realizations of m-sequences $(S_v(x) = Tr(x^v)$, the notation used in Sections 9.2 and 9.3), where $h(x) = Tr(x)$. We list these realizable pairs in Table 9.18. Thus, for n odd, all newly discovered binary 2-level autocorrelation sequences are Hadamard equivalent to m-sequences. Notice that Hadamard

Table 9.18. *Realizable pairs of m-sequences*

Hadamard relation, n odd	Realizable pair	Realization
$\widehat{f}(x) = \widehat{S}_k(x^t), t = \frac{k-1}{k}$	$(k, \frac{k-1}{k})$	Hyperoval seq.
$\widehat{N}(x) = \widehat{S_{2^k+1}}(x^{d^{-1}}), 3k \equiv 1 \pmod{n}$	$(2^k + 1, d^{-1})$	WG
$\widehat{C}_k(x) = \widehat{S}_3(x^t), t = \frac{2^k+1}{3}$	$(3, \frac{2^k+1}{3})$	Kasami power seq. B_k

equivalence is an equivalence relation. We now fix notation for all known classes of 2-level autocorrelation sequences for the rest of this section.

- PN, the set of the m-sequences,
- QR, the set of the quadratic residue sequences,
- $Hall$, the set of the Hall sextic residue sequences,
- GMW, the set of the GMW sequences (all types),
- O, the set consisting of the hyperoval sequences (Segre and Glynn types 1 and 2 ($G1$ and $G2$)),
- WG, the set of the WG sequences, and
- KPF, the set consisting of the Kasami power function sequences B_k, $k \leq$ $(n-1)/2$ and $\gcd(k, n) = 1$. (We separate 3-term and 5-term sequences from general B_k by denoting them as $T3$ and $T5$.)

Using this notation and Table 9.18, we have

$$PN \cup O \cup WG \cup KPF \subset H_{(Tr,Tr)}.$$

In other words, starting with a single m-sequence, we are able to obtain all sequences in the set of the left-hand side of the above inclusion by computing the second-order symmetric DHT $\widehat{f}(v, t)(x)$ where $f(x) = h(x) = Tr(x)$. In the following example, standard forms of known 2-level sequences of period 127 are those listed in Table 9.5 in Section 9.2.

Example 9.13 Let $p = 2$ and $n = 7$. Assume that $\mathbb{F}_{2^7}$ is defined by the primitive polynomial $x^7 + x + 1$. Let $h(x) = Tr(x)$ and $f(x) = Tr(x) \leftrightarrow \mathbf{b}$, an m-sequence, or $f(x) = Tr(x + x^9 + x^{13}) \leftrightarrow \mathbf{b}$, a 3-term sequence. We list some realizable pairs in Table 9.13. For example, for the case of $h(x) = Tr(x)$ and $f(x) = T3(x)$, we have

$$\frac{1}{128} \widehat{f}_h(13, 29)(\lambda) = (-1)^{T5(\lambda^{55})}, \ \lambda \in \mathbb{F}_{2^7},$$

where $T5(x) = Tr(x + x^5 + x^{21} + x^{13} + x^{29})$, the trace representation of a 5-term sequence of period 127. Therefore, for $n = 7$, the Hall sequences are also Hadamard equivalent to m-sequences.

An exhaustive search was performed to compute all sequences that are Hadamard equivalent to m-sequences for odd $n \leq 17$. Except for $n = 7$ (this case will be discussed further in the following subsection), for $9 \leq n \leq 17$, those which are Hadamard equivalent to m-sequences are the hyperoval, the WG, and the Kasami power function sequences. Interestingly, no previously unknown examples were found by this second-order symmetric DHT process for any odd $9 \leq n \leq 17$. In other words, the following result has been confirmed.

Table 9.19. *Some realizable pairs for* $n = 7$ *and* $h(x) = Tr(x)$

$f(x)$	(v, t)	Realization	Types
$Tr(x)$	$(3, 3)$	$1, 9, 13$	$T3 = G_1$
	$(11, 19)$	$1, 5, 13, 21, 29$	$T5 = G_2 = B_2$
	$(5, 11)$	$1, 3, 7, 19, 29$	WG
	$(5, 5)$	$9, 11, 21$	Hall!
$T3(x) =$ $Tr(x + x^9 + x^{13})$	$(13, 29)$	$3, 5, 15, 21, 55$	$T5(x^{55})$

$$PN \cup O \cup WG \cup KPF = H_{(Tr,Tr)} \text{ for } 9 \le n \le 17 \qquad (9.29)$$

However, it is unknown whether this result is true for $n \ge 19$.

9.4.4 Realization of the symmetric DHT

Let (v, t) be a realizable pair of $f(x)$ under the second-order symmetric DHT ($f(x) = h(x)$) where the realization is equal to $g(x)$. Gong and Golomb showed that in this case, there are either six realizable pairs for the realization or two realizable pairs. Moreover, the other five realizable pairs that produce $g(x)$ or a decimation of $g(x)$ are given by

$$
\begin{array}{ccc}
(v, t) & \longrightarrow \left(t, -(vt)^{-1}\right) & \longrightarrow \left(-(vt)^{-1}, v\right) \\
\downarrow & \downarrow & \downarrow \\
g(x) & g\left(x^{(vt)^{-1}}\right) & g\left(x^{-t^{-1}}\right)
\end{array}
$$

$$(9.30)$$

$$
\begin{array}{ccc}
\left(t^{-1}, v^{-1}\right) & \longrightarrow \left(v^{-1}, -vt\right) & \longrightarrow \left(-vt, t^{-1}\right) \\
\downarrow & \downarrow & \downarrow \\
g\left(x^{(vt)^{-1}}\right) & g(x) & g\left(x^{-t^{-1}}\right)
\end{array}
$$

If $v = t$ and $v^3 \equiv -1$, the above six realizable pairs degenerate into only two different pairs, that is, (v, v) and (v^{-1}, v^{-1}).

Example 9.14 Starting with a single m-sequence, we compute realizable pairs of m-sequences of period 127 by performing the second-order symmetric DHT (i.e., $f(x) = h(x) = Tr(x)$). In total, we obtain 20 realizable pairs, where each

Table 9.20. *A complete list of realizable pairs of the*
second-order symmetric DHT of m-sequences of
period 127

Realizable pair	Realization $g(x)$	Type
3-term		
(3, 3)	1, 9, 13	$T3(x)$
(3, 7)	1, 9, 15	$T3(x^{15})$
(7, 3)	19, 21, 31	$T3(x^{21})$
(43, 43)	1, 9, 15	$T3(x^{15})$
(43, 55)	1, 9, 13	$T3(x)$
(55, 43)	19, 21, 31	$T3(x^{21})$
5-term		
(3, 11)	3, 5, 9, 19, 29	$T5(x^5)$
(11, 19)	1, 5, 13, 21, 29	$T5(x)$
(19, 3)	1, 5, 7, 11, 31	$T5(x^{31})$
(13, 43)	1, 5, 13, 21, 29	$T5(x)$
(43, 47)	3, 5, 9, 19, 29	$T5(x^5)$
(47, 13)	1, 5, 7, 11, 31	$T5(x^{31})$
WG		
(5, 11)	1, 3, 7, 19, 29	$WG(x)$
(11, 15)	3, 7, 11, 19, 21	$WG(x^7)$
(15, 5)	1, 7, 9, 11, 23	$WG(x^{23})$
(13, 27)	3, 7, 11, 19, 21	$WG(x^7)$
(27, 9)	1, 3, 7, 19, 29	$WG(x)$
(9, 13)	1, 7, 9, 11, 23	$WG(x^{23})$
Hall		
(5, 5)	9, 11, 21	$Hall(x^{29})$
(27, 27)	9, 11, 21	$Hall(x^{29})$

of the classes of the 3-term sequences, 5-term sequences, and WG sequences are realized by six realizable pairs and the Hall sequences are realized by two realizable pairs. We list all these 20 realizable pairs in Table 9.20.

Note that $Hall(x^{29}) = Hall(x^{43}) = Hall(x^{55})$. From the exhaustive computation for $n = 7$, all known classes of 2-level autorrelation sequences except for the quadratic residue sequences are Hadamard equivalent to the class of m-sequences. Note that for each of the classes of the m-sequences, 3-term sequences, 5-term sequences, and WG sequences, there are 18 shift-distinct sequences. There are six shift-distinct Hall sequences and two shift-distinct quadratic residue sequences. For easy verification of the decimation-equivalent classes of these sequences, we also provide Table 9.21, in which each row

Table 9.21. *Decimation equivalence classes for n = 7*

3-term			5-term					WG					Hall		
1	9	13	1	5	21	13	29	1	3	7	19	29	5	27	63
3	27	29	3	15	63	29	47	3	9	21	23	47	15	13	31
5	43	3	5	19	29	3	9	5	15	13	63	9	19	1	47
7	63	55	7	13	5	55	19	7	21	11	3	19	13	31	15
9	13	47	9	43	31	47	7	9	27	63	11	7	43	29	55
11	15	1	11	55	13	1	3	11	5	27	21	3	55	43	29
13	47	21	13	3	19	21	63	13	29	55	15	63	3	7	23
15	1	9	15	23	47	9	27	15	43	29	31	27	23	3	7
19	11	15	19	63	9	15	43	19	23	3	47	43	63	5	27
21	31	19	21	29	15	19	23	21	63	5	9	23	29	55	43
23	5	43	23	31	27	43	1	23	11	9	7	1	31	15	13
27	29	7	27	1	55	7	21	27	13	31	5	21	1	47	19
29	7	63	29	9	23	63	31	29	47	19	43	31	9	21	11
31	19	11	31	7	1	11	5	31	55	43	13	5	7	23	3
43	3	27	43	11	7	27	13	43	1	47	55	13	11	9	21
47	21	31	47	27	11	31	55	47	7	23	1	55	27	63	5
55	23	5	55	21	3	5	15	55	19	1	29	15	21	11	9
63	55	23	63	47	43	23	11	63	31	15	27	11	47	19	1

QR
3, 5, 7, 23, 27, 29, 43, 55, 63
1, 9, 11, 13, 15, 19, 21, 31, 47

represents one decimation class. Note that the trace representation of each class in the first row is in standard form.

In the following, we show an extraordinary experimental result for the second-order symmetric DHT where $f(x) = h(x)$, taken from $\{T3, T5, WG, Hall, QR\}$ for $n = 7$. We present these realizable pairs and realizations in the following diagram, Figure 9.9 (in which 3T represents 3-term sequences, and 5T, 5-term sequences). In Figure 9.9, the double circles represent the pairs $(h(x), h(x))$. The numbers listed on the outer circle are the number of decimations for obtaining different decimation sequences, and the pair of numbers (v, t) are realizable pairs for the corresponding realization. For example, realizable pair $(3, 11)$ from the circle of m-sequences to the circle of 5-term sequences yields a 5-term sequence that is a 5-decimation from the standard 5-term sequence. Furthermore, a realizable pair $(5, 5)$ from the circle of 5-term sequences to the circle of the Hall sequences represents a realizable pair of the second symmetric DHT where $h(x) = T5(x)$, which realizes the Hall sequence. Both the Hall sequences and the 3-term sequences realize the quadratic residue sequences.

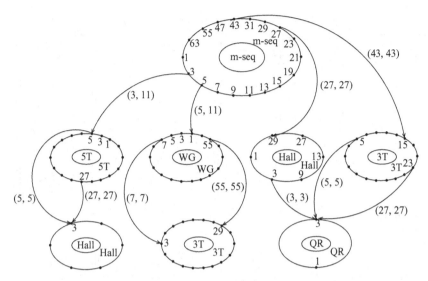

Figure 9.9. Operational flow of the second-order symmetric DHT for $n = 7$.

Figure 9.9 shows a very interesting phenomenon. For better visualization of the relationship among realizations from a single m-sequence, we simplify Figure 9.9 into a graph, Figure 9.10, where the vertices are the classes of 2-level autocorrelation sequences of period 127 (a total of six classes) and the edges are realizable pairs obtained by performing the second-order symmetric DHT on the class of the corresponding vertex. This graph clearly shows that all six classes are related by the second-order symmetric DHT, starting from a single m-sequence.

However, there is no simple explanation for this. Gong and Golomb (2001) verified that it is no longer true for $n = 17$, although at this length, there exist both Hall sextic residue sequences and quadratic residue sequences. One

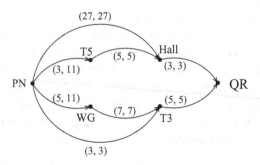

Figure 9.10. Graph of realizations of the second-order symmetric DHT for $n = 7$.

possible reason for this is that 17 is not the same type of number as 7, where

$$7 \equiv 3 \pmod 4 \text{ and } 17 \equiv 1 \pmod 4.$$

No other Mersenne prime exponent p satisfying both

$$p \equiv 3 \pmod 4 \text{ and } 2^p - 1 = 4a^2 + 27$$

is currently known.

Note

Multiple trace term sequences, including 3-term sequences 5-term sequences, and WG sequences were conjectured by No et al. in 1998 to be 2-level auto-correlation sequences. The trace representation of WG sequences obtained by Golomb, Gong, and Gaal was also presented in that paper. Previously, the conjecture on 3-term sequences appeared in Gong, Golomb, and Gaal (1997).

In 1998, Maschietti showed that the monomial hyperovals, Segre (1955) and Glynn types 1 and 2 (Glynn 1983), yield Hadamard difference sets with parameters $(2^n - 1, 2^{n-1} - 1, 2^{n-2} - 1)$. For more materials on hyperovals, see Hirschfeld (1975). Maschietti's proof involves a knowledge of hyperoval geometry. Soon after that, Xiang (1998) gave a short proof for part of the cases of hyperoval sequences by using the Hadamard transform and the Parseval equality. Dillon (1999) developed this method further and presented a short proof where he presents a proof for his Lemma 9.3, which is the proof we provide here for our Lemma 9.2. The linear spans of hyperoval sequences are given by Evan et al. (1999). For the Segre hyperoval sequences, Chang et al. (1999) presented another completely different proof, and from their proof, the trace representation of Segre hyperoval sequences is also obtained. (Because the Segre hyperoval sequences are in the same class as B_2, the trace representation of the Segre hyperoval sequences can also be obtained from (9.23) in Section 9.3.)

At almost the same time as the research on hyperoval sequences, No, Chung, and Yin (1998) found the Kasami power function representation of the WG sequences and verified their construction has 2-level autocorrelation for $n \leq 23$. By using this represenation, Dillon proved that this construction produces 2-level autocorrelation for the case of n odd. The proof given here for Lemma 9.3 is from Dillon (1999). Dobbertin (1999) conjectured B_k sequences have 2-level autocorrelation. In that paper, Dobbertin (2004) first proved that the Kasami power function is APN for n even, or equivalently, $x^d + (x + 1)^d$ is a 2-to-1 map on $\mathbb{F}_q$, so it implies that both $\sigma(x)$ (for the WG sequences) and $\Delta_k(x)$ are 2-to-1 maps on $\mathbb{F}_q$. Then, he determined the linear span and the trace representation of B_k sequences in that paper. Finally, Dillon and Dobbertin (2004) established that the B_k sequences have 2-level autocorrelation for both n odd and n even.

The 2-level autocorrelation of the WG sequences for the case of even n under the representation of the Kasami power function was also shown in that paper. For a proof of the result on MCM polynomials, see Cohen and Matthews (1994) for k odd and Dillon and Dobbertin (2004) for k even. Thus, by 2004, the validity of all conjectured 2-level autocorrelation sequences had been confirmed. For 3-term sequences, there are further interesting results, such as they form BCH codes of weight 5 and the dual code of this code has minimum distance 8, which were obtained by Chang et al. (2000). Gong and Golomb (1999) discovered that the Hadamard transform of the 3-term sequences is determined from one of the m-sequences and verified their result for $n \leq 23$ before Dillon and Dobbertin obtained the result of Lemma 9.4.

The iterative decimation Hadamard transform was introduced by Gong and Golomb (2002), which was inspired by the proofs of Lemmas 9.2, 9.3, and 9.4. The proof of the results shown by Eq. (9.30) was also obtained by them in Gong and Golomb (2002).

Baumert (1971) completed an exhaustive search for $n = 7$. The sequences D, E, and F listed in Table 9.5 were found by his search. The computer search for $n = 8$ was done in 1983 by Unjeng Cheng, and he reported two more sequences for $n = 8$, C, and D listed in Table 9.8. In 1992, when Dreier and Smith did an exhaustive search for $n = 9$, they found three more sequences: C, D, and E as listed in Table 9.14. Gaal and Golomb (2001) completed the exhaustive search for $n = 10$ and confirmed that the 10 classes of sequences listed in Table 9.22 in the Appendix are all the 2-level autocorrelation sequences for this length. Exhaustive computer searches to find all Hadamard ($2^n - 1$, $2^{n-1} - 1$, $2^{n-2} - 1$) cyclic difference sets are not currently feasible for any $n \geq 11$.

For $p > 2$, except for m-sequences and GMW sequences (four types), there are certain new constructions after 1997. The reader is referred to Helleseth, Kumar, and Martinson (2001), Arasu and Player (2003), Helleseth and Gong (2002), and No (2002). There are also some experimental results on new sequences obtained by applying the second-order decimation-Hadamard transform to ternary sequences (Gong, Helleseth, and Ludkovski (2001) and Lin (1998)).

Appendix: Known 2-level autocorrelation sequences of periods 1023, 2047, and 4095 and their trace representations

In the following tables, I represents a subset of the set consisting of all coset leaders modulo $2^m - 1$ such that $g(x) = \sum_{r \in I} Tr_1^m(x^r)$ where $m|n$. For $n = 12$,

Table 9.22. *A complete list of 2-level autocorrelation sequences of period 1023*

Types	Trace spectra	Number of terms	Linear span
m-seq	1	1	10
5-term sequence	1, 9, 73, 57, 121	5	50
WG-sequence	1, 3, 5, 7, 11, 13, 15, 35, 69, 71, 89, 105, 121	13	130
GMW constructions			
I for $\mathbb{F}_{2^5}$			
3	3, 17	2	20
5	5, 9	2	20
7	7, 19, 25, 69	4	40
11	11, 13, 21, 73	4	40
15	15, 23, 27, 29, 77, 85, 89, 147	8	80
1, 9, 7	1, 5, 7, 9, 19, 25, 69	7	70
3, 15, 11	3, 17, 11, 13, 21, 73, 15, 23, 27, 29, 77, 85, 89, 147	14	140
Total: 10 classes			

Table 9.23. *Known classes of 2-level autocorrelation sequences of period 2047:* $\mathbb{F}_{2^{11}}$ *defined by* $\alpha^{11} + \alpha^2 + 1$

Types	Trace spectra	Number of terms	Linear span
m-seq	1	1	11
3-term	1, 33, 49	3	33
5-term	1, 17, 137, 121, 143	5	55
WG	21, 23, 29, 35, 37, 41, 71, 89, 139, 165, 213, 307, 415	13	143
Segre =B2	1, 5, 13, 21, 53, 77, 85, 205, 213, 309, 333, 341, 413, 423, 469	15	165
Glynn 1	1, 5, 9, 13, 19, 37, 43, 67, 69, 137, 163, 211, 293	13	143
Glynn 2	1, 5, 13, 17, 29, 37, 49, 61, 69, 81, 93, 101, 113, 125, 139, 147, 151, 157, 171, 173, 183	21	231
B3	1, 5, 7, 9, 19, 25, 81, 169, 295	9	99
Total: 9 known classes			

Table 9.24. *Known classes of 2-level autocorrelation sequences ofperiod 4095*

Types	Trace spectra	Number of terms	Linear span
m-seq	1	1	12
B5	1, 33, 133, 159, 165, 637, 163, 661, 669, 421, 405, 667, 629, 621, 373	15	180
GMW constructions			
I			
3, $\mathbb{F}_{2^3}$	3, 17, 33, 5	4	48
7, $\mathbb{F}_{2^4}$	7, 67, 13, 37, 97, 133, 41, 73, 11	9	108
$\mathbb{F}_{2^6}$			
5	5, 17	2	24
11	11, 25, 137, 37	4	48
13	13, 41, 145, 19	4	48
23	23, 29, 275, 53, 149, 169, 281, 43	8	96
31	31, 61, 185, 59, 283, 181, 307, 55, 157, 425, 661, 173, 409, 299, 313, 47	16	192
$\mathbb{F}_{2^3} \subset \mathbb{F}_{2^6}$			
$(s_1, s_2) \to I$ with $\mathbb{F}_{2^6}$			
$(3, 5) \to 15, 11$	15, 57, 51, 177, 141, 165, 39, 291, 11, 25, 137, 37	12	144
$(3, 11) \to 3, 31$	3, 33, 31, 61, 185, 59, 283, 181, 307, 55, 157, 425, 661, 173, 409, 299, 313, 47	18	216
$(3, 13) \to 15, 1$	15, 57, 51, 177, 141, 165, 39, 291, 1	9	98
$(3, 23) \to 3, 13$	3, 33, 13, 41, 145, 19	6	72
$(3, 31) \to 15, 23$	15, 57, 51, 177, 141, 165, 39, 291, 23, 29, 275, 53, 149, 169, 281, 43	16	192
Total: 14 known classes			

for the case of the field chain of length 3, the pair (s_1, s_2) means the GMW function $g(x) = Tr_1^3(x^{s_1}) \circ Tr_3^6(x^{s_2})$, and I contains all the coset leaders modulo 63 in the expansion of $g(x)$. In this case s_1 has only one choice, $s_1 = 3$, and s_2 can be any coset leaders modulo 63 that are not equal to 1 and are relatively coprime to 63. Note that $Tr_1^3(x^3) \circ Tr_3^6(x) = Tr_1^6(x^3 + x^5)$. Thus, $Tr_1^3(x^3) \circ Tr_3^6(x^{s_2}) = Tr_1^6(x^{3s_2} + x^{5s_2})$, the exponents of which (corresponding coset leaders) are the numbers listed in Table 9.24. We list trace representations of 2-level sequences from the GMW constructions, multiple trace term construction, the hyperoval construction, and the Kasami power function construction if there are any. The finite fields used to compute the Glynn type 1 and type 2 hyperoval sequences are listed. For all the others, their trace representations are computed from the given formulae.

Exercises for Chapter 9

1. Design a 3-term sequence with period 31 where $f_o = x^5 + x^4 + x^3 + x + 1$, and provide the LFSR implementation for your design. Compare the resulting 3-term sequence with the quadratic residue sequence of period 31. What phenomenon do you find?

2. Let n be odd. A 3-term sequence $\mathbf{a}$ of period $2^n - 1$ is the sum of an m-sequence, say $\mathbf{s}$, which is constant-on-cosets, and its two decimated sequences, given as follows:

$$\mathbf{a} = \mathbf{s} + \mathbf{s}^{(q_1)} + \mathbf{s}^{(q_2)}$$

where $q_1 = 2^{(n-1)/2}$ and $q_2 = 2^{(n-1)/2} + 2^{(n+1)/2} + 1$.
   (a) Show that q_1^2 and q_2 are in the same coset modulo $2^n - 1$.
   (b) Prove that both q_1 and q_2 are coprime to $2^n - 1$. Therefore, a 3-term sequence is the sum of three m-sequences as specified above.
   (c) Let $f(x) = Tr(x + x^{q_1} + x^{q_2})$. Show that $f(x)$ is the trace representation of the 3-term sequence. For $n = 5$ and $n = 7$, the sets consisting of all coset leaders modulo 31 and 127 are given by

$$\Gamma_2(5) = \{0, 1, 3, 5, 7, 11, 15\}$$

and

$$\Gamma_2(7) = \{0, 1, 3, 5, 7, 9, 11, 13, 15, 19, 21, 23, 27, 29, 31, 43, 47, 55, 63\}$$

respectively. Compute the Hadamard transform $\widehat{f}(\lambda)$ of $f(x)$ in Example 9.1 for $\lambda = \alpha^i$ where $i \in \Gamma_2(n)$, $n = 5, 7$. Compute the ith element of a 3-decimation of the m-sequence $\mathbf{s}$ (that is, $\{s_{3i}\}$) where $i \in \Gamma_2(n)$, $n = 5, 7$. Compare these two lists for each of $n = 5, 7$.

3. Let $\mathbf{a} = 000100110101111$ be an m-sequence with period 15. Construct $\mathbf{b} = \{b_i\}$ whose elements are obtained by performing the trinomial decimation which is used to generate WG sequences from 5-term sequences. What do you get? Comment on your result.

4. For $n = 7$, using the decimation method, construct a 5-term sequence where one of the LFSRs has the characteristic polynomial $f(x) = x^7 + x^3 + 1$, and finding the corresponding WG-sequence using the trinomial decimation method. Provide the LFSR implementation for the 5-term sequence that you designed, and the lookup table (that is, the trinomial table) implementation for the WG sequence.

5. Show that the regular hyperoval $D(x^2)$ or any translate hyperoval $D(x^{2^i})$, where $\gcd(n, i) = 1$, $1 < n/2$ produces an m-sequence of period $2^n - 1$.

6. From Definition 9.1, the Kasami power function is defined as $f(x) = x^d$ where $d = 2^{2k} - 2^k + 1$ where $k < n$ and $\gcd(k, n) = 1$. Compute the Hadamard transform of $S_d(x) = Tr(x^d)$ for $n = 5$ and $n = 7$ for different choices of k. Using your data, verify Eq. (9.11).

7. A communication system employs a white noise source which has period in the interval [127, 2047], 2-level autocorrelation and linear span greater than 20. How many such designs can find? List parameters for each of your designs.
8. Verify the first order DHT values which are listed in Table 9.17, and check that the realizable pair (11, 11) produces a quadratic residue sequence. How many different ways can you generate a quadratic residue sequence of period 31?

10

Signal Sets with Low Crosscorrelation

In this chapter, we introduce constructions for signal sets with low crosscorrelation. These sequences have important applications in wireless CDMA communications. There are three classic constructions for signal sets with low correlation, namely, the Gold-pair construction, the Kasami (small) set construction, and the bent function signal set construction. In Section 10.1, we introduce some basic concepts and properties for crosscorrelation of sequences or functions, signal sets, and one-to-one correspondences among sequences, polynomial functions, and boolean functions. After that, three classic constructions will be presented in Sections 10.2, 10.3, and 10.4 respectively. With the development of new technologies, the demand for constraints on other parameters, such as linear spans of sequences, and the sizes of the signal sets has increased. Here, we will provide two examples of constructions that sacrifice ideal correlation in order to improve other properties, in Sections 10.5 and 10.6, respectively. One example is the interleaved construction for large linear spans, and the other is $\mathbb{Z}_4$ sequences to obtain large sizes of signal sets.

10.1 Crosscorrelation, signal sets, and boolean functions

In this section, we discuss some basic properties of crosscorrelation of sequences (some of them have been discussed in Chapter 1), refine the concept of signal sets, and develop the one-to-one correspondence between sequences and boolean functions. (Note that the one-to-one correspondence between sequences and functions is discussed in Chapter 6.)

We will keep the following notation in this section. Let p be any prime, n be a positive integer, $q = p^n$, and α be a primitive element in $\mathbb{F}_q$. Let $\mathbf{a} = \{a_i\}$

and $\mathbf{b} = \{b_i\}$ be two periodic sequences over $\mathbb{F}_p$ with periods $v = p^n - 1$ and t, where $t\,|\,v$, respectively; let $f(x)$ and $g(x)$ be their respective trace representations; that is, $a_i = f(\alpha^i)$ and $b_i = g(\alpha^i)$. The crosscorrelation between $\mathbf{a}$ and $\mathbf{b}$ (see Chapter 5) is defined by

$$C_{\mathbf{a},\mathbf{b}}(\tau) = \sum_{i=0}^{v-1} \omega^{a_{i+\tau} - b_i}, \ \tau = 0, 1, \ldots, \tag{10.1}$$

where ω is a primitive pth root of unity. Since $t\,|\,v$, we have

$$C_{\mathbf{a},\mathbf{b}}(\tau) = C_{\mathbf{a},\mathbf{b}}(\tau + kt), \ k = 0, 1, \ldots. \tag{10.2}$$

10.1.1 Basic properties of crosscorrelation

Property 10.1 *Let L be the left shift operator.*

(a) If both $\mathbf{a}$ and $\mathbf{b}$ are shifted by k, then their crosscorrelation does not change; that is,

$$C_{L^k\mathbf{a}, L^k\mathbf{b}}(\tau) = C_{\mathbf{a},\mathbf{b}}(\tau), 0 \leq k < v.$$

(b) Shift rule: the crosscorrelation of shifted versions of $\mathbf{a}$ and $\mathbf{b}$ is equal to the crosscorrelation of $\mathbf{a}$ and $\mathbf{b}$ up to some shift; that is,

$$C_{L^i\mathbf{a}, L^j\mathbf{b}}(\tau) = C_{\mathbf{a},\mathbf{b}}(\tau + i - j), 0 \leq i, j < v.$$

(c) Commutative rule:

$$C_{\mathbf{a},\mathbf{b}}(\tau) = \overline{C_{\mathbf{b},\mathbf{a}}(-\tau)}.$$

In other words, the crosscorrelation of $\mathbf{a}$ and $\mathbf{b}$ at τ is equal to the complex conjugate of the crosscorrelation of $\mathbf{a}$ and $\mathbf{b}$ at negative τ. In particular, if $p = 2$, then

$$C_{\mathbf{a},\mathbf{b}}(\tau) = C_{\mathbf{b},\mathbf{a}}(-\tau).$$

(Here both $\tau + j - i$ and $-\tau$ are reduced modulo v.)

(d) If both $\mathbf{a}$ and $\mathbf{b}$ are constant-on-cosets, so is their crosscorrelation function; that is,

$$C_{\mathbf{a},\mathbf{b}}(\tau) = C_{\mathbf{a},\mathbf{b}}(p\tau), \tau = 0, 1, \ldots.$$

Example 10.1 Let $\mathbf{a}$ and $\mathbf{b}$ be two binary sequences whose elements are given by

$$\mathbf{a} = 000100110101111 \text{ and } \mathbf{b} = 011.$$

Let T consist of different values in the set

$$\{C_{L^i(\mathbf{a}),L^j(\mathbf{b})}(\tau) \mid 0 \le \tau < 15, 0 \le i < 15, 0 \le j < 3\}.$$

Note that both $\mathbf{a}$ and $\mathbf{b}$ are constant-on-cosets. According to Property 10.1(b) and 10.1(d), to get the set T, we only need to compute $C_{\mathbf{a},\mathbf{b}}(\tau)$ for $\tau = 0, 1$. Since $C_{\mathbf{a},\mathbf{b}}(0) = -5$ and $C_{\mathbf{a},\mathbf{b}}(1) = 3$, we have $T = \{-5, 3\}$.

Next, we look at the crosscorrelation between the trace representations of these two sequences. Let α be a primitive element in $\mathbb{F}_{2^4}$ with minimal polynomial $x^4 + x + 1$. Then the trace representations of $\mathbf{a}$ and $\mathbf{b}$ are given by $f(x) = Tr_1^4(x) = x + x^2 + x^4 + x^8$ and $g(x) = Tr_1^2(x^5) = x^5 + x^{10}$, respectively. Thus, $C_{f,g}(\lambda)$, the crosscorrelation between f and g (defined in Section 8.4 of Chapter 8), takes the two values ± 4 for $\lambda \ne 0$. Precisely, we have

$$C_{f,g}(\lambda) = \sum_{x \in \mathbb{F}_{2^4}} (-1)^{Tr_1^4(\lambda x) + Tr_1^2(x^5)}$$

$$= 1 + \sum_{i=0}^{14} (-1)^{a_{i+\tau} + b_i}$$

$$= 1 + C_{\mathbf{a},\mathbf{b}}(\tau) \in \{\pm 4\}, \lambda = \alpha^\tau.$$

The following assertion follows immediately from the definitions of the shift operator and the decimation operator.

Property 10.2 *If $f(x)$ is the trace representation of $\mathbf{a}$, then $f(\alpha^j x)$ and $f(x^r)$ are the trace representations of $\mathbf{a}$ at shift j, $L^j(\mathbf{a})$, and the r-decimation of $\mathbf{a}$, $\mathbf{a}^{(r)}$, respectively; that is,*

$$L^j(\mathbf{a}) \leftrightarrow f(\alpha^j x), \tag{10.3}$$

$$\mathbf{a}^{(r)} \leftrightarrow f(x^r). \tag{10.4}$$

Recall that in Chapter 8 we introduced the notation $<\mathbf{a}, \mathbf{b}>$ to denote the dot product (see Chapter 1) of $\chi(\mathbf{a})$ and $\chi(\mathbf{b})$ where $\chi(\mathbf{x}) = (\omega^{x_0}, \dots, \omega^{x_{v-1}})$ where $\mathbf{x} = (x_0, \dots, x_{v-1}) \in \mathbb{F}_p^v$; that is,

$$<\mathbf{a}, \mathbf{b}> = (\chi(\mathbf{a}) \cdot \chi(\mathbf{b})) = \sum_{i=0}^{v-1} \chi(a_i) \chi^*(b_i) = \sum_{i=0}^{v-1} \omega^{a_i - b_i}.$$

Some relationships between this dot product and the crosscorrelation function of the sequences $\mathbf{a}$ and $\mathbf{b}$ are listed in Table 10.1 for easy reference.

Let f and g be two functions from $\mathbb{F}_{p^n}$ to F_p. The (Hamming) distance between f and g, denoted by $d(f, g)$, is defined as the number of $x \in \mathbb{F}_{p^n}$ for

Table 10.1. *Relationship of the dot product and
crosscorrelation*

$\mathbf{a}, \mathbf{b} \in \mathbb{F}_p^v, \tau \geq 0.$	
(a)	$C_{\mathbf{a},\mathbf{b}}(0) = <\mathbf{a}, \mathbf{b}>.$
(b)	$C_{\mathbf{a},\mathbf{b}}(\tau) = <L^{\tau}(\mathbf{a}), \mathbf{b}>.$
(c)	$C_{L^i\mathbf{a}, L^j\mathbf{b}}(\tau) = <L^{i+\tau}(\mathbf{a}), L^j(\mathbf{b})>, i, j \geq 0.$
(d)	$<\mathbf{a} + c, \mathbf{b} + d> = \omega^{d-c} <\mathbf{a}, \mathbf{b}>, c, d \in \mathbb{F}_p.$

which $f(x) \neq g(x)$; that is,

$$d(f, g) = |\{x \in \mathbb{F}_{p^n} \mid f(x) \neq g(x)\}|. \tag{10.5}$$

The (Hamming) weight of f, denoted by $w(f))$, is defined as the number of $x \in \mathbb{F}_{p^n}$ such that $f(x) \neq 0$; that is,

$$w(f)) = |\{x \in \mathbb{F}_{p^n} \mid f(x) \neq 0\}|. \tag{10.6}$$

Thus, we have

$$d(f, g) = w(f - g)); \tag{10.7}$$

that is, the distance between two functions f and g is equal to the weight of their difference. For the binary case, we have the following frequently used relationships between crosscorrelation and (Hamming) weight, or distance between sequences (or functions).

Property 10.3 *Let* $\mathbf{a} \leftrightarrow f(x)$ *and* $\mathbf{b} \leftrightarrow g(x)$ *be two binary sequences where one of them has period* $2^n - 1$. *Then*

$$w(L^{\tau}(\mathbf{a}) + \mathbf{b}) = 2^{n-1} - \frac{C_{\mathbf{a},\mathbf{b}}(\tau) + 1}{2}.$$

Equivalently,

$$C_{\mathbf{a},\mathbf{b}}(\tau) + 1 = 2^n - 2w(L^{\tau}(\mathbf{a}) + \mathbf{b}).$$

In the function version,

$$d(f(\lambda x), g(x)) = 2^{n-1} - \frac{1}{2}C_{f,g}(\lambda), \lambda = \alpha^{\tau}.$$

Note. These formulae are similar to Eq. (7.5) for autocorrelation in Section 7.2 of Chapter 7.

The next property indicates the crosscorrelation between $\mathbf{a}$ and a d-decimation of $\mathbf{b}$.

Property 10.4 *For* $\mathbf{a} \leftrightarrow f(x)$ *and* $\mathbf{b} \leftrightarrow g(x)$, *recall that* $\mathbf{b}$ *has period t, where* $t \mid (2^n - 1)$. *Let* $d > 1$ *satisfying,* $\gcd(d, t) = 1$. *Then*

$$C_{\mathbf{a}, \mathbf{b}^{(d-1)}}(\tau) = C_{\mathbf{b}, \mathbf{a}^{(d)}}(-d^{-1}\tau), \lambda = \alpha^{\tau}.$$

In particular, if $\mathbf{a} = \mathbf{b}$ *(so that* $f(x) = g(x)$*), then* $C_{\mathbf{a}, \mathbf{a}^{(d-1)}}(\tau) = C_{\mathbf{a}, \mathbf{a}^{(d)}}(-d^{-1}\tau)$.
Equivalently, in the function version,

$$C_{f(x), g(x^{d-1})}(\lambda) = C_{g(x), f(x^d)}(\lambda^{-d^{-1}}), \lambda \in \mathbb{F}_{2^n}.$$

In other words, the image of the crosscorrelation between the sequence $\mathbf{a}$, *and the* (d^{-1})*-decimation of* $\mathbf{b}$, $\mathbf{b}^{(d-1)}$, *is equal to the image of the crosscorrelation between* $\mathbf{a}$ *and the d-decimation of* $\mathbf{b}$, $\mathbf{b}^{(d)}$. *In particular, the image of the crosscorrelation between the sequence* $\mathbf{a}$ *and its* d^{-1}*-decimation is equal to the image of the crosscorrelation between* $\mathbf{a}$ *and its d-decimation.*

Proof. According to the definition of crosscorrelation,

$$\begin{aligned}
C_{\mathbf{a}, \mathbf{b}^{(d-1)}}(\tau) + 1 &= C_{f(x), g(x^{d-1})}(\lambda) \\
&= \sum_{x \in \mathbb{F}_{2^n}} (-1)^{f(\lambda x) + g(x^{d-1})} \\
&= \sum_{y \in \mathbb{F}_{2^n}} (-1)^{g(\lambda^{-d^{-1}}y) + f(y^d)} \quad (\text{set } x = \lambda^{-1} y^d) \\
&= C_{g(x), f(x^d)}(\lambda^{-d^{-1}}).
\end{aligned}$$

$\square$

10.1.2 Signal sets

We mentioned the concept of signal sets with low crosscorrelation toward the end of Section 5.1 of Chapter 5. Here we present it precisely.

Definition 10.1 *Let* $\mathbf{s}_j = (s_{j,0}, s_{j,1}, \ldots, s_{j,v-1})$, $0 \le j < r$, *be* r *shift-distinct p-ary sequences of period v. Let* $S = \{\mathbf{s}_0, \mathbf{s}_1, \ldots, \mathbf{s}_{r-1}\}$ *and*

$$\delta = \max |C_{\mathbf{s}_i, \mathbf{s}_j}(\tau)| \text{ for any } 0 \le \tau < v, 0 \le i, j < r, \tag{10.8}$$

where $\tau \neq 0$ *if* $i = j$. *The set* S *is said to be a* (v, r, δ) *signal set, and* δ *is referred to as the* maximum correlation *of* S. *We say that the set* S *has* low *crosscorrelation if* $\delta \le c\sqrt{v}$ *where* c *is a constant.*

A sequence in S is also called a signal from the point of view of engineering (see Proakis 1995). When we consider crosscorrelation between two sequences $\mathbf{s}_i$ and $\mathbf{s}_j$ in S, we simply write $C_{i,j}(\tau)$ for $C_{\mathbf{s}_i, \mathbf{s}_j}(\tau)$. We also denote by $C(S)$ the image of all the crosscorrelation functions of pairs of sequences in S and all the out-of-phase autocorrelation functions of sequences in S; that is,

> $C(S)$ is the set consisting of all distinct values in
> $\{C_{\mathbf{s}_i,\mathbf{s}_j}(\tau) \mid 0 \le \tau < v, 0 \le i, j < r, \tau \ne 0 \text{ if } i = j\}.$ (10.9)

Thus, $\delta = \max_{c \in C(S)} |c|$. Sometimes we refer to $C(S)$ as an image of the correlation of S.

Remark 10.1 The maximum correlation between any pair of the sequences in S is lower bounded, approximately, by the square root of the length of the sequences. (This was established by Welch in 1971.)

Example 10.2 Let **a** and **b** be shift-distinct m-sequences of period 7, say

$$\mathbf{a} = 1110100 \text{ and } \mathbf{b} = 1001011.$$

Let

$$\mathbf{s}_j = L^j(\mathbf{a}) + \mathbf{b}, 0 \le j < 7,$$

and $S = \{\mathbf{a}, \mathbf{b}\} \cup \{\mathbf{s}_j \mid 0 \le j < 7\}$. We may compute the sequences in S as follows.

$$S$$

	$\mathbf{b} = 1001011$
	$\mathbf{a} = 1110100$
$\mathbf{a} = 1110100$	$\mathbf{a} + \mathbf{b} = 0111111$
$L(\mathbf{a}) = 1101001$	$L(\mathbf{a}) + \mathbf{b} = 0100010$
$L^2(\mathbf{a}) = 1010011$	$L^2(\mathbf{a}) + \mathbf{b} = 0011000$
$L^3(\mathbf{a}) = 0100111$	$L^3(\mathbf{a}) + \mathbf{b} = 1101100$
$L^4(\mathbf{a}) = 1001110$	$L^4(\mathbf{a}) + \mathbf{b} = 0000101$
$L^5(\mathbf{a}) = 0011101$	$L^5(\mathbf{a}) + \mathbf{b} = 1010110$
$L^6(\mathbf{a}) = 0111010$	$L^6(\mathbf{a}) + \mathbf{b} = 1110001$

Then all the sequences in S are shift distinct. Since both **a** and **b** are m-sequences with the constant-on-cosets property, the image $C(S)$ of the correlation of S can be determined by the weight distribution of the sequences in $\{L^j(\mathbf{a}) + \mathbf{b} \mid j = 0, 1, 3\} \cup \{\mathbf{a}, \mathbf{b}\}$. There are three different values for weights of the sequences in this set, namely 6, 2, and 4. Thus $C(S) = \{-1, -5, 3\}$. Therefore, we have $\delta = 5$ and S is a $(7, 9, 5)$ signal set.

10.1.3 Signal sets from pairs of sequences

Assume that $\mathbf{a}$ and $\mathbf{b}$ are binary sequences, where the period of $\mathbf{a}$ is $v = 2^n - 1$ and the period of $\mathbf{b}$ is t with $t | v$. If S is constructed by

$$S = \{\mathbf{s}_j \, | 0 \le j < t\} \cup \{\mathbf{a}, \mathbf{b}\}, \, \mathbf{s}_j = L^j(\mathbf{a}) + \mathbf{b}, \tag{10.10}$$

then $C(S)$, the image of the correlation of S, can be computed by the following formulae:

$$C_{i,j}(\tau) = \langle L^\tau(\mathbf{s}_i), \mathbf{s}_j \rangle = \langle L^{\tau+i}(\mathbf{a}) + L^j(\mathbf{a}), \mathbf{b} + L^\tau(\mathbf{b}) \rangle, \tag{10.11}$$

or equivalently

$$C_{i,j}(\tau) + 1 = \sum_{x \in \mathbb{F}_q} \omega^{f(\alpha^{\tau+i}x) + f(\alpha^j x) + g(\alpha^\tau x) + g(x)}, \tag{10.12}$$

together with

$$C_{\mathbf{a},\mathbf{b}}(\tau), \, C_{\mathbf{a},\mathbf{s}_j}(\tau), \, C_{\mathbf{b},\mathbf{s}_j}(\tau), \, C_{\mathbf{a}}(\tau), \, C_{\mathbf{b}}(\tau). \tag{10.13}$$

(Here i and j may be equal.) Let $f_j(x) = f(\alpha^j x) + g(x)$. Then this is the trace representation of $\mathbf{s}_j, 0 \le j < t$. Thus, Eq. (10.12) becomes

$$C_{i,j}(\tau) + 1 = C_{f_i, f_j}(\alpha^\tau).$$

This is just the crosscorrelation of the functions f_i and f_j. In particular, for the following two cases, we can reduce the computation of the correlation of S dramatically.

Case 1. One of the sequences $\mathbf{a}$ and $\mathbf{b}$ is an m-sequence of period $2^n - 1$. We may assume that $\mathbf{a}$ is an m-sequence of period $2^n - 1$ with the trace representation $f(x) = Tr(x^r)$ where $\gcd(r, 2^n - 1) = 1$. Thus, the correlation of S reduces to computing

$$\Delta(\lambda, \beta) = \sum_{x \in \mathbb{F}_{2^n}} (-1)^{Tr(\lambda x^r) + g(\beta x) + g(x)}, \tag{10.14}$$

where $\lambda = 0$ or $\lambda = \alpha^i, \forall i \in \Gamma_2(n)$, the set of the coset leaders modulo $2^n - 1$, and for all $\beta \in \mathbb{F}_{2^n}$.

Case 2. Both $\mathbf{a}$ and $\mathbf{b}$ are m-sequences. We may assume that $f(x) = Tr(x)$ and $g(x) = Tr(x^d)$. (Note. Because the period t of $\mathbf{b}$ could be a proper factor of $2^n - 1$, d may not be coprime to $2^n - 1$. In other words, $\mathbf{b}$ could be an m-sequence of degree m for which $m | n$.) In this case, computation of the correlation of S can be further reduced to the computation of the crosscorrelation of the pair $\mathbf{a}$ and $\mathbf{b}$, or the Hadamard transform of $g(x)$, because $f(x) = Tr(x)$.

Table 10.2. $n = 5$, $\alpha^5 + \alpha^3 + 1 = 0$,
and $f(x) = Tr(x^d)$

d	$\widehat{f}(\lambda)$, $\lambda = \alpha^\tau$						
	0	1	3	5	7	11	15
3	-8	0	0	-8	8	8	0
5	-8	0	0	8	-8	8	0
7	-8	0	8	0	8	0	-8
11	-8	-8	0	8	0	0	8
15	12	8	-8	4	4	-4	0

In other words, we have

$$C(S) \subset \{C_{\mathbf{a},\mathbf{b}}(\tau) | \tau \in \Gamma_2(n)\} \cup \{-1\} \tag{10.15}$$
$$= \{\widehat{g}(\lambda) - 1 \,|\, \lambda = \alpha^\tau, \tau \in \Gamma_2(n)\} \cup \{-1\}.$$

We consolidate the above discussions into the following property.

Property 10.5 *Let S be constructed by Eq. (10.10).*

1. *The image of the crosscorrelation of S, $C(S)$, can be determined by Eq. (10.11) or Eq. (10.12) together with Eq. (10.13).*
2. *In particular, if one of **a** and **b** is an m-sequence of period $2^n - 1$, or if both of them are m-sequences, then their respective images of correlation can be computed by Eq. (10.14) or Eq. (10.15), respectively.*

For the case that both sequences are m-sequences, using Property 10.5 together with Property 10.3, the prototypes of $C(S)$ are determined by the weight distribution of the sequences in S. Thus, to compute $C(S)$, it suffices to compute the weight distribution of the sequences in S. For example, the values of the weight of the sequences in S in Example 10.1 belong to $\{2, 4, 6\}$. Thus $C(S)$ can be computed from these values using Property 10.3.

Example 10.3 For $n = 5$ and $n = 7$, let $\mathbf{a} \leftrightarrow Tr(x)$ and $\mathbf{b} \leftrightarrow f(x) = Tr(x^d)$ where $\gcd(d, 2^n - 1) = 1$. Because both $\mathbf{a}$ and $\mathbf{b}$ are m-sequences, according to Property 10.5, we only need to compute $C_{\mathbf{a},\mathbf{b}}(\tau) = \widehat{f}(\alpha^\tau) - 1, \forall \tau \in \Gamma_2(n)$ where $n = 5$ or $n = 7$, as shown in Tables 10.2 and 10.3.

Case 1. $n = 5$. We have

$$C_{\mathbf{a},\mathbf{b}}(\tau) \in \{-1, -1 \pm 8\}, \text{ for } d = 3, 5, 7, \text{ and } 11$$

and

$$C_{\mathbf{a},\mathbf{b}}(\tau) \in \{-1, -1 \pm 8, -1 \pm 4, -1 + 12\}, \text{ for } d = 15.$$

Table 10.3. $n = 7$, $\alpha^7 + \alpha + 1 = 0$, and $f(x) = Tr(x^d)$

d	$\widehat{f}(\lambda), \lambda = \alpha^\tau$																		
	0	1	3	5	7	9	11	13	15	19	21	23	27	29	31	43	47	55	63
3	16	0	0	0	16	0	0	16	0	-16	-16	0	-16	0	-16	16	16	0	16
5	16	0	0	0	16	0	0	-16	0	-16	-16	0	16	0	16	16	16	0	-16
7	-40	0	-16	0	-8	16	0	-8	-16	24	8	16	-8	0	8	8	-8	0	8
9	16	0	0	0	16	0	0	-16	0	16	8	-16	-16	0	-16	16	16	0	-16
11	16	0	16	16	16	-16	0	16	0	16	16	0	0	-16	16	0	16	0	0
13	16	0	0	16	-16	16	0	0	0	0	-16	-16	0	16	-16	0	-16	-16	16
15	16	16	16	-16	0	-16	-16	0	0	16	0	0	-16	16	16	16	0	0	0
19	-40	0	16	0	-8	-16	0	-8	-16	-8	8	16	24	0	8	8	-8	0	8
21	-40	0	-16	0	-8	0	0	8	16	8	8	-16	24	16	-8	-8	-8	16	8
23	16	16	0	0	0	0	0	0	16	-16	16	-16	16	0	16	0	-16	-16	0
27	16	-16	-16	-16	0	0	-16	0	16	16	0	16	0	16	-16	-16	0	0	16
29	16	0	0	0	16	16	16	16	-16	16	0	0	0	0	-16	0	0	0	-8
31	-40	-16	0	0	8	24	16	16	-8	-16	8	0	8	-8	16	0	-8	8	-8
43	16	-16	16	0	0	-16	16	-16	16	0	0	-16	0	0	16	16	0	8	16
47	-40	0	-16	-8	-8	8	0	16	16	8	8	8	0	-16	-8	0	24	8	-8
55	-40	16	-8	8	8	-8	-16	-8	0	16	8	0	0	-8	24	-16	8	16	0
63	-12	0	-8	16	-12	-8	16	20	-8	4	12	8	4	0	12	-20	4	-16	-4

Case 2. $n = 7$. We have

$$C_{\mathbf{a},\mathbf{b}}(\tau) \in \{-1, -1 \pm 16\}, \text{ for } d \in \{3, 5, 9, 11, 13, 15, 23, 27, 29, 43\}.$$
(10.16)

Example 10.4 Let $\mathbf{a}$ and $\mathbf{b}$ be the m-sequence and the quadratic residue sequence in Example 9.11 in Section 9.4; that is,

$\mathbf{a} = 1000010101110110001111100110100 \quad \leftrightarrow \quad f(x) = Tr(x)$
$\mathbf{b} = 1001001000011101010001111011011 \quad \leftrightarrow \quad g(x) = Tr(x + x^5 + x^7)$

where $a_i = Tr(\alpha^i)$ and $b_i = g(\alpha^i)$ where $\alpha^5 + \alpha^3 + 1 = 0$. Let

$$S = \{\mathbf{s}_j = L^j(\mathbf{a}) + \mathbf{b}|0 \le j < 31\} \cup \{\mathbf{a}, \mathbf{b}\}.$$

From Property 10.5, the prototype of $C(S)$ is determined by

$$\Delta(\lambda, \beta) = \sum_{x \in \mathbb{F}_{2^5}} (-1)^{Tr(\lambda x) + g(\beta x) + g(x)},$$

where $\lambda = 0$ or $\lambda = \alpha^i, \forall i \in \Gamma_2(5)$, the set of all the coset leaders modulo 31, and $\forall \beta \in \mathbb{F}_{2^5}$. Because there are seven coset leaders modulo 31, we need to compute $8 \times 32 = 256$ inner products of vectors of length 31. Note that if $\lambda = 0$, then $\Delta(0, \beta)$ is the autocorrelation of $g(x)$, which is equal to 0 if $\beta \ne 1$. Thus, we only need to compute $\Delta(\alpha^i, \alpha^j)$, for all $i \in \Gamma_2(5) = \{0, 1, 3, 5, 7, 11, 15\}$ and j with $0 \le j < 31$. These values are shown in Table 10.4. Thus, we obtain

$$C(S) = \{-1, -1 \pm 8, -1 \pm 16\} \implies \delta = 17.$$

Therefore, S is a (31, 33, 17) signal set.

10.1.4 One-to-one correspondence between sequences and boolean functions

In Section 6.4, we introduced the one-to-one correspondence between sequences and polynomial functions in terms of the trace representations of the sequences. Here, we revisit this relation and extend it to boolean functions. Recall that we introduced the following notation: $\mathcal{S}_2$, the set of all binary sequences with period $N|(2^n - 1)$; and $\mathcal{F}_2$, the set of all (polynomial) functions from $\mathbb{F}_{2^n}$ to $\mathbb{F}_2$.

A boolean function is a function of n variables from $\mathbb{F}_2^n$ to $\mathbb{F}_2$. An algebraic normal form of a boolean function of n variables is given by

$$g(x_0, \ldots, x_{n-1}) = \sum a_{i_1, \ldots, i_t} x_{i_1} \cdots x_{i_t}, \quad a_{i_1, \ldots, i_t} \in \mathbb{F}_2,$$
(10.17)

Table 10.4. *Evaluation of* $\Delta(\alpha^i, \alpha^j)$

i	0	1	3	5	7	11	15	i	0	1	3	5	7	11	15
j								j							
0	0	0	0	0	0	0	0	16	0	−8	0	−8	0	0	−8
1	0	0	0	8	0	8	0	17	8	0	0	0	0	8	8
2	0	8	0	0	−8	0	0	18	0	0	8	8	−8	0	0
3	8	8	8	0	0	8	−8	19	8	8	8	8	0	0	8
4	0	8	0	0	16	8	−8	20	0	0	−8	0	8	8	0
5	0	0	8	−8	0	8	8	21	−8	0	0	0	−8	8	0
6	8	0	−8	−8	−8	0	8	22	−8	0	0	8	0	8	8
7	8	0	−8	8	8	0	8	23	0	8	0	0	0	0	−8
8	0	8	0	0	8	0	0	24	8	0	8	0	0	0	0
9	0	0	0	0	0	8	0	25	8	−8	0	0	8	−8	−8
10	0	−16	0	8	0	0	0	26	−8	8	0	8	0	−16	0
11	−8	8	0	−8	0	0	0	27	0	0	16	−8	8	−8	8
12	8	8	−8	0	0	−8	8	28	8	0	0	8	0	0	−8
13	−8	0	8	8	0	0	0	29	0	0	0	−8	0	0	0
14	8	0	0	0	−8	0	0	30	0	−8	8	0	8	0	8
15	0	0	0	8	0	0									

where the sum runs through all the t-subsets $\{i_1, \ldots, i_t\} \subset \{0, 1, \ldots, n-1\}$. The degree of the boolean function g is the largest t for which $a_{i_1,\ldots,i_t} \neq 0$. We now denote by $\mathcal{B}_2$ the set of all boolean functions of n variables. We will establish that there exist one-to-one correspondences among these three sets:

$$\mathcal{S}_2 \leftrightarrow \mathcal{F}_2 \leftrightarrow \mathcal{B}_2.$$

We have seen the one-to-one correspondence between $\mathcal{S}_2$ and $\mathcal{F}_2$; that is, the sequences and the polynomial functions in Chapter 6. Applying the Lagrange interpolation formula to a given boolean function $g(x_0, \ldots, x_{n-1}) \in \mathcal{B}_2$ (see Eq. (6.10) in Section 6.4), we can determine its polynomial representation $f(x)$ as follows:

$$f(x) = \sum_{i=0}^{2^n-1} d_i x^i, \quad \text{where}$$
$$d_0 = g(0, \ldots, 0),$$
$$d_i = \sum_{x \in \mathbb{F}_{2^n}^*} g(x_0, \ldots, x_{n-1}) x^{-i}, \, 1 \le i \le 2^n - 1, \, x = \sum_{i=0}^{n-1} x_i \alpha_i$$

(10.18)

Next, we will show how to obtain a boolean representation from a polynomial representation of a function from $\mathbb{F}_{2^n}$ to $\mathbb{F}_2$ in terms of the linear space structure of the finite field $\mathbb{F}_{2^n}$. Let $\{\alpha_0, \ldots, \alpha_{n-1}\}$ be a basis of $\mathbb{F}_{2^n}$ over $\mathbb{F}_2$, denoted

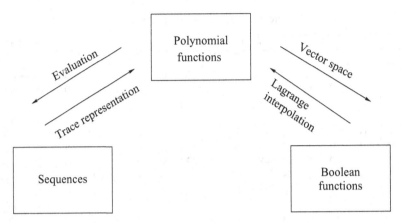

Figure 10.1. One-to-one correspondences among $\mathcal{S}_2$, $\mathcal{F}_2$, and $\mathcal{B}_2$.

by $\mathbb{F}_{2^n} = <\alpha_0, \ldots, \alpha_{n-1}>$, and let α be a primitive element of $\mathbb{F}_{2^n}$. For any $x \in \mathbb{F}_{2^n}$, we can represent x as

$$\rho : x = x_0\alpha_0 + x_1\alpha_1 + \cdots + x_{n-1}\alpha_{n-1} \leftrightarrow \mathbf{x} = (x_0, \ldots, x_{n-1}), x_i \in \mathbb{F}_2.$$

Thus ρ is an isomorphism between $\mathbb{F}_{2^n}$ and $\mathbb{F}_2^n$. Hence, for $f(x) \in \mathcal{F}$, we have

$$f(x) = f\left(\sum_{i=0}^{n-1} x_i\alpha_i\right) = g(x_0, \ldots, x_{n-1});$$

that is,

$$\kappa_\rho : f(x) \to g(x_0, \ldots, x_{n-1}), \tag{10.19}$$

which is a bijective map from $\mathcal{F}_2$ to $\mathcal{B}_2$ induced by ρ. Therefore, a conversion from a polynomial function to a boolean function is given by

$$\boxed{\begin{aligned} g(x_0, \ldots, x_{n-1}) = f\left(\sum_{i=0}^{n-1} x_i\alpha_i\right), \quad \text{where} \\ \mathbb{F}_{2^n} = <\alpha_0, \ldots, \alpha_{n-1}>. \end{aligned}} \tag{10.20}$$

The boolean function $g(x_0, x_1, \ldots, x_{n-1})$ has degree r where r is the maximum Hamming weight $w(k)$ as k runs through all the exponents in the trace terms of f. This value is also called the algebraic degree of $f(x)$.

Thus Eq. (10.18) gives a bijective map from $\mathcal{B}_2$ to $\mathcal{F}_2$ that is the inverse of Eq. (10.20). These one-to-one correspondences among those sets are shown in Figure 10.1

Example 10.5 Let $n = 3$, and let $\mathbb{F}_{2^3}$ be defined by the primitive polynomial $t(x) = x^3 + x + 1$, with α a root of $t(x)$ in $\mathbb{F}_{2^3}$. Thus $\{1, \alpha, \alpha^2\}$ is a basis for

$\mathbb{F}_{2^3}$ over $\mathbb{F}_2$. Let $f(x) = Tr(x^3)$. Then this is the trace representation of the m-sequence $\mathbf{a} = 1110100$. For any $x \in F_{2^3}$, we write $x = x_0 + x_1\alpha + x_2\alpha^2$. From the conversion formula (10.20),

$$
\begin{aligned}
f(x) = Tr(x^3) &= Tr((x_0 + x_1\alpha + x_2\alpha^2)^3) \\
&= Tr(x_0 + x_1 + x_2 + x_1x_2) + Tr((x_1 + x_0x_1 + x_0x_2)\alpha) \\
&\quad + Tr((x_0x_1 + x_2)\alpha^2) \\
&= (x_0 + x_1 + x_2 + x_1x_2) + (x_1 + x_0x_1 + x_0x_2)Tr(\alpha) \\
&\quad + (x_0x_1 + x_2)Tr(\alpha^2) \\
&= x_0 + x_1 + x_2 + x_1x_2.
\end{aligned}
$$

The last identity is obtained by noting that $Tr(\alpha) = Tr(\alpha^2) = 0$. Therefore, we have the following one-to-one correspondences:

$$\mathbf{a} = 1110100 \leftrightarrow Tr(x^3) \leftrightarrow x_0 + x_1 + x_2 + x_1x_2.$$

10.1.5 Walsh transform of boolean functions

Definition 10.2 *The* Walsh transform *of a boolean function f is defined by*

$$\widehat{f}(\mathbf{w}) = \sum_{\mathbf{x} \in \mathbb{F}_2^n} (-1)^{\mathbf{w} \cdot \mathbf{x} + f(\mathbf{x})}, \quad \mathbf{w} \in \mathbb{F}_2^n. \tag{10.21}$$

(Here we omit the bracket for the dot product of $\mathbf{w}$ and $\mathbf{x}$ for simplicity.)

In the following, we will derive a conversion between the Walsh transforms of boolean functions and the Hadamard transforms of their corresponding polynomial functions. We denote by $a(\mathbf{x}) = \mathbf{w} \cdot \mathbf{x} = \sum_{i=0}^{n-1} w_i x_i$ a linear boolean function, where $\mathbf{w} = (w_0, w_1, \ldots, w_{n-1}) \in \mathbb{F}_2^n$, a constant vector. Let $f(x)$ and $a(x)$ be the polynomial representations of $f(\mathbf{x})$ and $a(\mathbf{x})$, respectively. (We will use the same notation for both boolean functions and their corresponding polynomial forms when the context is clear.) The following property is immediate.

Property 10.6 *(a) There exists some $\lambda \in \mathbb{F}_{2^n}$ such that*

$$a(\mathbf{x}) = Tr(\lambda x). \tag{10.22}$$

(b) The Hadamard transform of $f(x)$ and the Walsh transform of $f(\mathbf{x})$ have the following relation:

$$\widehat{f}(\mathbf{w}) = \widehat{f}(\lambda), \mathbf{w} \in \mathbb{F}_2^n, \lambda \in \mathbb{F}_{2^n}, \tag{10.23}$$

where $\mathbf{w} \cdot \mathbf{x} = Tr(\lambda x)$. □

For example, with α in Example 10.5, using $Tr(\alpha^{2^i}) = 0$ and $Tr(\alpha^{3 \cdot 2^i}) = 1$, $i = 0, 1, 2$, we have

$$Tr(x) = Tr(x_0 + x_1\alpha + x_2\alpha^2)$$
$$= x_0 Tr(1) + x_1 Tr(\alpha) + x_2 Tr(\alpha^2)$$
$$= x_0 = \mathbf{w} \cdot \mathbf{x} \text{ where } \mathbf{w} = (1, 0, 0) \text{ and } \mathbf{x} = (x_0, x_1, x_2),$$
$$Tr(\alpha^3 x) = Tr(x_0\alpha^3 + x_1\alpha^4 + x_2\alpha^5)$$
$$= x_0 Tr(\alpha^3) + x_1 Tr(\alpha^4) + x_2 Tr(\alpha^5)$$
$$= x_0 + x_2 = \mathbf{w} \cdot \mathbf{x} \text{ where } \mathbf{w} = (1, 0, 1) \text{ and } \mathbf{x} = (x_0, x_1, x_2).$$

10.2 Odd case: Gold-pair signal sets and their generalization

In this section, we introduce the Gold-pair construction and its generalization to binary signal sets with parameters $(2^n - 1, 2^n + 1, 2^{(n+1)/2} + 1)$ where n is odd. We keep the notation that $n = 2m + 1$, an odd number, and α is a primitive element in $\mathbb{F}_{2^n}$.

10.2.1 Gold-pair construction

Construction: Select d from the following list:

(a) $d = 2^k + 1$ (the Gold decimation), $\gcd(k, n) = 1$, and $k \le \frac{n-1}{2}$.
(b) $d = 2^{2k} - 2^k + 1$ (the Kasami (large set) decimation), $\gcd(k, n) = 1$, and $k \le \frac{n-1}{2}$.
(c) $d = 2^{\frac{n-1}{2}} + 3$ (the Welch decimation).
(d) $d = 2^{2k} + 2^k - 1$ (the Niho decimation) where

$$k = \begin{cases} \frac{n-1}{4} & \text{if } n \equiv 1 \pmod 4 \\ \frac{3n-1}{4} & \text{if } n \equiv 3 \pmod 4. \end{cases}$$

(e) Inverse of d, for d in each of the above four cases.

For $0 \le j < 2^n - 1$, let $\mathbf{s}_j = \{s_{j,i}\}$ be a binary sequence whose elements are given by

$$s_{j,i} = Tr(\alpha^j \alpha^i + \alpha^{di}), i = 0, 1, \ldots, 2^n - 2.$$

Then $\mathbf{s}_j$ is called a Gold-pair sequence. Let $s_{2^n-1} = \mathbf{a} = \{Tr(\alpha^i)\}$ and $\mathbf{s}_{2^n} = \mathbf{b} = \mathbf{a}^{(d)}$. The set given by

$$S(d) = \{\mathbf{s}_j \mid 0 \le j \le 2^n\}$$

is said to be a Gold-pair (signal) set.

Note that both **a** and **b** are m-sequences, and $\mathbf{s}_j$ is a sum of **a** at shift j and **b** for $0 \le j < 2^n - 1$; that is,

$$\mathbf{s}_j = L^j \mathbf{a} + \mathbf{b}, 0 \le j < 2^n - 1.$$

Thus $S(d)$ has $2^n + 1$ shift-distinct sequences. Let $f(x) = Tr(x^d)$. According to Property 10.5, the prototype of crosscorrelation of $S(d)$, that is, $C(S(d))$, is determined by $C_{\mathbf{a},\mathbf{b}}(\tau)$ where τ runs through $\Gamma_2(n)$, the set of all the coset leaders modulo $2^n - 1$. In other words, we have

$$C(S(d)) \subset \{ \widehat{f}(\lambda) - 1 \mid \lambda = \alpha^\tau, \tau \in \Gamma_2(n) \}.$$

In the following, we show how to compute $C_{\mathbf{a},\mathbf{b}}(\tau)$, or equivalently, $\widehat{f}(\lambda)$, for the original Gold case $d = 2^k + 1$.

Theorem 10.1 *(Gold, 1968) Let n be odd and $d = 2^k + 1$ with $\gcd(k, n) = 1$. Then*

$$\widehat{f}(\lambda) = C_{\mathbf{a},\mathbf{b}}(\tau) + 1 = \begin{cases} 0 & \Longleftrightarrow \quad Tr(\lambda) = 0 \\ \pm 2^{(n+1)/2} & \Longleftrightarrow \quad Tr(\lambda) = 1, \end{cases}$$

where $\lambda = \alpha^\tau \in \mathbb{F}_{2^n}$.

Proof. Here we present a proof given by L. Welch instead of the original proof of Gold. Welch's proof is unpublished, but it is well known as the squaring method. Note that

$$\widehat{f}(\lambda) = \sum_{x \in \mathbb{F}_q} (-1)^{Tr(\lambda x) + Tr(x^d)} \quad (\text{set } q = 2^n), \tag{10.24}$$

and by squaring (10.24), we have

$$\widehat{f}^2(\lambda) = \sum_{x, w \in \mathbb{F}_q} (-1)^{Tr(\lambda x) + Tr(\lambda(x+w)) + Tr(x^d) + Tr((x+w)^d)}.$$

Substituting $d = 2^k + 1$, and then using the trace function identity $Tr(wx^{2^k}) = Tr(w^{2^{-k}}x)$, we get

$$\widehat{f}^2(\lambda) = \sum_{x, w \in \mathbb{F}_q} (-1)^{Tr(\lambda w) + Tr(w^d) + Tr(x^{2^k}w + xw^{2^k})}$$

$$= \sum_{w \in \mathbb{F}_q} (-1)^{h(w)} \sum_{x \in \mathbb{F}_q} (-1)^{Tr\left((w^{2^{-k}} + w^{2^k})x\right)}$$

$$(\text{set } h(w) = Tr(\lambda w) + Tr(w^d))$$

$$= 2^n \sum_{w \in L} (-1)^{h(w)}$$

where

$$L = \{w \in \mathbb{F}_q \mid w^{2^{-k}} + w^{2^k} = 0\}.$$

Since $\gcd(2k, n) = \gcd(k, n) = 1$, we have

$$w^{2^{-k}} + w^{2^k} = 0 \implies w^{2^{-k}} = w^{2^k} \implies w^{2^{2k}} = w$$

$$\implies w \in \mathbb{F}_{2^{2k}} \cap \mathbb{F}_{2^n} = \mathbb{F}_{2^{\gcd(2k,n)}} = \mathbb{F}_2 \implies L = \mathbb{F}_2.$$

Notice that $h(0) = 0$ and $h(1) = Tr(\lambda) + 1 \in \{0, 1\}$. Therefore

$$\widehat{f}^2(\lambda) = 2^n \sum_{w \in \mathbb{F}_2} (-1)^{h(w)} = 2^n \left(1 + (-1)^{Tr(\lambda)+1}\right). \tag{10.25}$$

If $Tr(\lambda) + 1 = 1 \implies Tr(\lambda) = 0, \lambda \in \mathbb{F}_q$, then Eq. (10.25) gives $\widehat{f}^2(\lambda) = 0$. Hence

$$\widehat{f}(\lambda) = 0 \iff Tr(\lambda) = 0. \tag{10.26}$$

If $Tr(\lambda) + 1 = 0, \lambda \in \mathbb{F}_q \implies Tr(\lambda) = 1$, then

$$\widehat{f}^2(\lambda) = 2^n(1 + 1) = 2^{n+1} \implies \widehat{f}(\lambda) = \pm 2^{(n+1)/2}.$$

Therefore,

$$\widehat{f}(\lambda) = \pm 2^{(n+1)/2} \iff Tr(\lambda) = 1. \tag{10.27}$$

The assertion follows from Eqs. (10.26) and (10.27). $\square$

Corollary 10.1 *We use the same notation as in the Gold-pair construction where $f(x) = Tr(x^d)$ $(n = 2m + 1)$.*

1. $S(d)$ is a $(2^n - 1, 2^n + 1, 2^{m+1} + 1)$ signal set.
2. The crosscorrelation of any pair of sequences in $S(d)$ or out-of-phase autocorrelation of any sequence in S is 3-valued and belongs to the set $\{-1, -1 \pm 2^{m+1}\}$.
3. In $\{\widehat{f}(\lambda) \mid \lambda \in \mathbb{F}_q\}$, 0 occurs 2^{n-1} times and $\pm 2^{m+1}$ (combined) occurs 2^{n-1} times. Precisely, the frequencies for these values are given by

$\widehat{f}(\lambda)$	frequency
0	2^{n-1}
2^{m+1}	$2^{n-2} + 2^{m-1}$
-2^{m+1}	$2^{n-2} - 2^{m-1}$

Table 10.5. *Exponents for the Gold-pair constructions*

n	k	Gold exp. $d = 2^k + 1$	Inverse d^{-1}	Kasami exp. $d = 2^{2k} - 2^k + 1$	Inverse d^{-1}
5	1	3	11		
	2	5	7	$13 \in C_{11}$	3
7	1	3	43		
	2	5	27	13	11
	3	9	15	$57 \in C_{23}$	29
9	1	3	171		
	2	5	103	13	59
	4	17	31	$241 \in C_{47}$	87
11	1	3	683		
	2	5	411	13	315
	3	9	231	57	413
	4	17	365	$241 \in C_{143}$	43
	5	33	63	$993 \in C_{95}$	151

Note. For the Kasami exponent d with $3k \equiv 1 \pmod{n}$, the Hadamard transform of $f(x) = Tr(x^d)$ has a result similar to Theorem 10.1, which appeared as identity (9.11) at the beginning of Section 9.3, when we discussed the Kasami power function construction of 2-level autocorrelation sequences.

Example 10.6 Let $n \in \{5, 7, 9, 11\}$. From the Gold-pair construction, we compute decimation values d for all cases, as shown in Tables 10.5 and 10.6. In these tables, C_i represents the coset containing the leader i modulo $2^n - 1$, and the columns under d^{-1} list the leaders of the cosets containing d^{-1} instead of d^{-1} itself. For the Kasami exponent, the case $k = 1$ is omitted from the list of the Kasami exponents since $2^{2k} - 2^k + 1 = 2^k + 1 = 2 + 1 = 3$ when $k = 1$, which degenerates to a Gold exponent. For each d in Table 10.5, $S(d)$ is a signal set with the features shown below.

n	Parameters	$C(S(d))$	δ
5	$(31, 33, 9)$	$\{-1, -9, 7\}$	9
7	$(127, 129, 17)$	$\{-1, -17, 15\}$	17
9	$(511, 513, 33)$	$\{-1, -33, 31\}$	33
11	$(2047, 2049, 65)$	$\{-1, -65, 63\}$	65

Remark 10.2 For $n = 5$ and 7, from Example 10.3, the exponents shown in Tables 10.5 and 10.6 are all the decimation sequences having 3-valued

Table 10.6. *Exponents for the Gold-pair constructions* (cont.)

n	Welch exp. d $2^{(n-1)/2} + 3$	Inverse d^{-1}	k	Niho exp. d $2^{2k} + 2^k - 1$	Inverse d^{-1}
5	7	5	$\frac{n-1}{4} = 1$	5	7
7	11	13	$\frac{3n-1}{4} = 5$	$39 \in C_{29}$	23
9	19	27	$\frac{n-1}{4} = 2$	19	27
11	35	117	$\frac{3n-1}{4} = 8$	$287 \in C_{249}$	107

crosscorrelation with $Tr(x)$. In fact, the four cases of d in the Gold-pair construction are all the known cases whose crosscorrelations with $Tr(x)$ are 3-valued and belong to $\{-1, -1 \pm 2^{m+1}\}$. (No other examples have been found by computer search where decimation does not belong to one of these four cases or their inverses.)

10.2.2 Randomness profile and implementation

The randomness profile of the Gold-pair construction ($n = 2m + 1$)

1. Each sequence has period $2^n - 1$.
2. There are $2^n + 1$ shift-distinct sequences in $S(d)$.
3. Crosscorrelation of any pair of sequences in $S(d)$ or out-of-phase autocorrelation of any sequence in $S(d)$ is 3-valued and belongs to $\{-1, \pm 2^{m+1}\}$. In other words, the image of the correlation of $S(d)$ is given by $C(S(d)) = \{-1, \pm 2^{m+1}\}$ for all d in the list.
4. Together with shift-equivalent sequences, from $S(d)$ there are $2^{2n} - 1$ different nonzero sequences.
5. There are $2^{n-1} + 1$ balanced sequences in $S(d)$. In particular, for all the Gold exponents and one Kasami exponent, the balanced sequences can be determined by

$$\mathbf{s}_j \text{ balanced}$$
$$\Longleftrightarrow \begin{cases} Tr(\alpha^j) = 0 & \text{for the Gold case } d = 2^k + 1 \\ Tr\left(\alpha^{(2^k+1)j}\right) = 0 & \text{for the Kasami case } d = 2^{2k} - 2^k + 1 \\ & \text{where } 3k \equiv 1 (\text{mod } n). \end{cases}$$

Together with $\mathbf{a}$, this constitutes a total of $2^{n-1} + 1$ balanced sequences.
6. Each sequence in $S(d)$ has linear span $2n$ except for $\mathbf{a}$ and $\mathbf{b}$, which have linear span n.
7. For a fixed α, a primitive element of $\mathbb{F}_q$, the numbers of the Gold-pair signal sets for each type of d are given by

$n > 9$	Gold exp.	Kasami exp.	Welch exp.	Niho exp.	All inverses
	$\phi(n)/2$	$\phi(n)/2 - 1$	1	1	$\phi(n) + 1$
Total			$2\phi(n) + 2$		

In the following, we discuss the implementation of these signal sets.

The LFSR implementation

1. Select odd n, and $t(x)$, a primitive polynomial of degree n over $\mathbb{F}_2$, as the characteristic polynomial of LFSR1.
2. Select d from the construction, and compute the minimal polynomial of α^d as the characteristic polynomial of LFSR2.
3. We obtain all the sequences in S by employing different initial states of LFSR1, as shown in Figure 10.2

Example 10.7 Design of a $(31, 33, 9)$ Gold-pair signal set.

Method 1. The LFSR implementation

1. Select $t(x) = x^5 + x^3 + 1$, a primitive polynomial over $\mathbb{F}_2$ of degree 5, as the characteristic polynomial of LFSR1.
2. Select $d = 1 + 2^2 = 5$ and compute the minimal polynomial of α^5 where $\mathbb{F}_{2^5}$ is defined by $t(\alpha) = 0$. This gives

$$v(x) = x^5 + x^4 + x^3 + x + 1.$$

   Use $v(x)$ as the characteristic polynomial of LFSR2.
3. Fixing the initial state of one of these two LFSRs and varying the other, we obtain all 33 sequences in $S(5)$, as shown in Figure 10.3.

Method 2. Software implementation

1. Select $f(x) = x^5 + x^3 + 1$, a primitive polynomial over $\mathbb{F}_2$ of degree 5, to generate the sequence **a**:

$$\mathbf{a} = 1000010101110110001111100110100.$$

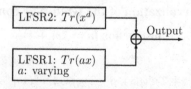

Figure 10.2. LFSR implementation of a Gold-pair generator.

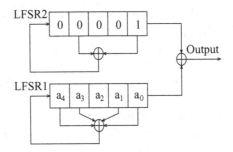

Figure 10.3. LFSR implementation of a $(31, 33, 9)$ Gold-pair generator.

2. Pick $d = 1 + 2^2 = 5$ and perform the 5-decimation operation on $\mathbf{a}$. We obtain $\mathbf{b} = \mathbf{a}^{(5)} = \{b_i\}$ where $b_i = a_{5i}$, as shown below:

$$\mathbf{b} = 1110110011100001101010010001011.$$

3. Compute

$$\mathbf{s}_i = L^i(\mathbf{a}) + \mathbf{b}, 0 \le i < 31.$$

Together with $\mathbf{a}$ and $\mathbf{b}$, there are 33 sequences in $S(5)$ that are presented in Table 10.7.

Randomness profile of $S(5)$:

1. Period 31.
2. There are 33 shift-distinct sequences in $S(5)$.
3. Crosscorrelation of any two sequences in S and out-of-phase autocorrelation of any sequence in S takes the three values -1, -9, and 7.
4. $S(5)$ is a $(31, 33, 9)$ signal set.
5. There are 17 balanced sequences in $S(5)$, including $\mathbf{a}$, $\mathbf{b}$, and those framed indices in Table 10.7; that is, $\mathbf{s}_i$ is balanced for $i \in C_1 \cup C_3 \cup C_{15}$. The other 16 sequences have as their respective weights either 12 or 20.
6. Linear span 10 except for $\mathbf{a}$ and $\mathbf{b}$, which have linear span 5.

10.2.3 Generalization of the Gold-pair construction

Construction ($n = 2m + 1$)

Let

$$T = \{g(x) + Tr_1^n(\beta x), \beta \in \mathbb{F}_{2^n}\} \cup \{Tr_1^n(x)\} \tag{10.28}$$

Table 10.7. s_i in $(31, 33, 9)$ signal set

i	s_i
0	0110100110010111100101110111111
1	1110011000001101110101011100010
2	1111100100111001010100001011001
3	1100011101010000010110100101111
4	1011101110000010010011111000011
5	0100001000100110011001000011011
6	1011000101101110001100110101010
7	0101011111111101001110110010010
8	1001101011011111100000000001110
9	0000000010011101011110110000001
10	0011010000011000000011010011110
11	0101110100010010111000010100000
12	1000111100000111001110011011100
13	0010101100101100100010000100101
14	0110001101111011111010111010110
15	1111001111010101001011000110000
16	1101001010001000101000111111101
17	1001000000110011101111001100111
18	0001010101000101100000101010011
19	0001111110101001111111100111010
20	0000101001110001000001111101000
21	0010000111000000111101001001100
22	0111011010100011000100100000100
23	1101100001100100110111110010100
24	1000010111101011010001010110101
25	0011111011110100011100011110111
26	0100100011001010000110001110010
27	1010010010110110110010101111000
28	0111110001001111011011101101101
29	1100110110111100001001101000110
30	1010111001011010101101100010001

and $S(g)$ be the set consisting of sequences whose trace representations are functions in T (evaluated at α). If

$$g(x) = \sum_{i=1}^{m} Tr_1^{n_i}(x^{1+2^i}),$$

where $n_i | n$, the size of the coset containing $1 + 2^i$, then $S(g)$ is a $(2^n - 1, 2^n + 1, 2^{m+1} + 1)$ signal set. Furthermore, crosscorrelation of any pair of the sequences in $S(g)$ or out-of-phase autocorrelation of any sequence in $S(g)$ is 3-valued and belongs to $\{-1, -1 \pm 2^{(n+1)/2}\}$. (Boztas and Kumar, 1994.)

Together with the case $g(x) = Tr(x^d)$ where d is taken from the Gold-pair construction, these are all the known constructions for $(2^n - 1, 2^n + 1, 2^{(n+1)/2} + 1)$ signal sets for n odd.

Example 10.8 Let $n = 5$, let α be the primitive element in Example 10.7, and let $g(x)$ be constructed via the above construction. Then $g(x) = Tr(x^3 + x^5)$. Let

$$a_i = Tr(\alpha^i)$$
$$b_i = g(\alpha^i), i = 0, 1, \ldots, 30.$$

Then

$$S(g) = \{L^i(\mathbf{a}) + \mathbf{b} \mid 0 \le i < 31\} \cup \{\mathbf{a}, \mathbf{b}\}$$

forms a $(31, 33, 9)$ signal set.

10.3 Even case: Kasami (small) signal sets and their generalization

In this section, we present constructions for binary signal sets with parameters $(2^n, 2^{n/2}, 2^{n/2} + 1)$ where n is even. We use the following notation in this section: $n = 2m$, $q = 2^n$, α is a primitive element in $\mathbb{F}_{2^n}$, $v = 2^m - 1$, and $d = 2^m + 1$.

10.3.1 Kasami (small) signal sets

Construction

Let $\mathbf{s}_\lambda = \{s_{\lambda,i}\}$ be a binary sequence whose elements are given by

$$s_{\lambda,i} = f_\lambda(\alpha^i), i = 0, 1, \ldots, \tag{10.29}$$

where

$$f_\lambda(x) = Tr_1^m \left(Tr_m^n(x^2) + \lambda x^d \right), \lambda \in \mathbb{F}_{2^m}, x \in \mathbb{F}_{2^n}. \tag{10.30}$$

A signal set S consists of $\mathbf{s}_\lambda$ for all $\lambda \in \mathbb{F}_{2^m}$; that is,

$$S = \{\mathbf{s}_\lambda \text{ such that } \lambda \in \mathbb{F}_{2^m}\}. \tag{10.31}$$

$\mathbf{s}_\lambda$ is said to be a Kasami (small set) sequence and S a Kasami (small) signal set. Note that $f_\lambda(x)$ is the trace representation of $\mathbf{s}_\lambda$.

Theorem 10.2 (Kasami, 1969) *S is a $(2^n, 2^m, 2^m + 1)$ $(n = 2m)$ signal set. Moreover, the crosscorrelation of any pair of sequences in S or out-of-phase autocorrelation of any sequence in S is 3-valued and belongs to $\{-1, -1 \pm 2^m\}$.*

To prove Theorem 10.2, we first derive the interleaved structure for the sequences in S. This structure also gives a simple proof for the unbalanced property of the Kasami small set sequences in the next subsection. The following assertions, related to finite fields and trace functions, will be used in several places.

Assertions:

1. $d^2 \equiv 2d$ (mod $2^n - 1$).
2. Let $\beta = \alpha^d$. Then β is a primitive element in $\mathbb{F}_{2^m}$.
3. For any $y \in \mathbb{F}_{2^m}$, $x \in \mathbb{F}_{2^n}$, $Tr_m^n(xy) = y Tr_m^n(x)$.

We avoid using double subscripts by renaming the sequences in S. Thus, let $\mathbf{u} = \{u_i\} = \mathbf{s}_\lambda \in S$. For $k = id + j, 0 \le i < v, 0 \le j < d$,

$$
\begin{aligned}
u_k = u_{id+j} &= Tr_1^m \left(Tr_m^n(\alpha^{2(id+j)}) + \lambda \alpha^{d(id+j)} \right) \\
&= Tr_1^m \left(\alpha^{2di} Tr_m^n(\alpha^{2j}) + \lambda \alpha^{2di} \alpha^{dj} \right) \text{ (by Assertions 1–3)} \\
&\implies u_{id+j} = Tr_1^m(\beta^{2i} t_j(\lambda)),
\end{aligned}
$$

where

$$
t_j(\lambda) = Tr_m^n(\alpha^{2j}) + \lambda \alpha^{dj}, \lambda \in \mathbb{F}_{2^m}. \tag{10.32}
$$

Thus, we have established the following result.

Lemma 10.1 *For* $\mathbf{u} = \{u_i\} = \mathbf{s}_\lambda \in S$,

$$
u_{id+j} = Tr_1^m(\beta^{2i} t_j), 0 \le i < v, 0 \le j < d,
$$

where $t_j = t_j(\lambda)$ *is defined in Eq. (10.32). Thus* $\mathbf{s}_\lambda$ *is a* (v, d) *interleaved sequence with the base sequence* $\mathbf{a} = \{a_i\}$ *whose elements are given by*

$$
a_i = Tr_1^m(\beta^{2i}), i = 0, 1, \ldots
$$

and the shift sequence $\mathbf{e} = (e_0, e_1, \ldots, e_{d-1})$ *is determined by* $t_j(\lambda)$ *in Eq. (10.32). In other words, each sequence in S can be arranged into a $v \times d$ array for which the jth $(0 \le j < d)$ column is given by $L^{e_j}(\mathbf{a})$, where e_j is determined by $t_j = \beta^{e_j}$ if $t_j \neq 0$. Otherwise, the jth column is the zero sequence.*

We shall also call $(t_0, t_1, \ldots, t_{d-1})$ a phase vector of $\mathbf{u}$ when it is regarded as a (v, d) interleaved sequence, because it determines phase shifts of column sequences of $\mathbf{u}$ with respect to $\mathbf{a}$.

Proof of Theorem 10.2. We will use an approach similar to the one in Section 8.1 of Chapter 8. For $\mathbf{s}_\eta$ and $\mathbf{s}_\lambda$ in S, considering the $v \times d$ arrays formed from $L^\tau(\mathbf{s}_\eta)$ and $\mathbf{s}_\lambda$, respectively, let A_j and B_j be their jth columns. Then

$$
A_j = \{Tr_1^m(\beta^{2i} t_{j+\tau}(\eta))\}_{i=1}^{v-1} \text{ and } B_j = \{Tr_1^m(\beta^{2i} t_j(\lambda))\}_{i=0}^{v-1}, 0 \le j < d,
$$

where $t_j(z)$ is defined by Eq. (10.32). Thus $C_{\eta,\lambda}(\tau)$, the crosscorrelation of s_η and s_λ, is equal to the sum of the inner products of A_j and B_j, $0 \le j < d$; that is,

$$C_{\eta,\lambda}(\tau) = \sum_{j=0}^{d-1} <A_j, B_j> .$$

Thus, it suffices to prove that there are at most two identical corresponding columns in their respective matrix forms. This is equivalent to saying that there are at most two values of j among $0 \le j < d$ such that

$$t_{j+\tau}(\eta) = t_j(\lambda). \tag{10.33}$$

In the following, we will reduce Eq. (10.33) to a quadratic equation. Considering Eq. (10.32), let $x = \alpha^j$, so we can rewrite $t_j(z)$ as:

$$
\begin{aligned}
t_j(z) &= Tr_m^n(x^2) + zx^d = x^2 + x^{2 \cdot 2^m} + zx^{2^m+1} \\
&= x^2(1 + x^{2(2^m-1)} + zx^{2^m-1}) \\
\Longrightarrow t_j(z) &= x^2(1 + y^2 + zy), \quad \text{where } y = x^{2^m-1}. \tag{10.34}
\end{aligned}
$$

Substituting Eq. (10.34) into Eq. (10.33),

$$t_{j+\tau}(\eta) = x^2(\alpha^{2\tau} + \alpha^{\tau 2^{m+1}} y^2 + \eta\alpha^{\tau d} y) = x^2(1 + y^2 + \lambda y) = t_j(\lambda).$$

Simplifying, we obtain a quadratic equation

$$a + by^2 + cy = 0, \tag{10.35}$$

where

$$a = 1 + \alpha^{2\tau}, b = 1 + \alpha^{\tau 2^{m+1}}, \text{ and } c = \lambda + \eta\alpha^{\tau d}$$

where $\tau \neq 0$. Because Eq. (10.35) is a quadratic equation over $\mathbb{F}_{2^n}$, it has at most two solutions y_i, $i = 1, 2$ in $\mathbb{F}_{2^n}$. For each of the y_i, there exists at most one $x = \alpha^j$, $0 \le j < d$ such that

$$y_i = x^{2^m-1}.$$

Therefore, there are at most two values of j among $0 \le j < d$ such that Eq. (10.33) is true. Thus

$$C_{\eta,\lambda}(\tau) = \begin{cases} -1 - 2^m & \text{if (10.33) has no solutions} \\ -1 & \text{if (10.33) has one solution} \\ -1 + 2^m & \text{if (10.33) has two solutions,} \end{cases}$$

which completes the proof. $\square$

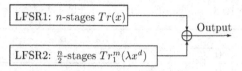

Figure 10.4. LFSR implementation of a Kasami (small) signal set generator.

Note that $\mathbf{s}_0 = \{Tr_1^n(\alpha^{2i})\} = \{Tr_1^n(\alpha^i)\}$ is an m-sequence of period $2^n - 1$. Thus, for $\lambda \neq 0$, we have

$$\mathbf{s}_\lambda = \mathbf{s}_0 + L^k \mathbf{a} \leftrightarrow Tr(x + \lambda x^d), \lambda = \beta^k, 0 \le k < 2^m - 1.$$

Hence, the Kasami (small) signal set can be written as

$$S = \{\mathbf{s}_{\beta^k} \mid 0 \le k < 2^m - 1\} \cup \{\mathbf{s}_0\}.$$

Profile of the randomness of Kasami (small) signal sets ($n = 2m$)
1. The Kasami (small) signal set S is a $(2^n - 1, 2^m, 2^m + 1)$ signal set.
2. Crosscorrelation of any two sequences in S or out-of-phase autocorrelation of any sequence in S is 3-valued and belongs to

$$\{-1, -1 \pm 2^m\}.$$

3. Imbalance range: $[1, 2^m + 1]$. (In fact, each sequence in S, except for $\mathbf{s}_0$, has weight either $2^{n-1} + 2^{m-1}$ or $2^{n-1} - 2^{m-1}$. We will show this result in the next subsection after we introduce a construction for generalized Kasami sequences.)
4. $\mathbf{a} = \{Tr_1^m(\beta^i)\}$ is an m-sequence of period $2^m - 1$. (Note that $Tr_1^m(x) = Tr_1^m(x^2)$.)
5. Linear span: $3n/2$, except for $\mathbf{s}_0$ whose linear span is n.

The LFSR implementation of the Kasami (small) signal sets is shown in Figure 10.4, from which all the sequences in S can be generated by varying the initial state of the short LFSR.

We will use an example to illustrate the properties of the Kasami signal sets discussed above.

Example 10.9 Design a $(63, 8, 9)$ Kasami signal set.

1. Pick $n = 6$, and $t(x) = x^6 + x + 1$ a primitive polynomial over $\mathbb{F}_2$; take α a root of $t(x)$ in $\mathbb{F}_{2^6}$, and $d = 9$. Let $\beta = \alpha^9$. Then β is a primitive element of $\mathbb{F}_{2^3}$.
2. Compute the minimal polynomial $v(x)$ of β, $v(x) = x^3 + x^2 + 1$ (or look-up table in Appendix C in Chapter 3).

Table 10.8. *Image of $t_j(\lambda) = x^2 + x^{16} + \lambda x^9, \forall \lambda \in \mathbb{F}_{2^3}^*$*

$\lambda = \beta^i$		$t_j(\lambda)$ for $x = \alpha^j$								
i	j	0	1	2	3	4	5	6	7	8
0		0	2	4	∞	1	2	∞	2	2
1		1	1	6	1	6	0	4	4	1
2		2	5	2	6	5	6	2	0	5
3		3	0	∞	5	2	3	1	∞	0
4		4	3	3	2	4	5	5	1	3
5		5	∞	1	4	0	1	0	6	∞
6		6	4	0	0	∞	∞	3	5	4

From Eq. (10.32), we have

$$t_j(\lambda) = x^2 + x^{16} + \lambda x^9, \, x = \alpha^j, 0 \le j < d,$$

whose values are shown in Table 10.8. In the ith row under the label i, the jth entry, $j = 0, \ldots, 8$ is the exponent r such that $t_j(\beta^i) = \beta^r$ where $x = \alpha^j$. By convention, if $t_j(\beta^i) = 0$, then the exponent is listed as ∞. (Note that $t_j(\beta^i) \in \mathbb{F}_{2^3}$.)

Table 10.8 directly determines the weight distribution of the sequences in S. In other words, for $i \in \{0, 3, 5, 6\}$, there are two zero columns in the 7×9 array form $\mathbf{s}_{\beta^i}$, and the other seven columns are shifts of the m-sequence $\mathbf{a} = 1110100$. For $i \in \{1, 2, 4\}$, all columns in the array are shifts of $\mathbf{a}$. Therefore, we have

$$w(\mathbf{s}_{\beta^i}) = \begin{cases} 7 \cdot 2^2 = 28 & \text{for } i \in \{0, 3, 5, 6\} \\ 9 \cdot 2^2 = 36 & \text{for } i \in \{1, 2, 4\}. \end{cases}$$

Thus, no sequences in S are balanced (except for the m-sequence $\mathbf{s}_0$). This is an interesting phenomenon, but was a puzzle in the literature for a long time. (We will give a proof of this result in a more general setting in the next subsection.) All eight Kasami sequences in S are shown in Table 10.9, represented as $(7, 9)$ interleaved sequences where the shift sequence of $\mathbf{s}_{\beta^i}$ is given by the $(i + 1)$st row under the label i in Table 10.8. For example, for $\mathbf{s}_1$, the phase vector is given by the first row under the label i in Table 10.8:

$$(0, 2, 4, \infty, 1, 2, \infty, 2, 2) \leftrightarrow (1, \beta^2, \beta^4, 0, \beta, \beta^2, 0, \beta^2, \beta^2)$$

and the first two column vectors of $\mathbf{s}_1$ are given by

$$\{Tr_1^3(\beta^{2i})\}_{i=0}^6 = 1110100$$

$$\{Tr_1^3(\beta^2\beta^{2i})\}_{i=0}^6 = 1101001.$$

Table 10.9. *Sequences in the (63, 8, 9) Kasami signal set*

$s_0 =$	$s_1 =$	$s_\beta =$	$s_{\beta^2} =$
0 0 0 0 0 1 0 0 0	1 1 1 0 1 1 0 1 1	1 1 0 1 0 1 1 1 1	1 0 1 0 0 0 1 1 0
0 1 1 0 0 0 1 0 1	1 1 0 0 0 1 0 1 1	0 0 1 0 1 1 0 0 0	1 1 1 1 1 1 1 1 1
0 0 1 1 1 1 0 1 0	1 0 1 0 0 0 0 0 0	0 0 0 0 0 1 1 1 0	0 1 0 0 1 0 0 1 1
0 0 1 1 1 0 0 1 0	0 1 0 0 1 1 0 1 1	1 1 0 1 0 0 0 0 1	1 1 1 0 1 0 1 0 1
0 1 0 1 1 0 1 1 1	1 0 0 0 1 0 0 0 0	1 1 1 1 1 1 0 0 1	0 0 0 1 0 1 0 1 0
0 1 1 0 0 1 1 0 1	0 0 1 0 1 0 0 0 0	1 1 1 1 1 0 1 1 1	0 1 0 1 1 1 0 0 1
0 1 0 1 1 1 1 1 1	0 1 1 0 0 1 0 1 1	0 0 1 0 1 0 1 1 0	1 0 1 1 0 1 1 0 0

$s_{\beta^3} =$	$s_{\beta^4} =$	$s_{\beta^5} =$	$s_{\beta^6} =$
0 1 0 0 1 0 1 0 1	1 0 0 1 1 0 0 1 0	0 0 1 1 1 1 1 0 0	0 1 1 1 0 0 0 0 1
0 1 0 1 1 0 0 0 1	0 0 0 1 0 1 1 0 0	1 0 0 0 1 0 1 1 0	1 0 1 1 0 0 0 1 0
1 1 0 1 0 1 0 0 1	1 1 1 0 1 1 1 0 1	1 0 0 1 1 0 1 0 0	0 1 1 1 0 0 1 1 1
1 0 0 1 1 1 1 0 0	0 1 1 1 0 1 1 1 1	1 0 1 0 0 1 0 0 0	0 0 0 0 0 0 1 1 0
1 1 0 0 0 1 1 0 1	0 1 1 0 0 0 0 1 1	0 0 1 0 1 1 1 1 0	1 0 1 1 0 0 1 0 0
0 0 0 1 0 0 1 0 0	1 0 0 0 1 1 1 1 0	1 0 1 1 0 1 0 1 0	1 1 0 0 0 0 0 1 1
1 0 0 0 1 1 0 0 0	1 1 1 1 1 0 0 0 1	0 0 0 1 0 0 0 1 0	1 1 0 0 0 0 1 0 1

The LFSR implementation of the (63, 8, 9) Kasami signal set is shown in Figure 10.5, from which all the sequences in S can be generated by varying initial states of the 3-stage LFSR.

10.3.2 Generalization of the Kasami (small) signal sets

From the proof of Theorem 10.2, the result that $C(S) = \{-1, -1 \pm 2^{n/2}\}$ depends only on the 2-level autocorrelation property of the column sequences,

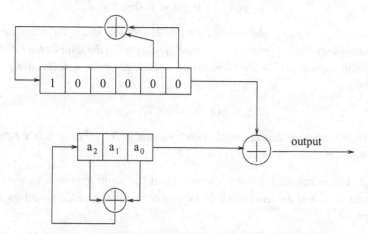

Figure 10.5. LFSR implementation of the (63, 8, 9) Kasami signal set.

where the sequences in S are regarded as (v, d) interleaved sequences. Thus, the trace function $Tr_1^m(x)$ employed in the Kasami (small) set construction can be replaced by any orthogonal function from $\mathbb{F}_{2^m}$ to $\mathbb{F}_2$.

Construction

Let $g(x) : \mathbb{F}_{2^m} \to \mathbb{F}_2$ be an orthogonal function (i.e., the evaluation of $g(x)$ is a 2-level autocorrelation sequence of period $2^m - 1$), and let $\mathbf{s}_\lambda = \{s_{\lambda,i}\}_{i \geq 0}$ whose elements are given by

$$s_{\lambda,i} = f_\lambda(\alpha^i), i = 0, 1, \ldots, \text{ where} \tag{10.36}$$
$$f_\lambda(x) = g\left(Tr_m^n(x^2) + \lambda x^d\right), \lambda \in \mathbb{F}_{2^m}, x \in \mathbb{F}_{2^n}. \tag{10.37}$$

So, $f_\lambda(x)$ is the trace representation of $\mathbf{s}_\lambda$. A signal set $S(g)$ consists of $\mathbf{s}_\lambda$ for all $\lambda \in \mathbb{F}_{2^m}$; that is,

$$S(g) = \{\mathbf{s}_\lambda \text{ such that } \lambda \in \mathbb{F}_{2^m}\}. \tag{10.38}$$

$S(g)$ is said to be a generalized Kasami (small) signal set.

Theorem 10.3 $S(g)$ *is a* $(2^n, 2^m, 2^m + 1)$ $(n = 2m)$ *signal set for any orthogonal function g. Furthermore,* $C(S(g)) = \{-1, -1 \pm 2^m\}$; *that is, the crosscorrelation of any pair of sequences in $S(g)$ or out-of-phase autocorrelation of a sequence in $S(g)$ is 3-valued and belongs to $C(S(g))$.*

A proof of Theorem 10.3 follows directly from the following lemma.

Lemma 10.2 *Let* $\mathbf{u} = \{u_i\} = \mathbf{s}_\lambda \in S(g)$, *defined above.*

(a) *The elements of* $\mathbf{u} = \{u_i\}$ *are given by*

$$u_{id+j} = g(\beta^{2i} t_j), 0 \leq i < v, 0 \leq j < d,$$

where $t_j = t_j(\lambda)$ *is defined in Eq. (10.32). Thus $\mathbf{s}_\lambda$ is a (v, d) interleaved sequence where the phase vector is $(t_0, t_1, \ldots, t_{d-1})$ (the same as the Kasami small sequences) and the elements of the base sequence* $\mathbf{a} = \{a_i\}$ *are given by*

$$a_i = g(\beta^{2i}), i = 0, 1, \ldots.$$

(b) *The sequence $\mathbf{s}_0$ has the trace representation $g(Tr_m^n(x^2))$, so it is a 2-level autocorrelation sequence of period $2^n - 1$.*

Proof. The assertion 1 follows directly from the same approach as used in Lemma 10.1, and the assertion 2 is a direct consequence of Construction I in Chapter 8. $\qquad\qquad\square$

Note that $\mathbf{a} \leftrightarrow g(x)$ is a 2-level autocorrelation sequence of period $v = 2^m - 1$. Thus, a generalized Kasami sequence can be constructed from the array form of the corresponding Kasami sequence by replacing the base sequence $\{Tr_1^m(\beta^{2j})\}_{j\geq 0}$ by $\mathbf{a}$ while the phase vector is kept unchanged. Note that if we choose $g(x) = Tr_1^m(x^2)$, then $S(g)$ is the Kasami (small) set. In the following, we will establish the weight distribution of the sequences in $S(g)$ for a general 2-level autocorrelation function $g(x)$.

Theorem 10.4 *With the above notation,*

$$w(\mathbf{s}_\lambda) = 2^{n-1} \pm 2^{m-1}, \forall \lambda \in \mathbb{F}_{2^m}^*.$$

In other words, each generalized Kasami sequence in $S(g)$, except for $\mathbf{s}_0$ (including the Kasami case) has weight either $2^{n-1} + 2^{m-1}$ or $2^{n-1} - 2^{m-1}$. Thus, except for the m-sequence $\mathbf{s}_0$, no sequences in $S(g)$ are balanced.

Proof. For $\lambda \neq 0$, we consider the sequence $\mathbf{s}_\lambda$, regarded as a (v, d) interleaved sequence. From Lemma 10.2(1) and the proof of Theorem 10.3, we only need to show that the quadratic equation

$$t_j(\lambda) = x^2(1 + y^2 + \lambda y) = 0, \quad \text{where } y = x^v, \tag{10.39}$$

has either exactly two solutions or no solutions for which

$$x = \alpha^j, 0 \leq j < d. \tag{10.40}$$

Note that Eq. (10.39) has solutions in $\mathbb{F}_{2^n}$ if and only if $Tr_1^n(\lambda^{-2}) = 0$. Because $\lambda \in \mathbb{F}_{2^m}$, $Tr_1^n(\lambda^{-2}) = Tr_1^n(\lambda^{-1}) = Tr_1^m(\lambda^{-1}Tr_m^n(1)) = 0$ from $Tr_m^n(1) = 1 + 1 = 0$. Thus the quadratic equation (10.39) always has two solutions in $\mathbb{F}_{2^n}$. Let $y_i, i = 1, 2$ be these two solutions. If there exists $j : 0 \leq j < d$ such that $y_1 = \alpha^{jv}$, then y_1 is a solution which satisfies Eq. (10.40). In this case,

$$y_1 y_2 = 1 \Longrightarrow y_2 = y_1^{-1} = \alpha^{-jv} = \alpha^{j'v}, \quad \text{where } 0 < j' = d - j < d.$$

Thus y_2 is also a solution that satisfies Eq. (10.40). Therefore, if one of these two solutions of Eq. (10.39) satisfies Eq. (10.40), so does the other. Consequently, we have

$$w(\mathbf{s}_\lambda) = \begin{cases} 2^{n-1} + 2^{m-1} & \text{if both solutions satisfy Eq. (10.40)} \\ 2^{n-1} - 2^{m-1} & \text{if neither of the solutions satisfy Eq. (10.40)}. \end{cases}$$

$\square$

Example 10.10 With $n = 6$, α, and $\beta = \alpha^9$ in Example 10.9, let $g(x) = Tr_1^3(x^3)$. Thus, for $\mathbf{s}_{\beta^i}$, the jth column sequence is given by

$$\{Tr_1^3((\beta^{2i}t_j)^3)\}_{i=0}^6,$$

where $(t_0, t_1, \ldots, t_8)$ is given by the ith row in Table 10.8. In other words, s_{β^i} can be obtained from Table 10.9 by replacing the base sequence $\mathbf{a} = 1110100$ by $\mathbf{a}^{(3)} = 1001011$ while the shifts are retained. This is the same as we did for the construction of GMW sequences using the interleaved approach. According to Theorem 10.4, no sequences in $S(g)$ (except for s_0) are balanced. This can be verified by Table 10.8, because there are either no ∞'s in each row or there are exactly two ∞'s in each row.

10.3.3 Profile of the randomness of $S(g)$ and implementation

A profile of the randomness of the generalized Kasami signal sets is shown as follows.

Profile of $(2^n, 2^m, 2^m + 1)$ signal set $S(g)$

Period	$2^n - 1$
Number of sequences in S	2^m
Cross/out-of-phase autocorrelation	$\{-1, -1 \pm 2^m\}$
Distribution of 0-1	Each non-m-sequence has weight $2^{n-1} \pm 2^{m-1}$ Imbalance range: $[1, 2^m + 1]$
Linear span LS	(a) $g(x) = Tr_1^m(x)$, $LS = \frac{3}{2}n$ (b) $g(x) = Tr_1^m(x^r)$, $\gcd(r, 2^m - 1) = 1$, $LS \geq m2^{w(r)}$ (c) $g(x)$ from Chapters 2, 8 and 9, linear span can be made large

Example 10.11 The following examples illustrate the generalized Kasami signal sets for $n = 6, 8,$ and 10.

(a) Let $n = 6$, and let α be the same as in Example 10.9. Then there are only two sequences of period 7 with 2-level autocorrelation. Thus there are only two choices for $g(x)$:

$$g(x) = Tr_1^3(x) \text{ or } g(x) = Tr_1^3(x^3),$$

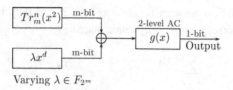

Varying $\lambda \in F_{2^m}$

Figure 10.6. Galois configuration for implementation of generalized Kasami signal sets.

which produce two generalized Kasami signal sets (including the Kasami case), as shown in Examples 10.9 and 10.10, respectively. In other words, we have

$$S_1 = S(Tr_1^3(x)) = \{Tr_1^3(Tr_3^6(x) + \lambda x^9) \mid \lambda \in \mathbb{F}_{2^3}\}, \text{ and}$$

$$S_2 = S(Tr_1^3(x^3)) = \{Tr_1^3 \left((Tr_3^6(x) + \lambda x^9)^3\right) \mid \lambda \in \mathbb{F}_{2^3}\}.$$

Both sets produce $(63, 8, 9)$ signal sets. The linear span of sequences in S_1 is 9, or 6 for $\lambda = 0$; and the linear span of sequences in S_2 is 21, or 12 for $\lambda = 0$. The latter follows from the expansion:

$$Tr_1^3 \left((Tr_3^6(x) + \lambda x^9)^3\right) = Tr_1^6(x^3 + (1 + \lambda^2)x^5 + \lambda x^{13}) + Tr_1^3(\lambda^3 x^{27}).$$

(b) $n = 8$. There are two shift-distinct 2-level autocorrelation sequences of period 15. Both are m-sequences. Thus $g(x) \in \{Tr_1^4(x), Tr_1^4(x^7)\}$, and $S(g)$ is a $(255, 16, 17)$ signal set.

(c) $n = 10$. There are six shift-distinct m-sequences and two shift-distinct quadratic residue sequences of period 31. Thus, there are eight $(1023, 32, 33)$ generalized Kasami signal sets in all.

A generalized Kasami (small) signal set can be implemented by the Galois configuration, as shown in Figure 10.6.

10.4 Even case: Bent function signal sets

In this section, using bent functions, we present another construction of $(2^n - 1, 2^m, 2^m + 1)$ signal sets for $n = 2m$ where m is even.

10.4.1 Bent functions

A bent function is a boolean function in n variables, say $f(x_0, \ldots, x_{n-1})$, whose Walsh transform has constant magnitude; that is,

$$\widehat{f}(\mathbf{w}) = \sum_{\mathbf{x} \in \mathbb{F}_2^n} (-1)^{\mathbf{w} \cdot \mathbf{x} + f(\mathbf{x})} = \pm\sqrt{q}, \ \forall \mathbf{w} \in \mathbb{F}_2^n \ (q = 2^n). \quad (10.41)$$

From Property 10.6, a boolean function is bent if and only if the Hadamard transform of its corresponding polynomial satisfies

$$\widehat{f}(\lambda) = \sum_{x \in \mathbb{F}_{2^n}} (-1)^{Tr(\lambda x) + f(x)} = \pm \sqrt{q}, \quad \forall \lambda \in \mathbb{F}_{2^n}. \tag{10.42}$$

Note that $\widehat{f}(\lambda)$ is an integer. Thus, bent functions only exist when n is even. In the following, we will use both representations for boolean functions. There are two general constructions for bent functions, as presented below. Again take $n = 2m$, $v = 2^m - 1$, $d = 2^m + 1$, and α a primitive element in $\mathbb{F}_{2^n}$. Let $t(x) = x^2 + c_1 x + c_0$, $c_i \in \mathbb{F}_{2^m}$ be the minimal polynomial of α over $\mathbb{F}_{2^m}$. Thus, we can write $\mathbb{F}_{2^n} = \{x + \alpha y \mid x, y \in \mathbb{F}_{2^m}\}$.

Construction I (McFarland) for Bent Functions	
Boolean Form	Polynomial Form
$f(\mathbf{x}, \mathbf{y}) = \mathbf{x} \cdot \psi(\mathbf{y}) + g(\mathbf{y})$ where	$f(z) = x\psi(y) + g(y)$ where
$\mathbf{x}, \mathbf{y} \in \mathbb{F}_2^m$,	$z = x + \alpha y, x, y \in \mathbb{F}_{2^m}$
$g(y_0, \ldots, y_{m-1})$, a boolean $\mathbb{F}_2^m \to \mathbb{F}_2$	g, a function $\mathbb{F}_{2^m} \to \mathbb{F}_2$
$\psi(y_0, \ldots, y_{m-1})$, a permutation of $\mathbb{F}_2^m$	$\psi(y)$, a permutation of $\mathbb{F}_{2^m}$

Construction II (Dillon) for Bent Functions
Polynomial Form $f(x) : \mathbb{F}_{2^n} \to \mathbb{F}_2, d = 2^m + 1$
$f(0) = 0, a_i = f(\alpha^i), 0 \le i < 2^n - 1$ such that
$a_{kd+i} = a_i$, and $w((a_0, a_1, \ldots, a_{d-1})) = 2^{m-1}$

Construction II, given here, is slightly different from Dillon's original construction. We provide a proof below.

Property 10.7 *The trace representation of any binary sequence* $(a_0, a_1, \ldots, a_{d-1})$ *with weight* 2^{m-1} *is bent. Furthermore, the least period of the sequence is equal to* d.

Proof. We will use the interleaved structure of m-sequences to establish this lemma. For Construction II, let $\mathbf{a} = \{a_i\}_{i=0}^{2^n - 2}$. We arrange $\mathbf{a}$ into a (v, d)

array:

$$A = \begin{bmatrix} a_0 & a_1 & \cdots & a_{d-1} \\ a_d & a_{d+1} & \cdots & a_{d+(d-1)} \\ a_{2d} & a_{2d+1} & \cdots & a_{2d+(d-1)} \\ \vdots & & & \\ a_{(v-1)d} & a_{(v-1)d+1} & \cdots & a_{(v-1)d+(d-1)} \end{bmatrix} ;$$

then

$$A = \begin{bmatrix} R \\ R \\ \vdots \\ R \end{bmatrix} = [A_0, A_1, \ldots, A_{d-1}],$$

where $R = (a_0, a_1, \ldots, a_{d-1})$ is the first row vector of A where $w(R) = 2^{m-1}$, and the A_j's are the column vectors of A where A_j is either the zero sequence or the constant 1 sequence. In other words, all row vectors of A are identical and the weight of the row vector is 2^{m-1}. By the construction of R or $\mathbf{a}$, $f(x)$ is the trace representation of R. Next, we write $\mathbf{b} = \{b_i\}$, where $b_i = Tr(\lambda \alpha^i)$, which can be arranged into a (v, d) array:

$$B = \begin{bmatrix} b_0 & b_1 & \cdots & b_{d-1} \\ b_d & b_{d+1} & \cdots & b_{d+(d-1)} \\ b_{2d} & b_{2d+1} & \cdots & b_{2d+(d-1)} \\ \vdots & & & \\ b_{(v-1)d} & b_{(v-1)d+1} & \cdots & b_{(v-1)d+(d-1)} \end{bmatrix} = [B_0, B_1, \ldots, B_{d-1}].$$

According to Theorem 5.2 in Chapter 5, there is only one column in B that is the zero sequence, and the other 2^m columns are shift-equivalent m-sequences of period $2^m - 1$. So, every nonzero column of B has weight 2^{m-1}. Thus, the Hadamard transform of $f(x)$ is determined by the number of 1's in entries of the sum of the matrices A and B. In the following, we will show how to count this number. Without loss of generality, we may assume that the first column of B is the zero sequence. Thus, we have the following two cases.

Case 1. $a_0 = 0$. In the matrix $A + B$, there are 2^{m-1} columns that have weight 2^{m-1} and 2^{m-1} columns have weight $2^{m-1} - 1$ (those obtained from the complementing of B_j). Using a similar result to Property 10.4, we have

$$\widehat{f}(\lambda) = 1 + \{2^{2m} - 1 - 2[2^{m-1} \cdot 2^{m-1} + 2^{m-1} \cdot (2^{m-1} - 1)]\} = 2^m.$$

Case 2. $a_0 = 1$. In the matrix $A + B$, the first column vector has weight $2^m - 1$. For the rest of the 2^m column vectors, $2^{m-1} + 1$ columns have weight 2^{m-1}, and the other $2^{m-1} - 1$ columns have weight $2^{m-1} - 1$. Thus, we have

$$\widehat{f}(\lambda) = 1 + \{2^{2m} - 1 - 2[2^m - 1 + (2^{m-1} + 1) \cdot 2^{m-1}$$
$$+ (2^{m-1} - 1) \cdot (2^{m-1} - 1)]\} = -2^m.$$

Therefore, for any $\lambda \neq 0$, we have

$$\widehat{f}(\lambda) = \begin{cases} 2^m & \text{if } B_0 = \mathbf{0} \text{ and } a_0 = 0 \\ -2^m & \text{if } B_0 = \mathbf{0} \text{ and } a_0 = 1. \end{cases}$$

If $\lambda = 0$, then $\widehat{f}(0)$ is determined by the number of ones in A. In other words, we have

$$\widehat{f}(0) = 1 + [2^{2m} - 1 - 2w(a_0, a_1, \ldots, a_{2^n - 2}]$$
$$= 1 + [2^{2m} - 1(2^m - 1) \cdot 2^{m-1}] = 2^m.$$

Thus, $f(x)$ is bent. Since the Hamming weight of R is a power of 2, the least period of $\mathbf{a}$ is equal to d.   □

Example 10.12 Let $n = 4$, and let α be a primitive element of $\mathbb{F}_{2^4}$ with minimal polynomial $t(x) = x^4 + x + 1$.

Method 1. Using Construction I, let $\psi(y_0, y_1) = (y_0, y_1)$, a permutation of $\mathbb{F}_2^2$, and $g(y_0, y_1) = y_0$, a map from $\mathbb{F}_2^2$ to $\mathbb{F}_2$. Then

$$f(x_0, x_1, y_0, y_1) = (x_0, x_1) \cdot \psi(y_0, y_1) + g(y_0, y_1) = x_0 y_0 + x_1 y_1 + y_0$$

is a bent function from $\mathbb{F}_2^4$ to $\mathbb{F}_2$. The values of the function f and the Walsh spectrum of f are shown in Table 10.10 where $\mathbf{z} = (x_0, x_1, y_0, y_1) \in \mathbb{F}_2^4$ and $\mathbf{w} = (w_0, w_1, w_2, w_3) \in \mathbb{F}_2^4$.

Method 2. Using Construction II, let $\mathbf{a} = 00101$, which is a 3-decimation from $L(\mathbf{b})$ where $\mathbf{b} = 000100110101111 \leftrightarrow Tr(x)$. Thus the trace representation of $\mathbf{a} = 00101$ is given by $f(x) = Tr(\alpha x^3)$ where $a_i = f(\alpha^i)$, $i = 0, 1, 2, 3$, and 4. Note that

$$A = \begin{bmatrix} 00101 \\ 00101 \\ 00101 \end{bmatrix}.$$

According to Construction II, $f(x)$ is a bent function from $\mathbb{F}_{2^4}$ to $\mathbb{F}_2$. Furthermore,

$$\{\widehat{f}(\alpha^i)\}_{i=0}^{14} = (4, -4, 4, -4, 4, 4, -4, 4, -4, 4, 4, -4, 4, -4, 4) \text{ and } \widehat{f}(0) = 4.$$

Table 10.10. *A bent function from*
Construction I

z, w	$f(\mathbf{z})$	$\widehat{f}(\mathbf{w})$
0 0 0 0	0	4
1 0 0 0	0	−4
0 1 0 0	0	4
1 1 0 0	0	−4
0 0 1 0	1	4
1 0 1 0	0	4
0 1 1 0	1	4
1 1 1 0	0	4
0 0 0 1	0	4
1 0 0 1	0	−4
0 1 0 1	1	−4
1 1 0 1	1	4
0 0 1 1	1	4
1 0 1 1	0	4
0 1 1 1	0	−4
1 1 1 1	1	−4

Let $\{1, \alpha, \alpha^2, \alpha^3\}$ be a basis of $\mathbb{F}_{2^4}$. Then the boolean form of $f(x)$ relative to this basis is given by

$$f(x_0, x_1, x_2, x_3) = Tr\left(\alpha(x_0 + x_1\alpha + x_2\alpha^2 + x_3\alpha^3)^3\right).$$

Simplifying, we obtain

$$f(x_0, x_1, x_2, x_3) = x_0x_1 + x_0x_2 + x_0x_3 + x_1x_2 + +x_1x_3 + x_2x_3 + x_2.$$

Thus we have the following one-to-one correspondences:

$$00101 \leftrightarrow Tr(\alpha x^3) \leftrightarrow x_0x_1 + x_0x_2 + x_0x_3 + x_1x_2 + +x_1x_3 + x_2x_3 + x_2.$$

Example 10.13 Let $f(x) = Tr_1^m(x^d)$. For $\eta \in \mathbb{F}_{2^n}^*$, we have

$$\widehat{f}(\eta) = \sum_{x \in \mathbb{F}_{2^n}} (-1)^{Tr(\eta x) + Tr_1^m(x^d)}$$

$$= \sum_{x \in \mathbb{F}_{2^n}} (-1)^{Tr(y) + Tr_1^m(\lambda y^d)},$$

where $y = \eta x$, $\lambda = \eta^d \in \mathbb{F}_{2^m}^*$, and $Tr(x) = Tr_1^n(x)$. Thus $\widehat{f}(\eta) = 2^n - 2w(\mathbf{s}_\lambda)$, $\lambda \neq 0$, where $\mathbf{s}_\lambda$ is a sequence in the Kasami (small) set ($\lambda \neq 0$). According to Theorem 10.4, $w(\mathbf{s}_\lambda) = 2^{n-1} \pm 2^{m-1}$. Therefore,

$$\widehat{f}(\eta) = \pm 2^m, \eta \in \mathbb{F}_{2^n}^*.$$

Note that for $\eta = 0$, $T = \{Tr_1^m(x^d) \mid x \in \mathbb{F}_{2^n}\}$ consists of d copies of the m-sequence $\{Tr_1^m(\beta^j)\}_{j=0}^{2^m-2}$ where β is a primitive element in $\mathbb{F}_{2^m}$. Therefore, the number of 1's in T is given by

$$(2^m + 1)2^{m-1} = 2^{n-1} + 2^{m-1} \implies \hat{f}(0) = \sum_{x \in \mathbb{F}_{2^n}} (-1)^{Tr_1^m(x^d)} = -2^m.$$

Thus $f(x) = Tr_1^m(x^d)$ or $f(x) = Tr(\eta x) + Tr_1^m(x^d)$, $\forall \eta \in \mathbb{F}_{2^n}^*$ is bent. For example, if $n = 4$, then $Tr_1^2(x^5) \leftrightarrow 011$ is a bent function from $\mathbb{F}_{2^4}$ to $\mathbb{F}_2$.

Remark 10.3 From Proposition 6.7 in Chapter 6, $f(x)$ is bent if and only if the additive autocorrelation of $f(x)$, $V_f(w)$, is equal to 0 for $w \neq 0$ and equal to 2^n for $w = 0$. In other words,

$$f(x) \text{ bent} \iff$$

$$V_f(w) = \sum_{x \in \mathbb{F}_{2^n}} (-1)^{f(x+w)+f(x)} = \begin{cases} 2^n & \text{if } w = 0 \\ 0 & \text{if } w \neq 0. \end{cases} \quad (10.43)$$

Thus, bent functions produce binary sequences with 2-level additive autocorrelation, given by Eq. (10.43); that is, all out-of-phase autocorrelation values are equal to zero. From Eq. (10.43) and the Hadamard matrix property (see Section 2.4 in Chapter 2), a bent function yields a $2^n \times 2^n$ Hadamard matrix (see Dillon (1975) for details).

10.4.2 Bent function signal sets

Construction

For $n = 2m$ where m is even, let $f(\mathbf{x})$ be a bent function in m variables and set

$$f_{\mathbf{z}}(\mathbf{x}) = f(\mathbf{x}) + \mathbf{z} \cdot \mathbf{x}, \quad \mathbf{x}, \mathbf{z} \in \mathbb{F}_2^m. \quad (10.44)$$

Let $\mathbf{s_z} = \{s_{\mathbf{z},i}\}$ whose elements are given by

$$s_{\mathbf{z},i} = f_{\mathbf{z}}(x_{0,i}, x_{1,i}, \ldots, x_{m-1,i}) + Tr_1^n(\sigma_0\alpha^i), i = 0, 1, \ldots, \quad (10.45)$$

where

$$x_{j,i} = Tr_1^n(\eta_j\alpha^i), 0 \leq j < m, \text{ and } \sigma_0 \notin \mathbb{F}_{2^m}, \quad (10.46)$$

where $\{\eta_0, \eta_1, \ldots, \eta_{m-1}\}$ is a basis of $\mathbb{F}_{2^m}/F_2$. In other words, we have

$$s_{\mathbf{z},i} = f(\mathbf{x}_i) + \mathbf{z} \cdot \mathbf{x}_i + Tr_1^n(\sigma_0\alpha^i), i = 0, 1, \ldots, \quad (10.47)$$

where $\mathbf{x}_i = (x_{0,i}, x_{1,i}, \ldots, x_{m-1,i})$. A signal set $S(f)$ is given by

$$S(f) = \{\mathbf{s_z} \mid \mathbf{z} \in \mathbb{F}_2^m\}. \quad (10.48)$$

Then $S(f)$ is a $(2^n - 1, 2^m, 2^m + 1)$ signal set.

In the following, we will derive the trace representation of $\mathbf{s_z}$, that is, the polynomial form of Eq. (10.45). Let $\{\beta_0, \ldots, \beta_{m-1}\}$ be the dual basis of $\{\eta_0, \ldots, \eta_{m-1}\}$. From Theorem 3.13 in Chapter 3, any element y in $\mathbb{F}_{2^m}$ can be represented as

$$y = \sum_{j=0}^{m-1} y_j \beta_j, \, y_j \in \mathbb{F}_2, \tag{10.49}$$

where

$$y_j = Tr_1^m(\eta_j y), 0 \le j < m. \tag{10.50}$$

This relation aids in finding trace representations of the sequences in $S(f)$.

Property 10.8 *For any $\mathbf{z} \in \mathbb{F}_2^m$, there exists a unique $\lambda \in \mathbb{F}_{2^m}$ such that*

$$\mathbf{s}_{\mathbf{z},i} = f\left(Tr_m^n(\alpha^i)\right) + Tr_1^n\left((\lambda + \sigma_0)\alpha^i\right), i = 0, 1, \ldots. \tag{10.51}$$

Proof. Let $x = \alpha^i \in \mathbb{F}_{2^n}$. Again, we avoid using double subscripts by renaming $\mathbf{x}_i$. Thus, let $\mathbf{x} = (x_0, \ldots, x_{m-1}) = \mathbf{x}_i$. According to Eq. (10.46), we have $x_j = Tr_1^n(\eta_j x), 0 \le j < m$. Note that $Tr_m^n(x) \in \mathbb{F}_{2^m}$, denoted by $y = Tr_m^n(x)$. Consequently,

$$
\begin{aligned}
Tr_1^m(\eta_j y) &= Tr_1^m(\eta_j Tr_m^n(x)) = Tr_1^m(Tr_m^n(\eta_j x)) \text{ (since } \eta_j \in \mathbb{F}_{2^m}) \\
&= Tr_1^n(\eta_j x) \text{ (by the transitivity of the trace function)} \\
&\Longrightarrow Tr_1^m(\eta_j y) = Tr_1^n(\eta_j x). \tag{10.52}
\end{aligned}
$$

Thus, we have

$$
\begin{aligned}
Tr_m^n(x) = y &= \sum_{j=0}^{m-1} Tr_1^m(\eta_j y)\beta_j \\
&= \sum_{j=0}^{m-1} Tr_1^n(\eta_j x)\beta_j \text{ (by Eq. (10.52)).}
\end{aligned}
$$

Hence,

$$\mathbf{x} = (x_0, x_1, \ldots, x_{m-1}) \leftrightarrow Tr_m^n(x). \tag{10.53}$$

Therefore, from the above one-to-one correspondence and Property 10.6, there exists $\lambda \in \mathbb{F}_{2^m}$ such that

$$\mathbf{z} \cdot \mathbf{x} = Tr_1^m(\lambda Tr_m^n(x)) = Tr_1^n(\lambda x). \tag{10.54}$$

Substituting Eq. (10.53) and Eq. (10.54) into Eq. (10.45), we obtain

$$s_{\mathbf{z},i} = f(Tr_m^n(\alpha^i)) + Tr_1^n((\lambda + \sigma_0)\alpha^i), \, i = 0, 1, \ldots, \, \lambda \in \mathbb{F}_{2^m}. \quad \square$$

According to Property 10.8, we have the following polynomial form construction for bent function signal sets.

Construction of bent function signal sets in polynomial form
Let
- $f(x) : \mathbb{F}_{2^m} \to \mathbb{F}_2$, bent,
- $f_\lambda(x) = f(Tr^n_m(x)) + Tr^n_1((\lambda + \sigma_0)x)$, $\lambda \in \mathbb{F}_{2^m}$, $\sigma_0 \in \mathbb{F}_{2^n} \setminus \mathbb{F}_{2^m}$,
- $\mathbf{s}_\lambda = \{\mathbf{s}_{\lambda,i}\}$, where $\mathbf{s}_{\lambda,i} = f_\lambda(\alpha^i)$, $i = 0, 1, \ldots,$ and
- $S = \{\mathbf{s}_\lambda \mid \lambda \in \mathbb{F}_{2^m}\}$.
Then S is a $(2^n - 1, 2^{n/2}, 2^{n/2} + 1)$ signal set.

Property 10.9 *Every sequence in S is balanced.*

Proof. In order to prove a sequence $\mathbf{s}_\lambda$ is balanced, it suffices to show that

$$\sum_{x \in \mathbb{F}_{2^n}} (-1)^{f_\lambda(x)} = 0, \forall \lambda \in \mathbb{F}_{2^m}. \tag{10.55}$$

Note that $Tr^n_m(x)$ is the trace representation of an m-sequence over $\mathbb{F}_{2^m}$ of degree 2. According to Definition 5.5 and Theorem 5.7 in Section 5.6 of Chapter 5, $Tr^n_m(x)$ satisfies the 2-tuple balance property; that is, any pair $(\lambda, \mu) \in \mathbb{F}_{2^m}$ occurs once in the following set:

$$\{(Tr^n_m(x), Tr^n_m(\gamma x)) \mid x \in \mathbb{F}_{2^n}\}$$

when $\gamma \notin \mathbb{F}_{2^m}$. We now use this property to derive Eq. (10.55). Let $\gamma = \lambda + \sigma_0$. Then $\sigma_0 \notin \mathbb{F}_{2^m}$ and $\lambda \in \mathbb{F}_{2^m} \implies \gamma \notin \mathbb{F}_{2^m}$. Using the 2-tuple balance property, we have

$$\sum_{x \in \mathbb{F}_{2^n}} (-1)^{f_\lambda(x)} = \sum_{x \in \mathbb{F}_{2^n}} (-1)^{f(Tr^n_m(x)) + Tr^m_1(Tr^n_m(\gamma x))}$$

$$= \sum_{\lambda, \mu \in \mathbb{F}_{2^m}} (-1)^{f(\lambda) + Tr^m_1(\mu)}$$

$$= \sum_{\lambda \in \mathbb{F}_{2^m}} (-1)^{f(\lambda)} \sum_{\mu \in \mathbb{F}_{2^m}} (-1)^{Tr^m_1(\mu)} = 0.$$

The last identity follows from $\sum_{\mu \in \mathbb{F}_{2^m}} (-1)^{Tr^m_1(\mu)} = 0$. $\qquad\square$

A profile of the randomness of bent function signal sets is presented in Table 10.11. Bent function signal sets can be implemented by the Galois configuration, as shown in Figure 10.7. Compared with the generalized Kasami

Table 10.11. *Profile of bent function signal sets for* $n = 4t$

Period	$2^n - 1$
Number of Sequences in S	$2^{n/2}$
Cross/out-of-phase autocorrelation	$\{-1, -1 \pm 2^{n/2}\}$
Distribution of 0-1	balanced for all sequences in S
Linear span	$\leq \sum_{i=1}^{n/4} \binom{n}{i}$

signal sets, a 2-level autocorrelation function $g(x)$ is replaced by a bent function $f(x)$. Furthermore, each sequence in a bent function signal set is a sum of two sequences. One is given by $f(Tr_m^n(x))$ and the other by an m sequence with the trace representation $Tr((\lambda + \sigma_0)x)$ where $\lambda \in \mathbb{F}_{2^m}$ and $\sigma_0 \notin \mathbb{F}_{2^m}$. This structure results in a completely different 0-1 distribution from the Kasami or generalized Kasami signal sets.

Example 10.14 Let $n = 8$. We will use the same parameters as in Example 8.5 of Chapter 8 in Section 8.2; that is, $\mathbb{F}_{2^8}$ is defined by the primitive polynomial $c(x) = x^8 + x^4 + x^3 + x^2 + 1$, α a root of $c(x)$, and $\beta = \alpha^d$ $(d = 17)$ a primitive element in $\mathbb{F}_{2^4}$ with minimal polynomial $t(x) = x^4 + x + 1$.

1. Pick $f(x)$ to be the bent function from Method 2 in Example 10.12; that is,

$$f(x) = Tr_1^4(\beta x^3) \leftrightarrow \{f(\beta^i)\}_{i=0}^{14} = 001010010100101 = \mathbf{d}.$$

2. Set $\sigma_0 = \alpha \notin \mathbb{F}_{2^4}$, and

$$f_\lambda(x) = f(Tr_4^8(x)) + Tr_1^8((\lambda + \alpha)x), \lambda \in \mathbb{F}_{2^4}.$$

Then we have

$$\begin{aligned} f_\lambda(x) &= Tr_1^4(\beta Tr_4^8(x)^3) + Tr_1^8((\lambda + \alpha)x) \\ &= Tr_1^8(\beta x^3 + \beta^8 x^9 + (\lambda + \alpha)x). \end{aligned}$$

Note that this representation of $f_\lambda(x)$ gives linear span 24 for each sequence in S constructed below.

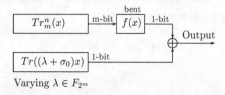

Varying $\lambda \in F_{2^m}$

Figure 10.7. Galois configuration for implementation of bent function signal sets.

3. Set

$$\mathbf{s}_{\lambda,i} = f_\lambda(\alpha^i), i = 0, 1, \ldots,$$

and

4. $S = \{\mathbf{s}_\lambda \mid \lambda \in \mathbb{F}_{2^4}\}$.

We can easily get 16 sequences in S from Example 8.5 in Chapter 8, where we constructed the GMW sequence of period 255 using the interleaved method. Let $\mathbf{c} = \{c_i\}$ with $c_i = Tr_1^8(\alpha^i)$ be that m-sequence of period 255, regarded as a (15, 17) interleaved sequence. We reproduce it here:

$$\mathbf{c} \leftrightarrow A = A(\mathbf{a}, \mathbf{e}) = \begin{bmatrix} 000001000111100010 \\ 010111000000011001 \\ 001001101111001000 \\ 001010110111010110 \\ 010110000111111011 \\ 011110101111010001 \\ 000011011001111110 \\ 011100110001011010 \\ 001000100101010 \\ 011101111011001111 \\ 011111101001100111 \\ 010100011000001111 \\ 010101011111100101 \\ 000010011111111100 \\ 001011110001010100 \end{bmatrix} = \begin{bmatrix} A_0 \\ A_1 \\ \vdots \\ A_{14} \end{bmatrix} = [B_0, B_1, \ldots, B_{16}],$$

where its base sequence and shift sequence are given by

$$\mathbf{a} = \{Tr_1^4(\beta^i)\}_{i=0}^{14} = 000100110101111$$

$$\{Tr_4^8(\alpha^j)\}_{j=0}^{16} \leftrightarrow \mathbf{e} = (\infty, 2, 4, 2, 8, 12, 4, 0, 1, 9, 9, 14, 8, 5, 0, 3, 2).$$

(Note that the j entry in $\mathbf{e}$ is the exponent r of β for which $\beta^r = Tr_4^8(\alpha^j)$, and A_i and B_j are the ith row vector and the jth column vector of A, respectively.) Let $\mathbf{u} = \{u_i\}$ whose elements are given by

$$u_i = f(Tr_4^8(\alpha^i)), i = 0, 1, \ldots.$$

Then

$$\mathbf{s}_{\beta^j} = \mathbf{u} + L^{17j}(\mathbf{c}) + L(\mathbf{c}), 0 \le j < 15, \text{ and } \mathbf{s}_0 = \mathbf{u} + L(\mathbf{c}),$$

where L is the (left) shift operator. Note that $\mathbf{u}$ is a $(15, 17)$ interleaved sequence with the base sequence $\mathbf{d} = 001010010100101 \leftrightarrow f(x)$ and the same shift sequence $\mathbf{e}$ as the m-sequence $\mathbf{c}$. Thus, we have

$$\mathbf{u} \leftrightarrow A(\mathbf{d}, \mathbf{e}) = \begin{bmatrix} 011101100111100001 \\ 000010001000100010010 \\ 010101010100001101 \\ 001000101111100000 \\ 000010010001000011110 \\ 011101100111100001 \\ 000010001000100010010 \\ 010101010100001101 \\ 001000101111100000 \\ 000010010001000011110 \\ 011101100111100001 \\ 000010001000100010010 \\ 010101010100001101 \\ 001000101111100000 \\ 000010010001000011110 \end{bmatrix}$$

where the jth column is the sequence $L^{e_j}(\mathbf{d})$. Therefore, the array form of $\mathbf{s}_{\beta^j}, 0 \le j < 15$, is given by

$$A(\mathbf{d}, \mathbf{e}) + \begin{bmatrix} A_j \\ A_{j+1} \\ \vdots \\ A_{14} \\ A_0 \\ \vdots \\ A_{j-1} \end{bmatrix} + [B_1, B_2, \ldots, B_{16}, L(B_0)],$$

where $L(B_0)$ is the shifted sequence from B_0 by applying the (left) shift operator. In other words, by shifting row vectors in the array form of the m-sequence $\mathbf{c}$, we obtain all sequences in S except for $\mathbf{s}_0$. A profile of the randomness of S is given in Table 10.12, and the implementation is shown in Figure 10.8.

10.5 Interleaved construction of signal sets

In this section, we present a binary signal set with parameters $(v^2, v + 1, 2v + 3)$ in terms of interleaved sequences. Here we have two choices, namely

Table 10.12. *Profile of $S(f)$ with the bent function*
$$f(x) = Tr_1^4(\beta x^3) \ (n = 2 \cdot 4)$$

Period	$2^8 - 1 = 255$
Number of Sequences in S	16
Cross/out-of-phase autocorrelation	$\{-1, -1 \pm 16\}$
Distribution of 0-1	128 1's and 127 0's for each sequence in S
Linear span	24
	(note: possible maximum linear span is 32 for this period)

$v = 2^n - 1$ or v a prime. This is an example of signal set design where the optimum correlation of the signal set is sacrificed to achieve large linear spans.

10.5.1 Constructions of $(v^2, v + 1, 2v + 3)$ signal sets

Procedure 1
1. Choose $\mathbf{a} = (a_0, a_1, \ldots, a_{v-1})$ and $\mathbf{b} = (b_0, b_1, \ldots, b_{v-1})$, two binary sequences of period v with 2-level autocorrelation.
2. Pick $\mathbf{e} = (e_0, e_1, \ldots, e_{v-1})$, an integer sequence whose elements are taken from $\mathbf{Z}_v$, the set consisting of integers modulo v.

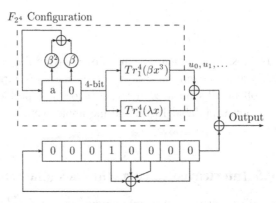

Figure 10.8. The Galois configuration for implementation of the (255, 16, 17) bent function signal set in Example 10.14.

3. Construct $\mathbf{u} = (u_0, u_1, \ldots, u_{v^2-1})$, a (v, v) interleaved sequence whose jth column sequence is given by $L^{e_j}(\mathbf{a})$.
4. Set

$$\mathbf{s}_j = (s_{j,0}, s_{j,1}, \ldots, s_{j,v^2-1}), 0 \le j < v$$

whose elements are defined by

$$s_{j,i} = u_i + b_{j+i} \text{ or } \mathbf{s}_j = \mathbf{u} + L^j(\mathbf{b}), 0 \le j < v.$$

5. A signal set S is defined as

$$S = \{\mathbf{s}_j \mid j = 0, 1, \ldots, v - 1\} \cup \{\mathbf{u}\}.$$

If the shift sequence $\mathbf{e}$ satisfies the following difference condition:

$$\boxed{\begin{array}{l} \text{for each } 1 \le s < v, \text{ the differences} \\ e_j - e_{j+s}, 0 \le j < v - s \text{ are all distinct} \end{array}} \tag{10.56}$$

or equivalently,

$$\boxed{|\{e_j - e_{j+s} \mid 0 \le j < v - s\}| = v - s, \text{ for all } 1 \le s < v} \tag{10.57}$$

then S is a $(v^2, v + 1, 2v + 3)$ signal set. Moreover, the crosscorrelation of any two sequences in S or the out-of-phase autocorrelation of any sequence in S belongs to the set $\{1, -v, v + 2, 2v + 3, -2v - 1\}$.

In the following, we show two methods for construction of sequences over $\mathbb{Z}_v$ satisfying Eq. (10.56) (or equivalently, Eq. (10.57)).

Construction A

(Shortened GMW Construction): $(v^2, v + 1, 2v + 3)$ signal set where $v = 2^m - 1$. Let $n = 2m$ and $d = 2^m + 1$. Select α a primitive element in $\mathbb{F}_{2^n}$, and set $\beta = \alpha^d$, a primitive element in $\mathbb{F}_{2^m}$.

1. *Short* 2-level autocorrelation sequence $\mathbf{b}$: Select $\mathbf{b}$ a binary 2-level autocorrelation sequence of period v.
2. *Long* 2-level autocorrelation sequence $\mathbf{c}$: Choose $g(x)$ an orthogonal function from $\mathbb{F}_{2^m}$ to $\mathbb{F}_2$, and let $\mathbf{c} = \{c_i\}$ whose elements are given by

$$c_i = g(Tr^n_m(\alpha^i)), i = 0, 1, \ldots.$$

Then $\mathbf{c}$ is a 2-level autocorrelation sequence of period $2^n - 1$ from the GMW construction (see Chapter 8). Thus, a $v \times d$ array from $\{c_i\}$ has the following form:

$$
\mathbf{c} =
\begin{bmatrix}
0 & c_1 & c_2 & \cdots & c_{d-2} & c_{d-1} \\
0 & c_{d+1} & c_{d+2} & \cdots & c_{d+d-2} & a_{2d-1} \\
\vdots & & & & & \\
0 & c_{(v-1)d+1} & c_{(v-1)d+2} & \cdots & c_{(v-1)d+d-2} & d_{vd-1}
\end{bmatrix}
\qquad (10.58)
$$

with "cut" markers above the first and last columns.

3. Interleaved sequence $\mathbf{u}$: Let U be a matrix obtained from $\mathbf{c}$ by deleting the first and the last columns, that is, deleting (c_{id}, c_{id+d-1}), $i = 0, 1, \ldots$. Then U yields a (v, v) interleaved sequence $\mathbf{u}$ of period v^2 whose shift sequence satisfies Eq. (10.57). Note that U has the base sequence given by $\{g(\beta^i)\}$, a 2-level autocorrelation sequence of period v, and the shift sequence $\mathbf{e} = (e_0, \ldots, e_{v-1})$ whose elements are given by

$$
e_j = e \iff Tr_m^n(\alpha^{j+1}) = \beta^e, \, 0 \le j < v. \qquad (10.59)
$$

Here it is not necessary to actually compute e_j.

4. Set

$$
\mathbf{s}_j = \mathbf{u} + L^j(\mathbf{b}), \, 0 \le j < v
$$

and $S = \{\mathbf{s}_j \mid 0 \le j < v\}$. Then S is a $(v^2, v+1, 2v+3)$ signal set.

Simple case

We may select $g(x) = Tr_1^m(x)$ and $\mathbf{b}$ an arbitrary binary m-sequence of period $2^m - 1$. In this case, $\mathbf{c}$ is an m-sequence of period $2^n - 1$. The signal set given by Construction A shortens $\mathbf{c}$ by deleting (c_{di}, c_{id+d-1}), $i = 0, 1, \ldots$ to obtain the interleaved sequence $\mathbf{u}$ with shift sequence satisfying Eq. (10.57). This is a case similar to the Kasami (small) signal set. But here the short sequence $\mathbf{b}$ can be any m-sequence of period $2^m - 1$. (Note. In the Kasami (small) set case, the short m-sequence is completely determined by $\mathbf{c}$, which is $\mathbf{c}^{(d)}$, a d-decimation of $\mathbf{c}$ where $d = 2^m + 1$.)

To simplify notation, we may sometimes use the same symbol for an interleaved sequence and for its array form.

Example 10.15 Construct a $(49, 8, 17)$ signal set. Let $n = 6 \implies m = 3$ and $v = 7$. We will use the procedure of the simple case to construct a $(49, 8, 17)$ signal set.

1. Choose **b** $= 1001011$, an m-sequence of period 7.
2. We select the same m-sequence **c** of period 63 as in Example 10.9 for the Kasami signal set $(63, 8, 9)$; that is,

$$\mathbf{c} = \begin{bmatrix} \boxed{0} & 0000100 & \boxed{0} \\ \boxed{0} & 1100010 & \boxed{1} \\ \boxed{0} & 0111101 & \boxed{0} \\ \boxed{0} & 0111001 & \boxed{0} \\ \boxed{0} & 1011011 & \boxed{1} \\ \boxed{0} & 1100110 & \boxed{1} \\ \boxed{0} & 1011111 & \boxed{1} \end{bmatrix}$$

3. By deleting the first column and the last column of **c**, which are framed in the above array, the resulting array gives a $(7, 7)$ interleaved sequence **u** which satisfies Eq. (10.57); that is,

$$\mathbf{u} = \begin{bmatrix} 0000100 \\ 1100010 \\ 0111101 \\ 0111001 \\ 1011011 \\ 1100110 \\ 1011111 \end{bmatrix}$$

4. Set $\mathbf{s}_j = \mathbf{u} + L^j(\mathbf{b})$: $j = 0, 1, \ldots, 6$ and $S = \{\mathbf{s}_j \mid 0 \le j < 7\} \cup \{\mathbf{u}\}$.
   Note that adding $L^j(\mathbf{b})$, **b** at shift j, to the $(7, 7)$ interleaved sequence **u** is equivalent to complementing those columns in **u** for which the bits in $L^j(\mathbf{b})$ are 1's. We illustrate this idea precisely below. Let B_j denote a $v \times v$ matrix where the top row of the matrix is $L^j(\mathbf{b})$ and the rest of the rows are identical to the top row. For example, we have

$$B_0 = \begin{bmatrix} 1001011 \\ 1001011 \\ 1001011 \\ 1001011 \\ 1001011 \\ 1001011 \\ 1001011 \end{bmatrix}$$

Thus,

$$\mathbf{s}_j = \mathbf{u} + B_j, 0 \le j < 7.$$

We can easily write out the elements of $\mathbf{s}_0$ and $\mathbf{s}_1$ from the above arrays **u** and B_0; that is,

$$\mathbf{s}_0 = \mathbf{u} + B_0 = \begin{bmatrix} 0000100 \\ 1100010 \\ 0111101 \\ 0111001 \\ 1011011 \\ 1100110 \\ 1011111 \end{bmatrix} + \begin{bmatrix} 1001011 \\ 1001011 \\ 1001011 \\ 1001011 \\ 1001011 \\ 1001011 \\ 1001011 \end{bmatrix} = \begin{bmatrix} 1001111 \\ 0101001 \\ 1110110 \\ 1110010 \\ 0010000 \\ 0101101 \\ 0010100 \end{bmatrix}$$

$$\mathbf{s}_1 = \mathbf{u} + B_1 = \begin{bmatrix} 0000100 \\ 1100010 \\ 0111101 \\ 0111001 \\ 1011011 \\ 1100110 \\ 1011111 \end{bmatrix} + \begin{bmatrix} 0010111 \\ 0010111 \\ 0010111 \\ 0010111 \\ 0010111 \\ 0010111 \\ 0010111 \end{bmatrix} = \begin{bmatrix} 0010011 \\ 1110101 \\ 0101010 \\ 0101110 \\ 1001100 \\ 1110001 \\ 1001000 \end{bmatrix}$$

There are four 1's in **b** and also in the shifts of **b**. Thus, there are $4 \cdot 3 + 3 \cdot 4 = 24$ 1's and 25 0's in $\mathbf{s}_j$. So, the $\mathbf{s}_j$'s are balanced except for **u**.

Profile of the $(49, 8, 17)$ signal set

1. Period 49.
2. Eight shift-distinct sequences.
3. Maximum magnitude of crosscorrelation is 17.
4. Crosscorrelation takes five values:

$$\{1, -7, 9, 17, -15\}.$$

5. Balance: 25 0's and 24 1's in one period of each sequence (except **u**).
6. Linear span: 24, except for **u** with linear span 21.

Construction B

$(p^2, p + 1, 2p + 3)$ signal set (p prime)

1. Choose **a** and **b** from the set consisting of the quadratic residue sequences modulo p and the Hall sextic residue sequences modulo p if those sequences exist for such p.
2. Choose α, a primitive element of $\mathbb{F}_p$.
3. Compute

$$e_j = \alpha^j \in \mathbb{F}_p, \ 0 \leq j < p.$$

The rest of the steps are the same as in Procedure 1.

Example 10.16 Let $p = 11$, a prime. Choose

$$\mathbf{a} = \mathbf{b}^{(2)} = (11011100010) \text{ and } \mathbf{b} = (10100011101),$$

where $\mathbf{b}$ is a quadratic residue sequence of period 11. Thus, $\mathbf{a}$ and $\mathbf{b}$ are two shift-distinct quadratic residue sequences of period 11. Since 2 is a primitive element of $\mathbb{F}_{11}$, we then compute $e_j \equiv 2^j \pmod{11}$, $0 \le j < 11$, as shown below:

$$\mathbf{e} = (1, 2, 4, 8, 5, 10, 9, 7, 3, 6, 1).$$

From $\mathbf{a}$ and $\mathbf{e}$ we can construct the interleaved sequence $\mathbf{u}$ whose jth column sequence is $L^{e_j}(\mathbf{a})$:

$$\mathbf{u} = \begin{bmatrix} 10101010101 \\ 01110100100 \\ 11000111101 \\ 11010010011 \\ 10011101001 \\ 00100111010 \\ 00011110110 \\ 01111011000 \\ 10110001111 \\ 01001001110 \\ 11101100011 \end{bmatrix}.$$

Therefore

$$\mathbf{s}_j = \mathbf{u} + L^j(\mathbf{b}), \text{ and } S = \{\mathbf{s}_j \mid 0 \le j < 11\} \cup \{\mathbf{u}\},$$

which is a (121, 11, 25) signal set. We can easily obtain any sequence in S from $\mathbf{u}$ and $\mathbf{b}$. For example, $\mathbf{s}_0$ is obtained by complementing the jth column of $\mathbf{u}$ at those j's such that $b_j = 1$, that is, complementing columns of $j \in \{0, 1, 6, 7, 8, 10\}$. Thus, we have

$$\mathbf{s}_0 = \mathbf{u} + \mathbf{b} = \begin{bmatrix} 00001001000 \\ 11010111001 \\ 01100100000 \\ 01110001110 \\ 00111110100 \\ 10000100111 \\ 10111101011 \\ 11011000101 \\ 00010010010 \\ 11101010011 \\ 01001111110 \end{bmatrix}$$

It can be verified that a profile of the randomness of S is as follows:

1. Period 121.
2. Twelve shift-distinct sequences in S.
3. Maximum magnitude of crosscorrelation is equal to 25.
4. Crosscorrelation takes five values:

$$\{1, -11, 13, 25, -23\}.$$

5. Each sequence in S, except for $\mathbf{u}$, is balanced; that is, there are 61 0's and 60 1's in each period of the sequence.
6. The linear span is equal to 120 for $\mathbf{s}_j$ and 110 for $\mathbf{u}$.

10.5.2 Profile of the interleaved constructions A and B

1. S is a $((2^m - 1)^2, 2^m + 1, 1 + 2^{m+1})$ signal set and a $(p^2, p + 1, 2p + 3)$ signal set, from Constructions A and B respectively.
2. Crosscorrelation of any pair of sequences in S or out-of-phase autocorrelation takes five values:

$$\{1, -v, v + 2, 2v + 3, -2v - 1\},$$

where $v = 2^m - 1$ for Construction A and $v = p$ for Construction B.
3. Each sequence in S except for $\mathbf{u}$ has $(v^2 + 1)/2$ zeros and $(v^2 - 1)/2$ ones. In other words, all the sequences in S except for $\mathbf{u}$ satisfy the balance property.
4. We denote the linear span of a sequence $\mathbf{s}$ by $LS(\mathbf{s})$.
   (a) For $v = 2^m - 1$ in Construction A, the linear span of any sequence in S is lower-bounded by

   $$LS(\mathbf{s}_j) > (2^m - 1)LS(\mathbf{a})/2 + LS(\mathbf{b}) \text{ and } LS(\mathbf{u}) > (2^m - 1)LS(\mathbf{a})/2$$

   when $LS(\mathbf{a}) \geq n$. Otherwise, $LS(\mathbf{s}_j) = m2^m$ and $LS(\mathbf{u}) = m(2^m - 1)$.
   (b) For $v = p$, if both $\mathbf{a}$ and $\mathbf{b}$ are quadratic residue sequences, then the linear span of a sequence in S is given by

   $$LS(\mathbf{s}_j) = \begin{cases} \frac{p^2 - 1}{2} & \text{for } p \equiv 7 \pmod 8 \\ p^2 - 1 & \text{for } p \equiv 3 \pmod 8 \end{cases}$$

   and

   $$L(\mathbf{u}) = \begin{cases} \frac{p(p-1)}{2} & \text{for } p \equiv 7 \pmod 8 \\ p(p - 1) & \text{for } p \equiv 3 \pmod 8. \end{cases}$$

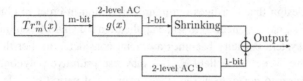

Figure 10.9. Galois configuration of $((2^m - 1)^2, 2^m + 1, 1 + 2^{m+1})$ signal sets.

10.5.3 Implementation

Construction A, which generates $((2^m - 1)^2, 2^m + 1, 1 + 2^{m+1})$ signal sets, can be implemented, as shown in Figure 10.9, by using two binary sequence generators with 2-level autocorrelation of period $2^n - 1$ ($n = 2m$, as a long sequence) and $2^m - 1$ (as a short sequence) respectively, together with a shrinking operation: deleting two consecutive bits for each $2^m + 1$ consecutive bits from the 2-level binary sequence of period $2^n - 1$, where one of these two bits corresponds to a zero column of the array form of the long sequence.

For example, the signal set given in Example 10.15 can be implemented as in Figure 10.10.

Construction B can be implemented via prestorage of both the shift sequence **e** and the quadratic residue sequence **a** (here we choose **b** = **a**) for small p (for example, $p < 2^{25}$).

10.6 $\mathbb{Z}_4$ signal sets

In this section, we will introduce the design of signal sets where the elements of sequences are taken from $\mathbb{Z}_4 = \{0, 1, 2, 3\}$ (mod 4). This is an example of

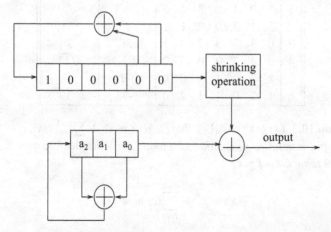

Figure 10.10. Implementation of the (47, 8, 17) signal set in Example 10.15.

signal set design that sacrifices optimum correlation in order to obtain large sizes of the signal sets, needed in some applications of CDMA communication where the system capacity becomes a crucial consideration. For the case of m-sequences over a finite field $\mathbb{F}_q$, we may use primitive polynomials over $\mathbb{F}_q$ of degree n to generate m-sequences over $\mathbb{F}_q$ of period $q^n - 1$. Here we intend to generate sequences over $\mathbb{Z}_4$ with period $2^n - 1$ in terms of known results on binary m-sequences of period $2^n - 1$. This leads to the investigation of constructions of basic irreducible polynomials over $\mathbb{Z}_4$ in terms of primitive polynomials over $\mathbb{F}_2 = \mathbb{Z}_2$. (In this section, we will use $\mathbb{Z}_2$ for $\mathbb{F}_2$ to emphasize the connection with $\mathbb{Z}_4$). In the following, we first present an algorithm for construction of basic irreducible polynomials over $\mathbb{Z}_4$ and then show a design for $\mathbb{Z}_4$ signal sets with parameters $(2^n - 1, r, \delta)$.

The crosscorrelation of two $\mathbb{Z}_4$ sequences $\mathbf{a}$ and $\mathbf{b}$ of period v is defined by

$$C_{\mathbf{a},\mathbf{b}}(\tau) = \sum_{k=0}^{v} \omega^{a_{k+\tau} - b_k}, \quad 0 \leq \tau < v,$$

where $\omega = -\sqrt{-1}$, a primitive 4th root of unity ($i = \sqrt{-1}$ by convention). We have the following map from $\mathbb{Z}_4$ to the complex number field:

j	0	1	2	3
ω^j	1	$-i$	-1	i

For example, with $\mathbf{a} = (0, 2, 2, 3, 1, 1, 0)$ and $\mathbf{b} = (3, 2, 2, 1, 2, 3, 1)$, we have

τ			$L^\tau(\mathbf{a}) - \mathbf{b}$					$C_{\mathbf{a},\mathbf{b}}(\tau)$	$|C_{\mathbf{a},\mathbf{b}}(\tau)|$
0	1	0	0	2	3	2	3	i	1
1	3	0	1	0	3	1	3	$2 + i$	$\sqrt{5}$
2	3	1	3	0	2	1	1	i	1
3	0	3	3	3	2	3	1	$3i$	3
4	2	3	2	3	0	3	2	$-2 + 3i$	$\sqrt{13}$
5	2	2	2	1	0	0	0	$-i$	1
6	1	2	0	1	1	2	0	$-3i$	3

Definition 10.3 *Let $g(x) \in \mathbb{Z}_4[x]$, that is, $g(x) = x^n + \sum_{i=0}^{n-1} c_i x^i$, $c_i \in \mathbb{Z}_4$, be a monic polynomial. $g(x)$ is said to be* monic basic irreducible *over $\mathbb{Z}_4$ if the modulo 2 reduction of $g(x)$,*

$$\overline{g}(x) = x^n + \sum_{i=0}^{n-1} (c_i \bmod 2) x^i,$$

is a monic irreducible polynomial over $\mathbb{Z}_2$.

Example 10.17 Let

$$g(x) = x^5 + 2x^4 + x^3 + 3 \in \mathbb{Z}_4[x].$$

Then

$$\overline{g}(x) = x^5 + x^3 + 1$$

is primitive over $\mathbb{Z}_2$, and therefore it is irreducible over $\mathbb{Z}_2$. Thus $g(x)$ is a basic irreducible polynomial over $\mathbb{Z}_4$.

10.6.1 Algorithm for finding basic irreducible polynomials over $\mathbb{Z}_4$

Algorithm 10.1 ALGORITHM FOR FINDING BASIC IRREDUCIBLE POLYNOMIALS OVER $\mathbb{Z}_4$

Input: $f(x)$, *a primitive polynomial over $\mathbb{Z}_2$ of degree n.*
Output: $g(x)$, *a basic irreducible polynomial over $\mathbb{Z}_4$ of degree n.*

Procedure(f, g):

1. *Set $h(x) = f(x)$ and regard $h(x)$ as a polynomial over $\mathbb{Z}_4$.*
2. *Compute $(-1)^n h(x)h(-x)$, which will be found to be a polynomial of degree n in x^2 over $\mathbb{Z}_4$:*

$$g(x^2) = (-1)^n h(x)h(-x).$$

3. Return $g(x)$
4. Quit.

Example 10.18 Select $f(x) = x^5 + x^3 + 1$, a primitive polynomial over $\mathbb{Z}_2$.

Procedure(f, g):

1. Set $h(x) = f(x)$ and regard $h(x)$ as a polynomial over $\mathbb{Z}_4$.
2. Compute

$$\begin{aligned}
(-1)^5 h(x)h(-x) &= -(x^5 + x^3 + 1)(-x^5 - x^3 + 1) \\
&= x^{10} + 2x^8 + x^6 + 3 \\
&= (x^2)^5 + 2(x^2)^4 + (x^2)^3 + 3.
\end{aligned}$$

3. Return $g(x) = x^5 + 2x^4 + x^3 + 3$, a basic irreducible polynomial over $\mathbb{Z}_4$.
4. Quit.

This is in fact the basic irreducible polynomial in Example 10.17.

10.6.2 $\mathbb{Z}_4$ signal sets of $S(t), t = 0, 1,$ and 2

Algorithm 10.2 ALGORITHM FOR GENERATING $\mathbb{Z}_4$ FAMILIES $S(0)$, $S(1)$, AND $S(2)$

Input: $f(x)$, *a primitive polynomial over $\mathbb{Z}_2$ of degree n.*
Output: $S(t)$, $\mathbb{Z}_4$ *families, $t = 0, 1, 2$.*

Procedure(f, $S(0)$, $S(1)$, $S(2)$):

1. *Apply Algorithm 10.1 to $f(x)$ for computing a basic irreducible polynomial*
   $g(x) = x^n - \sum_{i=0}^{n-1} g_i x^i$, $g_i \in \mathbb{Z}_4$.
2. *Generate $\mathbb{F}_{2^n}$ by $f(\alpha) = 0$, and compute the minimal polynomials of α^3 and α^5 over $\mathbb{Z}_2$:*

$$f_{\alpha^3}(x) = x^n + \sum_{i=0}^{n-1} l_i x^i, l_i \in \mathbb{Z}_2,$$

$$f_{\alpha^5}(x) = x^n + \sum_{i=0}^{n-1} k_i x^i, k_i \in \mathbb{Z}_2.$$

3. *Randomly select initial states:*

$$(a_0, a_1, \ldots, a_{n-1}), a_i \in \mathbb{Z}_4,$$
$$(b_0, b_1, \ldots, b_{n-1}), b_i \in \mathbb{Z}_2,$$
$$(c_0, c_1, \ldots, c_{n-1}), c_i \in \mathbb{Z}_2.$$

4. *Generate a quaternary sequence $\mathbf{a} = \{a_i\}$ by $g(x)$:*

$$a_{k+n} = \sum_{i=0}^{n-1} g_i a_{i+k}, k = 0, 1, \ldots, \text{ in } \mathbb{Z}_4,$$

and two binary sequences $\mathbf{b} = \{b_i\}$ and $\mathbf{c} = \{c_i\}$ by $f_{\alpha^3}(x)$ and $f_{\alpha^5}(x)$ respectively:

$$b_{k+n} = \sum_{i=0}^{n-1} l_i b_{i+k}, k = 0, 1, \ldots, \text{ in } \mathbb{Z}_2,$$

$$c_{k+n} = \sum_{i=0}^{n-1} k_i c_{i+k}, k = 0, 1, \ldots, \text{ in } \mathbb{Z}_2.$$

5. *Compute a quaternary sequence $\mathbf{s} = \{s_i\}$:*

$$s_i = a_i + 2u b_i + 2v c_i, i = 0, 1, \ldots, \text{ in } \mathbb{Z}_4,$$

where u and v belong to $\mathbb{Z}_2$.

Table 10.13. *Parameters of $\mathbb{Z}_4$ families*

Family	Period v (or length)	Size r	δ
$S(0)$	$2^n - 1$	$v + 2$	$\sqrt{v+1} + 1$
$S(1)$	$2^n - 1$	$\geq v^2 + 3v + 2$	$2\sqrt{v+1} + 1$
$S(2)$	$2^n - 1$	$\geq v^3 + 4v^2 + 5v + 2$	$4\sqrt{v+1} + 1$

6. *Return*

$S(0) = \{\mathbf{s} | u = 0, v = 0,$ *for all initial states of* $\mathbf{a}\}$;

$S(1) = \{\mathbf{s} | u = 1, v = 0,$ *for all initial states of* $\mathbf{a}$ *and* $\mathbf{b}\}$;

$S(2) = \{\mathbf{s} | u = 1, v = 1,$ *for all initial states of* $\mathbf{a}$, $\mathbf{b}$ *and* $\mathbf{c}$ $\}$.

7. *Quit*

We present the sizes of these signal sets and their maximum correlation in Table 10.13.

Example 10.19 Compute $\mathbb{Z}_4$ families, $S(i), i = 0, 1, 2$ for $n = 5$.

Input: $f(x) = x^5 + x^3 + 1$, a primitive polynomial over $\mathbb{Z}_2$ of degree $n = 5$.

Output: $S(t)$, $\mathbb{Z}_4$ families, $t = 0, 1, 2$.

Procedure$(f, S(0), S(1), S(2))$

1. Apply Algorithm 10.1 to $f(x)$, to get the basic irreducible polynomial $g(x) = x^5 + 2x^4 + x^3 + 3 = x^5 - (2x^4 + 3x^3 + 1)$.
2. Generate $\mathbb{F}_{2^5}$ by $\alpha^5 + \alpha^3 + 1 = 0$, and compute the minimal polynomials of α^3 and α^5 over $\mathbb{Z}_2$:

$$f_{\alpha^3}(x) = x^5 + x^3 + x^2 + x + 1,$$
$$f_{\alpha^5}(x) = x^5 + x^4 + x^3 + x + 1.$$

3. Arbitrarily select initial states:

$$(a_0, a_1, a_2, a_3, a_4) = (1, 3, 0, 0, 0), a_i \in \mathbb{Z}_4,$$
$$(b_0, b_1, b_2, b_3, b_4) = (1, 0, 0, 0, 0), b_i \in \mathbb{Z}_2,$$
$$(c_0, c_1, c_2, c_3, c_4) = (1, 0, 0, 0, 0), c_i \in \mathbb{Z}_2.$$

4. Generate a quaternary sequence $\mathbf{a} = \{a_i\}$ by $g(x)$:

$$\mathbf{a} = 13000111112033032012202121 01330,$$

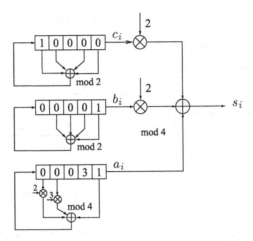

Figure 10.11. LFSR implementation of $\mathbb{Z}_4$ sequence $\{s_i\}$ for $n = 5$.

where

$$a_{k+5} = 2a_{k+4} + 3a_{k+3} + a_k, k = 0, 1, \ldots, \text{ in } \mathbb{Z}_4,$$

and two binary sequences $\mathbf{b} = \{b_i\}$ and $\mathbf{c} = \{c_i\}$ by $f_{\alpha^3}(x)$ and $f_{\alpha^5}(x)$:

$$\mathbf{b} = 10000101101010001110111111001001,$$

where

$$b_{k+5} = b_{k+3} + b_{k+2} + b_{k+1} + b_k, k = 0, 1, \ldots, \text{ in } \mathbb{Z}_2, \text{ and}$$
$$\mathbf{c} = 00001101010010001011111101100111,$$

where

$$c_{k+n} = c_{k+4} + c_{k+3} + c_{k+1} + c_k, k = 0, 1, \ldots, \text{ in } \mathbb{Z}_2.$$

5. Compute quaternary sequences:

$$\mathbf{a} = 1300011111203303201220212101330 \in S(0);$$
$$\mathbf{a} + 2\mathbf{b} = 3300031331001303023202030103330 \in S(1);$$
$$\mathbf{a} + 2\mathbf{b} + 2\mathbf{c} = 3300211133003303221020232303112 \in S(2).$$

Now $s_i = a_i + 2b_i + 2c_i$, $i = 0, 1, \ldots, \{s_i\}$ is a $\mathbb{Z}_4$ sequence, whose LFSR implementation is shown in Figure 10.11.

The parameters of these signal sets for $n = 5$ are shown below.

Parameters for $n = 5$

Family	Period v (or length)	Size r	δ
$S(0)$	31	33	$4\sqrt{2}+1$
$S(1)$	31	1057	$8\sqrt{2}+1$
$S(2)$	31	≥ 33792	$16\sqrt{2}+1$

Next we will give an example of a $\mathbb{Z}_4$ signal set $S(2)$ whose parameters are taken from the specification of the scrambling sequences in the 3G standard (the third generation of mobile communications).

Example 10.20 $S(2)$ **Design for** $n = 8$

1. Select $n = 8$ and $f(x) = x^8 + x^5 + x^3 + x^2 + 1$, primitive over $\mathbb{Z}_2$.
2. Apply Algorithm 10.1 to $f(x)$ to obtain a basic irreducible polynomial over $\mathbb{Z}_4$ as follows:

$$g(x) = x^8 - (3x^5 + x^3 + 3x^2 + 2x + 3).$$

3. Generate $\mathbb{F}_{2^8}$ by $f(\alpha) = 0$, and compute the minimal polynomials of α^3 and α^5 over $\mathbb{Z}_2$ (or use the look-up table in Appendix C in Chapter 3):

$$f_{\alpha^3}(x) = x^8 + x^7 + x^5 + x + 1,$$
$$f_{\alpha^5}(x) = x^8 + x^7 + x^5 + x^4 + 1.$$

4. Randomly select initial states:

$$(a_0, a_1, \ldots, a_7), a_i \in \mathbb{Z}_4,$$
$$(b_0, b_1, \ldots, b_7), b_i \in \mathbb{Z}_2,$$
$$(c_0, c_1, \ldots, c_7), c_i \in \mathbb{Z}_2.$$

5. Generate a quaternary sequence $\mathbf{a} = \{a_i\}$ using $g(x)$:

$$a_{k+8} = 3a_{k+5} + a_{k+3} + 3a_{k+2} + 2a_{k+1} + 3a_k, k = 0, 1, \ldots, \text{ in } \mathbb{Z}_4,$$

and two binary sequences $\mathbf{b} = \{b_i\}$ and $\mathbf{c} = \{c_i\}$ by $f_{\alpha^3}(x)$ and $f_{\alpha^5}(x)$ respectively:

$$b_{k+8} = b_{k+7} + b_{k+5} + b_{k+1} + b_k, k = 0, 1, \ldots, \text{ in } \mathbb{Z}_2,$$
$$c_{k+8} = c_{k+7} + c_{k+5} + c_{k+4} + c_k, k = 0, 1, \ldots, \text{ in } \mathbb{Z}_2.$$

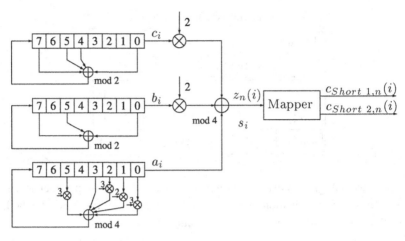

Figure 10.12. LFSR implementation of $\mathbb{Z}_4$ Signal Sets for $n = 8$.

6. Compute quaternary sequences $\mathbf{s} = \{s_i\}$:

$$\mathbf{s} = \mathbf{a} + 2\mathbf{b} + 2\mathbf{c} \text{ in } \mathbb{Z}_4$$

for all intial states of $\mathbf{a}$, $\mathbf{b}$, and $\mathbf{c}$. This gives the family $S(2)$, whose parameters are given by

Parameters for an $S(2)$ Design, $n = 8$

$\mathbb{Z}_4$ Family	Period v	Size r	δ
$S(2)$	255	≥ 16842752	65

The LFSR implementation of $\mathbb{Z}_4$ sequence $\{s_i\}$ for $n = 8$ is shown in Figure 10.12.

Note

For sequences with low correlation, Helleseth and Kumar (1998) have an excellent chapter in the *Handbook of Coding Theory*. For the lower bound on correlation, see Welch (1974) and also Sidelnikov (1978). For the Gold-pair construction, the Gold case was discovered by Gold in 1968, and a proof for Corollary 10.1 for the Kasami (large set) case was given by Kasami (1969) or a later version in 1971. "Kasami Case (large set)" $2^{2k} - 2^k + 1$, may have been first observed by L. Welch, whose result circulated without journal publication. Short proofs for the Kasami decimation $d = 2^{2k} - 2^k + 1$ were given by Dillon (see the note for Chapter 9) and Dobbertin (1999). Welch's and Niho's cases were two old, long-standing conjectures in 1970 and in 1972 and were recently proved by Canteaut, Charpin, and Dobbertin (2000) and Hollmann and Xiang (2001). The generalization of the Gold-pair construction was discovered by Boztas and Kumar in 1992. (A result similar to Theorem 10.1 for the Kasami exponent with $3k \equiv 1 \pmod{n}$ was stated in Chapter 9).

For n even, the Kasami (small) set construction was found by Kasami (1969). For the generalized Kasami construction, the case where $g(x) = Tr_1^n(x^r)$ was generalized by No and Kumar in 1989 and are called No sequences; the case where $g(x)$ is an arbitrary orthogonal function was partially generalized by No et al. in 1997, and this general form appeared in Gong (2002). Bent functions were named by O. Rothaus 1976. Bent function signal sets were constructed by Olsen, Scholtz, and Welch in 1982, and the linear spans of these sequences were studied by Kumar (1983, 1988). The proofs for correlation and linear span of the bent functions signal sets can be found in these papers. For various constructions of bent functions, see Rothaus (1976), McFarland (1973), Dillon (1975), and Carlet (1994).

For the even case $n = 2m$, there are two more constructions. One is the Kerdock code construction with parameters $(2^n - 1, 2^{n-1}, 1 + 2^{n/2})$; see McWilliams Sloane (1977) and Helleseth and Kumar (1998). The other chooses the decimation $d = 2^m + 2^{(m+1)/2} + 1$ or $d = 2^{m+1} + 3$ where m is odd. For $f(x) = Tr(x^d)$, the Hadamard transform of f has exactly three values: $0, \pm 2^{m+1}$. Furthermore, 0 occurs $3 \cdot 2^{n-2}$ times, 2^{m+1} occurs $2^{n-3} + 2^{m-2}$ times, and -2^{m+1} occurs $2^{n-3} - 2^{m-2}$ times. This gives signal sets with parameters $(2^n - 1, 2^n + 1, 1 + 2^{m+1})$, which is not as good as the (generalized) Kasami case or the bent case. These two decimations were conjectured in Niho's thesis (1972) and proved by Cusick and Dobbertin in 1996.

For the interleaved construction of signal sets with parameters $(v^2, v + 1, 2v + 3)$, the case where $v = 2^n - 1$ and the column sequences are m-sequences was constructed in 1995 by Gong. Paterson (1998) extended this to $v = p$ (binary case) where the two short sequences **a** and **b** are identical and found another construction for the shift sequences satisfying the difference condition (10.57) using a special class of MDS codes. This investigation was further extended to $v = p^n - 1$ (p-ary case) and $v = p$ (binary case) where the two short sequences can be different by Gong (2002). The results using interleaved constructions A and B and the results on the linear spans of interleaved sequences for both binary and nonbinary cases, presented here, can be found in Gong (2002). (For the linear spans of quadratic residue sequences, see Ding, Helleseth, and Shan (1998).) $\mathbb{Z}_4$ signal sets were constructed by Kumar, Helleseth, Calderbank, and Hammons in their IEEE Information Theory Society best research paper of 1996. The results introduced here are from that paper. The construction of linear recursive sequences over rings was investigated as early as the 1930s by Ward (1933). Research along this line has flourished since the end of the 1980s.

For the p-ary case where $p > 2$, the status of the generalization from binary cases to p-ary cases is as follows. The Gold-pair construction has been extended to the Gold type decimation $d = \frac{1}{2}(p^k + 1)$ and the Kasami-Welch-Trachtenberg type decimation to $d = p^{2k} - p^k + 1$ $(\gcd(k, n) = 1)$ by Helleseth (1976, 1999) and Trachtenberg (1970). Unlike the binary case, here n can be even. For the Welch and Niho decimations, the Welch decimation was extended to the ternary case by Dobbertin et al. (2001), but not to the general p-ary case. The ternary Niho case was also conjectured by Dobbertin et al. Bent function signal sets for the p-ary case, $p > 2$, were found by Kumar and Moreno (1991). For the generalized Kasami (small) set, generalizing the binary case to the p-ary case with $p > 2$ still remains open.

Exercises for Chapter 10

1. Compute all sequences in a $(31, 33, 9)$ Gold-pair signal set by choosing different **a**. Give the LFSR implementation for your design.
2. There are 50 users in an indoor wireless mobile communication network system. The system requires that
   (a) the scrambling sequence (binary) used by each user is shift distinct from the other users;
   (b) each sequence is balanced with length 127;
   (c) the maximal crosscorrelation between any two of these sequence is 17.
   Design a signal set that satisfies these requirements.
3. Design a Kasami set with parameters $(63, 8, 9)$ by using a different design from the example shown in the text.
   (a) Give the LFSR implementation.
   (b) How many shift sequences are in this Kasami set?
   (c) Compute the 0-1 distribution for each signal in the Kasami set and crosscorrelation for one pair of the signals.
4. A CDMA system needs to employ a signal set with a period of at least 1024, and the maximum value of the crosscorrelation of the signal set is less than 80. How many such designs are there? Give the parameters for each of these designs.
5. Randomly choose four binary m-sequences of period 31, and compute the crosscorrelation of each pair of these m-sequences. What is the smallest maximum crosscorrelation value for all pairs?
6. Can you give a nontrivial bound for the maximum crosscorrelation of any pair of binary m-sequences of degree n?
7. Let

$$\mathbf{e} = (2, 4, 2, 8, 12, 4, 0, 1, 9, 9, 14, 8, 5, 0, 3).$$

Then **e** satisfies the difference condition (10.56). Using **e** as the exponent sequence, construct an interleaved signal set with parameters (225, 16, 33). Give the sequence **u** and one of the sequences in this signal set (not **u**) in their matrix forms.

8. Design an interleaved signal set with parameters (49, 9, 17).
   (a) Give an LFSR implementation for your design.
   (b) Compare your design to a Kasami signal set having similar parameters.
   (c) For any pair of sequences in your interleaved signal set, find the shifts that yield the maximum crosscorrelation value 17.

9. Research Problem: For an interleaved signal set with parameters $(v^2, v + 1, 2v + 3)$, the crosscorrelation of any pair of the sequences in the signal set or the out-of-phase autocorrelation of any sequence in the signal set will be reduced to the set $\{1, -v, v + 2\}$ if the shift sequence $\mathbf{e} = (e_0, e_1, \ldots, e_{v-1})$ satisfies the following condition: for all $1 \leq s < v$,

$$|\{e_j - e_{j+s} \mid 0 \leq j < v - s\} \cup \{e_{v-s+j} - e_j - 1 \mid 0 \leq j < s\}| = v.$$

Does such a vector **e** exist? For small values of v, exhaustively search for such vectors **e**.

11

Correlation of Boolean Functions

The one-to-one correspondences among sequences, polynomial functions, and boolean functions serve as bridges for connections between sequence design with good correlation and constructions of boolean functions with strong cryptographic properties. These materials will be discussed in this chapter. It is worth pointing out that the cryptographic properties of boolean functions discussed in the literature are related to the Hadamard transform of functions (fundamental theory of linear cryptanalysis) and the convolution of functions (fundamental theory of differential cryptanalysis).

To construct pseudorandom sequences with large linear span, a natural way is to apply boolean functions to a set of LFSRs (these LFSRs may be equal). The resulting sequences are called filtering function sequences if the LFSRs are equal or combinatorial function sequences if they are distinct. From the one-to-one correspondence between boolean functions in n variables and polynomial functions from $\mathbb{F}_{2^n}$ to $\mathbb{F}_2$, as introduced in Section 10.1 of Chapter 10, the known constructions, including GMW sequences (all four types), Gold-pair sequences, Kasami and generalized Kasami sequences, and bent function sequences can be considered special cases of this general construction.

For cryptographic applications, from known pairs of plaintext and ciphertext, one can obtain segments of the output sequences of the filtering/combinatorial generators. Given this fact, an attacker tries to find initial states of these LFSRs by computing various correlations of the output sequences in order to get the entire output sequences. To prevent this type of attack, one of the effective ways is to employ boolean functions with certain low correlation properties. Another application of boolean functions in cryptology is employed in block cipher models for providing confusion and diffusion between plaintext and ciphertext. All the desired properties for boolean functions, such as nonlinearity, correlation immunity and resiliency, and propagation, are related to the Walsh transform (or equivalently the Hadamard transform) and the convolution of functions. In this chapter, we introduce these concepts and provide examples obtained from sequences with good correlation.

11.1 Invariants, resiliency, and nonlinearity

For a boolean function $f(x_0, \ldots, x_{n-1}) : \mathbb{F}_2^n \to \mathbb{F}_2$, we consider a classification of $\mathcal{B}$, the set consisting of all boolean functions in n variables. Let G consist of the following operations on the variables $(x_0, \ldots, x_{n-1})$ or on f: (a) all permutations on $\mathbf{x} = (x_0, x_1, \ldots, x_{n-1})$; (b) complementation on any subset of the components of $\mathbf{x}$; and (c) complementation on functions. Let

$$c_{\mathbf{w}} = \sum_{\mathbf{x} \in \mathbb{F}_2^n} (f(\mathbf{x}) + \mathbf{w} \cdot \mathbf{x}), \tag{11.1}$$

where the addition in $f(\mathbf{x}) + \mathbf{w} \cdot \mathbf{x}$ is the addition in $\mathbb{F}_2$; i.e., the sum is reduced modulo 2 and the $\sum$ is ordinary integer summation. Thus, $c_{\mathbf{w}}$ is equal to the (Hamming) weight of the function $f(\mathbf{x}) + \mathbf{w} \cdot \mathbf{x}$, defined as the number of $\mathbf{x} \in \mathbb{F}_2^n$ such that $f(\mathbf{x}) + \mathbf{w} \cdot \mathbf{x} = 1$, or equivalently, the distance between f and the linear function $\mathbf{w} \cdot \mathbf{x}$:

$$c_{\mathbf{w}} = w(f(\mathbf{x}) + \mathbf{w} \cdot \mathbf{x}) = d(f, \mathbf{w} \cdot \mathbf{x}) = |\{\mathbf{x} \in \mathbb{F}_2^n \mid f(\mathbf{x}) + \mathbf{w} \cdot \mathbf{x} = 1\}|.$$

$$\tag{11.2}$$

Golomb introduced the following invariance properties of boolean functions under G in 1959. These results were also collected in Chapter 8 of Golomb (1967). We define $R_{\mathbf{w}} = \max\{c_{\mathbf{w}}, 2^n - c_{\mathbf{w}}\}$, $\mathbf{w} \in \mathbb{F}_2^n$. Then, we have the following concepts about invariants $R_{\mathbf{w}}$ with respect to the operators of G.

Definition 11.1 (GOLOMB, 1959) *With the above notation,*

- *R_0 is invariant under G (this is equal to the smaller of the weight of f or the weight of the complement of f). It is called the 0-order invariant of f.*
- *The set $\{R_{\mathbf{w}} \mid \mathbf{w} \in \mathbb{F}_2^n, w(\mathbf{w}) = k\}$, when it is rearranged in descending order, is invariant under G. The $R_{\mathbf{w}}$'s in descending order are called the kth-order invariants of f, $1 \le k \le n$.*

Example 11.1 Let $n = 4$ and $f(x_0, x_1, x_2, x_3) = x_0 x_1 + x_2 + x_3$. We compute the truth table of f, $c_{\mathbf{w}}$ and the invariants $R_{\mathbf{w}}$ ($\mathbf{w} = (w_0, w_1, w_2, w_3)$), shown in Table 11.1.

Thus the 0-order invariant is equal to 8, first-order invariants $(8, 8, 8, 8)$, second-order invariants $(12, 8, 8, 8, 8, 8)$, third-order invariants $(12, 12, 8, 8)$,

Table 11.1. *Invariants of* $x_0 x_1 + x_2 + x_3$

(x_0, x_1, x_2, x_3)	$f(\mathbf{x})$	(w_0, w_1, w_2, w_3)	$c_{\mathbf{w}}$	$R_{\mathbf{w}}$
0 0 0 0	0	0 0 0 0	8	8
1 0 0 0	0	1 0 0 0	8	8
0 1 0 0	0	0 1 0 0	8	8
1 1 0 0	1	0 0 1 0	8	8
0 0 1 0	1	0 0 0 1	8	8
1 0 1 0	1	1 1 0 0	8	8
0 1 1 0	1	1 0 1 0	8	8
1 1 1 0	0	0 1 1 0	8	8
0 0 0 1	1	1 0 0 1	8	8
1 0 0 1	1	0 1 0 1	8	8
0 1 0 1	1	0 0 1 1	4	12
1 1 0 1	0	1 1 1 0	8	8
0 0 1 1	0	1 1 0 1	8	8
1 0 1 1	0	1 0 1 1	4	12
0 1 1 1	0	0 1 1 1	4	12
1 1 1 1	1	1 1 1 1	12	12

and fourth-order invariant 12. All kth-order invariants are invariant under the operators of G.

Golomb described a method for computing the invariants in terms of the Walsh coefficients of f. In our notation, this method is equivalent to the following assertion (by Property 10.3 of Section 10.1 in Chapter 10):

$$c_{\mathbf{w}} = 2^{n-1} - \frac{1}{2}\widehat{f}(\mathbf{w}), \mathbf{w} \in \mathbb{F}_2^n. \tag{11.3}$$

In other words, the invariants can be computed via the Walsh transform. Furthermore, we have the following relation:

$$c_{\mathbf{w}} = 2^{n-1} \iff \widehat{f}(\mathbf{w}) = 0. \tag{11.4}$$

Siegenthaler (1984) investigated the so-called correlation immunity of boolean functions in order to construct boolean functions used in combinatorial generators that can resist correlation attack.

Definition 11.2 (SIEGENTHALER, 1984) *A boolean function $f(\mathbf{x})$ in n variables is kth-order correlation immune if each k-subset K of $\{0, \dots, n-1\}$, $Z = f(\mathbf{x})$, considered as a random variable over $\mathbb{F}_2$, is independent of all x_i for $i \in K$ where the x_i's are considered as random variables over $\mathbb{F}_2$ taking the values 0 or 1 with equal probability. Furthermore, if f is balanced and kth-order correlation immune, then f is said to be kth-order resilient.*

Computation of the conditional probabilities for which boolean functions are kth-order correlation immune is equivalent to counting the number of 1's of Z when $\mathbf{x}$ takes values in the restricted subspaces of $\mathbb{F}_2^n$ of dimension $n - k$. However, these numbers are related to the number of 1's of Z on subspaces of $\mathbb{F}_2^n$ of dimension $n - 1$, that is, the subspaces defined by any linear combination of r coordinates x_i, $1 \leq r \leq k$. Therefore, kth-order correlation immunity (or resiliency) can be determined by the invariants $c_{\mathbf{w}}$ or equivalently by the Walsh transform of f. We discuss this in detail below.

Lemma 11.1 *Let* $Z = f(\mathbf{x})$ *be a boolean function in* n *variables. For a* k-subset K of $\{0, 1, \ldots, n - 1\}$, *we write* $K = \{j_1, \ldots, j_k\}$, *and let* $y_i = x_{j_i}$, $i = 1, \ldots, k$. *Then* Z *is independent of* $\mathbf{y} = (y_1, \ldots, y_k)$ *if and only if* Z *is independent of any linear combination of the* y_i's, *i.e.,* $\mathbf{w} \cdot \mathbf{y} = \sum_{i=1}^{k} w_i y_i$ *for every nonzero constant vector* $\mathbf{w} = (w_1, \ldots, w_k)$, $w_i \in \mathbb{F}_2$.

Proof. The assertion can be stated using the following probabilities:

$$P\{Z \mid y_1 = t_1, \ldots, y_k = t_k\} = P\{Z\}, \text{ for any } t_i \in \mathbb{F}_2$$

if and only if

$$P\{Z \mid \mathbf{w} \cdot \mathbf{y} = c\} = P\{Z\}, \text{ for all nonzero } \mathbf{w} \in \mathbb{F}_2^k, \text{ and } c \in \mathbb{F}_2.$$

This result can be established by mathematical induction. The necessary condition is clear. We only need to show that it is sufficient.

Case 1. $k = 2$. The method used below is an elementary method for counting the 0-1 distribution of binary sequences, that is, counting the following numbers:

1. The number of $\mathbf{x} \in \mathbb{F}_2^n$ for which $Z = c$ given that $y_1 = i$ and $y_2 = j$, with $c, i, j \in \mathbb{F}_2$, is denoted by $M_c(i, j)$.
2. The number of $Z = c$ given that $y_1 + y_2 = b$, $b \in \mathbb{F}_2$, is denoted by $N_c(b)$.
3. The number of $Z = c$ given that $y_k = i$, $i \in \mathbb{F}_2$ for $k = 1, 2$, is denoted by $T_{c,y_k}(i)$.

The conditional probability of Z given that $y_1 = i$ and $y_2 = j$ is given by

$$P\{Z = c \mid y_1 = i, y_2 = j\} = \frac{M_c(i, j)}{2^{n-2}}.$$

Thus, the random variable Z is independent of (y_1, y_2) if and only if these conditional probabilities are equal for any assignment of (i, j), that is, for a fixed c,

$$M_c(i, j) = u_c, \text{ a constant for all } (i, j). \tag{11.5}$$

This can be derived from the relationships among the three types of numbers defined above together with the assumption that Z is independent of $y_1 + y_2$, y_1, and y_2, as presented below.

Identities (without any restriction)

$$M_1(1, 0) + M_1(0, 1) = N_1(1)$$
$$M_1(0, 0) + M_1(1, 1) = N_1(0)$$
$$M_1(0, 0) + M_1(0, 1) = T_{1, y_1}(0)$$
$$M_1(1, 0) + M_1(1, 1) = T_{1, y_1}(1)$$
$$M_1(0, 0) + M_1(1, 0) = T_{1, y_2}(0)$$
$$M_1(0, 1) + M_1(1, 1) = T_{1, y_2}(1).$$

By the assumption that Z is independent of $y_1 + y_2$, y_1, and y_2, we have the following three additional identities:

$$P\{Z = 1 \mid y_1 + y_2 = b\} = \tfrac{N_1(b)}{2^{n-1}} \implies N_1(0) = N_1(1),$$
$$P\{Z = 1 \mid y_1 = i\} = \tfrac{T_{1, y_1}(i)}{2^{n-1}} \implies T_{1, y_1}(0) = T_{1, y_1}(1),$$
$$P\{Z = 1 \mid y_2 = i\} = \tfrac{T_{1, y_2}(i)}{2^{n-1}} \implies T_{1, y_2}(0) = T_{1, y_2}(1).$$

(Here we set $c = 1$ to simplify the proof, but for the case $c = 0$, these identities are exactly the same except that all the subscript 1's are replaced by 0's.) From these nine identities, by simple manipulation, it follows that Eq. (11.5) is true.

The above nine identities can be visualized in Figure 11.1, which gives a short proof for the assertion. In Figure 11.1, the vertices are $M_c(i, j)$, rewritten as w, x, y, and z. Each edge denotes the sum of its two endpoint vertices. Thus, these are the six identities listed above without restriction. Z being independent of y_1, and independent of y_2, implies that the sums of pairs of opposite edges

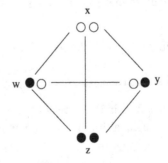

Figure 11.1. Graph of the conditional probabilities $P\{Z \mid y_1 = i, y_2 = j\}$.

are equal, and Z being independent of $y_1 + y_2$ implies that the sums of the two diagonal edges are equal. Thus, we have

$$\left. \begin{array}{l} w + z = x + y \\ w + x = y + z \\ w + y = x + y \end{array} \right\} \implies w = x = y = z.$$

Case 2. Assume that the sufficient condition is true for $k < r$. For $k = r$, let $S = (Z, y_3, \ldots, y_r)$. In a way similar to that used for $k = 2$, we can show that if S is independent of y_1, y_2, and $y_1 + y_2$, then S is independent of (y_1, y_2). In other words, we only need to replace all the subscript 1's in the above nine identities by $\mathbf{c} = (c_0, c_3, \ldots, c_r)$ where the $c_i \in \mathbb{F}_2$ are constant, from which the result follows. $\qquad \square$

From this lemma and the definition of correlation immunity, it is immediate that a boolean function $Z = f(\mathbf{x})$ is kth-order correlation immune if and only if Z is independent of $\mathbf{w} \cdot \mathbf{x}$ for all $\mathbf{w} \in \mathbb{F}_2^n$ such that $1 \le w(\mathbf{w}) \le k$. In other words, f is kth-order correlation immune if and only if

$$P\{Z \mid \mathbf{w} \cdot \mathbf{x} = 1\} = P\{Z \mid \mathbf{w} \cdot \mathbf{x} = 0\}, \tag{11.6}$$

$$\text{for all } \mathbf{w} \in \mathbb{F}_2^n \text{ such that } 1 \le w(\mathbf{w}) \le k.$$

From this assertion, we can establish a relation between correlation immunity and invariants of boolean functions.

Property 11.1 *A boolean function f is kth-order correlation immune if and only if*

$$c_{\mathbf{w}} = 2^{n-1}, \text{ for all } \mathbf{w} \in \mathbb{F}_2^n \text{ such that } 1 \le w(\mathbf{w}) \le k, \tag{11.7}$$

or equivalently,

$$\widehat{f}(\mathbf{w}) = 0, \text{ for all } \mathbf{w} \in \mathbb{F}_2^n \text{ such that } 1 \le w(\mathbf{w}) \le k. \tag{11.8}$$

Proof. We follow a similar approach to the proof for Lemma 11.1. For $Z = f(\mathbf{x})$, and a fixed $\mathbf{w} = (w_0, \ldots, w_{n-1}) \in \mathbb{F}_2^n$, we denote

$$N_{ij} = |\{\mathbf{x} \in \mathbb{F}_2^n \mid Z = i \text{ and } \mathbf{w} \cdot \mathbf{x} = j\}|, i, j \in \mathbb{F}_2.$$

Note that there are 2^{n-1} vectors $\mathbf{x} \in \mathbb{F}_2^n$ such that $\mathbf{w} \cdot \mathbf{x} = j$, $j \in \mathbb{F}_2$, where $\mathbf{w} \ne \mathbf{0}$. Therefore, the N_{ij}'s satisfy the following relations.

$$N_{00} + N_{10} = 2^{n-1}, \tag{11.9}$$

$$N_{01} + N_{11} = 2^{n-1}. \tag{11.10}$$

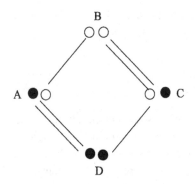

Figure 11.2. Graph of the conditional probabilities $P\{Z|\mathbf{w}\cdot\mathbf{x}\}$.

On the other hand, the conditional probability of Z given that $\mathbf{w}\cdot\mathbf{x} = j$ is given by:

$$P\{Z = i \mid \mathbf{w}\cdot\mathbf{x} = j\} = \frac{N_{ij}}{2^{n-1}}. \tag{11.11}$$

Therefore, from Eq. (11.6), f is kth-order correlation immune if and only if for all $\mathbf{w}\in \mathbb{F}_2^n$ with $1 \leq w(\mathbf{w}) \leq k$,

$$\begin{aligned} P\{Z = 1 \mid \mathbf{w}\cdot\mathbf{x} = 0\} = P\{Z = 1 \mid \mathbf{w}\cdot\mathbf{x} = 1\} &\Longrightarrow N_{10} = N_{11}. \\ P\{Z = 0 \mid \mathbf{w}\cdot\mathbf{x} = 0\} = P\{Z = 0 \mid \mathbf{w}\cdot\mathbf{x} = 1\} &\Longrightarrow N_{00} = N_{01}. \end{aligned} \tag{11.12}$$

Combining this with Eqs. (11.9) and (11.10), f is kth-order correlation immune if and only if

$$c_{\mathbf{w}} = N_{10} + N_{01} = 2^{n-1}, \text{ for all } \mathbf{w}\in \mathbb{F}_2^n \text{ with } 1 \leq w(\mathbf{w}) \leq k.$$

From the relation $\widehat{f}(\mathbf{w}) = 2^n - 2c_{\mathbf{w}}$, the last assertion follows. □

Similar to the proof for Lemma 11.1, the above identities can be visualized in Figure 11.2, which yields a short proof for the assertion. In Figure 11.2, the vertices are $N_{i,j}$, rewritten as A, B, C, and D. An edge with a single line denotes that the sum of the two connecting vertices is equal to 2^{n-1}, and an edge with double lines denotes that the two endpoint vertices are equal. The latter follows from Z being independent of $\mathbf{w}\cdot\mathbf{x}$. Note that $c_{\mathbf{w}} = A + C$. Thus, we have

$$\left.\begin{aligned} A + B = C + D &= 2^{n-1} \\ A &= D \\ B &= C \end{aligned}\right\} \Longrightarrow c_{\mathbf{w}} = A + C = 2^{n-1}.$$

Property 11.1 implies that Z is independent of any k input variables if and only if the probability that Z is equal to any linear combination of r input variables is $1/2$ for all $1 \leq r \leq k$. This could be viewed as Z being indistinguishable from all linear combinations of r input variables with $1 \leq r \leq k$. Perhaps this view may help to formulate a security notion for symmetric cryptography.

Note. We omitted proofs of Lemma 7.1 and Theorem 7.3 in Chapter 7. Right now, the reader can easily work them out by applying a counting method similar to the one used for Lemma 11.1 or Property 11.1.

If $f(x)$ is balanced, then $c_0 = 2^{n-1} \iff \widehat{f}(0, \ldots, 0) = 0$. Thus f is kth-order resilient if and only if

$$c_{\mathbf{w}} = 2^{n-1}, \text{ for all } \mathbf{w} \in \mathbb{F}_2^n \text{ such that } 0 \leq w(\mathbf{w}) \leq k, \qquad (11.13)$$

or equivalently,

$$\widehat{f}(\mathbf{w}) = 0, \text{ for all } \mathbf{w} \in \mathbb{F}_2^n \text{ such that } 0 \leq w(\mathbf{w}) \leq k. \qquad (11.14)$$

Example 11.2 For the boolean function f in Example 11.1, since $c_{\mathbf{w}} = 8$ for all $\mathbf{w} \in \mathbb{F}_2^4$ such that $0 \leq w(\mathbf{w}) \leq 1$, it follows that f is first-order resilient.

Siegenthaler described the degree d of f, the order of correlation immunity or resiliency, k, and the number of variables, n, as satisfying the following relationships:

$d + k \leq n$	if f is kth-order correlation immune
$d + k \leq n - 1$	if f is kth-order resilient and $k < n - 1$, and
$d = 1$	if $k = n - 1$.

Note that the only function that achieves $(n - 1)$th-order resiliency is the linear function

$$x_0 + \cdots + x_{n-1}$$

and its complement.

Nonlinearity of a boolean function $f(\mathbf{x})$, $\mathbf{x} = (x_0, \ldots, x_{n-1}) \in \mathbb{F}_2^n$, is a measure of how far f is from all affine functions.

Definition 11.3 (NONLINEARITY) *Nonlinearity of a boolean function $f(\mathbf{x})$ in n variables is defined as*

$$N_f = \min_{\{\mathbf{w} \in \mathbb{F}_2^n, c \in \mathbb{F}_2\}} d(f(\mathbf{x}), \mathbf{w} \cdot \mathbf{x} + c), \qquad (11.15)$$

where $d(f, g)$ denotes the Hamming distance between f and g.

According to Eq. (11.2), the nonlinearity of f is equal to the minimum value of the invariants; that is,

$$N_f = \min_{\{w \in \mathbb{F}_2^n\}} \min\{c_w, 2^n - c_w\}. \tag{11.16}$$

For the function f in Example 11.1, since the minimal value of the set consisting of $\min\{c_w, 16 - c_w\}$ for all $w \in \mathbb{F}_2^4$ is equal to 4, the nonlinearity of f is equal to 4. From Eqs. (11.2) and (11.3), the nonlinearity of f can be characterized by the Walsh transform when a boolean form is adopted or the Hadamard transform if a polynomial form is adopted. In other words,

$$N_f = 2^{n-1} - \frac{1}{2} \max_{w \in \mathbb{F}_2^n} |\widehat{f}(w)| \tag{11.17}$$

or equivalently,

$$N_f = 2^{n-1} - \frac{1}{2} \max_{\lambda \in \mathbb{F}_{2^n}} |\widehat{f}(\lambda)|, \tag{11.18}$$

where $f(x)$ is the polynomial form of the boolean function $f(x_0, \ldots, x_{n-1})$.

Remark 11.1 Let $a \leftrightarrow Tr(x)$ and $b \leftrightarrow f(x)$. Then the nonlinearity of the function f is determined by the maximum magnitude of the crosscorrelation between a and b. If we define $S = \{L^j(a) + b \mid 0 \le j < t\}$ where t is the period of b, then the computation for the nonlinearity of f is much easier than for determining the maximum correlation δ of S, because from Property 10.5 in Section 10.1 of Chapter 10,

$$\delta = \max_{\beta \in \mathbb{F}_{2^n}} (2^n - 2N_{f_\beta(x)}),$$

where $f_\beta(x) = f(\beta x) + f(x)$. Thus the nonlinearity of $f(x)$ is equal to the nonlinearity of $f_0(x)$, which is just one case out of 2^n cases for δ. Computation costs for these two quantities are equal only for $f(x) = Tr(x^d)$ where $\gcd(d, 2^n - 1) = 1$, that is, when $f(x)$ is the trace presentation of an m-sequence.

In the following example, we show some functions with large nonlinearity obtained from sequences with good correlation.

Example 11.3 (a) $n = 2m$. Let f be a bent function in n variables. Since

$$\widehat{f}(\lambda) = \pm 2^m, \forall \lambda \in \mathbb{F}_{2^n},$$

the nonlinearity of f is given by

$$N_f = A(n) = 2^{n-1} - 2^{m-1}. \tag{11.19}$$

(b) $n = 2m + 1$. Let $f(x) = Tr(x^d)$ where d is taken from the Gold-pair construction. In this case, $\widehat{f}(\lambda)$, the Hadamard transform of f, belongs to the set P, as defined by

$$P = \{0, \pm 2^{m+1}\} \tag{11.20}$$

for all $\lambda \in \mathbb{F}_{2^n}$. Therefore, the nonlinearity of f is given by

$$N_f = B(n) = 2^{n-1} - 2^m. \tag{11.21}$$

(c) $n = 2m + 1$. Let $f(x)$ be taken from the 2-level autocorrelation functions constructed in Chapter 9. From Lemmas 2, 3, and 4 in Chapter 9, we have

$$\widehat{f}(\lambda) \in P \text{ if } f(x) = G_1, WG(x), \text{ or } C_k.$$

Therefore $N_f = B(n)$.

Some comments on bounds on the nonlinearity of boolean functions
(1) For n even, according to Parseval's identity, the value $A(n)$, defined by (11.19), is maximal. Thus, bent functions have maximum distance from all affine functions. However, bent functions are not balanced.
(2) For n odd, note that if the crosscorrelation between f and g belongs to P, then the maximum magnitude of the values in P is 2^{m+1}. According to Parseval's identity, 2^{m+1} is minimal among the magnitudes of all the 3-valued crosscorrelation functions. In this case, we say that the set P is preferred, and the crosscorrelation function of $f(x)$ and $g(x)$ is preferred. For the case $g(x) = Tr(x)$, the crosscorrelation function of $f(x)$ and $Tr(x)$ is equal to the Hadamard transform of $f(x)$. So, in this case, it is equivalent to saying that the Hadamard transform of f is preferred. Thus the parameter $\delta = 1 + 2^{m+1}$ for the Gold-pair signal sets is minimal among all the signal sets having 3-valued correlation.

However, for n odd, the nonlinearity given by $B(n)$, defined by Eq. (11.21), is not maximal. There is a well-known example found by Patterson and Wiedemann (1983) for $n = 15$ whose nonlinearity is larger than $B(15)$, as shown below in the trace representations

$$f_1(x) = \sum_{r \in I} Tr(x^r), I = \{651, 1519, 7595\}, \text{ and}$$

$$f_2(x) = \sum_{r \in I} Tr(x^r), I = \{217, 651, 1085, 2387, 3255\},$$

where $\mathbb{F}_{2^{15}}$ is defined by $x^{15} + x + 1$. The nonlinearity and weight of these two functions are given by

$$N_{f_i} = 16276 > B(15) = 16256 \text{ and } w(f_i) = 16492, i = 1, 2.$$

Note that both of these functions are not balanced, although their nonlinearity is larger than $B(15)$. For odd n, no systematic methods are yet known which are not from concatenations of linear functions, for constructing functions from $\mathbb{F}_{2^n}$ to $\mathbb{F}_2$ (or equivalently, boolean functions in n variables) such that their nonlinearity is greater than $B(n)$.

Remark 11.2 From Property 11.1, kth-order correlation immunity (or resiliency) and nonlinearity of boolean functions are determined by $c_{\mathbf{w}}$ or equivalently by the invariants $R_{\mathbf{w}}$. Because the invariants (in descending order or without any order) are invariant under the operators of G, kth-order correlation immunity (or resiliency) and the nonlinearity of boolean functions are invariant under the operators of G.

11.2 Dual functions and resiliency

In this section, we present a condition such that boolean functions with the preferred Hadamard transform possess first-order resiliency. To establish this result, we introduce a new function, called a dual function, determined by the Hadamard transform.

Definition 11.4 Let $f : \mathbb{F}_{2^n} \to \mathbb{F}_2$. Its dual function, denoted by $\sigma_f(x)$, is defined as

$$\sigma_f(\lambda) = \begin{cases} 0 & \text{if } \widehat{f}(\lambda) = 0 \\ 1 & \text{if } \widehat{f}(\lambda) \neq 0. \end{cases} \tag{11.22}$$

Example 11.4 Let $n = 5$ and $\mathbb{F}_{2^5}$ be defined by $\alpha^5 + \alpha^3 + 1 = 0$.

(a) $f(x) = Tr(x^3)$. We compute:

$$i = 0, \ldots, 30$$

$\widehat{f}(\alpha^i)$	-8	0	0	0	-8	0	8	0	-8	-8	0	8	8	0	0	0	-8	8	-8	8	8	0	0	8	8	0	8	0	0
$\sigma_f(\alpha^i)$	1	0	0	0	1	0	1	0	1	1	0	1	1	0	0	0	1	1	1	0	0	1	1	0	1	0	0

Since the maximum magnitude of the Hadamard transform is equal to 8, the nonlinearity of f is given by $N_f = 16 - \frac{1}{2} \cdot 8 = 12$. This exponent is a Gold-pair exponent, so, by Theorem 10.1 of Chapter 10 (or directly computing the discrete Fourier transform of σ_f), we have $\sigma_f(x) = Tr(x)$.

(b) $f(x) = Tr(x^{15})$. The evaluation of the dual function of $f(x)$ is given as follows:

$$i = 0, \ldots, 30$$

$\widehat{f}(\alpha^i)$	12 88−884 −8 484 4−4−8−4408 −84 4 4−4−4 0 −8 4−4 0 400
$\sigma_f(\alpha^i)$	1 11 111 1 1111 1 1 1 1101 11 1 1 1 1 0 1 1 1 0 100

The maximum magnitude of the Hadamard transform is equal to 12. Thus, the nonlinearity of f is given by $16 - \frac{1}{2} \cdot 12 = 10$. Applying the DFT, we get $\sigma_f(x) = Tr(x^5 + x^7 + x^{11})$.

Example 11.5 The dual function of a bent function f is a constant function. In other words, we have

$$f \text{ is bent} \implies \sigma_f(x) = 1, \forall x \in \mathbb{F}_{2^n}.$$

Note that if the Hadamard transform of a function f is preferred, such as the function $Tr(x^3)$ in Example 11.4, again, from the Parseval identity

$$\sum_\lambda \widehat{f}(\lambda)^2 = 2^{2n},$$

we see that $\widehat{f}(\lambda) \neq 0$ for 2^{n-1} of the λ's. Therefore the dual functions of those functions are always balanced. Another remarkable property of functions with preferred Hadamard transforms is that their dual functions completely determine first-order resiliency. We state this result without proof.

Fact 11.1 *Assume that the Hadamard transform of f is preferred. Then there exists some basis such that a boolean form of f under this basis is first-order resilient if and only if $\sigma_f(x)$ is not a linear function; that is, $\sigma_f(x) \neq Tr(\beta x)$ for any $\beta \in \mathbb{F}_{2^n}$.*

From the Hadamard transforms of 2-level autocorrelation sequences (Lemmas 9.2, 9.3, and 9.4 or the results summarized in Table 9.18 of Chapter 9) and the Gold-pair sequences in Section 10.1, we can easily get the dual functions of those functions, which are presented in Tables 11.2 and 11.3. The polynomial functions in these two tables are all the known polynomial functions with preferred Hadamard transforms for any odd n. Thus, the nonlinearity of such functions is equal to $B(n)$. But there are many examples found by computer search whose Hadamard transforms are also preferred. The following example is one of these.

Table 11.2. *n odd, functions with 2-level AC, $N_f = B(n)$, preferred $\hat{f}$, and first-order resiliency*

Functions	Dual Functions	Sequences
$Tr(x^d)$	$Tr(x^{2^k+1})$ nonlinear	for d: Kasami exp. with $3k \equiv 1 \pmod n$ for d: the rest of Kasami exponents, Welch and Niho cases
$WG(x)$	$Tr(x^{d^{-1}})$	WG sequences in Section 9.1 of Chapter 9
$G_1(x)$	$Tr(x^{\frac{k-1}{k}})$	Glynn type 1 sequences in Section 9.2 of Chapter 9
$C_k(x)$	$Tr(x^{2^k+1})$	Sequences from the Kasami power function construction in Section 9.3 of Chapter 9 including Segre C_2, 3-term, and 5-term sequences
$Tr_{F_1/F_0}(x^{s_1}) \circ \cdots$ $\circ Tr_{F_i/F_{i-1}}(x^{s_i})$ $\circ \cdots \circ$ $Tr_{F_r/F_{r-1}}(x^{s_r}),$ $s_i = \|F_{i-1}\|^{t_i} + 1$	nonlinear	GMW sequences (type 2) of Chapter 8

Example 11.6 Let $n = 7$, let $\mathbb{F}_{2^7}$ be defined by $\alpha^7 + \alpha + 1$, and let $f(x) = Tr(x + x^7 + x^{29})$. In Table 11.4, we present the Hadamard transform of $f(x)$ and the autocorrelation of $f(x)$ for $x = \alpha^i$ for all the coset leaders i modulo 127. By computation, we have the dual of $f(x)$, $\sigma_f(x) = Tr(x^5 + x^{13} + x^{19} + x^{21} + x^{29})$, which is nonlinear. We also calculate the autocorrelation of $g(x) = \sigma_f(x)$ and the Hadamard transform of $g(x)$, as shown in Table 11.4.

From Table 11.4, we have $\hat{f}(\lambda), \hat{g}(\lambda) \in \{0, \pm 16\}, \forall \lambda \in \mathbb{F}_{2^7}$. Thus, both the Hadamard transform of $f(x)$ and the Hadamard transform of $g(x)$, the dual of $f(x)$, are preferred. Hence, both of them have nonlinearity 56. From Fact 11.1, the boolean form of f under some basis is first-order resilient. Let $\mathbf{a} = \{a_i\}$ and

Table 11.3. *n odd, functions with $N_f = B(n)$ and preferred $\hat{f}$*

Functions	Dual Functions	Sequences
$\sum_{i=1}^{(n-1)/2} c_i Tr(x^{1+2^i}),$ $c_i \in \mathbb{F}_2$	$Tr(x)$ linear linear	(1) for the Gold type exponent: $c_i = 1$ for one of i's (2) for $c_i = 1$ for all i's (3) for $c_i = 1$ for any subset of i's if $n = p$ prime and 2 is a primitive element in $\mathbb{F}_p$

Table 11.4. *Example of sequences with preferred correlation*

i	0	1	3	5	7	9	11	13	15	19	21	23	27	29	31	43	47	55	63
$f(\alpha^i)$	1	1	0	1	0	0	1	0	1	0	1	0	1	0	0	1	1	0	1
$\widehat{f}(\alpha^i)$	16	0	−16	16	−16	0	−16	16	0	−16	16	16	0	0	0	0	0	0	16
$C_f(\alpha^i)$	128	−8	8	8	−16	0	8	−8	−16	−8	16	−8	−8	8	8	16	8	0	−8
$g(\alpha^i)$	1	1	0	0	0	0	0	1	1	1	1	0	0	1	0	1	0	1	1
$\widehat{g}(\alpha^i)$	16	0	16	−16	−16	0	16	16	0	−16	16	−16	0	0	0	0	0	0	16
$C_g(\alpha^i)$	128	−8	−8	8	0	0	0	8	0	−8	8	8	−8	0	−8	8	8	0	−8

$\mathbf{b} = \{b_i\}$ where $a_i = f(\alpha^i)$ and $b_i = \sigma_f(\alpha^i)$, $i = 0, \ldots, 126$. Then the elements of $\mathbf{a}$ and $\mathbf{b}$ are given by

$$\mathbf{a} = \{a_i\}_{i=0}^{126} = 1\,1\,1\,0\,1\,1\,0\,0\,1\,0\,1\,1\,0\,0\,0\,1\,1\,0\,0\,0\,1\,1\,1\,0\,0\,0\,0$$
$$1\,0\,0\,1\,0\,1\,1\,0\,0\,0\,1\,0\,0\,1\,1\,1\,1\,1\,1\,0\,1\,0\,1\,0\,1\,0\,1$$
$$1\,0\,0\,0\,0\,0\,1\,1\,0\,1\,1\,0\,1\,0\,0\,1\,0\,1\,0\,0\,1\,0\,0\,1\,0\,0\,1$$
$$0\,1\,0\,1\,1\,1\,1\,1\,1\,1\,0\,0\,0\,1\,1\,0\,0\,1\,1\,0\,0\,1\,0\,0\,0\,1\,1$$
$$1\,0\,0\,1\,0\,1\,0\,0\,0\,1\,0\,1\,1\,0\,1\,1\,0\,1\,1$$

$$\mathbf{b} = \{b_i\}_{i=0}^{126} = 1\,1\,1\,0\,1\,0\,0\,0\,1\,0\,0\,0\,0\,1\,0\,1\,1\,0\,0\,1\,0\,1\,0\,0\,0\,1\,1$$
$$0\,0\,1\,1\,0\,1\,0\,0\,1\,0\,1\,1\,1\,0\,1\,1\,1\,0\,1\,0\,0\,0\,0\,1\,0\,1\,1$$
$$0\,1\,0\,0\,1\,1\,1\,0\,0\,1\,1\,0\,0\,0\,0\,0\,1\,1\,0\,1\,1\,0\,1\,0\,1\,0\,0$$
$$1\,1\,1\,1\,1\,1\,0\,0\,0\,1\,1\,0\,1\,0\,1\,0\,0\,0\,1\,1\,0\,0\,0\,1\,1\,1\,0$$
$$0\,1\,1\,1\,0\,1\,0\,0\,1\,0\,1\,1\,1\,0\,0\,1\,0\,1\,1$$

Because $C_f(\alpha^i) \in \{0, \pm 8, \pm 16\}$ and $C_g(\alpha^i) \in \{0, \pm 8\}$, the autocorrelation functions of $\mathbf{a}$ and $\mathbf{b}$ are given as follows:

$$C_{\mathbf{a}}(\tau) \in \{-1, -9, 7, -17, 15\}, \text{ and } C_{\mathbf{b}}(\tau) \in \{-1, -9, 7\}, \tau \neq 0.$$

11.3 Dual functions, additive autocorrelation, and the propagation property

In this subsection, we will use the connection between the Hadamard transform and the convolution transform discussed in Section 6.7 of Chapter 6 to show that the additive autocorrelations of some functions in Table 11.2 are preferred, that is, belong to $\{0, \pm 2^{\frac{n+1}{2}}\}$, and have *first-order propagation*, which will be defined below. Recall that the additive autocorrelation of a polynomial function $f : \mathbb{F}_{2^n} \to \mathbb{F}_2$ is defined by (see Chapter 6)

$$V_f(w) = \sum_{\mathbf{x} \in \mathbb{F}_{2^n}} (-1)^{f(\mathbf{x}+w)+f(\mathbf{x})}, \; w \in \mathbb{F}_{2^n}. \tag{11.23}$$

For the boolean form of f, we have

$$V_f(\mathbf{w}) = \sum_{\mathbf{x} \in \mathbb{F}_2^n} (-1)^{f(\mathbf{x}+\mathbf{w})+f(\mathbf{x})}, \ \mathbf{w} \in \mathbb{F}_2^n. \tag{11.24}$$

Definition 11.5 *A boolean function f in n variables is said to have kth-order propagation if*

$$V_f(\mathbf{w}) = 0, \ \text{for all } \mathbf{w} \in \mathbb{F}_2^n \ \text{such that } 1 \le w(\mathbf{w}) \le k.$$

Functions with preferred Hadamard transforms have several interesting phenomena.

Lemma 11.2 *Let n be odd. If the Hadamard transform of $f(x)$ is preferred, then*

$$V_f(w) = -\widehat{\sigma_f}(w), \ w \ne 0, \ w \in \mathbb{F}_q;$$

that is, the additive autocorrelation of f is equal to the negative of the Hadamard transform of the dual of f.

Proof. From Proposition 6.7 (Section 6.7, Chapter 6), we have

$$V_f(w) = \frac{1}{q} \sum_{\lambda \in \mathbb{F}_q} (-1)^{Tr(w\lambda)} \widehat{f}^2(\lambda) \ (q = 2^n).$$

Because the Hadamard transform of f is preferred, $\widehat{f}(\lambda)^2 = 2^{n+1} \sigma_f(\lambda)$. Substituting this into the above equation, we get

$$V_f(w) = 2 \sum_{\lambda} (-1)^{Tr(w\lambda)} \sigma_f(\lambda).$$

By noting that $2\sigma_f(\lambda) = 1 - (-1)^{\sigma_f(\lambda)}$, we obtain

$$V_f(w) = \sum_{\lambda} (-1)^{Tr(w\lambda)} - \sum_{\lambda} (-1)^{Tr(w\lambda)+\sigma_f(\lambda)}.$$

This is equal to $-\widehat{\sigma_f}(w)$ when $w \ne 0$ and to 2^n when $w = 0$. $\qquad \square$

In the following, we determine which functions in Table 11.2 have the preferred additive autocorrelation function.

Theorem 11.1 *Let*

$$d = 2^{2k} - 2^k + 1 \ where \ 3k \equiv 1 \ (mod \ n). \tag{11.25}$$

Then the additive autocorrelations of the functions listed in the following table are preferred, that is, 3-valued, and belong to P. (All d or k in the table satisfy Eq. (11.25).)

f	$\sigma_f(x)$
$Tr(x^d)$	$Tr(x^{2^k+1})$
$WG(x)$	$Tr(x^{d^{-1}})$
$C_3(x)$	$Tr(x^3)$
$C_k(x)$, 5-term	$Tr(x^{d^{-1}})$

Proof. (a) $f(x) = Tr(x^d)$ where d satisfies Eq. (11.25). From Corollary 10.1 (Section 10.2, Chapter 10), we see that $\widehat{\sigma_f}(w)$ is preferred. Applying Lemma 11.2, the additive autocorrelation of f is preferred.

(b) $f(x) = WG(x)$. Because the Hadamard transform of $Tr(x^d)$ is preferred, according to Property 10.4 (Section 10.1, Chapter 10), so too is $Tr(x^{d^{-1}})$, the dual of f.

(c) $f(x) = C_k(x)$ where $k = 3$ or $3k \equiv 1 \pmod{n}$. Note that $k = 3 \Longrightarrow \sigma_f(x) = Tr(x^3)$ (see the dual of f in Table 11.2). Thus, this is the same case as Case 1. So, the assertion is true. For the case $3k \equiv 1 \pmod{n}$, note that

$$d(2^k + 1) \equiv 3 \pmod{2^n - 1}. \tag{11.26}$$

Therefore,

$$\sigma_f(x) = Tr(x^{d^{-1}}).$$

This is the dual of the WG function. Thus the result is true.   □

Theorem 11.1 presents a very interesting result on the Kasami exponent d with $3k \equiv 1 \pmod{n}$. We restate it as follows. Recall that $Tr(t(x))$ is defined as the trace representation of the 5-term sequences in Construction B in Section 9.1 of Chapter 9. Thus

$$C_k(x) = Tr(t(x^{2^k+1})) \text{ and } WG(x) = Tr(t(x + 1) + 1), \tag{11.27}$$

where $3k \equiv 1 \pmod{n}$. From Theorem 11.1 and the proof of Theorem 11.1, the dual functions of $C_k(x)$ and $WG(x)$ are equal and are given by $Tr(x^{(2^k+1)/3})$ because $d^{-1} = (2^k + 1)/3$.

In the following, we investigate the propagation property for the functions in Theorem 11.1. Let $g(x) = \sigma_f(x)$ where f is taken from the list in Theorem 11.1. We consider two cases.

1. $f(x) = Tr(x^d)$ or $f(x) = C_3(x)$. In this case, we have $g(x) = Tr(x^{2^k+1})$ or $g(x) = Tr(x^3)$. Applying Theorem 10.1 (Section 10.2, Chapter 10), the dual of $g(x)$ is the linear function $Tr(x)$ for both cases. According to Fact 11.1,

there exist no bases of $\mathbb{F}_{2^n}/\mathbb{F}_2$ such that a boolean form of $g(x)$ is first-order resilient.

2. $f(x)$ is taken as $C_k(x)$ or $WG(x)$, given by Eq. (11.27). Then $g(x) = Tr(x^{d^{-1}})$ for both cases. Thus, we have

$$\widehat{g}(\lambda) = \widehat{S_d}(\lambda^{-d^{-1}}) \quad (S_d(x) = Tr(x^d))$$

$$\implies \quad \left. \begin{array}{l} \sigma_g(x) = Tr(x^{-d^{-1}(2^k+1)}) \\[2ex] -d^{-1}(2^k+1) \not\equiv 2^j \pmod{2^n - 1} \end{array} \right\} \implies \sigma_g(x) \text{ is nonlinear.}$$

Thus, $g(x)$ is first-order resilient, so that $f(x)$ has first-order propagation.

We summarize the above discussions, together with the other properties in Table 11.2, in the following corollary.

Corollary 11.1 *Any function* $f(x) \in \{C_k(x), WG(x)\}$ *satisfies the following properties where* $3k \equiv 1 \pmod{n}$.

(a) *f has 2-level autocorrelation.*

(b) *Nonlinearity $N_f = B(n) = 2^{n-1} - 2^{(n-1)/2}$.*

(c) *$\widehat{f}(\lambda)$ is preferred, that is, belongs to P. Or equivalently, the crosscorrelation of $f(x)$ and $Tr(x)$ is preferred.*

(d) *f is first-order resilient under some basis.*

(e) *The additive autocorrelation function of f is preferred.*

(f) *f has first-order propagation under some basis.*

Example 11.7 Let $n = 7$. Then $k = 5 \implies n - k = 2 \implies 2^{n-k} + 1 = 5$, and $t(x) = x + x^5 + x^{21} + x^{13} + x^{29}$. Thus

$$C_5(x) = Tr(t(x^{2^2+1})) = Tr(x^5 + x^{19} + x^{29} + x^3 + x^9)$$
$$WG(x) = Tr(t(x + 1) + 1) = Tr(x + x^3 + x^7 + x^{19} + x^{29}).$$

Both $C_5(x)$ and $WG(x)$ have the following properties:

(a) Orthogonal or 2-level autocorrelation.

(b) Nonlinearity $N_f = 56$.

(c) Hadamard transform is preferred, that is, belongs to $\{0, \pm 16\}$.

(d) First-order resiliency under some basis.

(e) The additive autocorrelation function is preferred, that is, belongs to $\{0, \pm 16\}$.

(f) First-order propagation under some basis.

From the proof of Theorem 11.1, we obtain a sufficient condition for additive autocorrelation being preferred. We illustrate this process in the following figure. In other words, if the Hadamard transforms of both f and the dual of f are preferred, so is the additive autocorrelation of f.

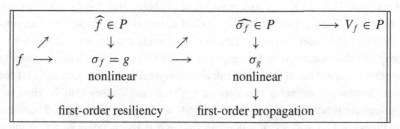

Note that the additive autocorrelation of the function f in Example 11.6 is preferred, because the Hadamard transforms of both f and its dual are preferred. In the following table, we summarize the classes of functions that we discuss in this section, where each class has one more property added to it than in the class above it.

(a) $f \to \widehat{f} \in P$	functions in Tables 11.2 and 11.3
(a), and (b) f first-order resiliency	functions in Table 11.2
(a), (b), and (c) $f \to V_f \in P$	functions in Theorem 11.1
(a), (b), (c), and (d) f 1-propagation	WG and $C_k(x)$ where $3k \equiv 1 (\mathrm{mod}\ n)$

Remark 11.3 Correlation immunity (or resiliency) and nonlinearity of a function from $\mathbb{F}_{2^n}$ to $\mathbb{F}_2$ are invariant under the operators of G (see Remark 11.2). It can be easily verified that the additive autocorrelation of a function is also invariant under G. For $f : \mathbb{F}_{2^n} \to \mathbb{F}_2$, we denote by $G \cdot f$ the class of functions obtained by performing the operators of G on f: $G \cdot f = \{\tau(f) \mid \tau \in G\}$. Then $G \cdot f$ preserves the correlation immunity or resiliency, nonlinearity, additive autocorrelation, and propagation of the function f. An interesting result is that the sequences having the functions in $G \cdot f$ as their trace representations are shift distinct. In other words, let $\mathbf{a} \leftrightarrow \tau(f)$ and $\mathbf{b} \leftrightarrow \eta(f)$ where $\tau(f), \eta(f) \in G \cdot f$, $\tau \neq \eta$. Then $\mathbf{a}$ and $\mathbf{b}$ are shift distinct. This is a remarkable phenomenon relating classification of boolean functions in n variables and polynomial functions from $\mathbb{F}_{2^n}$ to $\mathbb{F}_2$.

Historical note

For boolean functions in n variables, or equivalently polynomial functions from $\mathbb{F}_{2^n}$ to $\mathbb{F}_2$, the concepts of correlation immunity, resiliency, and nonlinearity of boolean functions were investigated by Golomb in 1959 under the terminology of invariants of boolean functions. (Invariants of a boolean function are the distances between the boolean function and all affine functions; see Definition 11.1.) He was the first to study boolean functions using the Walsh transform. This work was also included in his book *Shift Register Sequences*, Chapter VIII. Siegenthaler (1984) studied correlation immunity for the construction of boolean functions that can resist correlation attacks, and the degree property also appeared in that paper. The definition of correlation immunity presented here is taken from Siegenthaler's original definition. Lemma 11.1 for general random variables was derived by Xiao and Massey (1988). Here we presented a proof via counting 0-1 distributions of binary sequences. The result of Property 11.1 states that a boolean function is independent of any k input variables if and only if the distance between the function and a linear combination of any r input variables is equal to 2^{n-1} for all $1 \leq r \leq k$, or equivalently, the Walsh transform is equal to zero at nonzero input vectors whose Hamming weights are less than or equal to k. This is a simple consequence of Golomb's work on the invariants (i.e., the distances) (1959, 1982), which was motivated by the search for nonlinear shift register sequences that would resist cryptographic attacks based on correlation, but his Jet Propulsion Laboratory work on this subject (1957–1959), anticipating much of Siegenthaler (1984) and Xiao and Massey (1988), has not yet been declassified. The result of Property 11.1 was rediscovered by Xiao and Massey in 1988. Since then, it has become a popular definition for correlation immunity and resiliency of boolean functions in the literature.

For n even, $A(n) = 2^{n-1} - 2^{(n/2)-1}$ is the maximum nonlinearity that a boolean function in n variables can achieve. However, for n odd, $B(n) = 2^{n-1} - 2^{(n-1)/2}$ is not maximal for the nonlinearities of all the boolean functions in n variables. Maximum nonlinearity of boolean functions is equal to the covering radius of a Reed-Muller code in a coding context. For $n = 15$, the two examples found by Patterson and Wiedemann (1983, 1990) were from this viewpoint. The nonlinearity of these two functions is equal to 16,276, which is larger than $B(15) = 16,256$. Note that these two functions can be modified to be balanced and still have nonlinearity larger than $B(15)$ (see Maitra and Sarkar 2000). (The trace representations of Patterson and Wiedemann's functions were computed by K.M. Khoo.) That the polynomial functions from GMW type 2 sequences, given in Table 11.2, have the preferred Hadamard transform was proved by Klapper (1996) and by Games (1987) for the GMW type 1 case (GMW type 1 is a special case of GMW type 2; see Chapter 8). The quadratic functions in polynomial form listed in Table 11.3 for Case (3) were shown by Khoo, Gong, and Stinson (2002). A proof of Fact 11.1 was given by Zheng and Zhang (1999) using boolean forms and by Gong and Youssef (2002) using polynomial forms. An algorithm for finding such bases was also provided in

Gong and Youssef (2002). The propagation property of boolean functions as well as some bounds on resiliency and nonlinearity of boolean functions have been discussed in recent years by a number of researchers Nyberg (1991), Dawson and Tavares (1991), Maitra and Sarkar (2000), Zheng and Zhang (2001), Canteaut et al. (2001), and Carlet (2002). The resiliency property of boolean functions has a connection to the combinatorial objects called orthogonal arrays (see Camion (1992)).

Exercises for Chapter 11

1. Calculate the invariants C_w and the R_w's for each of the following boolean functions:
   (a) $f(x_0, x_1, x_2) = x_0 x_1 + x_0 x_2 + x_1 x_2$.
   (b) $f(x_0, x_1, x_2, x_3) = x_0 x_2 + \bar{x}_1 x_3 + \bar{x}_2 x_3$.
   (c) $f(x_0, x_1, x_2, x_3) = x_0 x_1 + x_0 x_2 + x_0 x_3 + x_1 x_2 + x_1 x_2 + x_2 x_3$.
   (d) $f(x_0, x_1, x_2, x_3, x_4) = x_0 + x_1 x_2 + \bar{x}_0 x_3 x_4$.
   Note. In these problems, the bar over a letter ($\bar{x}$) denotes complementation; and the plus sign ($+$) indicates modulo 2 addition (that is, $GF(2)$ addition, and not the logic OR operation).
2. Constructing nonlinear boolean functions:
   (a) Find a third-order correlation-immune boolean function of five binary variables that is nonlinear.
   (b) Find a fourth-order correlation-immune boolean function of seven binary variables that is nonlinear.

12

Applications to Radar, Sonar, Synchronization, and CDMA

12.1 Overview

In the first three of the applications mentioned in the title of this chapter, one of the objectives (often the major objective) is to determine a point in time with great accuracy. In radar and sonar, we want to determine the round-trip time from transmitter to target to receiver very accurately, because the one-way time (half of the round-trip time) is a measure of the distance to the target (called the range of the target).

The simplest approach would be to send out a pure impulse of energy and measure the time until it returns. The ideal impulse would be virtually instantaneous in duration, but with such high amplitude that the total energy contained in the pulse would be significant, much like a Dirac delta function. However, the Dirac delta function not only fails to exist as a mathematical function, but it is also unrealizable as a physical signal. Close approximations to it – very brief signals with very large amplitudes – may be valid mathematically, but are impractical to generate physically. Any actual transmitter will have an upper limit on peak power output, and hence a short pulse will have a very restricted amount of total energy: at most, the peak power times the pulse duration. More total energy can be transmitted if we extend the duration; but if we transmit at uniform power over an extended duration, we do not get a sharp determination of the round-trip time. This dilemma is illustrated in Figure 12.1.

In the presence of noise, the time of the peak of the autocorrelation function of the extended pulse becomes difficult to determine with high precision.

It is here that clever combinatorial mathematics comes to the rescue. By using a suitably coded pulse pattern, or coded signal pattern, we extend the duration, thereby increasing the total transmitted energy as much as desired, while still maintaining a sharp spike in the autocorrelation function to mark the round-trip

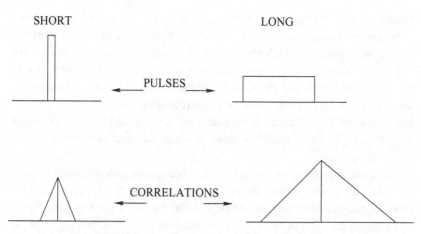

Figure 12.1. The shorter the pulse, the sharper the autocorrelation function.

propagation time with great precision. Such a technique was successfully used in 1961 by the Jet Propulsion Laboratory to bounce a radar signal off the surface of Venus and detect it back on earth. Not only was that the first successful radar ranging of another planet in the solar system, but the distance thus measured made it possible to improve the accuracy of the astronomical unit (the mean radius of the earth's orbit around the sun and the basic yardstick for distances within the solar system) by three orders of magnitude; see Butrica (1996).

12.2 Types of signals and correlations

A monostatic radar is one that has transmitter and receiver at the same location, which requires that the transmitter is turned off while the signal is being received. A bistatic radar is one that has two separate antennas, often at a considerable distance from one another, for transmitting and receiving. With a bistatic radar, it is possible, in principle, to be transmitting all the time.

A pulse radar is one that transmits a succession of pulses, not necessarily uniformly spaced, but that is turned off between pulses. A CW (continuous wave) radar is one which stays on the air for an extended period of time, transmitting a sine wave, usually with phase modulation added. A frequency hop radar is one that jumps from one to another of a finite set of frequencies, according to some predetermined pattern, thus combining certain features of the pulse radar and the CW radar.

For correlation purposes, most radar patterns are either finite (the transmitter is turned off before and after the pattern is sent) or periodic (the same basic pattern is repeated periodically a number of times). If a CW radar is modulated in both phase and amplitude, the resulting signal can be modeled as a sequence of complex numbers, which also have both phase and amplitude. If only the amplitude varies, the sequence of complex numbers becomes a sequence of real numbers. If only the phase varies, the sequence of complex numbers reduces to a sequence of values on the unit circle of the complex plane. The real line intersects the unit circle at the points $+1$ and -1, and binary modulation by $+1$ and -1 can be regarded either as phase modulation or as amplitude modulation.

The returning signal can be correlated either against an ideal model of itself or against some other signal (often called a complementary signal or sequence), especially designed to highlight some specific feature of the returning signal. In all cases, the ratio of the value of correlation for $\tau = 0$, that is, when the signal is aligned with itself, versus the maximum value of the correlation for $\tau \neq 0$, is a measure of the clarity with which range can be measured in a noisy environment.

If there is a relative motion (either toward or away) between the transmitter and the target, then the returning radar or sonar signal is shifted not only in time, but also in frequency. The frequency shift (Doppler shift) is proportional to the time derivative of range – that is, to the velocity of approach or separation between communicator and target. The two-dimensional autocorrelation function of the signal, in the time and frequency domain, is called its ambiguity function, and the ideal shape of such a function is a spike or an inverted thumb-tack (or drawing pin, in the United Kingdom). The reader may also consult Moreno (1999) for more information about the ambiguity function.

12.3 Barker sequences

Barker (1953) asked, for what lengths L do binary sequences of $+1$'s and -1's, $\{a_j\}_{j=1}^{L}$, exist, with finite, unnormalized autocorrelation $K(\tau)$, defined by

$$K(\tau) = \sum_{j=1}^{L-\tau} a_j a_{j+\tau}, \qquad (12.1)$$

bounded by 1 in absolute value for $\tau \neq 0$? That is, the requirement is

$$|K(\tau)| \leq 1 \text{ for } 1 \leq |\tau| \leq L - 1.$$

Barker gave examples having the following lengths L:

L	Sequence
1	$+1$
2	$+1, +1$
3	$+1, +1, -1$
4	$+1, +1, -1, +1$
5	$+1, +1, +1, -1, +1$
7	$+1, +1, +1, -1, -1, +1, -1$
11	$+1, +1, +1, -1, -1, -1, +1, -1, -1, +1, -1$
13	$+1, +1, +1, +1, +1, -1, -1, +1, +1, -1, +1, -1, +1$

Turyn and Storer (1961) showed that there are no other Barker sequences for *odd* lengths $L > 13$. It is still unproven that even-length Barker sequences with $L > 4$ do not exist, though this is generally believed. It has been shown (see ·Turyn (1968)) that even-length Barker sequences for $L \geq 4$ give rise to circulant Hadamard matrices of order L, as described in Section 2 of Chapter 2.

12.4 Generalized Barker sequences

Golomb and Scholtz (1965) defined a *generalized Barker sequence* of length L to be a sequence $\{a_j\}_{j=1}^{L}$ of complex numbers on the unit circle (that is, $|a_j| = 1$ for $1 \leq j \leq L$) with finite unnormalized autocorrelation $K(\tau)$, defined by

$$K(\tau) = \sum_{j=1}^{L-\tau} a_j a_{j+\tau}^*, \tag{12.2}$$

satisfying

$$|K(\tau)| \leq 1 \text{ for } 1 \leq |\tau| \leq L - 1. \tag{12.3}$$

In Golomb and Scholtz (1965), a group of $4L^2$ transformations on the sequence $\{a_j\}_{j=1}^{L}$ is identified which preserves the Barker property (12.3), and examples of generalized Barker sequences with $L \leq 16$ are given. Subsequently, examples have been found for all lengths $L \leq 49$. Lengths beyond $L = 20$ have mostly been found by members of Hans Dieter Lüke's group at Aachen University of Technology. The examples for $L > 36$ are due to Brenner (1998), who is a member of the Aachen group.

With $K(\tau)$ defined as in Eq. (12.2), $K(0) = L$, and hence the normalized correlation $C(\tau)$ for a generalized Barker sequence must satisfy

$$|C(\tau)| = \frac{1}{L}|K(\tau)| \leq \frac{1}{L} \text{ for all } 1 \leq |\tau| \leq L - 1. \tag{12.4}$$

This condition becomes increasingly difficult to satisfy as L increases, and it is not known whether any generalized Barker sequences exist for large values of L. However, for applications to radar, conditions considerably weaker than Eq. (12.4) are still quite useful. For example, if $\{a_j\}_{j=1}^{L}$ is a sequence for which $K(\tau)$, as defined in (12.2), satisfies

$$|K(\tau)| \le cL^{\frac{1}{2}}, 1 \le |\tau| \le L - 1,$$

then the normalized correlation $C(\tau)$ satisfies

$$|C(\tau)| = \frac{1}{L}|K(\tau)| \le cL^{-\frac{1}{2}}, 1 \le |\tau| \le L - 1. \tag{12.5}$$

Such a family of complex sequences, existing for every length $L = n^2$, using nth roots of unity as the terms of the sequence, was described by Frank (1963). Another family of sequences satisfying Eq. (12.5), for every integer length $L \ge 1$, using Lth roots of unity as the terms, was described in Zhang and Golomb (1993).

12.5 Huffman's impulse-equivalent pulse trains

The generalized Barker sequences of the previous section correspond to phase modulation on a sinusoidal radar carrier signal, where the modulation lasts for L time intervals, changing from each interval to the next, and where the radar transmitter is turned off before and after the signal consisting of the L phase-modulated intervals.

Huffmann (1962) considered the corresponding problem for an amplitude modulated radar signal and imposed an even stronger restriction than Barker's on the out-of-phase values of the autocorrelation:

$$K(\tau) = \sum_{j=1}^{L-\tau} a_j a_{j+\tau} = 0 \text{ for } 1 \le |\tau| \le L - 2,$$

where all the terms a_j of the sequence $\{a_j\}_{j=1}^{L}$ are real. As usual,

$$K(0) = \sum_{j=1}^{L} a_j^2 = \sum_{j=1}^{L} |a_j|^2$$

is the total energy of the signal, and

$$K(L - 1) = a_1 a_L = K(-(L - 1))$$

cannot be 0 if the sequence truly has length L.

As a general method of studying the finite autocorrelation function of the sequence $\{a_0, a_1, a_2, \ldots, a_n\}$ of complex numbers of length $L = n + 1$, we consider the two associated polynomials

$$P(x) = a_0 x^n + a_1 x^{n-1} + \cdots + a_{n-1} x + a_n = a_0 \prod_{i=1}^{n} (x - r_i)$$

and

$$Q(x) = x^n P\left(\frac{1}{x}\right) = a_0 + a_1 x + \cdots + a_n x^n = a_0 \prod_{i=1}^{n} (1 - r_i x),$$

where the r_i, in general, are complex numbers.

It is readily seen that

$$P(x)Q^*(x) = \sum_{\tau=-n}^{n} K(\tau) x^{n+\tau}, \tag{12.6}$$

where $K(\tau) = \sum_{j=0}^{n-\tau} a_j a_{j+\tau}^*$ and $K(-\tau) = \sum_{j=\tau}^{n} a_j a_{j-\tau}^* = K^*(\tau)$ for $\tau \geq 0$. Here $K(0) = \sum_{j=0}^{n} |a_j|^2 = E$ is the total energy in the signal; and $K(n) = a_0 a_n^*$, $K(-n) = a_0^* a_n = K^*(n)$. We normalize the sequence by requiring $|K(n)| = |K(-n)| = 1$. If $K(n) = \eta$ with $|\eta| = 1$, then $K(-n) = \eta^*$. We will be particularly interested in the case that η is real, and hence either $K(n) = K(-n) = 1$ or $K(n) = K(-n) = -1$. With Huffman's additional restriction, namely that $K(\tau) = 0$ for $1 \leq |\tau| \leq n - 1$, Eq. (12.6) becomes

$$P(x)Q^*(x) = \sum_{\tau=-n}^{n} K(\tau) x^{n+\tau} = \eta x^{2n} + E x^n + \eta^*, \; |\eta| = 1,$$

and we will focus on the two cases

$$x^{2n} + E x^n + 1 = P(x)Q^*(x) \tag{12.7}$$

or

$$-x^{2n} + E x^n - 1 = P(x)Q^*(x). \tag{12.8}$$

In Eq. (12.7), we can write

$$x^{2n} + E x^n + 1 = (x^n + R^n)\left(x^n + \frac{1}{R^n}\right), \, R > 0,$$

where $2 < E = R^n + \frac{1}{R^n} < \infty$. In Eq. (12.8), we can write

$$-x^{2n} + E x^n - 1 = -(x^n - R^n)\left(x^n - \frac{1}{R^n}\right), \, R > 0,$$

where again $2 < E = R^n + \frac{1}{R^n} < \infty$.

Hence, for Eq. (12.7), the roots of $P(x)Q^*(x) = 0$ are the complex roots of $(x^n + R^n)(x^n + \frac{1}{R^n}) = 0$, which are $\{R\alpha^t : t = 1, 3, 5, \ldots, 2n - 1\}$ and $\{\frac{1}{R\alpha^t} : t = 1, 3, 5, \ldots, 2n - 1\}$, where $\alpha = e^{\pi i/n}$. Similarly, for Eq. (12.8), the roots of $P(x)Q^*(x) = 0$ are the complex roots of $(x^n - R^n)(x^n - \frac{1}{R^n}) = 0$, which are $\{R\beta^u : u = 1, 2, \ldots, n\}$ and $\{\frac{1}{R}\beta^u : u = 1, 2, \ldots, n\}$, where $\beta = e^{2\pi i/n}$.

The sequence $\{a_j\}_{j=0}^N$ is a sequence of real numbers if and only if all coefficients of $P(x)$ (and hence also of $Q(x)$) are real, which occurs if and only if the subset of the $2n$ roots of $P(x)Q^*(x) = 0$ that are roots of $P(x)$ is closed with respect to complex conjugation.

An example of such a sequence $\{a_j\}$ of length $L = n + 1 = 5$ is $\{1, -1, \frac{1}{2}, 1, 1\}$, with $P(x) = x^4 - x^3 + \frac{1}{2}x^2 + x + 1$, $Q(x) = x^n P\left(\frac{1}{x}\right) = 1 - x + \frac{1}{2}x^2 + x^3 + x^4$, and $P(x)Q^*(x) = x^8 + 4\frac{1}{4}x^4 + 1 = (x^4 + 4)(x^4 + \frac{1}{4})$. The roots of $x^4 + 4 = 0$ are $\{1 + i, 1 - i, -1 + i, -1 - i\}$, and the roots of $x^4 + \frac{1}{4} = 0$ are $\{\frac{1+i}{2}, \frac{1-i}{2}, \frac{-1+i}{2}, \frac{-1-i}{2}\}$. The roots of $P(x) = 0$ are $\{1 + i, 1 - i, \frac{-1+i}{2}, \frac{-1-i}{2}\}$, verified by

$$(x - (1 + i))(x - (1 - i))\left(x - \frac{-1+i}{2}\right)\left(x - \frac{-1-i}{2}\right)$$

$$= (x^2 - 2x + 2)\left(x^2 + x + \frac{1}{2}\right) = x^4 - x^3 + \frac{1}{2}x^2 + x + 1.$$

Also, for the sequence $\{1, -1, \frac{1}{2}, 1, 1\}$, the autocorrelation $K(\tau)$ is given by $K(0) = 4\frac{1}{4}$, $K(1) = K(2) = K(3) = 0$, and $K(4) = 1$.

Long Huffman sequences are useful to the extent that the magnitudes of the terms are approximately equal, so that the transmitted energy is distributed fairly uniformly throughout the broadcast interval. Methods for achieving this type of uniformity are not adequately understood for the general case and may not exist.

12.6 Pulse patterns and optimal rulers

A pulse radar is able to send one or more pulses of radio-frequency energy toward a target. It is a convenient and rather realistic assumption to require all the pulses to be the same in both duration and amplitude. The signal design problem then reduces to devising patterns of these identical pulses, so that the autocorrelation function of the pattern is as impulse-like as possible.

The problem is usually restated as follows: For each positive integer n, what is the shortest length $L = L(n)$ for which there is a sequence $\{a_1, a_2, \ldots, a_n\}$ with $0 = a_1 < a_2 < \cdots < a_n = L$, such that the set of $\binom{n}{2}$ differences $\{a_j - a_i\}$, with $1 \le i < j \le n$, are all distinct?

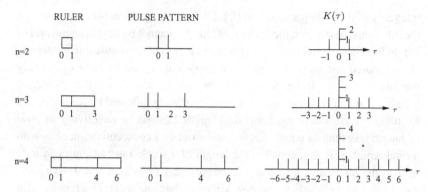

Figure 12.2. Perfect rulers for $n = 2, n = 3$, and $n = 4$, with the corresponding pulse radar patterns and their unnormalized autocorrelation functions.

The model underlying this restatement is the following. At each integer a_i in the sequence, there is a pulse of brief duration and unit amplitude. Thus, the sequence corresponds to a pulse pattern with n pulses, spread out over a total duration of L (or, more precisely L plus one pulse duration) with the unnormalized autocorrelation function $K(\tau)$ satisfying $K(0) = n$, $0 \leq K(\tau) \leq 1$ for $1 \leq \tau \leq L$, and $K(\tau) = 0$ for $\tau > L$. (The distinctness of the differences $\{a_j - a_i\}$ guarantees $K(\tau) \leq 1$ for all τ, $|\tau| \geq 1$.) Finding the shortest length $L = L(n)$ for the sequence achieves the desired signal parameters with the shortest duration for the pulse pattern.

The sequence model is also described in terms of a certain class of rulers (the measuring devices, not the monarchs or autocrats).

A ruler of length L has only n marks on it, at integer positions $a_1, a_2, \ldots, a_n$, where $a_1 = 0$ and $a_n = L$ are the two endpoints of the ruler. If every integer distance d, $1 \leq d \leq L$, can be measured in one and only one way as a distance between two of the n marks, then the ruler is called a perfect ruler. For a perfect ruler, $L = \binom{n}{2}$, because there are exactly $\binom{n}{2}$ distances between the n marks, and these must be some permutation of $\{1, 2, 3, \ldots, L\}$. In Figure 12.2, we see perfect rulers for $n = 2, 3$, and 4, with the corresponding radar pulse patterns and their autocorrelation functions $K(\tau)$.

Unfortunately, for $n > 4$, there are no perfect rulers.

Theorem 1 *For $n > 4$, no perfect rulers exist.*

Proof. A ruler with n marks has $n - 1$ intervals between marks. For a perfect ruler, these $n - 1$ intervals must all have distinct positive integer lengths, and the sum of these lengths must be $L = \binom{n}{2}$. Hence these intervals must be (in some order) $1, 2, 3, \ldots, n - 1$, because any other set of $n - 1$ distinct positive integers will have a larger sum than $\binom{n}{2}$. Because all distances between marks on the ruler must be distinct, the interval of length 1 cannot be next to an interval of length less than or equal to $n - 2$ (because the sum of two consecutive intervals is a measured distance of the ruler, and every length from 1 to $n - 1$ is already measured, as a single interval). This can only be achieved if the interval of length 1 is at one end of the ruler and is immediately followed by the interval of length $n - 1$. Similarly, the interval of length 2 cannot be next to any interval of length less than or equal to $n - 3$ (to avoid two consecutive intervals with a total length $\leq n - 1$, equaling the length of a single interval), nor can it be next to the interval of length $n - 2$ (because $2 + (n - 2) = n = 1 + (n - 1)$, and the distance n would be measured in more than one way). This requires that the interval of length 2 must also be at an end of the ruler and must also be next to the interval of length $n - 1$. But then the entire ruler consists of only three intervals: $1, n - 1, 2$, which means there are only four marks altogether, and $n = 4$. (Note that with $n = 4$, there really is a perfect ruler with the consecutive intervals of lengths 1, 3, 2.)   □

There are two obvious ways to relax the requirement on a perfect ruler to get objects that exist for all n. A covering ruler with n marks and length L measures every distance from 1 to L, as a distance between two marks on the ruler, in at least one way; whereas a spanning ruler with n marks and length L measures every distance from 1 to L, as a distance between two marks on the ruler, in at most one way. The interesting combinatorial problems are to determine the longest covering ruler with n marks, and the shortest spanning ruler with n marks, for each positive integer n. Both of these problems have long histories in the combinatorial literature. However, the application to pulse radar involves only finding the shortest spanning ruler for each n. (Martin Gardner (1983) termed these objects Golomb rulers, a name which seems subsequently to have been widely adopted.)

The behavior of $L(n)$ as a function of n, for the shortest spanning ruler, is quite erratic in detail, although it appears that asymptotically $L(n) \sim n^2$ as $n \to \infty$. The value of $L(n)$ has been determined by exhaustive computer search for all $n \leq 25$. Lengths 15 and 16 were done by Shearer (1990), lengths 17 and 18 by Sibert (1993), and length 19 by Dollas, Rankin, and McCracken (1998). In addition to left-right reversal of the ruler, these rulers are not unique for several of the smaller values of n. One example of a spanning ruler of length $L(n)$, for each $n \leq 25$, is shown in Table 12.1. For $n \geq 20$, a distributed worldwide computer effort has been underway for several years, so that $L(20) = 283$, $L(21) = 333$, $L(22) = 356$, $L(23) = 372$, $L(24) = 425$, and $L(25) = 480$ are now known (see Hayes (1998)).

Table 12.1. *Table of the shortest spanning rulers*

n	$L(n)$	m	Sequence of Marks
2	1	1	0, 1
3	3	1	0, 1, 3
4	6	1	0, 1, 4, 6
5	11	2	0, 1, 4, 9, 11
6	17	4	0, 1, 4, 10, 12, 17
7	25	5	0, 1, 4, 10, 18, 23, 25
8	34	1	0, 1, 4, 9, 15, 22, 32, 34
9	44	1	0, 1, 5, 12, 25, 27, 35, 41, 44
10	55	1	0, 1, 6, 10, 23, 26, 34, 41, 53, 55
11	72	2	0, 1, 4, 13, 28, 33, 47, 54, 64, 70, 72
12	85	1	0, 2, 6, 24, 29, 40, 43, 55, 68, 75, 76, 85
13	106	1	0, 2, 5, 25, 37, 43, 59, 70, 85, 89, 98, 99, 106
14	127	1	0, 5, 28, 38, 41, 49, 50, 68, 75, 92, 107, 121, 123, 127
15	151	1	0, 6, 7, 15, 28, 40, 51, 75, 89, 92, 94, 121, 131, 147, 151
16	177	1	0, 1, 4, 11, 26, 32, 56, 68, 76, 115, 117, 134, 150, 163, 168, 177
17	199	1	0, 8, 31, 34, 40, 61, 77, 99, 118, 119, 132, 143, 147, 182, 192, 194, 199
18	216	1	0, 2, 10, 22, 53, 56, 82, 83, 89, 98, 130, 148, 153, 167, 188, 192, 205, 216
19	246	1	0, 1, 6, 25, 32, 72, 100, 108, 120, 130, 153, 169, 187, 190, 204, 231, 233, 242, 246
20	283	1	0, 24, 30, 43, 55, 71, 75, 89, 104, 125, 127, 162, 167, 189, 206, 215, 272, 275, 282, 283
21	333	1	0, 4, 23, 37, 40, 48, 68, 78, 138, 147, 154, 189, 204, 238, 250, 251, 256, 277, 309, 331, 333
22	356	1	0, 1, 9, 14, 43, 70, 106, 122, 124, 128, 159, 179, 204, 223, 253, 263, 270, 291, 330, 341, 353, 356
23	372	1	0, 6, 22, 24, 43, 56, 95, 126, 137, 146, 172, 173, 201, 213, 258, 273, 281, 306, 311, 355, 365, 369, 372
24	425	1	0, 9, 33, 37, 38, 97, 122, 129, 140, 142, 152, 191, 205, 208, 252, 278, 286, 326, 332, 353, 368, 384, 403, 425
25	480	1	0, 12, 29, 39, 72, 91, 146, 157, 160, 161, 166, 191, 207, 214, 258, 290, 316, 354, 372, 394, 396, 431, 459, 467, 480

In Table 12.1, the quantity m is the number of inequivalent rulers of length $L(n)$, which are shortest spanning rulers with n marks. Only one of each set of m rulers is listed explicitly.

Minimum spanning rulers have another, very different application in radio astronomy. In radio astronomy, only receiving antennas are used, and the spacing between two antennas generates a difference in the time a signal is received, which is used in making interferometry measurements on the signal. If several antennas are constructed along a straight line, they should be spaced along that line in the pattern of a spanning ruler, to get maximum interferometry information by having all the arrival time differences distinct.

The properties of these rulers also play an important role in x-ray diffraction crystallography. In this technique, one attempts to determine the bonding angles of a crystal by shining x-rays at it and observing the diffraction patterns that emerge. What in fact is measurable by this method is the differences of the bonding angles, from which one wishes to reconstruct the bonding angles themselves. This leads to an inverse problem: given the autocorrelation function $K(\tau)$, what is the set of possible signals which might have produced it? For several decades, crystallographers relied on a "theorem" of S. Piccard that asserted (in our terminology): "If two spanning rulers have the same autocorrelation function $K(\tau)$, the two rulers are either identical or mirror images of each other." A counterexample was found in Bloom and Golomb (1977) for 6-mark rulers: $\{0, 1, 4, 10, 12, 17\}$ and $\{0, 1, 8, 11, 13, 17\}$, which are in fact shortest 6-mark spanning rulers. This generalizes to a two-parameter family of counterexamples, all involving 6-mark spanning rulers. No counterexamples with fewer than 6-marks are possible, and none with more than 6-marks are known. (There are partial results that suggest that counterexamples may occur only in the case $n = 6$; see Yovanof and Golomb (1998). It is known that no other counterexamples occur for $n \leq 12$.) Very recently, A. Bekir (2004) gave a proof that the only counterexamples to S. Piccard's "Theorem" occur for six-mark rulers.

For additional information on constructions based on perfect circular rulers, see Golomb and Taylor (2002).

12.7 Perfect circular rulers from cyclic projective planes

A systematic construction for very good spanning rulers (though not necessarily best spanning rulers) with k marks, and length not exceeding $(k - 1)^2$, can be obtained from cyclic (v, k, λ)-designs with $v = n^2 + n + 1, k = n + 1, \lambda = 1$ (called finite projective planes), which are known to exist whenever n is a prime or a power of a prime. Think of the cyclic $(v, k, 1)$ design D as a circle of circumference v, with k marks at the k positions d_i, for all $d_i \in D$.

On this circle, every integer arc length from 1 to $v - 1$ can be found in one and only one way as a distance between two marks. Remove the longest arc between two consecutive marks, and straighten out what is left. This will be a spanning ruler with k marks and length at most $v - k = n^2 = (k - 1)^2$.

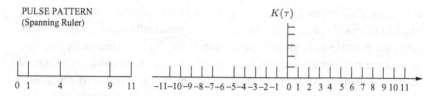

Figure 12.3. Radar pulse pattern, and its autocorrelation, for $n = 5$ pulses.

For example, $D = \{3, 6, 7, 12, 14\}$ is a $(21, 5, 1)$ cyclic difference set, and the longest arc (modulo 21) is from 14 to 3, which we remove. We then look at $D - 3 = \{0, 3, 4, 9, 11\}$, and a ruler with five marks, at $0, 3, 4, 9$ and 11, is a shortest spanning ruler, measuring every length from 1 to 11 inclusive except for length 10. (This is the other shortest spanning ruler with five marks, inequivalent to the one shown in Figure 12.3.)

Figure 12.4 shows that the perfect circular ruler of circumference 13, corresponding to the $(v, k, \lambda) = (13, 4, 1)$ cyclic difference set. Here, $n = k - \lambda = 3$ is a "multiplier"; and the difference set $D = \{0, 2, 5, 6\}$ (mod 13) is a union of the cyclotomic cosets $\{0\}$ and $\{2, 6, 5\}$.

Note that our geometric model uses a circle of circumference v, with marks on this circumference at the points $d_1, d_2, \ldots, d_k$, where these are measured arc lengths from an arbitrary starting point. By the cyclic difference set property for

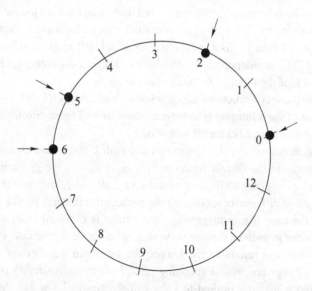

Figure 12.4. Perfect circular ruler modulo 13, from the $(v, k, \lambda) = (13, 4, 1)$ cyclic difference set $D = \{0, 2, 5, 6\}$ (mod 13).

$\lambda = 1$, the differences $d_j - d_i$, with $i \neq j$, take on each value from 1 to $v - 1$, modulo v, exactly once. Therefore, arcs (or equivalently central angles) of the circle are found in each integer length from 1 to $v - 1$, each in exactly one way, as the circumferential distances between two marks on the circle. (The central angles of the circle, as determined by pairs of marks on the circumference, take on every value $2\pi s / v$ for $s = 1, 2, 3, \ldots, v - 1$.) It is in this sense that cyclic projective planes correspond not only to cyclic $(v, k, 1)$ difference sets, but also to perfect circular rulers.

The known constructions for cyclic $(v, k, 1)$ difference sets are of $k - 1 = n$ being of the form $q = p^m$ with p prime and m a positive integer. By the Bruck-Ryser-Chowla theorem, these finite projective planes are known *not* to exist for infinitely many other values of n. However, there are also infinitely many values of n for which the existence of a plane of order n remains an open question. (See Hall (1986).)

There are two ways to go from a perfect circular ruler to a good (though not necessarily best) spanning ruler. As already noted, one way is to remove the longest arc between consecutive marks from a perfect circular ruler on $k = n + 1$ marks and straighten out what is left to get a spanning ruler on k marks. (Among all $(v, k, 1)$ cyclic difference sets for given k, select the one whose longest arc between consecutive marks is greatest to get as short a spanning ruler as possible by this method.)

The other way to go from a perfect circular ruler to a good spanning ruler is to cut the circular ruler at one of the marks and then straighten out the entire circle to get a spanning ruler. A perfect circular ruler with k marks and circumference $v = k^2 - k + 1$ then yields a spanning ruler with $k + 1$ marks and length $L = k^2 - k + 1$. This is independent of which $(v, k, 1)$ cyclic difference set is chosen and of which of the k marks is used to cut the circle.

Because these constructions are only known to exist when $k - 1 = q = p^m$, the spacing of such integers q influences the utility of these constructions for obtaining bounds on $L(k)$ for all values of k.

More generally, a circular spanning ruler with k marks and circumference v has k marks at the integer positions $0 \leq g_1 < g_2 < \cdots < g_k$ such that the differences $g_j - g_i$ are all distinct modulo v for all $i \neq j$, without the requirement that these differences include all the consecutive integers from 1 to $v - 1$. It is still the case that a straight spanning ruler is obtained from a circular spanning ruler in each of the two ways described above for the case of perfect circular rulers and removing the longest arc results in the shortest spanning ruler from the given circular spanning ruler. For circular spanning rulers, the circumference v is not limited to values of the form $n^2 + n + 1$. Moreover, the correspondence is now bidirectional, because, given any straight spanning ruler of length L, a circular spanning ruler of circumference $\leq 2L + 1$, can be obtained by adjoining a closing arc of length $\leq L + 1$. (If the closing arc has length $L + 1$, it is easily seen that all measured arcs on the circle with lengths less than the circumference of $2L + 1$ will be distinct.) An interesting open question is to determine the smallest circular spanning ruler with k marks that corresponds to the shortest straight spanning ruler with k marks.

There is a long-standing conjecture that $L(n) < n^2$ for all $n \geq 2$. From the table for $L(n)$ with $2 \leq n \leq 25$, we find that $\frac{L(n)}{n^2} < 0.8$ throughout this range. When a perfect circular ruler with n marks exists, the corresponding straight spanning ruler will always have $L(n) < n^2$, but the gaps between consecutive values of such integers n, from known constructions, become arbitrarily large.

Because there are $\binom{n}{2}$ measured distances on an n-mark ruler, $L(n) \geq \binom{n}{2}$ for all $n \geq 2$, and the inequality is strict for $n > 4$ (the last perfect ruler). For example, it is not difficult to show that $\frac{L(n)}{n^2} > \frac{7}{12}$ for all large values of n. The challenging open problems are to determine

$$\liminf_{n \to \infty} \frac{L(n)}{n^2} \quad \text{and} \quad \limsup_{n \to \infty} \frac{L(n)}{n^2}.$$

These two values are equal if and only if $\lim_{n \to \infty} \frac{L(n)}{n^2}$ exists, in which case all three values are equal.

12.8 Two-dimensional pulse patterns

John Costas (1984) proposed the following problem: We wish to design an $n \times n$ frequency hop pattern, for radar or sonar, using n consecutive time intervals $t_1, t_2, \ldots, t_n$ and n consecutive frequencies $f_1, f_2, \ldots, f_n$, where some permutation of the n frequencies is assigned to the n consecutive time slots. Moreover, this should be done in such a way that, if two frequencies f_i and $f_{i+\tau}$ occur at the two times t_j and t_{j+s}, then there is no $i', i' \neq i$, where the two frequencies $f_{i'}$ and $f_{i'+\tau}$ occur at times $t_{j'}$ and $t_{j'+s}$. This constraint corresponds to an ideal, or thumb-tack ambiguity function for the frequency hop pattern.

We may represent the frequency hop pattern by an $n \times n$ permutation matrix (a_{ij}), where $a_{ij} = 1$ if and only if frequency f_i is used at time t_j, otherwise $a_{ij} = 0$. The extra condition is that the $\binom{n}{2}$ vectors connecting the n positions in the matrix where 1's are located are all distinct as vectors: no two vectors are the same in both magnitude and slope. One may visualize a dot at each position where $a_{ij} = 1$. When the pattern is shifted in both time (horizontally) and frequency (vertically), any dot can be brought into coincidence with any other dot. However, the extra Costas condition is that no such shift (other than the identity, which is no shift at all) will bring two dots into coincidence with two other dots.

Costas succeeded, initially, in finding examples, by exhaustive computer search, only for $n \leq 12$. However, several systematic constructions for these Costas arrays are now known, giving examples for arbitrarily large values of n. All of these systematic constructions are based on the existence of primitive roots in finite fields. Three such constructions are the following:

1. *The Welch construction*, for $n = p - 1$ and $p - 2$, p prime. Let g be a primitive root modulo p. The dots of the permutation matrix occur at the locations (i, g^i) for $1 \leq i \leq p - 1$, giving a Costas array of order $n = p - 1$.

   Because $g^{p-1} \equiv 1 \pmod{p}$, there is a dot at $(p - 1, g^{p-1})$, which is at a corner of the matrix. Removing the row and column of this dot leaves a Costas array of order $n = p - 2$.

2. *The Lempel construction* for $n = q - 2$, q a prime power. Let α be any primitive element in $\mathbb{F}_q$. The dots of the permutation matrix occur at the locations (i, j) whenever $\alpha^i + \alpha^j = 1$ in $\mathbb{F}_q$, $1 \leq i, j \leq n - 2$. (This always produces a symmetric matrix.)

3. *The Golomb construction*, for $n = q - 2$ and $q - 3$, q a prime power. Let α and β be any two primitive elements in $\mathbb{F}_q$. The dots of the permutation matrix occur at the locations (i, j) whenever $\alpha^i + \beta^j = 1$ in $\mathbb{F}_q$, $1 \leq i, j \leq n - 2$. (The special case when $\alpha = \beta$ is the Lempel construction.)

It has been shown that for all $q > 2$, the field $\mathbb{F}_q$ contains primitive elements α and β (not necessarily distinct) with $\alpha + \beta = 1$; see Moreno (1999). Using such α and β in the Golomb construction, since $\alpha^1 + \beta^1 = 1$, we have $(1, 1)$ as the location of a dot in the construction. Removing the top row and left column of the matrix leaves a $(q - 3) \times (q - 3)$ Costas array.

For proofs that these three constructions must yield Costas arrays, see Golomb (1984). For additional variants, and the way they yield examples of Costas arrays for many values of $n < 360$, see Golomb and Taylor (1984). Since 1984, the smallest values of n for which no examples of Costas arrays are known have been $n = 32$ and $n = 33$.

The complete enumeration of Costas arrays through $n = 13$ was reported in Golomb and Taylor (1984). Subsequent values have been found by Silverman ($n \leq 16$), and O. Moreno ($n \leq 23$), and $n = 24$ in a recent report by Beard, Erickson, and Russo (2004) leading to the tabulation of $C(n)$, the total number of $n \times n$ permutation matrices that are Costas arrays; the reduced number $c(n)$, where two Costas arrays which differ only by one of the eight dihedral symmetries of the square are not considered distinct; and $s(n)$, the number of inequivalent arrays that have diagonal symmetry, all of which are shown in Table 12.2. For $n > 2$, these three quantities are linearly related by $C(n) = 8c(n) - 4s(n)$. A newer article by Beard *et al.* (2005) extends the enumeration to $n = 25$ and gives a lower bound for $n = 26$.

Table 12.2. *The number of Costas arrays for*
$n \leq 26$; *where* $C(n)$ *is the total number,* $c(n)$
is the reduced number, and $s(n)$ *is the number*
of symmetric Costas arrays of order n

n	$C(n)$	$c(n)$	$s(n)$
2	2	1	1
3	4	1	1
4	12	2	1
5	40	6	2
6	116	17	5
7	200	30	10
8	444	60	9
9	760	100	10
10	2160	277	14
11	4368	555	18
12	7852	990	17
13	12828	1616	25
14	17252	2168	23
15	19612	2467	31
16	21104	2648	20
17	18276	2294	19
18	15096	1892	10
19	10240	1283	6
20	6464	810	4
21	3536	446	8
22	2052	259	5
23	872	114	10
24	200	25	0
25	88	12	2
26*	56	8	2

It is quite possible that $C(n) = 0$ for some, or even for infinitely many, values of n. However, the three general constructions listed above guarantee that $C(n) > 0$ for arbitrarily large values of n. For the Welch construction above, since $2^{16} + 1 = 65,537$ is prime, $C(2^{16}) \geq \phi(2^{16}) = 2^{15} = 33,768$, and $\limsup_{n \to \infty} C(n) = \infty$.

12.9 Periodic modulation

For CW radar, biphase modulation based on a periodic binary sequence can be used if the sequence has a two-valued correlation:

$$C(\tau) = \frac{1}{P} \sum_{i=1}^{p} a_i a_{i+\tau} = \begin{cases} 1, & \tau \equiv 0 \,(\mathrm{mod}\,P) \\ \alpha, & \tau \not\equiv 0 \,(\mathrm{mod}\,P), \end{cases}$$

---

* The enumeration for $n = 26$ may not be complete.

where P is the period of the binary sequence $\{a_i\}$, whose terms are either $+1$ or -1.

While m-sequences (maximum-length linear binary shift register sequence, mentioned earlier) are the most popular choice to generate the sequence $\{a_i\}$, the other cyclic Paley-Hadamard sequences (corresponding to the cyclic Hadamard difference sets described in Section 7.1 of Chapter 7) give the same behavior for $C(\tau)$, namely

$$C(\tau) = \begin{cases} 1, & \tau \equiv 0 \,(\mathrm{mod}\,P), \\ \frac{-1}{P}, & \tau \not\equiv 0 \,(\mathrm{mod}\,P). \end{cases}$$

More generally, one may start with any cyclic difference set with parameters $v, k,$ and λ (see Hall (1986) and Jungnickel and Pott (1999) for standard terminology) and generate a binary sequence of period v, containing a "$+1$" k times and a "-1" $v - k$ times. Then, if this sequence is compared to a phase shift of itself with $\tau \not\equiv 0 \,(\mathrm{mod}\,v)$, $a_i a_{i+\tau} = (+1)(+1) = +1$ a total of λ times per period; $a_i a_{i+\tau} = (+1)(-1) = -1$ a total of $k - \lambda$ times per period; $a_i a_{i+\tau} = (-1)(+1) = -1$ a total of $k - \lambda$ times per period; and therefore $a_i a_{i+\tau} = (-1)(-1) = +1$ a total of $v - \lambda - 2(k - \lambda) = v - 2k + \lambda$ times per period. Hence, for such a sequence

$$C(\tau) = \frac{1}{v} \sum_{i=1}^{v} a_i a_{i+\tau} = \begin{cases} 1, & \tau \equiv 0 \,(\mathrm{mod}\,v), \\ 1 - \frac{4(k-\lambda)}{v}, & \tau \not\equiv 0 \,(\mathrm{mod}\,v). \end{cases} \tag{12.9}$$

It is especially favorable to configure the system in such a way that $C(\tau) = 0$ for all $\tau \not\equiv 0 \,(\mathrm{mod}\,v)$. However, this will not happen in Eq. (12.9) for $v > 4$, because v will be odd and cannot equal $4(k - \lambda)$. There are several modifications that can be made to get $C(\tau) = 0$ for $\tau \neq 0$:

(i) Instead of using the values $+1$ and -1, the sequence $\{a_i\}$ could consist of $+1$ and $-b$, for some real $b \geq 0$.

(ii) Instead of using the values $+1$ and -1, the sequence $\{a_i\}$ could consist of $+1$ and $e^{i\phi}$, for some phase angle $\phi \neq 0, \pi$.

(iii) The transmitted sequence $\{a_i\}$ can consist of $+1$'s and -1's, but the reference sequence against which it is correlated can consist of $+1$ and $-b$, for some real $b \geq 0$.

The mathematical consequences of each of these three possible modifications are explored in Golomb (1992). Possibility (ii) seems especially promising for practical radars.

12.10 The application to CDMA wireless technology

A cellular telephone system is designed to serve a large number of users who may be in the same geographical area and may be attempting to talk at the same time. The term cellular refers to the fact that a very large service area is divided up into cells, where each cell has a relay tower to pick up and forward calls from its local region. For mobile users (e.g., someone talking on a cell phone while driving), the system must provide a way to hand off this user from one cell to another, preferably without the user being aware that this transition is even occurring. (It is called a "soft handoff" if the user is communicating with both the old and the new base station before the old one is released.)

With simultaneous users in a limited geographical region, the obvious problem is interference of the calls with each other. The earliest proposed systems addressed this problem either by assigning different radio frequencies to the various users (called frequency division multiple access, or FDMA) or by assigning different time slots (necessarily recurring soon enough to maintain call continuity) to different users (called time division multiple access, or TDMA). However, by nearly universal agreement, a superior approach, used in some second generation cell phone systems and planned or already implemented in virtually all third generation (3G) systems, is known as code division multiple access, or CDMA for short.

The basic idea of CDMA is to use a wide frequency band common to all the users, whose signals are made distinguishable by assigning mutually uncorrelated code modulation patterns to the various users. Communication systems for point-to-point military communications using this basic type of code modulation had been in use since the 1960s and were known as direct sequence spread spectrum systems, but CDMA was invented to handle the multiuser situation with many telephones transmitting simultaneously.

We have previously mentioned a number of ways to obtain fairly large sets of binary sequences with mutually (pairwise) low crosscorrelation values. However, the first CDMA standard to be implemented was the IS-95 standard, introduced by QUALCOMM, INC., where the modulation was based on the use of m-sequences. Specifically, if a particular m-sequence of period $2^n - 1$ is used, each of its $2^n - 1$ cyclic shifts are very nearly uncorrelated with one another. It is these shifted versions of the same underlying sequence that provide the set of mutually noninterfering signals in the IS-95 standard and in many of its successors, such as CDMA2000. (Actually, with a spreading sequence of period $2^{42} - 1$, true orthogonality is not achieved over the short integration period involved. The cross-correlation achieved in practice averages on the order of $1/\sqrt{m}$, where m is the number of terms in the integration period.) Observe that for this approach to work, it is necessary to maintain overall communication *coherence*. If the signals of different users are allowed to drift randomly in time, some of the advantage of the low correlation may be lost. (The IS-95 and CDMA 2000 systems synchronize base stations, using GPS, so that when a handoff occurs the new base station is prepared, whereby the user terminal acquires immediately, without a prolonged search. The individual users have different–known–offsets, but due to position and velocity they may drift among themselves by as much as several sequence terms, or "chips." This is also accommodated by a very brief search. Such imperfections in the relative timing do not noticeably affect cross-correlation between different users, which is on the order of $1/\sqrt{m}$ because of partial period correlation.)

Note that the basic idea will work in exactly the same way if the m-sequence is replaced by any other binary sequence corresponding to a cyclic Hadamard difference set. Thus, the many basically different constructions described in this book for cyclic Hadamard difference sets can be viewed as alternative ways to implement CDMA code modulation. While linear shift register sequences are extremely easy to generate, advances in computer technology have rendered this advantage relatively unimportant.

In the European standard for 3G, called WCDMA (for wide-band CDMA) or UMTS, the main difference is that the base stations are not mutually synchronized by GPS. Hence each user terminal must fully resynchronize with a new base station when it enters its sphere of influence. (Once it is synchronized, it operates essentially like CDMA2000, including soft handoffs.)

It would be conceivable to field a fully non-coherent CDMA system, but this would entail many new difficulties. The cyclic shifts of a single binary sequence would no longer guarantee that the users would be mutually distinguishable. Instead, sets of sequences, like the Gold codes, which have favorable autocorrelation individually and favorable mutual crosscorrelation, would be recommended. In view of the Welch-Sidelnikov bound on simultaneous correlation (Welch 1974 and Sidelnikov 1978), the coherent system using cyclic shifts of an m-sequence should outperform non-coherent systems constrained as to both autocorrelation and crosscorrelation. (The terms coherent and non-coherent normally refer to *phase*, but in this section we use them with reference to *chip position* in the modulating sequence.) It is unlikely that anyone would choose to implement a non-coherent system for CDMA. (In recent years, an extensive literature has grown up around signal sets with simultaneous auto- and cross-correlation constraints called optical orthogonal codes, from one of the early suggested applications to optical communications.)

To field a practical CDMA wireless telephony system, many additional technical problems must be solved. For example, the information to be conveyed must be incorporated into the signal. The soft-handoff issue must be addressed. Signal fading and blockage due to obstacles in the environment must be dealt with. Interference from sources other than competing cell-phone calls must be overcome. These engineering issues are beyond the scope of the present book. However, for an excellent account by one of the leading pioneers of QUALCOMM's original CDMA system, see Viterbi (1995).

The multiple access aspect of CDMA refers to the possibility of a number of users operating in the same signaling environment at the same time, without drowning out each other's signals. A family of combinatorial designs that we named Tuscan squares in Golomb and Taylor (1985), originally introduced for frequency-hop multiple access applications, has already become the subject of an extensive combinatorial literature.

Signal design problems in communications almost invariably correspond to interesting combinatorial problems. Conversely, almost every major family of combinatorial designs can be interpreted as the solution to a family of signal design problems.

Exercises for Chapter 12

(These exercises provide examples of constructions that were mentioned but not adequately illustrated in the text of Chapter 12.)

I. Polyphase Sequences with Good Autocorrelation

Compute both the periodic and the aperiodic autocorrelation of each of the following sequences. (Remember that with complex numbers, correlation is computed as a type of Hermitian dot product.)

1. Binary (i.e, two-phase) length 11 sequence:

$$\{+1, +1, +1, -1, -1, -1, +1, -1, -1, +1, -1\}.$$

(Barker sequence (Barker 1953).)

2. Sextic (i.e., six-phase) length 9 sequence:

$$\{1, 1, \epsilon, \epsilon, \epsilon^5, \epsilon^4, \epsilon, \epsilon^3, 1\} = \{1, 1, \epsilon, \epsilon, -\epsilon^2, -\epsilon, \epsilon, -1, 1\}$$

where $\epsilon = e^{2\pi i/6}$. (Generalized Barker sequence (Golomb and Scholtz, 1965).)

3. Nine-phase length 9 sequence:

$$\{\alpha^0, \alpha^1, \alpha^3, \alpha^6, \alpha^{10}, \alpha^{15}, \alpha^{21}, \alpha^{28}, \alpha^{36}\} = \{1, \alpha, \alpha^3, \alpha^6, \alpha, \alpha^6, \alpha^3, \alpha, 1\},$$

where $\alpha = e^{2\pi i/9}$. (Golomb-Chu sequence (Chu 1972, and Zhang and Golomb 1993).)

4. Four-phase length 16 sequence:

$$\{i^0, i^0, i^0, i^0; i^0, i^1, i^2, i^3; i^0, i^2, i^4, i^6; i^0, i^3, i^6, i^9\}$$
$$= \{1, 1, 1, 1, 1, i, -1, -i, 1, -1, 1, -1, 1, -i, -1, i\},$$

where $i = e^{2\pi i/4} = \sqrt{-1}$. (Frank sequence (Frank 1963).)

II. Perfect Circular Rulers and Short Spanning Rulers

5. If $D = \{d_1, d_2, \ldots, d_k\}$ is a (v, k, λ) cyclic difference set (i.e., $d_i - d_j$ takes every nonzero value exactly λ times modulo v for $1 \leq i, j \leq k$), show that $D' = aD + b = \{ad_1 + b, ad_2 + b, \ldots, ad_k + b\}$ is again a (v, k, λ) cyclic difference set, for all integers a with $(a, v) = 1$, and all integers b, where all arithmetic is modulo v. (D and every D' of this type are called equivalent difference sets.)

6. Verify that $D_0 = \{0, 1, 11, 19, 26, 28\}$ is a $(31, 6, 1)$ cyclic difference set. Show that D_0 is equivalent to a difference set formed as the union of two 3-element cyclotomic cosets modulo 31, relative to the multiplier $m = k - \lambda = 5$.

7. Among all difference sets equivalent to D_0 of the previous problem, find one (call it D_1) that corresponds to the 6-mark perfect circular ruler of circumference 31 having the longest arc between any two of its adjacent marks. (For D_0, this longest arc, between "1" and "11", has length 10, which can be improved upon.)

8. Removing the longest arc from D_1 of the previous problem, and straightening out what remains, where are the marks on the resulting spanning ruler? Is this the shortest possible 6-mark spanning ruler?

III. Costas Array Constructions

9. **Welch construction:** Plot the points $(i, 6^i)$, modulo 11, for $1 \leq i \leq 10$, on a 10×10 grid, and verify that the result is a 10×10 Costas array. Show how to modify this to obtain a 9×9 Costas array and then an 8×8 Costas array.

10. **Lempel construction:** Plot the points (i, j) such that $2^i + 2^j \equiv 1 \pmod{13}$, for $1 \leq i, j \leq 11$, on an 11×11 grid, and verify that the result is an 11×11 Costas array, symmetric in the $i = j$ axis. Show how to modify this to obtain a 10×10 (symmetric) Costas array. (Note that this 10×10 array is not equivalent to the 10×10 array in Problem 9 with respect to any symmetry of the 10×10 square.)

11. **Golomb-Lempel construction:** Plot the points (i, j) such that $7^i + 11^j \equiv 1 \pmod{17}$, for $1 \leq i, j \leq 15$, on a 15×15 grid, and verify that the result is a 15×15 Costas array. Show how to modify this to obtain a 14×14 Costas array.

12. **Golomb-Lempel construction:** Plot the points (i, j) such that $\alpha^i + \beta^j = 1$, where α and β are elements of $GF(2^4)$ with α a root of $x^4 + x + 1 = 0$ and $\beta = \alpha^4$, for $1 \leq i, j \leq 14$, on a 14×14 grid, and verify that the result is a 14×14 Costas array. Show how to modify this to obtain a 13×13 Costas array and then a 12×12 Costas array.

Note: For proofs of the general validity of these constructions for Costas arrays, see Golomb (1984).

IV. Huffman Sequences

13. Find all the complex roots of $x^8 - 2^4 = 0$ and of $x^8 - 2^{-4} = 0$.

14. Use your result from the previous problem to construct a Huffman sequence of nine real amplitudes, as described in Section 12.5, and verify that your sequence has the required autocorrelation function.

Bibliography

K. T. Arasu and K. J. Player. A new family of cyclic difference sets with singer parameters in characteristic three, *Designs, Codes and Cryptography* **28** (2003), pp. 75–91.

R. H. Barker. Group synchronization of binary digital systems, *Communication Theory (Proceedings of the Second London Symposium on Information Theory)*, London, Butterworths, 1953, pp. 273–287.

L. D. Baumert. *Cyclic Difference Sets*, Lecture Notes in Mathematics, Vol. 182, Berlin, Springer-Verlag, 1971.

L. D. Baumert, S. W. Golomb, and Marshall Hall, Jr. Discovery of an Hadamard matrix of order 92, *Bull. Amer. Math. Soc.* **11** (1962), pp. 237–238.

J. K. Beard, K. Erickson, and J. C. Russo. Combinatoric collaboration on Costas arrays and radar applications, *Proceedings of the 2004 IEEE Radar Conference*, Piscataway, NJ, USA, 2004, pp. 260–265.

J. K. Beard, K. G. Erickson, J. C. Russo, M. C. Monteleone, and M. T. Wright. Costas array generation and search methodology, submitted for publication, 2005.

H. Beker and F. Piper. *Cipher Systems*, John Wiley and Sons, New York, 1982.

A. Bekin. On the nonexistence of additional counterexamples to S. Piccard's Theorem, Ph.D. Dissertation, University of Southern California, Los Angeles, December 2004.

E. R. Berlekamp. *Algebraic Coding Theory*, New York, McGraw-Hill, 1968.

B. C. Berndt and R. J. Evans. Sums of Gauss, Jacobi and Jacobsthal, *J. Number Theory* **11** (1979), pp. 349–398.

B. C. Berndt, R. J. Evans, and K. S. Williams. *Gauss and Jacobi Sums*, New York, John Wiley, Interscience, 1998.

T. Beth, D. Jungnickel, and H. Lenz. *Design Theory*, Cambridge, Cambridge University Press, Vol. 1, 2nd Edition 1999, p. 305.

H. S. Black. *Modulation Theory,* New York, Van Nostrand, 1953.

G. S. Bloom and S. W. Golomb. Applications of numbered, undirected graphs, *Proceedings of the IEEE* **65** (1977), pp. 562–570.

S. Boztas and P. V. Kumar. Binary sequences with Gold-like correlation but larger linear span, *IEEE Trans. Inform. Theory* **40** (1994), pp. 532–537.

A. R. Brenner. Polyphase Barker sequences up to length 45 with small alphabets, *Electronics Letters* **34** (1998), pp. 1576–1577.

A. J. Butrica. *To See the Unseen: A History of Planetary Radar Astronomy*, National Aeronautics and Space Administration, Washington DC, 1996.

P. Camion, C. Carlet, P. Charpin, and N. Sendrier. On correlation-immune functions, *Advances in Cryptology-Crypto'91*, Lecture Notes in Computer Science, Vol. 576, Berlin, Springer-Verlag, 1992, pp. 86–100.

A. Canteaut, C. Carlet, P. Charpin, and C. Fontaine. On cryptographic properties of the cosets of $R(1, m)$, *IEEE Trans. Inform. Theory* 47 (2001), pp. 1494–1513.

A. Canteaut, P. Charpin, and H. Dobbertin. Binary m-sequences with three-valued cross-correlation: a proof of Welch's conjecture, *IEEE Trans. Inform. Theory* 46 (2000), pp. 4–8.

C. Carlet. A larger class of cryptographic Boolean functions via a study of the Maiorana-McFarland construction, *Advances in Cryptology-Crypto'2002*, Lecture Notes in Computer Science, Vol. 2442, Berlin, Springer-Verlag, 2002, pp. 549–564.

C. Carlet. Two new classes of bent functions, *Advances in Cryptology-Eurocrypt'93*, Lecture Notes in Computer Science, Vol. 765, Berlin, Springer-Verlag, 1994, pp. 77–101.

C. Carlet, P. Charpin, and V. Zinoviev. Codes, bent functions and permutations suitable for DES-like cryptosystems, *Designs, Codes and Cryptography* 15 (1998), pp. 125–156.

A. H. Chan and R. A. Games. On the quadratic spans of de Bruijn sequences, *IEEE Trans. Inform. Theory* 36 (1990), pp. 822–829.

A. H. Chan, R. A. Games, and E. L. Key. On the complexities of de Bruijn sequences, *J. Combin. Theory* 33 (1982), pp. 233–246.

A. C. Chang, P. Gaal, S. W. Golomb, G. Gong, T. Helleseth, and P. V. Kumar. On a conjectured ideal autocorrelation sequence and a related triple-error correcting cyclic code, *IEEE Trans. Inform. Theory* 46 (2000), pp. 680–686.

A. C. Chang, S. W. Golomb, G. Gong, and P. V. Kumar. On the linear span of ideal autocorrelation sequences arising from the Segre hyperoval, *Sequences and their Applications – Proceedings of SETA'98, Discrete Mathematics and Theoretical Computer Science*, London, Springer-Verlag, 1999.

U. J. Cheng. Exhaustive construction of (255, 127, 63)-cyclic difference sets, *Jounal of Combinatorial Theory, Series A* 33 (1983), pp. 115–125.

D. C. Chu. Polynominal codes with good periodic correlation properties, *IEEE Trans. Inform. Theory* 18 (1972), pp. 531–532.

H. B. Chung and J. S. No. Linear span of extended sequences and cascaded GMW sequences, *IEEE Trans. Inform. Theory* 45 (1999), pp. 2060–2065.

S. D. Cohen and R. W. Matthews. A class of exceptional polynomials, *Transactions of the American Mathematical Society* 345 (1994), pp. 897–909.

J. P. Costas. A study of a class of detection waveforms having nearly ideal range-doppler ambiguity properties, *Proc. IEEE* 72 (1984), pp. 996–1009.

T. W. Cusick and H. Dobbertin. Some new three-valued crosscorrelation functions for binary m-sequences, *IEEE Trans. Inform. Theory* 42 (1996), pp. 1238–1240.

Z. D. Dai, G. Gong, and D. F. Ye. Decompositions of cascaded GMW functions, *Science in China (Series A)* 44 (2001), pp. 709–717.

M. H. Dawson and S. E. Tavares. An expanded set of s-box design criteria based on information theory and its relation to differential-like attacks, *Advances in*

Cryptology-Eurocrypt'91, Lecture Notes in Computer Science, Vol. 547, Berlin, Springer-Verlag, 1991, pp. 352–367.

J. F. Dillon. Multiplicative difference sets via additive characters, *Designs, Codes and Cryptography*, **17** (1999), pp. 225–236.

J. F. Dillon. Elementary Hadamard difference sets, in *Proc. Sixth S-E Conf. Comb. Graph Theory and Comp.*, F. Hoffman *et al.* (Eds.), Winnipeg Utilitas Math, 1975, pp. 237–249.

J. F. Dillon and H. Dobbertin. New cyclic difference sets with Singer parameters, *Finite Fields and Their Applications*, **10** (2004), pp. 342–389.

C. S. Ding, T. Helleseth, and W. J. Shan. On the linear complexity of Legendre sequences, *IEEE Trans. Inform. Theory* **44** (1998), pp. 1276–1278.

S. S. Ding. *Linear Feedfack Shift Register Sequences*, Shanghai, Scientific and Technology Publications, 1982, Chap. 1. (In Chinese)

H. Dobbertin. Another proof of Kasami's theorem, *Designs, Codes, and Cryptography*, **17** (1999), pp. 177–180.

H. Dobbertin. Construction of bent functions and balanced Boolean functions with high nonlinearity, *Proceedings of Fast Software Encryption, Second International Workshop*, Berlin, Springer-Verlag, 1995, pp. 61–74.

H. Dobbertin. Kasami power functions, permuation polynomials and cyclic difference sets, *Difference sets, sequences and their correlation properties (Bad Windsheim, 1998), NATO Adv. Sci. Inst. Ser. C Math. Phys. Sci.* **542** (1999), pp. 133–158.

H. Dobbertin, T. Helleseth, P. V. Kumar, and H. Martinsen. Ternary m-sequences with three-valued cross-correlation functions: new decimations of Welch and Niho type, *IEEE Trans. Inform. Theory* **47** (2001), pp. 1473–1481.

A. Dollas, W. T. Rankin, and D. McCracken. A new algorithm for Golomb ruler derivation and proof of the 19 mark ruler, *IEEE Trans. Inform. Theory* **44** (1998), pp. 379–382.

R. B. Dreier and K. W. Smith. Exhaustive determination of (511, 255, 127) cyclic difference sets, manuscript, 1991.

T. Etzion and A. Lempel. Construction of de Bruijn sequences of minimal complexity, *IEEE Trans. Inform. Theory* **30** (1984), pp. 705–709.

R. Evan, H. D. L. Hollman, C. Krattenthaler, and Q. Xiang. Gauss sums, Jacobi sums and p-ranks of cyclic difference sets, *Journal of Combinatorial Theory, Series A* **87** (1999), pp. 74–119.

R. Fano. *Transmission of Information*, Cambridge, MIT Press, 1961.

Fibonacci, *Liber Abaci*, 1202. English translation by L. E. Sigler, Springer-Verlag, New York, 2002.

R. L. Frank. Polyphase codes with good nonperiodic correlation properties, *IEEE Trans. Inform. Theory* **9** (1963), pp. 43–45.

P. Gaal and S. W. Golomb. Exhaustive determination of (1023, 511, 255)-cyclic difference sets, *Mathematics of Computation* **70** (2001), pp. 357–366.

R. Gallager. *Information Theory and Reliable Communication*, New York, John Wiley and Sons, 1968.

R. A. Games. An algebraic construction of sonar sequences using M-sequences, *SIAM J. Algebraic Discrete Methods* **8** (1987), pp. 753–761.

R. A. Games. Crosscorrelation of M-sequences and GMW-sequences with the same primitive polynomial, *Discrete Appl. Math.* **12** (1985), pp. 139–146.

R. A. Games. A generalized recursive construction for de Bruijn sequences, *IEEE Trans. Inform. Theory* **29** (1983), pp. 843–850.

R. A. Games. The geometry of m-sequences: three-valued crosscorrelations and quadratics in finite projective geometry, *SIAM J. Algebraic Discrete Methods* **7** (1986), pp. 43–52.

R. A. Games. The geometry of quadrics and correlations of sequences, *IEEE Trans. Inform. Theory* **32** (1986), pp. 423–426.

R. A. Games. There are no de Bruijn sequences of span n with complexity $2^{n-1} + n + 1$, *J. Combin. Theory, Ser. A* **34** (1983), pp. 248–251.

R. A. Games and A. H. Chan. A fast algorithm for determining the complexity of a binary sequence with period 2^n, *IEEE Trans. Inform. Theory* **29** (1983), pp. 144–146.

M. Gardner. *Wheels, Life and Other Mathematical Amusements*, New York, W. H. Freeman, 1983, Chap. 15, pp. 152–165.

D. G. Glynn. Two new sequences of ovals in finite Desarguesian planes of even order, *Lecture Notes in Mathematics*, Vol. 1036, Berlin, Springer-Verlag, 1983, pp. 217–229.

R. Gold. Maximal recursive sequences with 3-valued recursive cross-correlation functions, *IEEE Trans. Inform. Theory* **14** (1968), pp. 154–156.

S. W. Golomb. Algebraic constructions for Costas arrays, *Journal Comb. Theory (A)* **37** (1984), pp. 13–21.

S. W. Golomb. Irreducible polynomials, synchronization codes, primitive necklaces, and the cyclotomic algebra, *Combinatorial Mathematics and Its Applications*, R. C. Bose and T. A. Dowling (Eds.), Chapel Hill, University of North Carolina, 1969, pp. 358–370.

S. W. Golomb. On the classification of balanced binary sequences of period $2^n - 1$, *IEEE Trans. Inform. Theory* **26** (1980), pp. 730–732.

S. W. Golomb. On the classification of Boolean functions, *IEEE Trans. Inform. Theory* **5** (1959), pp. 176–186. Also appears in Golomb (1982), Chap. VIII.

S. W. Golomb. *Remarks on Orthogonal Sequences*, The Glenn L. Martin Company, Baltimore, MD, July 28, 1954. (Memo)

S. W. Golomb. *Sequences with Randomness Properties*, Baltimore, Glenn L. Martin Company, 1955.

S. W. Golomb. *Shift Register Sequences*, San Francisco, Holden-Day, 1967, revised edition, Laguna Hills, CA, Aegean Park Press, 1982.

S. W. Golomb. Two-valued sequences with perfect periodic autocorrelation, *IEEE Trans. Aerospace Electronic Systems* **28** (1992), pp. 383–386.

S. W. Golomb and G. Gong. Periodic sequences with the "trinomial property," *IEEE Trans. Inform. Theory* **45** (1999), pp. 1276–1279.

S. W. Golomb, R. E. Peile, and R. A. Scholtz. *Basic Concepts in Information Theory and Coding*, New York, Plenum Press, 1994.

S. W. Golomb and R. A. Scholtz, Generalized Barker sequences, *IEEE Trans. Inform. Theory* **11** (1965), pp. 533–537.

S. W. Golomb and H. Taylor. Constructions and properties of Costas arrays, *Proc. IEEE* **72** (1984), pp. 1143–1163.

S. W. Golomb and H. Taylor. Cyclic projective planes, perfect circular rulers, and good spanning rulers, in *Sequences and Their Applications, Discrete Mathematics and Theoretical Computer Science − Proceedings of SETA'01*, V. Kumar, T. Helleseth, and K. Yang (Eds.), London, Springer-Verlag, 2002, pp. 166–181.

S. W. Golomb and H. Taylor. Tuscan squares — A new family of combinatorial designs, *Ars Combinatoria* **20**-B (1985), pp. 115–132.

G. Gong. New Designs for signal sets with low cross-correlation, balance property and large linear span: $GF(p)$ Case, *IEEE Trans. Inform. Theory* **48** (2002), pp. 2847–2867.

G. Gong. Runs, component sequences and vector sequences of m-sequences over Galois fields, *Acta Electronica Silica* **14** (1986), pp. 94–100.

G. Gong. q-ary cascaded GMW sequences, *IEEE Trans. Inform. Theory* **42** (1996), pp. 263–267.

G. Gong. *Sequence Analysis*, Lecture Notes for Course CO739x, University of Waterloo, http://www.cacr.math.uwaterloo.ca/~ggong, Winter 1999, Chapter 5.

G. Gong. Theory and applications of q-ary interleaved sequences, *IEEE Trans. Inform. Theory* **41** (1995), pp. 400–411.

G. Gong, Z. D. Dai, and S. W. Golomb. Enumeration and criterion for cyclically shift-distinct q-ary GMW sequences of period $q^n - 1$, *IEEE Trans. Inform. Theory* **46** (2000), pp. 474–484.

G. Gong, A. Di Porto, and W. Wolfowicz. Galois linear group sequences, *La Comunicazione, Note Recensioni Notizie* **XLII** (1993), pp. 83–89.

G. Gong, P. Gaal, and S. W. Golomb. A suspected infinite class of cyclic Hadamard difference sets, *Proceedings of 1997 IEEE Information Theory Workshop*, July 6-12, 1997, Longyearbyen, Svalbard, Norway.

G. Gong and S. W. Golomb. Binary sequences with two-level autocorrelation, *IEEE Trans. Inform. Theory* **45** (1999), pp. 692–693.

G. Gong and S. W. Golomb. The decimation-Hadamard transform of two-level autocorrelation sequences, *IEEE Trans. Inform. Theory* **48** (2002), pp. 853–865.

G. Gong and S. W. Golomb. Hadamard transforms of three term sequences, *IEEE Trans. Inform. Theory* **45** (1999), pp. 2059–2060.

G. Gong and S. W. Golomb. Personal communications, 2001.

G. Gong and S. W. Golomb. Transform domain analysis of DES, *IEEE Trans. Inform. Theory* **45** (1999), pp. 2065–2073.

G. Gong, T. Helleseth, and M. Ludkovski. New ternary 2-level autocorrelation sequences and realizable pairs, submitted to *IEEE Trans. Inform. Theory*. A preliminary version (titled as New families of ideal 2-level autocorrelation ternary sequences from second order DHT) appeared in *Pre-proceedings of the second International Workshop in Coding and Cryptography*, January 8–12, 2001, Paris, pp. 345–354.

G. Gong and K. M. Khoo. Additive autocorrelation of resilient boolean functions, the *Proceedings of Tenth Annual Workshop on Selected Areas in Cryptography, Lecture Notes in Computer Science*, Vol. 3006, Berlin, Springer-Verlag, 2004, pp. 275–290.

G. Gong and H. Y. Song. Two-tuple balance of bon-binary sequences with ideal two-level autocorrelation, abstract in the *Proceedings of 2003 IEEE International Symposium on Information Theory (ISIT2003)*, June 29–July 4, Yokohama, Japan, pp. 404. Full version, Technical Report CORR 2002-20, University of Waterloo, 2002.

G. Gong and A. Youssef. Cryptographic properties of the Welch-Gong transformation sequence generators, *IEEE Trans. Inform. Theory* **48** (2002), pp. 2837–2846.

B. Gordon, W. H. Mill, and L. R. Welch. Some new difference sets, *Canadian J. Math.* **14** (1962), pp. 614–625.

M. Hall, Jr. *Combinatorial Theory*, 2nd ed., New York, Wiley-Interscience, 1986.

M. Hall, Jr. A Survey of Difference Sets, *Proc. Am. Math. Soc.* **7** (1956), pp. 975–986.

G. H. Hardy and E. M. Wright. *An Introduction to the Theory of Numbers*, 5th ed., New York, Oxford University Press, 1980.

B. Hayes. Computing science: Collective wisdom, *American Scientist* **86** (1998), pp. 118–122.

T. Helleseth. Correlation of m-sequences and related topics, *Sequences and their Applications – Proceedings of SETA '98*, Discrete Mathematics and Theoretical Computer Science, Berlin, Springer-Verlag, 1999, pp. 49–66.

T. Helleseth. Some results about the cross-correlation functions between two maximal length linear sequences, *Discrete Math.* **16** (1976), pp. 209–232.

T. Helleseth and G. Gong. New nonbinary sequences with ideal two-level autocorrelation, *IEEE Trans. Inform. Theory* **48** (2002), pp. 2868–2872.

T. Helleseth and P. V. Kumar. Codes and sequences over Z_4, in A. Pott *et al.* (Eds.), *Difference Sets, Sequences and Their Correlation Properties (Bad Windsheim, 1998)*, NATO Adv. Sci. Inst. Ser. C, Math. Phys. Sci. **542** (1999).

T. Helleseth and P. V. Kumar. Sequences with low correlation, in *Handbook of Coding Theory*, V. Pless and C. Huffman (Eds.), New York, Elsevier Science, 1998, pp. 1765–1853.

T. Helleseth, P. V. Kumar, and H. Martinsen. A new family of ternary sequences with ideal two-level autocorrelation function, *Designs, Codes and Cryptography* **23** (2001), pp. 157–166.

T. Herlestam. On functions of linear shift register sequences, *Advances in Cryptology-Eurocrypt'85*, Lecture Notes in Computer Science, Vol. 219, Berlin, Springer-Verlag, 1985, pp. 119–129.

J. W. P. Hirschfeld. Ovals in a Desarguesian plane of even order, *Ann. Mat. Pure Appl.* **102** (1975), pp. 75–89.

J. W. P. Hirschfeld. *Projective Geometries over Finite Fields*, London, Oxford University Press, 2nd ed., 1998.

H. D. L. Hollmann and Q. Xiang. A proof of the Welch and Niho conjectures on cross-correlations of binary m-sequences, *Finite Fields and Their Applications* **7** (2001), pp. 253–286.

X. D. Hou. q-ary bent functions constructed from chain rings, *Finite Fields and Their Applications* **4** (1998), pp. 55–61.

D. A. Huffman. The generation of impulse-equivalent pulse trains, *IEEE Trans. Inform. Theory* **8** (1962), pp. 10–16.

K. Ireland and M. Rosen. *A Classical Introduction to Modern Number Theory*, New York, Springer-Verlag, 2nd ed., 1991.

D. Jungnickel and A. Pott. Difference sets: an introduction, *Difference sets, Sequences and Their Correlation Properties (Bad Windsheim, 1998)*, A. Pott *et al.* (Eds.), NATO Adv. Sci. Inst. Ser. C, Math. Phys. Sci. **542** (1999).

T. Kasami. Weight distributions of Bose-Chaudhuri-Hocquenghem codes, *Combinatorial Mathematics and Its Applications*, R. C. Bose and T. A. Dowling (Eds.), University of North Carolina, Chapel Hill, NC, 1969, pp. 335–357.

T. Kasami. The weight enumerators for several classes of subcodes of the 2nd-order Reed-Muller codes, *Information and Control* **18** (1971), pp. 369–394.

H. Kharaghani and B. Tayfeh-Rezaie. A Hadamard matrix of order 428, preprint, June 2004.

K. M. Khoo, G. Gong, and D. R. Stinson. A new family of Gold-like sequences, *Proceedings of the 2002 IEEE International Symposium on Information Theory*, Lausanne, Switzerland, June 30–July 5, 2002, p. 81.

J. H. Kim and H. Y. Song. Existence of cyclic Hadamard difference sets and its relation to binary sequences with ideal autocorrelation, *Journal of Communications and Networks* 1 (1999), pp. 14–18.

A. Klapper. d-form sequences: families of sequences with low correlation values and large linear spans, *IEEE Trans. Inform. Theory* 41 (1995), pp. 423–431.

A. Klapper. Large families of sequences with near-optimal correlations and large linear span, *IEEE Trans. Inform. Theory* 42 (1996), pp. 1241–1248.

A. Klapper, A. H. Chan, and M. Goresky. Cascaded GMW sequences, *IEEE Trans. Inform. Theory* 39 (1993), pp. 177–183.

A. Klapper, A. H. Chan, and M. Goresky. Cross-correlations of linearly and quadratically related geometric sequences and GMW sequences, *Discrete Appl. Math.* 46 (1993), pp. 1–20.

P. V. Kumar. Frequency-hopping code sequence designs having large linear span, *IEEE Trans. Inform. Theory* 34 (1988), pp. 146–151.

P. V. Kumar. *On bent sequences and generalized bent functions*, Ph. D. Thesis, University of Southern California, Los Angeles, 1983.

P. V. Kumar, T. Helleseth, A. R. Calderbank, and A. R. Hammons. Large families of quaternary sequences with low correlation, *IEEE Trans. Inform. Theory* 42 (1996), pp. 579–592.

P. V. Kumar and O. Moreno. Prime-phase sequences with periodic correlation properties better than binary sequences, *IEEE Trans. Inform. Theory* 37 (1991), pp. 603–616.

R. Lidl and H. Niederreiter. *Finite Fields*, Encyclopedia of Mathematics and its Applications, Vol. 20, Reading, MA, Addison-Wesley, 1983. (Revised version, Cambridge University Press, 1997.)

A. H. Lin. *From cyclic Hadamard difference sets to perfectly balanced sequences*, Ph.D. thesis, University of Southern California, Los Angeles, 1998.

E. Lucas. Sur La Récherche des grand nonbres première, *Assoc. Francais pour l'Avanc. des Scièpces* 5 (1876), pp. 61–68.

H. D. Lüke. *Korrelationssignale*, Springer-Verlag, Berlin, 1992.

I. G. MacDonald. *Symmetric Functions and Hall Polynomials*, Oxford, Oxford University Press, 3rd ed., 1999.

S. Maitra and P. Sarkar. Modifications of Patterson-Wiedemann functions for cryptographic applications, *IEEE Trans. Inform. Theory* 48 (2002), pp. 278–284.

S. Maitra and P. Sarkar. Nonlinearity bounds and constructions of resilient boolean functions, *Advances in Cryptology-Crypto '2000*, Lecture Notes in Computer Science, Vol. 1880, Berlin, Springer-Verlag, 2000, pp. 515–532.

R. W. Marsh. Table of irreducible polynomials over GF(2) through degree 19, *U.S. Department of Commerce*, Office of Technical Services, Washington D.C., 1957.

A. Maschietti. Difference sets and hyperovals, *Designs, Codes and Cryptography* 14 (1998), pp. 89–98.

J. L. Massey. Shift-register synthesis and BCH decoding, *IEEE Trans. Inform. Theory* **15** (1969), pp. 81–92.

G. L. Mayhew and S. W. Golomb. Linear spans of modified de Bruijn sequences, *IEEE Trans. Inform. Theory* **36** (1990), pp. 1166–1167.

R. J. McEliece. *Finite Fields for Computer Scientists and Engineers*, The Kluwer International Series in Engineering and Computer Science, Vol. 23, Boston, Kluwer Academic, 1986.

R. L. McFarland. A family of noncyclic difference sets, *Journal of Combinatorial Theory, Series A* **15** (1973), pp. 1–10.

R. L. McFarland and B. F. Rice. Translates and multipliers of abelian difference sets, *Proc. Amer. Math. Soc.* **68** (1978), pp. 375–379.

F. J. McWilliams and N. J. A. Sloane. *Theory of Error-Correcting Codes*, Amsterdam, North-Holland, 1977 (revised, 1991).

W. Meier and O. Staffelbach. Fast correlation attacks on stream ciphers, *Advances in Cryptology-Eurocrypt'88*, Lecture Notes in Computer Science, 1988, Vol. 330, pp. 301–314.

A. J. Menezes, P. C. van Oorschot, and S. A. Vanstone. *Handbook of Applied Cryptography*, Boca Raton, LA, CRC Press, 1996.

O. Moreno. Survey of results on signal patterns for locating one or multiple targets, in A. Pott *et al.* (Eds.), *Difference Sets, Sequences and Their Correlation Properties (Bad Windsheim, 1998)*, NATO Adv. Sci. Inst. Ser. C, Math. Phys. Sci. **542** (1999).

Y. Niho. *Multi-valued cross-correlation functions between two maximal linear recursive sequences*, Ph.D. Thesis, University of Southern California, Los Angeles, USCEE report 409, 1972.

J. S. No. Generalization of GMW sequences and No sequences, *IEEE Trans. Inform. Theory* **42** (1996), pp. 260–262.

J. S. No. p-ary unified sequences: p-ary extended d-form sequences with ideal autocorrelation property, *IEEE Tran. Inform. Theory* **48** (2002), pp. 2540–2546.

J. S. No, H. B. Chung, and M. S. Yun. Binary pseudorandom sequences of period $2^m - 1$ with ideal autocorrelation generated by the polynomial $z^d + (z + 1)^d$, *IEEE Trans. Inform. Theory* **44** (1998), pp. 1278–1282.

J. S. No, S. W. Golomb, G. Gong, H. K. Lee, and P. Gaal. Binary pseudorandom sequences of period $2^n - 1$ with ideal autocorrelation, *IEEE Trans. Inform. Theory* **44** (1998), pp. 814–817.

J. S. No and P. V. Kumar. A new family of binary pseudo-random sequences having optimal periodic correlation properties and larger linear span, *IEEE Trans. Inform. Theory* **35** (1989), pp. 371–379.

J. S. No, H. K. Lee, H. B. Chung, H. Y. Song, and K. C. Yang. Trace representation of Legendre sequences of Mersenne prime period, *IEEE Trans. Inform. Theory* **42** (1996), pp. 2254–2255.

J. S. No, K. Yang, H. G. Chung, and H. Y. Song. New constructions for families of binary sequences with optimal correlation properties, *IEEE Trans. Inform. Theory* **43** (1997), pp. 1596–1602.

K. Nyberg. Perfect nonlinear S-boxes, *Advances in Cryptology-Eurocrypt'91*, Lecture Notes in Computer Science, Vol. 547, Berlin, Springer-Verlag, 1991, pp. 378–386.

J. D. Olsen, R. A. Scholtz, and L. R. Welch. Bent-function sequences, *IEEE Trans. Inform. Theory* **28** (1982), pp. 858–864.

O. Ore. Contributions to the Theory of Finite Fields, *Trans. Am. Math. Soc.* **36** (1934), pp. 243–274.

R. E. A. C. Paley. On Orthogonal Matrices, *J. Math. Phys.* **12** (1933), pp. 311–320.

W. J. Park and J. J. Komo. Relationships between m-sequences over $GF(q)$ and $GF(q^m)$, *IEEE Trans. Inform. Theory* **35** (1989), pp. 183–186.

K. G. Paterson. Binary sequence sets with favorable correlations from difference sets and MDS codes, *IEEE Trans. Inform. Theory* **44** (1998), pp. 172–180.

N. J. Patterson and D. H. Wiedemann. Correction to the covering radius of the $(2^{15}, 16)$ Reed-Muller code is at least 16276, *IEEE Trans. Inform. Theory* **36** (1990), p. 443.

N. J. Patterson and D. H. Wiedemann. The covering radius of the $(2^{15}, 16)$ Reed-Muller Code is at least 16276, *IEEE Trans. Inform. Theory* **29** (1983), pp. 354–356.

O. Perron. Bemerkungen über die verteilung der quadratischen reste, *Math. Zeitschrift* **56** (1952), pp. 122–130.

J. G. Proakis. *Digital Communications*, New York, McGraw-Hill, 3rd ed., 1995.

L. Rédei. *Algebra*, Germany, Leipzig, Geest & Portig, 1959; Oxford, New York, Pergamon Press, 1st English edition, 1967.

O. S. Rothaus. On "bent" functions, *Journal of Combinatorial Theory, Series A* **20** (1976), pp. 300–305.

R. A. Rueppel. *Analysis and Design of Stream Ciphers*, Berlin, Springer-Verlag, 1986.

H. J. Ryser. *Combinatorial Mathematics*, Carus Mathematical Monographs, No. 14, Math. Assoc. of America, Washington DC, 1963.

R. A. Scholtz and L. R. Welch. GMW sequences, *IEEE Trans. Inform. Theory* **30** (1984), pp. 548–553.

B. Segre. Ovals in a finite projective plane, *Canadian J. Math.* **7** (1955), pp. 414–416.

B. Segre and U. Bartocci. Ovali ed altre curve nei piani di Galois di caratteristica due, *Acta Arith.* **8** (1971), pp. 423–449.

I. Selin. *Detection Theory*, Princeton, NJ, Princeton University Press, 1965.

E. S. Selmer. *Linear Recurrence Relations over Finite Fields*, Bergan, Norway, Department of Mathematics, University of Bergen, 1966.

C. E. Shannon. A mathematical theory of communication, *Bell System Technical Journal* **27** (1948), pp. 379–423 (Part I), July (1948), pp. 623–656 (Part II), October.

J. B. Shearen. Some new optimum Golomb rulers, *IEEE Trans. Inform. Theory* **36** (1990), pp. 183–184.

D. A. Shedd and D. V. Sarwate. Construction of sequences with good correlation properties, *IEEE Trans. Inform. Theory*, Vol. 25, No. 1, January 1979, pp. 94–97.

W. O. Sibert, in a letter to H. Taylor and S. Golomb, April 1993. (This reference appeared in the paper of Dollas, Rankin and McCracken 1998.)

V. M. Sidelnikov. On mutual correlation of sequences, *Probl. Kybern.* **24** (1978), pp. 537–545.

T. Siegenthaler. Correlation-immunity of nonlinear combining functions for cryptographic applications, *IEEE Trans. Inform. Theory* **30** (1984), pp. 776–780.

J. Singer. A theorem in finite projective geometry and some applications to number theory, *Trans. Amer. Math. Soc.* **43** (1938), pp. 377–385.

H. Y. Song and S. W. Golomb. On the existence of cyclic Hadamard difference sets, *IEEE Trans. Inform. Theory* **40** (1994), pp. 1266–1268.

R. G. Stanton and D. A. Sprott. A family of difference sets, *Canad. J. Math.* **10** (1958), pp. 73–77.

H. Stichtenoth. *Algebraic Function Fields and Codes*, Berlin, Springer-Verlag, 1993.

H. M. Trachtenberg. *On the cross-correlation functions of maximal linear sequences*, Ph.D. Thesis, University of Southern California, Los Angeles, 1970.

R. Turyn. Sequences with small correlation, in Henry B. Mann (Ed.), *Error Correcting Codes,* New York, John Wiley & Sons, 1968, pp. 195–228.

R. J. Turyn. Character sums and difference sets, *Pacific Journal of Mathematics* **15** (1965), pp. 319–346.

R. J. Turyn and J. Storer. On binary sequences, *Proceedings of the American Mathematical Society* **12** (1961), pp. 394–399.

B. L. van der Waerden. *Modern Algebra* (English Translation), New York, Frederick Ungar, 1953.

A. J. Viterbi. *CDMA − Principles of Spread Spectrum Communication*, Reading, MA, Addison-Wesley, 1995.

M. Ward. The arithmetical theory of linear recurring series, *Trans. Amer. Math. Soc.* **35** (1933), pp. 600–628.

M. Ward. The characteristic number of a sequence of integers satisfying a linear recursion relation, *Trans. Amer. Math. Soc.* **33** (1931), pp. 153–165.

L. R. Welch. Lower bounds on the minimum correlation of signals, *IEEE Trans. Inform. Theory* **20** (1974), pp. 397–399.

J. Williamson. Hadamard's determinant theorem and the sum of four squares, *Duke Math J.* **11** (1944), pp. 65–81.

Q. Xiang. On balanced binary sequences with two-level autocorrelation functions, *IEEE Trans. Inform. Theory* **44** (1998), pp. 3153–3156.

G. Z. Xiao and J. L. Massey. A spectral characterization of correlation-immune combining functions, *IEEE Trans. Inform. Theory* **34** (1988), pp. 569–571.

G. Yovanof and S. W. Golomb. The polynomial model in the study of counterexamples to S. Piccard's Theorem, *Ars Combinatoria* **48** (1998), pp. 43–63.

K. C. Zeng and M. X. Huang. On the linear syndrome method in cryptoanalysis, *Advances in Cryptology-Crypto'88,* S. Goldwasser (Ed.), Lecture Notes in Computer Science, Vol. 403, Berlin, Springer-Verlag, 1990, pp. 469–478.

N. Zhang and S. W. Golomb. Polyphase sequences with low autocorrelations, *IEEE Trans. Inform. Theory* **39** (1993), pp. 1085–1089.

Y. L. Zheng and X. M. Zhang. Improved upper bound on the nonlinearity of high order correlation immune functions, *Proceedings of Selected Areas in Cryptography 2000,* Lecture Notes in Computer Science, Vol. 2012, Berlin, Springer-Verlag, 2001, pp. 262–274.

Y. L. Zheng and X. M. Zhang. Relationships between bent functions and complementary plateaued functions, *Information and Communications Security (ICICS'99)*, Lecture Notes in Computer Science, Vol. 1787, Berlin, Springer-Verlag, 1999, pp. 60–75.

N. Zierler. Linear recurring sequences, *J. Soc. Indust. Appl. Math.* **7** (1959), pp. 31–48.

Index

Printed in the United States
By Bookmasters